ASTROPHYSICS AND SPACE SCIENCE LIBRARY

A SERIES OF BOOKS ON THE RECENT DEVELOPMENTS
OF SPACE SCIENCE AND OF GENERAL GEOPHYSICS AND ASTROPHYSICS
PUBLISHED IN CONNECTION WITH THE JOURNAL
SPACE SCIENCE REVIEWS

Editorial Board

J. E. BLAMONT, *Laboratoire d'Aeronomie, Verrières, France*

R. L. F. BOYD, *University College, London, England*

L. GOLDBERG, *Kitt Peak National Observatory, Tucson, Ariz., U.S.A.*

C. DE JAGER, *University of Utrecht, The Netherlands*

Z. KOPAL, *University of Manchester, England*

G. H. LUDWIG, *NASA Headquarters, Washington, DC, U.S.A.*

R. LÜST, *President Max-Planck-Gesellschaft zur Förderung der Wissenschaften, München, F.R.G.*

B. M. McCORMAC, *Lockheed Palo Alto Research Laboratory, Palo Alto, Calif., U.S.A.*

L. I. SEDOV, *Academy of Sciences of the U.S.S.R., Moscow, U.S.S.R.*

Z. ŠVESTKA, *University of Utrecht, The Netherlands*

VOLUME 110
PROCEEDINGS

ASTRONOMY WITH SCHMIDT-TYPE TELESCOPES

ASTRONOMY WITH SCHMIDT-TYPE TELESCOPES

PROCEEDINGS OF THE 78th COLLOQUIUM
OF THE INTERNATIONAL ASTRONOMICAL UNION,
ASIAGO, ITALY, AUGUST 30 – SEPTEMBER 2, 1983

Edited by

MASSIMO CAPACCIOLI

*Institute of Astronomy and Asiago Astrophysical Observatory,
University of Padova, Padova, Italy*

D. REIDEL PUBLISHING COMPANY

A MEMBER OF THE KLUWER ACADEMIC PUBLISHERS GROUP

DORDRECHT / BOSTON / LANCASTER

Library of Congress Cataloging in Publication Data

International Astronomical Union. Colloquium (78th : 1983 : Asiago, Italy)
Astronomy with Schmidt-type telescopes.

(Astrophysics and space science library ; v. 110)
Includes indexes.
1. Schmidt telescope—Congresses. I. Capaccioli, M.
II. Title. III. Series.
QB88.I58 1983   522'.2   84-4695
ISBN 90-277-1756-7

Published by D. Reidel Publishing Company,
P.O. Box 17, 3300 AA Dordrecht, Holland.

Sold and distributed in the U.S.A. and Canada
by Kluwer Academic Publishers,
190 Old Derby Street, Hingham, MA 02043, U.S.A.

In all other countries, sold and distributed
by Kluwer Academic Publishers Group,
P.O. Box 322, 3300 AH Dordrecht, Holland.

All Rights Reserved
© 1984 by D. Reidel Publishing Company, Dordrecht, Holland
No part of the material protected by this copyright notice may be reproduced or
utilized in any form or by any means, electronic or mechanical
including photocopying, recording or by any information storage and
retrieval system, without written permission from the copyright owner

Printed in The Netherlands

TABLE OF CONTENTS

EDITOR'S FOREWORD	xiii
LIST OF PARTICIPANTS	xv
WELCOME ADDRESS TO THE PARTICIPANTS L.Rosino, Chairman of the SOC	xix
ADDRESS BY THE GENERAL SECRETARY OF THE INTERNATIONAL ASTRONOMICAL UNION R.M.West	xxi
RESOLUTION	1
SCHMIDT TELESCOPES AS DISCOVERY INSTRUMENTS L.Woltjer	3
THE ESO SKY SURVEYS R.M.West	13
SKY SURVEYS WITH THE UK 1.2m SCHMIDT TELESCOPE R.D.Cannon	25
OBJECTIVE PRISM SURVEYS M.F.McCarthy	37
THE CASE LOW-DISPERSION NORTHERN SKY SURVEY P.Pesch, N.Sanduleak	53
THE DETECTION OF FAINT IMAGES AGAINST THE SKY BACKGROUND D.F.Malin	57
ENHANCEMENT OF FAINT IMAGES FROM UK SCHMIDT TELESCOPE PLATES B.W.Hadley	73
STELLAR PHOTOMETRY WITH AN AUTOMATED MEASURING MACHINE G.Gilmore	77
ANALYSIS OF IMAGES WITH THE APM SYSTEM AT CAMBRIDGE E.Kibblewhite, M.Bridgeland, P.Bunclark, M.Cawson, M.Irwin	89

REDUCTION TECHNIQUES - ESO FACILITIES
    P.Crane   99

MACHINE-PROCESSING OF OBJECTIVE-PRISM PLATES
AT THE ROYAL OBSERVATORY, EDINBURGH
    R.G.Clowes   107

THE AUTOMATED DETECTION OF VARIABLE OBJECTS ON SCHMIDT PLATES
    M.R.S.Hawkins   121

A FAINT GALAXY SURVEY FROM COSMOS MEASURES ON DEEP UKST PLATES
    H.T.MacGillivray, R.J.Dodd   125

VISUAL AND AUTOMATIC CLASSIFICATION OF GALAXY IMAGES
    A.Wirth   129

A SYSTEM FOR OBJECT DETECTION AND IMAGE CLASSIFICATION
ON PHOTOGRAPHIC PLATES
    M.L.Malagnini, M.Pucillo, P.Santin, G.Sedmak, G.L.Sicuranza   133

AUTOMATIC ANALYSIS OF OBJECTIVE PRISM SPECTRA
    P.Hewett, M.Irwin, P.Bunclark, M.Bridgeland,
    E.Kibblewhite, R.McMahon   137

REMARKS ON T-GRAIN TECHNOLOGY APPLIED TO ASTROPHOTOGRAPHY
    A.Maury   141

MICROSPOTS ON IIIa-J PLATES ('GOLD SPOT DISEASE')
    M.E.Sim   143

INTERNAL CALIBRATION OF ASTRONOMICAL PHOTOGRAPHS
    P.S.Bunclark, M.J.Irwin   147

INTERNATIONAL HALLEY WATCH WIDE FIELD NETWORK FOR LARGE-SCALE
PHENOMENA, CALIBRATION OF SCHMIDT PLATES USING STAR PROFILES
    D.A.Klinglesmith III, S.W.Rupp   155

THE 2020GM PDS MICRODENSITOMETER AT MUENSTER
    W.C.Seitter   159

M.A.M.A. PROJECT - A NEW MEASURING MACHINE IN PARIS
    J.Guibert, P.Charvin, P.Stoclet   165

IMAGE DETECTION SYSTEM FOR SCHMIDT PLATES
    H.Maehara, T.Yamagata   169

A NEW WIDE FIELD ELECTROGRAPHIC CAMERA AS AN OPTIMUM DETECTOR
FOR SCHMIDT-TYPE TELESCOPES
    P.J.Griboval, X.Z.Jia   173

TABLE OF CONTENTS

ASTRONOMICAL PERFORMANCES OF THE MEPSICRON,
A NEW LARGE AREA IMAGING PHOTON COUNTER
C.Firmani, L.Gutierrez, E.Ruiz, L.Salas,
G.F.Bisiacchi, F.Paresce ... 177

THE APPLICATION OF OPTICAL FIBRE TECHNOLOGY
TO SCHMIDT TELESCOPES
J.A.Dawe, F.G.Watson ... 181

IMAGES FROM LARGE SCHMIDT TELESCOPES
D.S.Brown, C.N.Dunlop, J.V.Major ... 185

THE DETERMINATION OF THE VIGNETTING FUNCTION
OF A SCHMIDT TELESCOPE
J.A.Dawe ... 193

THE HIPPARCOS IMAGE
J.-Y.Le Gall, M.Saisse ... 197

THE SCHMIDT TELESCOPE ON CALAR ALTO
K.Birkle ... 203

THE 50/70-CM SCHMIDT TELESCOPE
AT THE BULGARIAN NATIONAL ASTRONOMICAL OBSERVATORY
M.K.Tsvetkov ... 207

ON THE ASTRONOMICAL RESEARCH
WITH THE TORUN 60/90 CM SCHMIDT TELESCOPE
A.Woszczyk ... 211

JAPANESE ORBITING ULTRAVIOLET TELESCOPE PROJECT:
UVSAT WORKING GROUP REPORT
M.Iye ... 215

ASTROMETRY WITH SCHMIDT TELESCOPES
C.A.Murray ... 217

COMPILATION OF THE HIPPARCOS INPUT CATALOGUE-
AN EXTENSIVE USE OF SCHMIDT SKY SURVEYS
C.Turon-Lacarrieu ... 225

SMALL BODIES OF THE SOLAR SYSTEM
C.T.Kowal ... 229

WIDE-FIELD IMAGING OF HALLEY'S COMET DURING 1985-1986
USING SCHMIDT-TYPE TELESCOPES
J.C.Brandt, D.A.Klinglesmith III, M.B.Niedner,Jr., J.Rahe ... 233

OBJECTIVE PRISM SPECTROSCOPY OF THE TAIL OF COMET AUSTIN 1982 G
K.Jockers, L.G.Balázs ... 237

MISSING MATTER IN THE VICINITY OF THE SUN
  J.N.Bahcall   241

GALACTIC RESEARCH WITH SCHMIDT-TELESCOPES
  W.Becker   247

SPACE DISTRIBUTION OF RED GIANTS AND THE GALACTIC STRUCTURE
  K.Ishida   257

OBJECTIVE PRISM SURVEY OF THE OUTER GALACTIC HALO
  K.Ratnatunga, K.C.Freeman   261

A SURVEY FOR O-B STARS IN THE PUPPIS WINDOW
  D.J.Westpfahl,Jr.   265

STATISTICS OF A-TYPE STARS AS POSSIBLE INDICATOR OF STAR FORMATION
  L.G.Balazs   269

DISCOVERY OF NEW BRIGHT PECULIAR STARS OF THE NORTHERN SKY
  W.P.Bidelman   273

USE OF THE UK SCHMIDT FOR A PHOTOMETRIC INVESTIGATION
OF RED STARS IN THE ORION NEBULA MOLECULAR COMPLEX
  A.D.Andrews, B.McGee   279

A SURVEY OF NORTHERN BOK GLOBULES AND THE MON OB1/RI ASSOCIATION
FOR H-ALPHA EMISSION STARS
  K.Ogura, T.Hasegawa   283

PRESENT STATE OF THE WORK ON AUTOMATIC SPECTRAL CLASSIFICATION
AT TARTU
  V.Malyuto   287

LA MESURE DES VITESSES RADIALES AVEC UN PRISME OBJECTIF
ASSOCIE A UN TELESCOPE DE SCHMIDT
  Ch.Fehrenbach, R.Burnage   291

THE YOUNG OPEN CLUSTER NGC 2384
  S.M.Hassan   295

ASIAGO SCHMIDT SURVEYS OF VARIABLE STARS AND SUPERNOVAE
  L.Rosino   301

HAMBURG OBSERVATORY NORTHERN MILKY WAY SPECTRAL SURVEY
FOR EMISSION OBJECTS
  L.Kohoutek   311

STAR COUNTS
  R.G.Kron   315

TABLE OF CONTENTS

STELLAR POPULATION SYNTHESIS AND STAR COUNTS TO CONSTRAIN
THE GALACTIC STRUCTURE
    A.Robin, M.Creze   325

AN AUTOMATED METHOD OF GENERAL STAR COUNTS FOR DARK CLOUDS
    H.Ohtani   329

THE MAGELLANIC CLOUDS
    B.E.Westerlund   333

STAR COUNTS AND DYNAMICAL PARAMETERS IN THE SMC
    M.Kontizas, E.Kontizas   347

NARROW SPECTRAL RANGE OBJECTIVE-PRISM TECHNIQUE
APPLIED TO A SEARCH FOR SMALL MAGELLANIC CLOUD MEMBERS
    M.Azzopardi   351

PHOTOGRAPHIC PHOTOMETRY WITH SCHMIDT PLATES
OF STAR CLUSTERS IN THE SMC
    M.Kontizas   355

LUMINOSITY FUNCTIONS OF OLD GLOBULAR CLUSTERS IN THE SMC
    M.Kontizas, E.Kontizas   359

ELLIPTICITIES OF 'BLUE' AND 'RED' GLOBULAR CLUSTERS IN THE SMC
    E.Kontizas, D.Dialetis, Th.Prokakis, M.Kontizas   363

PHOTOMETRY OF EXTENDED SOURCES
    G.de Vaucouleurs   367

HIGH-RESOLUTION NARROW-FIELD VERSUS LOW-RESOLUTION WIDE-FIELD
OBSERVATIONS OF GALAXIES
    M.Capaccioli, E.Davoust, G.Lelièvre, J.-L.Nieto   379

GLOBAL STRUCTURE OF GALAXIES AND QUANTITATIVE CLASSIFICATION
    S.Okamura, K.Kodaira, M.Watanabe   383

SURFACE PHOTOMETRY OF PURE DISK GALAXIES
    C.Carignan   385

ELLIPTICAL GALAXIES WITH SHELLS
    D.Carter   389

BRAND NEW CANDIDATES FOR DEEP SCHMIDT PLATES
    M.Capaccioli, E.Davoust, G.Lelièvre, J.-L.Nieto   393

A SEARCH FOR 'YOUNG GALAXIES'
    R.McMahon, R.Terlevich, C.Hazard, M.Irwin, J.Melnick   395

A SURVEY OF SOUTHERN COMPACT AND BRIGHT NUCLEUS GALAXIES
    A.P.Fairall   397

OBJECTIVE-PRISM REDSHIFTS OF FAINT GALAXIES
    J.A.Cooke, B.D.Kelly, S.M.Beard, D.Emerson   401

OBJECTIVE-PRISM GALAXY REDSHIFTS
IN FIELDS AROUND THE SOUTH GALACTIC POLE
    Q.A.Parker, H.T.MacGillivray, R.J.Dodd,
    J.A.Cooke, S.M.Beard, B.D.Kelly, D.Emerson   405

SEARCHING FOR EMISSION-LINE GALAXIES
    T.D.Kinman   409

REDUCTION OF SLITLESS SPECTRA -
THE DETECTION OF FAINT EMISSION LINES
    H.-M.Adorf, H.-J.Röser   423

UV GALAXIES AND SUPERASSOCIATIONS
    E.Ye.Khachikian   427

THE CERRO EL ROBLE SAMPLE OF FAINT ULTRAVIOLET EXCESS OBJECTS
IN THE SOUTH GALACTIC POLE
    L.E.Campusano, C.Torres   433

A STUDY OF ULTRAVIOLET-EXCESS GALAXIES BASED ON THE KISO SURVEY
    B.Takase, T.Noguchi, H.Maehara   439

QUASI STELLAR OBJECTS AND ACTIVE GALACTIC NUCLEI
    C.Barbieri   443

DETECTION OF QSOs NEAR LARGE GALAXIES
    A.S.Pocock, J.C.Blades, M.V.Penston, M.Pettini   457

AN OPTICAL SURVEY FOR VERY FAINT QUASAR CANDIDATES
    D.Hamilton   461

ON THE CROSS-CORRELATION OF GALAXIES WITH UVX OBJECTS AND QSOS
    B.J.Boyle, T.Shanks, R.Fong   467

THE OPTICAL VARIABILITY OF 3C 446
    C.Barbieri, S.Cristiani, G.Romano   473

THE PHYSICAL NATURE OF THE BLUE OBJECTS
IN THE FIELD OF BD+15°2469 (VIRGO CLUSTER)
    L.Abati Erculiani, H.Lorenz   475

TABLE OF CONTENTS

OPTICAL IDENTIFICATIONS OF RADIO SOURCES WITH ACCURATE POSITIONS USING THE UKST IIIa-J PLATES
    A.Savage, D.L.Jauncey, M.J.Batty, S.Gulkis, D.D.Morabito, R.A.Preston   481

GALAXY COUNTS
    J.A.Tyson   489

FAINT GALAXY NUMBER COUNTS
    T.Shanks, P.R.F.Stevenson, R.Fong, H.T.MacGillivray   499

OBSERVATION OF INTERGALACTIC DUST BY SCHMIDT-TELESCOPES
    S.Marx   507

SUPERCLUSTERING AND SUBSTRUCTURES
    G.Chincarini   511

COSMOLOGY WITH THE SPACE SCHMIDT TELESCOPE - GALAXY COLORS AND COLOR DISTRIBUTIONS
    G.A.Bruzual   527

OPTICAL DESIGN WITH THE SCHMIDT CONCEPT
1) GROUND-BASED DEVELOPMENT
2) THE SPACE SCHMIDT PROJECT FOR THE 1990'S ?
    G.Lemaître   533

IMPLEMENTATION AND USE OF WIDE FIELDS IN FUTURE VERY LARGE TELESCOPES
    J.R.P.Angel   549

SCHMIDT ASTRONOMY AND THE SPACE TELESCOPE
    B.M.Lasker   563

GROUND BASED DEVELOPMENTS - REDUCTION SYSTEMS
    P.J.Grosbøl   571

SUMMARY
    L.Woltjer   577

INDEX OF AUTHORS   581

INDEX OF NAMES   583

INDEX OF ASTRONOMICAL OBJECTS   605

INDEX OF SUBJECTS   611

EDITOR'S FOREWORD

The idea of holding a colloquium on Schmidt telescopes (techniques and science) originated from the observation that, in the last ten years and in spite of the remarkable developments and achievements in this field of astronomical research, there had been no specific opportunity for the experts to meet together, make the point on the state of the art, discuss and coordinate future plans. Therefore, Prof.L.Rosino, one of the pioneers in the use of wide-field telescopes, driven also by the wish of honouring the over four decades of activity of the Asiago Observatory, proposed to the Executive Committee of the International Astronomical Union to sponsor a colloquium on 'Astronomy with Schmidt type telescopes' to be held at Asiago at the end of the summer of 1983. Details about the composition of the Scientific Organizing Committee and the sponsoring organizations are given in Prof.Rosino's 'Welcome to the Participants'. The granting of this proposal was the beginning of a number of headaches for the members of the Local Organizing Committee, R.Barbon, F.Ciatti, P.Rafanelli and myself. If, organizationwise, the colloquium was successful, this is truly due to the generous efforts of my colleagues of the SOC and to the efficient organization of the Linta Park, the hotel hosting the meeting.

During four days of constant rain (and with few breaks for some sober social events), the 120 participants from about 30 countries gave rise to a lively meeting, full of interesting contributes and ponderous lectures. All of the papers and posters but one were collected in this book.

Unfortunately, one of the highlights of the meeting, the 'Joint Discussion' leaded by R.Kron, was not included (except for a couple of bites scattered here and there), since in general the speakers did not provide the texts of their remarks. As an important remnant of the discussion, however, there is the 'Resolution' opening this volume. Another excellent contribution which is missed is the summary of the poster papers by R.Cannon. It was not requested for publication since the posters themselves are present in these proceedings.

After the addresses to the participants by Prof.Rosino, director of the Asiago Astrophysical Observatory, Dr.R.West on behalf of I.A.U., and Prof.L.Merigliano, Rector of the Padova University, the scientific work was opened by a lecture of Prof.L.Woltjer, who also took the hard task of summarizing the meeting. Except for the two papers of Prof.Woltjer, the order of most of the other contributions was modified according to editorial reasons. Therefore, the Table of Contents can not account for the chairmen of the sessions (Prof.J.Bahcall, Dr.V.Blanco, Dr.R.Cannon,

Prof.G.de Vaucouleurs, Prof.C.Fehrenbach, Prof.E.Khachikian, Dr.S.Marx, Prof.B.Takase and Prof.L.Woltjer) whose experienced guidance of the discussion is gratefully acknowledged.

In closing, I wish to thank all those who helped me in the preparation of this volume, particularly Drs. E.Held and R.Rampazzo.

M.CAPACCIOLI

LIST OF PARTICIPANTS

ABATI ERCULIANI L., Astronomical Observatory, Padova
ADORF H.-M., Max-Plank-Institut für Astronomie, Heidelberg
ALKSNIS A., Radioastrophysical Observatory, Riga
ANDREWS A.D., Armagh Astronomical Observatory
ANGEL J.R.P., Steward Observatory
BAHCALL J.N., Institute for Advanced Studies, Princeton
BALAZS L.G., Konkoly Observatory, Budapest
BARBIERI C., University of Padova
BARBON R., Astrophysical Observatory, Asiago
BECKER W., Astronomical Institute, Basel
BENACCHIO L., Astronomical Observatory, Padova
BENVENUTI P., IUE Observatory, VILSPA
BERTOLA F., University of Padova
BETTONI GALLETTA D., Astronomical Observatory, Padova
BIDELMAN W.P., Case Western Reserve University
BIRKLE K., Centro Astronomico Hispano-Alémân, Almeria
BLANCO V., Cerro Tololo Interamerican Observatory
BONOLI F., University of Bologna
BOYLE B.J., University of Durham
BRACCESI A., University of Bologna
BRUZUAL G., C.I.D.A., Merida
BUNCLARK P., Institute of Astronomy, Cambridge
BURNAGE R., Haute Provence Observatory
CAMPUSANO L.E., University of Santiago
CANNON R.D., Royal Observatory, Edinburgh
CAPACCIOLI M., University of Padova
CARIGNAN C., Kapteyn Astronomical Institute, Groningen
CARTER D., Mt. Stromlo and Siding Spring Observatories
CAWSON M., Institute of Astronomy, Cambridge
CHINCARINI G., University of Oklahoma, Norman
CIATTI F., Astrophysical Observatory, Asiago
CRANE P., European Southern Observatory
CRISTALDI S., Astronomical Observatory, Catania
CRISTIANI S., European Southern Observatory
DAVOUST E., Astronomical Observatory, Besançon
DAWE J.A., Royal Observatory, Edinburgh
DETTMAR R.J., Max-Plank-Institut für Radioastronomie, Bonn
DE VAUCOULEURS G., University of Texas, Austin
DODD R.J., Carter Observatory
D'ODORICO S., European Southern Observatory
DUERBECK H., University Observatory, Bonn
FAIRALL A.P., University of Cape Town

FEDERICI L., University of Bologna
FEHRENBACH C., Haute Provence Observatory
FIRMANI C., UNAM, Mexico City
GALLETTA G., Astronomical Observatory, Padova
GILMORE G., Royal Observatory, Edinburgh
GROSBØL P.J., European Southern Observatory
GUIBERT J., Observatory of Paris
HADLEY B.W., Royal Observatory, Edinburgh
HAMATSCHEK R., K. Zeiss, Jena
HAMILTON D., University of Chicago
HASSAN S.M., University of Riad
HAWKINS M., Royal Observatory, Edinburgh
HEWETT P., Institute of Astronomy, Cambridge
HUNT L., Astronomical Observatory, Florence
IIJIMA T., Astrophysical Observatory, Asiago
ISHIDA K., Astronomical Observatory, Tokyo
IYE M., European Southern Observatory
KHACHIKIAN E.Y., Astrophysical Observatory, Byurakan
KINMAN T.D., Kitt Peak National Observatory
KLINGLESMITH D.A.III, Goddard Space Flight Center, Greenbelt
KODAIRA K., Astronomical Observatory, Tokyo
KOHOUTEK L., Astronomical Observatory, Hamburg
KONTIZAS E., National Observatory, Athens
KONTIZAS M., National Observatory, Athens
KRON R.G., University of Chicago
KUNT D., Observatory of Meudon
LABHARDT L., University of Washington, Seattle
LASKER B.M., Space Telescope Science Institute, Baltimore
LAUBERTS A., European Southern Observatory
LEMAITRE G., Astronomical Observatory, Marseille
LODEN L.O., Astronomical Observatory, Uppsala
LONGO G., Astronomical Observatory, Naples
MACGILLIVRAY H.T., Royal Observatory, Edinburgh
MAHEARA H., Astronomical Observatory, Tokyo
MAFFEI P., University of Perugia
MALAGNINI M.L., Astronomical Observatory, Trieste
MALIN D.F., Anglo-Australian Observatory
MARX S., Tautenburg Observatory
MAURY A., Observatory of C.E.R.G.A.
McCARTHY M.F., Vatican Observatory
McMAHON R., Institute of Astronomy, Cambridge
MILLIKAN A., Eastman Kodak Research Laboratories
MOSS C., Vatican Observatory
NATSVLISHVILI R., Astrophysical Observatory, Abastumani
NIETO J.-L., Observatory of Pic du Midi
OGURA K., University of Tokyo
OHTANI H., University of Manchester
PARKER Q.A., University of St. Andrews
PESCH P., Warner and Swasey Observatory
POCOCK A.S., Royal Greenwich Observatory
POLLAS C., Observatory of C.E.R.G.A.

LIST OF PARTICIPANTS

POULAIN P., Observatory of Toulouse
RAFANELLI P., Astronomical Observatory, Padova
ROBIN A., Observatory of Besançon
ROMANO G., University of Padova
ROSINO L., University of Padova
SANDULEAK N., Warner and Swasey Observatory
SANTANGELO M., University of Pisa
SEDMAK G., University of Trieste
SEITTER W.C., University of Muenster
SHANKS T., University of Durham
SHI W., University of Manchester
SIM M.E., Royal Observatory, Edinburgh
TAKASE B., Astronomical Observatory, Tokyo
TSVETKOV M.K., Bulgarian National Astronomical Observatory
TURON-LACARRIEU C., Observatory of Paris
TYSON J.A., Bell Laboratories, Murray Hill
VAN CITTERS W., National Science Foundation, Washington
WEST R.M., European Southern Observatory
WESTERLUND B.E., Astronomical Observatory, Uppsala
WESTPFAHL D.J.Jr., Montana State University
WILD P., Astronomical Institute, Bern
WIRTH A., Harvard-Smithsonian Center for Astrophysics, Cambridge
WOLTJER L., European Southern Observatory
WOSZCYC A., University of Torun
ZAVATTI F., University of Bologna
ZEILINGER W., University of Wien
ZITELLI V., University of Bologna

WELCOME ADDRESS TO THE PARTICIPANTS

L. Rosino
Director of the Asiago Astrophysical Observatory
of the University of Padova and
Chairman of the Scientific Organizing Committee

Ladies and Gentlemen, dear Colleagues,

I am very pleased to welcome you at Asiago for this Second Conference on the Schmidt telescopes and their role in Astronomy. The previous Meeting was held at the Hamburg Observatory, the same place where Barnard Schmidt invented and built his new telescope. After twelve years of intense activity, it was time to promote a new Conference in order to make the point on the results hitherto achieved and discuss the prospectives of future work with Schmidt telescopes.

I like to express my gratitude to the Executive Committee of the International Astronomical Union for having accepted my proposal that a Colloquium would take place at Asiago, in the site and under the patronage of the University of Padova. After all, Padova has a privilege unique in the world: under her sky 374 years ago with his "canocchiale" Galileo started the exploration of the heavens, which is now continued, in a much larger scale, by the Schmidt telescopes. We shall be very glad and honoured to receive and to show our instruments, among which there are two Schmidt telescopes, to the Colleagues who wish to visit our Observatory.

Two powerful Schmidt telescopes have entered in operation in the southern emisphere after the Hamburg Conference: the ESO 1-m at La Silla in Chile and the UK 122-cm, equipped with an objective prism of the same size, at Siding Spring in Australia. They are both used to extend the Palomar Sky Atlas to the southern emisphere. Professor Woltjer, Dr. West, Dr. Cannon will certainly give us information on those surveys and on the programmes carried out with these and other Schmidt telescopes. What I can say is that I had the opportunity of examining some of the films of the survey now in distribution and I was impressed by the high quality of the images and the deepness of the sky photographs.

High sensitivity emulsions and new photographic techniques have in fact increased the efficiency of the Schmidt telescopes which can operate from the ultraviolet to the infrared. Moreover, several proposals have been made to put large Schmidt telescopes into orbit around the Earth so that they may scan the whole of the sky in the far ultraviolet.

Since a Schmidt telescope can produce a very large amount of material, techniques have been introduced for the automated fast reduction of wide field plates with the highest possible precision. The programme of this Colloquium indicates that all the topics, together with the prospectives of astrometry, spectroscopy and space research with Schmidt telescopes, will be amply discussed.

In conclusion, I express my gratitude to the Colleagues, some from very distant countries, who accepted to participate actively in this Conference with review papers, communications and posters. I am also grateful to the sponsoring organizations: the International Astronomical Union, the University of Padova, the Italian National Council of Researches and to the public and private institutions who have given their support and assistance to this Colloquium; in particulat to the Regione Veneta, the cities of Padova and Asiago, the Provincia of Vicenza, the Comunita' Montana, the Banca Cattolica del Veneto. My deepest thanks to all of them and to the Rector of the University of Padova who has accepted our invitation to be with us at the opening of this Conference. Finally, I am particularly thankful to the Colleagues of the Scientific Organizing Committee,

R.Cannon, K.Freeman, K.Henize, E.Khachikian, E.Kibblewhite, F.Macchetto, S.Marx, M.McCarthy, W.Sargent, A.Savage, J.Sersic, B.Takase, A.Tyson, R.West and B.Westerlund,

and to the members of the Local Organizing Committee,

M.Capaccioli, R.Barbon, F.Ciatti and P.Rafanelli,

who have taken care of the organization.

To all the participants I express my best wishes for a fruitful work.

ADDRESS BY THE GENERAL SECRETARY OF THE
INTERNATIONAL ASTRONOMICAL UNION

Richard M. West

It is a pleasure to extend warm greetings from the IAU Executive Committee to the organizers and to all participants in IAU Colloquium 78.

When the possibility about holding a meeting on "Astronomy with Schmidt-type Telescopes" was discussed at the time of the IAU General Assembly in Patras, Greece, August 1982, there was full concensus that this was a most timely subject and by accepting sponsorship, the IAU Executive Committee expressed its appreciation of the high level of the proposed programme.

Today, one year later, the great need for this meeting is underlined by the large number of participants and the many papers which have been announced. It is interesting to note that no less than six Commissions and two Working Groups of the IAU are responsible for the scientific programme, indicating a very broad support and a wide range of topics. Curiously enough, the person who has contributed most to this branch of astronomical science, and whose name figures in the title of the meeting, Bernhard Schmidt, was not a member of the IAU and carried out most of his revolutionary work without being known by the world-wide astronomical community. Perhaps this is because in those days, more than 50 years ago, only full-time, professional astronomers could become IAU members and because consultants were not yet co-opted in the IAU Commissions.

Looking through the programme, one is immediately impressed by the wide variety of topics. It may perhaps even be justified to ask what, for instance, comets and elliptical galaxies have to do with each other. However, this is typical in Schmidt astronomy since the photographical emulsion does not distinguish between the two: it shows everything which can be seen in a given sky area. Any astronomical photograph may reveal completely unforeseen objects, and everybody who scrutinizes such material may be lucky to benefit from what is commonly termed serendipity.

I am sure that the participants in this meeting will, in a similar way, learn much from areas outside their own field during the next few days, and I wish all of you a most fruitful and interactive meeting.

RESOLUTION

Participants in the IAU Colloquium No.78: 'Astronomy with Schmidt-Type Telescopes', in concordance with the IAU Commission 9 Working Group on Photographic Problems.
Acknowledging the impressive advances during the past decade resulting from the introduction of new emulsions of increased sensitivities and over a wide spectral range, and
Recalling the continued interest in our science shown by several manufacturers of photographic materials, in particular the Eastman Kodak Company and,
Noting the continuing and growing need for photographic emulsions in support of both ground-based and space astronomy, especially with wide-field, high resolution sky survey telescopes for which it is unlikely in the near future that panoramic electronic detectors will have the detector size and resolution available on photographic materials,
Urge all manufacturers to initiate and continue research with the aim of providing new emulsions with improved characteristics - for example quantum efficiency, sensitivity and spectral range.

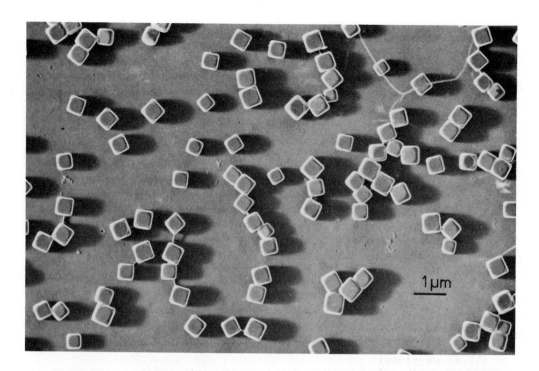
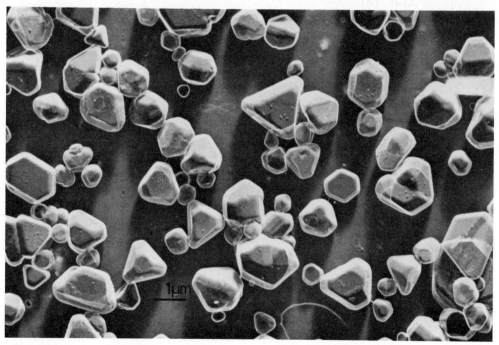

Photographic grains typical of those found in IIIa-J (upper) and (lower) 103a-O emulsions.

SCHMIDT TELESCOPES AS DISCOVERY INSTRUMENTS

L. Woltjer
European Southern Observatory
Karl-Schwarzschild-Strasse 2
D-8046 Garching b. München

INTRODUCTION

In 1930 B. Schmidt invented the telescope that bears his name. Eleven years later, there were already two dozen Schmidt telescopes with primaries of 25 cm or more in diameter (Dimitroff and Baker 1945). An invention spreading so rapidly must have fulfilled a basic need - the need for a large field with good image quality over a substantial wavelength range. Here I shall briefly review the role of the large Schmidt telescopes in contemporary astronomy.

SKY SURVEYS

Perhaps the most important contributions made by Schmidt telescopes is that they have provided for the first time very deep maps of the whole sky down to stellar magnitude 20 and beyond. In table 1 the large scale deep surveys are listed. The Palomar Sky Survey and the ESO Quick Blue Survey have been fully completed, the SRC Blue survey is nearly complete, and the others are in progress. The newer photographic emulsions have given significant gains in both resolution and limiting magnitude. As a consequence, the recent decision at Palomar to make a new survey of the northern sky is most welcome.

In addition to these large mapping surveys, also more specialized surveys have been undertaken at various observatories covering important parts of the sky: numerous surveys for blue stars (on the basis of U-B and/or U-V colour), near IR surveys, surveys with special colour filters for emission line objects, spectral surveys for the classification of stars or for the discovery of quasars, and surveys for variable stars, stars of large proper motion and asteroids. Surveys for highly polarized objects could also be of interest, but to date only a few experimental studies have been made.

Table 1

DEEP SKY SURVEYS

SURVEY	COLOUR	DEC	N	EMULSION	λ	m
PALOMAR	BLUE	>-33	935	103a-O	3400 - 4800	21
	RED	>-33	935	103a-E	6300 - 6600	20
ESO QUICK	BLUE	<-18	606	IIa-O	3700 - 5000	21
ESO/SRC	BLUE	<-17	606	IIIa-J	3800 - 5300	23
	RED	<-18	606	IIIa-F	6300 - 6800	22
UK SCHMIDT	BLUE	-18 — +3	288	IIIa-J	3800 - 5300	23
EQUATORIAL	RED	-18 — +3	288	IIIa-F	6300 - 6800	22

Note: N = number of fields
λ = estimated half power wavelength limits of the filter-plate combination
m = limiting magnitude on the best plates

Before proceeding with a discussion of the scientific areas in which these surveys are of importance, it is necessary to consider for a moment the relative roles of Schmidt telescopes and of large reflectors in the 4-m class. Of course, as has been pointed out frequently, the Schmidt surveys have led to the discovery of many "interesting" objects that subsequently have been studied with large telescopes in spectroscopic and imaging modes. But for many survey type studies the better scale of large telescopes would also be preferable.

As an example, in figure 1 a comparison is made between the images of the Sculptor dwarf irregular galaxy (Laustsen et al 1977) and its surroundings on the ESO Quick Blue Survey, the SCR blue (J) survey, the ESO red (F) survey, and on a IIIa-J plate taken with the ESO 3.6 m telescope. The greater depth of the IIIa-J in comparison with the IIa-O is very much in evidence, as is the superiority of the 3.6-m image in comparison with that obtained with the Schmidt

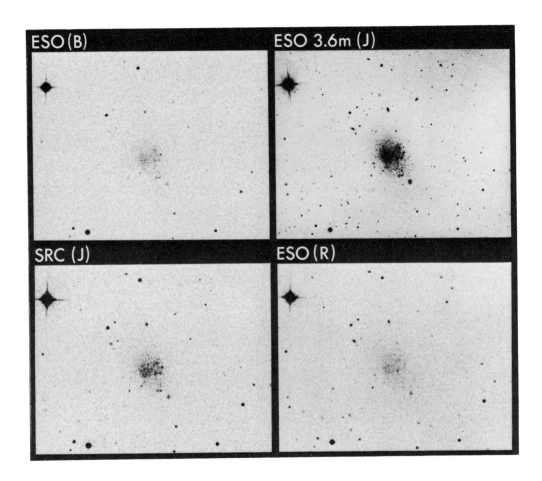

Figure 1: The Sculptor Dwarf Irregular Galaxy photographed with the ESO Schmidt in the Quick Blue Survey on IIa-O (ESO-B) and in the red survey on IIIa-F (ESO-R), with the UK Schmidt in the blue survey on IIIa-J (SRC-J) and with the ESO 3.6 m telescope on IIIa-J (ESO 3.6 m-J). The bright blue stars in the dwarf galaxy are clearly visible on the blue exposures. The surrounding foreground stars and background galaxies have on the average about equal images on the SRC(J) and ESO(R) plates.

telescopes. While the bright blue stars of the dwarf galaxy are comparatively faint on the red survey, it is interesting to note that almost all field stars visible on the red survey plate are visible on the blue one and vice versa. However, many more field stars are visible on the 3.6-m plates, and also the distinction between faint galaxies and stars is much clearer. Suppose then we have both a Schmidt and a large telescope available, what criterion should we apply to decide which of the two to use for a particular survey?

A crude answer may be based on the following considerations. The field of a large telescope is usually of the order of a square degree. In a typical deep survey, several plates have to be taken per field, and with the pressure on large telescope time it is difficult to extend such a survey much beyond 10 square degrees. To learn something about the objects being surveyed, one should be at least able to determine their number with some accuracy - perhaps to within 20 %. This would require 25 objects to be detected in the survey, corresponding to 2.5 objects per square degree or to a total number over the whole sky $N = 10^5$. While one may vary these numbers a bit, the conclusion is clear: if $N \gg 10^5$ it is reasonable to make a survey with a large telescope, but if $N \ll 10^5$ this is impossible, and a large field Schmidt is needed. Of course, this does not mean that one cannot use a Schmidt also in the former case, especially when no other telescope is available, but the results are likely to be less satisfactory if the objects are faint.

Another way of looking at the situation is to note that an all-sky survey in two colours with large telescopes would require about 10^5 plates corresponding to about 25,000 dark nights or about 250 telescope years. With around 10 large telescopes in the world engaged for 10 % of the dark time in such a survey, it would take 250 years to complete it. With five of the world's large Schmidts doing the same programme, a gain of a factor of $5 \times 30 = 150$ in time would be achieved - the factor of 30 arising from the larger field. As a result, such a survey could be completed in two years. If one were to add 3 spectral plates and 3 more colours per field, the time required would become four times larger. With a further 20 plates for searches for variables and for astrometry, 3500 years would be needed with the large telescopes and 25 years with the five Schmidts. What these figures show is that with large telescopes less than a few percent of the sky is likely to be imaged during the coming decades. On the other hand, though some selectivity is needed in Schmidt programmes, the present Schmidt telescopes in the world appear to be sufficient to deal with much of what is needed during the medium term future.

SCIENTIFIC PROGRAMMES WITH SCHMIDT TELESCOPES

In table 2 are listed some typical objects for which the all-sky number $N \lesssim 10^5$ and which consequently are in the prime domain of research with Schmidt telescopes.

Table 2

TYPICAL OBJECTS IN SCHMIDT ASTRONOMY

GALAXIES	B < 16	PLANETARIES
DWARF GALAXIES		H II REGIONS
PECULIAR GALAXIES		SUPERNOVA REMNANTS
QUASARS	B < 19	DARK NEBULAE
BL LAC		
CLUSTERS OF GALAXIES	z < 0.5	
SUPERCLUSTERS	z < 1	
PECULIAR STARS	B < 15	ASTEROIDS
VARIABLE STARS		COMETS
LOW METALLICITY STARS		
INTRINSICALLY FAINT STARS		

Relatively bright galaxies have been extensively searched for on the sky surveys by Zwicky et al. (1968), by Vorontsov-Velyaminov and Arhipova (1968), by Nilson (1973) and by Lauberts (1982). The catalogues made by these authors include a total of about 50,000 galaxies down to (uncertain) photographic magnitudes of about 15.5 or angular diameters of about 1 arc minute. These catalogues have served as a rich material for further studies with larger telescopes and also for the study of the distribution of galaxies in space.

About 1800 dwarf galaxies - difficult to detect because of low surface brightness - have been found on the sky surveys by Tully (see van Woerden 1980). Many of these have been subsequently studied by 21-cm radio techniques. Many "interacting", "peculiar" and "eruptive" galaxies have been found by Vorontsov-Velyaminov (1959), by Lauberts (1982), by Arp and Madore (unpublished), and by Zwicky (1971). Some of these have been studied in more detail also spectroscopically leading to concepts involving tidal interaction between galaxies and eruptive phenomena in the nuclei of galaxies. Seyfert galaxies and galaxies with active star formation have been found by Markarian (1969) on the basis of their blue continuum on objective prism plates.

In studies of the distribution of fainter galaxies on the Palomar survey, Abell (1958) and Zwicky and his associates (1968) found numerous clusters which have gained in importance with the discovery of much associated hot, X-ray emitting gas. It is now generally agreed that the distribution of clusters is not random, and this has led to the concept of superclusters as possibly the largest units in the universe.

Quasars were first discovered on the basis of the identification of radio sources. The subsequent discovery by Sandage (1965) that quasars without perceptible radio emission constitute a far more numerous class resulted from searches of Schmidt plates on the basis of U-B and U-V colours. Apart from the U-B excess which since that time has been found to be a good indicator for quasars with redshifts less than $z = 2.2$, quasars have been found on the basis of the detection of emission lines in low dispersion objective prism plates taken with Schmidt telescopes. A large survey for U-B excess objects covering 1/4 of the whole sky has recently been made by Schmidt and Green (1983). Apart from about a hundred quasars and Seyferts with $B < 16$, more than a thousand white dwarfs, cataclysmic variables and other blue stars were discovered.

The number of quasars as a function of magnitude increases very steeply due to cosmic evolution (factor of eight per magnitude) and, as a consequence, small variations in photometric characteristics of plates may lead to large variations in the numbers found. Nevertheless, the very large incompleteness (factor of about 3) in the objective prism surveys with the 24-inch Schmidt at Tololo as shown by comparison with data from the grism surveys with the 4-m telescope (Osmer 1980, Woltjer and Setti 1982) remains surprising.

The quasars provide a good example of the interplay between Schmidt telescopes and large reflectors. Virtually all information below $B = 19$ comes from the former, while the latter have provided most of the data for $B > 19$, the total number of quasars with $B < 19$ being about 10^5 in agreement with the discussion given previously.

Related to quasars are the BL Lac objects, which are characterized by large amplitude photometric variability and by strong polarization (5 - 40 %). Almost nothing is known about possible radio quiet BL Lac objects. Perhaps the polarization characteristics could be used as the basis for a systematic Schmidt survey.

Galactic structure is one of the classical areas of Schmidt astronomy. The work of McCuskey and others with the Case Schmidt, that of Becker with Palomar Schmidt plates, the work at Bergedorf and elsewhere have much contributed to our knowledge about the distribution of stars (see for example Blaauw and Schmidt 1965). Interstellar absorption has caused major problems in some of these studies. The ease with which 21-cm and other radio studies could bypass this problem has perhaps shifted the emphasis away from this type of work in the galactic plane, but it remains necessary to search for distant OB stars to provide distance estimates for features identified in the radio astronomical studies.

At higher galactic latitudes, the problems of interstellar absorption diminish. Currently, the work of Becker is being extended by Sandage down to fainter magnitudes with larger telescopes. Again, it is seen here how one passes from a first view with Schmidt telescopes to more detailed studies with large telescopes in small areas selected on the basis of the earlier Schmidt work. Partly for galactic structure studies, but mainly to find white dwarfs and other "blue stars", extensive surveys of stars with small or negative U-B or U-V colours have been made by Haro, Luyten, Richter and others with the Tonanzintla, Palomar and Tautenburg Schmidt telescopes. The change in emphasis in these surveys is noteworthy. While Luyten (1965) still could write that "unrecognized blue galaxies may also be included, but this is merely a calculated risk", at present many studies of this type are done with blue galaxies and quasars as the primary objective.

One of the fundamental problems in galactic structure is the determination of the faint end of the luminosity function. A common technique - although giving severely biased samples - is to look for objects of large proper motion on Schmidt plates. As shown by Luyten, accuracies of a few times 10^{-2} arc seconds per year may be obtained from repeat plates of the Palomar survey, and the publications of the Observatory of Minnesota contain numerous catalogues of high proper motion stars. Present evidence does not support the idea that low luminosity stars account for much of the "missing mass" near the sun. However, "degenerate black dwarfs", stars with too low a mass to achieve hydrogen burning, still could make a large contribution. As an example in the models of Staller and de Jong (1981), a case is described where black dwarfs account for the missing mass but where the all-sky number of 10^5 is reached only at V = 23. A proper motion survey on IV-N plates for such cool stars could be of interest.

Spectral surveys with objective prisms have led to the discovery of many interesting stars: WR, Ap, Am, Ba, S, Hα emission stars and others. In other studies, some such stars have been found by taking plates through specially chosen relatively narrow colour filters. Searches for variable stars, supernovae, etc. represent another common area of research with Schmidts.

Of particular interest in studies of galactic evolution are the first generations of stars to form. Do these have zero or very low metal abundances or were heavy elements already present at a very early stage ? The answer to this question must come from Schmidt telescope searches for probably rare "zero" metal stars. An interesting attempt to find such stars is being made by Shectmann and Preston at Palomar who use a filter isolating a small spectral interval around the Ca K line and an objective prism. The filter cuts down on the sky background and reduces the overlap of spectra; consequently, faint magnitudes may be reached.

Schmidt telescopes are also of much use in the study of interstellar matter. Planetary nebulae have been looked for by Abell, Kohoutek and others. It is interesting to note that in a list of 83 new planetaries discovered between 1978 and 1981 (Kohoutek 1983), there are still 15 objects found on the Palomar survey. Apparently, even this very extensively searched survey still has many objects to be discovered. H II regions and supernova remnants are another area of Schmidt astronomy: It would be difficult to study large objects like the Gum nebula or the Vela Supernova Remnant with small field instruments. Numerous dark nebulae have also been found on the sky surveys.

Finally, in the solar system, searches for asteroids on Schmidt plates are of importance. Of special interest are the earth crossing asteroids (Apollo, Amor, etc.) which may have had a significant impact on the appearance of the surface of the earth and on the evolution of life on earth. Special searches as well as serendipitous finds have now led to the discovery of 49 such objects (Shoemaker 1983). An entirely different area of Schmidt work concerns comets, where a large field is needed for proper imaging of the long tails.

THE FUTURE OF SCHMIDT TELESCOPES

From the preceding discussion it is clear that existing Schmidt telescopes still have many years of fruitful work in front of them. With regard to the longer term future, however, two issues appear to be of prime importance: detectors and a Schmidt telescope in space. With regard to the former, further improvements in photographic plates certainly would be important. The introduction of CCD-type detectors, however, could make a qualitative change. At present, CCDs are still small (few cm^2), but there does not appear to be, in principle, any objection to fitting the curved focal place of a Schmidt telescope with a mosaic of CCDs. Also the data handling problem probably would not be an obstacle a decade from now. One may wonder, however, whether pushing the Schmidt images a further 2-3 magnitudes fainter would justify the cost of such an approach. Schmidt astronomy has been particularly fruitful because until now most objects found with Schmidt telescopes could be studied in more detail and also spectroscopically with larger telescopes. This would no longer be the case in a Schmidt telescope that would go much deeper. Certainly, one can think of programmes to do at, say, V = 25 - for example an astrometric search for very faint black dwarfs, but the number of possible programmes seems rather limited.

More promising might be a Schmidt telescope in space; the access to the uv and near IR part of the spectrum could lead to many discoveries. In addition, the potentially better angular resolution and the somewhat reduced sky brightness might allow also deeper surveys in the optical domain with a good separation between galaxies, quasars and stars; if so, this might have interest in the study of the distribution of matter in the universe on the largest scale.

The success of such a Schmidt again depends on the development of a detector that can make use of the resolution achievable in space, and which at the same time covers a large area. The balance between field and angular resolution, both from the optical and the detector point of view, will need careful consideration.

REFERENCES

Abell, G.: 1958, Astrophys. J. Suppl. 3, pp. 211-288
Blaauw, A. and Schmidt, M. eds.: 1965, Stars and Stellar Systems V: Galactic Structure, U. of Chicago Press
Dimitroff, G.Z., Baker, J.G.: 1945, "Telescopes and Accessories", Blakiston
Kohoutek, L.: 1983, IAU Symposium 103, pp. 17-30
Lauberts, A.: 1982, The ESO/Uppsala Survey of the ESO (B) Atlas, ESO
Laustsen, S., Richter, W., van der Lans, J., West, R.M., and Wilson, R.N.: 1977, Astron. Astrophys. 54, pp. 639-640
Luyten, W.J.: 1965, Stars and Stellar Systems V: Galactic Structure, U. of Chicago Press, p. 394
Markarian, B.E.: 1969, Astrophys. 5, pp. 286-301
Nilson, P.: 1973, Uppsala General Catalogue of Galaxies, Uppsala Astr. Obs. Ann. 6
Osmer, P.S.: 1980, Astroph. J. Suppl. 42, pp. 523-540
Sandage, A.: 1965, Astroph. J. 141, pp. 1560-1578
Schmidt, M. and Green, R.F.: 1983, Astroph. J. 269, pp. 352-374
Shoemaker, E.M.: 1983, Ann. Rev. Earth Planet. Sci. 11, pp. 461-494
Staller, R.F. and de Jong, T.: 1981, Astron. Astrophys. 98, pp. 140-148
van Woerden, H.: 1980, ESO/ESA Workshop on Dwarf Galaxies, ESO, pp. 23-29
Vorontsov-Velyaminov, B.A.: 1959, Atlas and Catalogue of Interacting Galaxies, Moscow State University
Vorontsov-Velyaminov, B.A. and Arhipova, V.P.: 1968, Morfologiceskij Katalog Galaktik, Moscow State University
Woltjer, L. and Setti, G.: 1982, "Astrophysical Cosmology", Pont. Acad. Scient. Scripta Varia 48, pp. 293-313
Zwicky, F.: 1971, Catalogue of Selected Compact Galaxies and of Post-Eruptive Galaxies, F. Zwicky publ.
Zwicky, F., Herzog, E., Wild, P., Karpowicz, M. and Kowal, C.T.: 1968, Catalogue of Galaxies and of Clusters of Galaxies, Cal. Inst. of Technology, Pasadena

THE ESO SKY SURVEYS

Richard M. West
EUROPEAN SOUTHERN OBSERVATORY, Karl-Schwarzschild-Straße 2
D-8046 Garching bei München, FRG

ABSTRACT

Technical aspects of the production of photographic Sky Surveys and Atlasses are discussed. The need to optimize all factors, e.g. telescope site, telescope optics, mechanics and control system, hypersensitization, calibration, processing and reproduction is illustrated by the ESO 1m Schmidt telescope and the ESO Sky Surveys.

1. INTRODUCTION

Many photographic sky surveys have been undertaken during the past 100 years. Each of these constitutes a "snapshot" of a part of the sky, and thus provides 1) a <u>record</u> for later research, 2) an <u>inventory</u> of objects in that sky area and 3) <u>identification</u> of objects already known. The general expression is $I = I(\alpha, \delta, \lambda, t)$ which emphasizes the significance of the time parameter: Any sky survey provides a basis for comparison with earlier and future observations and therefore the possibility of discovering and studying <u>variable</u> phenomena. The best known modern photographic sky surveys are those undertaken at Palomar, at the European Southern Observatory and with the UK Schmidt telescope. References are, for instance, found in the papers by by Minkowski and Abell (1963), West (1972) and West and Schuster (1982).

My talk today has been preceded by Professor Woltjer's introduction, dealing with many astronomical aspects of the work with wide-angle telescopes, including the types of discoveries which are possible with such instruments. The next speaker, Dr. R.D. Cannon, will tell you about the highlights of research with the UK 48" Schmidt telescope. I therefore feel that I should not try to repeat these two speakers by going into detail about the rich discoveries made by the ESO 1m Schmidt Telescope and by attempting to summarize the many current astronomical research projects based on the ESO Southern Sky Surveys. It came to my mind that whereas we often talk about scientific research, we seldom realize and discuss the great technical difficulties in obtaining the high-quality photographic plates which are the basis for modern sky surveys. Thus, I have decided to concentrate mostly on the various phases of making such plates, starting with the telescope site and following through until the atlas copies have been distributed. Only thereafter can astronomers all over the world benefit from a survey and start making real astronomical use of it.

In describing the various problems which have been encountered with the ESO Schmidt and at the ESO Sky Atlas Laboratory over the years, I hope to impress upon those who decide to undertake similar projects and those who are just regular users of wide-angle telescopes, how important it is that every link in the chain is optimized. Whenever something goes wrong, the end product will suffer and will not contain the astronomical information which we expect it to carry. Talking with colleagues from other Schmidt telescopes, it has become abundantly clear to me that they have very similar problems, although this is perhaps not always told to the public in a loud voice. I shall therefore in this paper try to expose some of the major problems which were encountered during the production of the ESO Sky Surveys and, fortunately, in almost all cases solved in a satisfactory way.

2. ELEMENTS OF A SKY SURVEY

Chronologically, the elements which are necessary for the achievement of a successful and useful sky survey are the following: telescope site (altitude, climate, seeing); telescope (optical, mechanical and control systems); photographic material (availability, spectral sensitivity, speed, etc.); hypersensitization (methods, stability); exposure at telescope, calibration, processing, copying of original plates, protection and archiving of originals and copies and finally, distribution of the resulting atlasses. In what follows, we shall take a closer look at some of these points in connection with the ESO 1 m Schmidt telescope and the ESO(B) and ESO(R) Surveys.

2.1 Site and Telescope

The ESO La Silla site is probably one of the best on earth. In addition to a large number of clear nights (at least 2500 hours per year) it has a significant fraction of very good seeing of obvious benefit to any seeing limited observations, e.g. the taking of survey plates with the ESO 1 m Schmidt.

The ESO Convention of 5 October 1962 specifically mentions a Schmidt telescope of about 1.2 m aperture, but in the end a 1 m telescope was built as an improved copy of the Hamburg Schmidt telescope. The mirror is 162 cm, the focal length is 3065 mm (plate scale 67.45 arcsec/mm;

the field is 290 x 290 mm^2, or $5.°5 \times 5.°5$. The constructor was W. Strewinsky (who also built the Hamburg Schmidt) and the telescope entered into operation in 1972. It soon became clear that, although the ESO Schmidt was improved in several respects as compared to the Hamburg Schmidt, it was not able to take full advantage of the <u>very good</u> seeing at La Silla, the desire to make <u>long exposures</u> in the dark La Silla sky, and the recent availability of <u>high resolution</u> emulsions. These three factors together put very high demands on the optical and mechanical characteristics of the telescope, expressed through the need to achieve near perfect guiding over several hours.

Much of the time during the past 10 years has been spent in significantly improving the performance of the ESO Schmidt telescope in order to make it possible to obtain deep survey plates of near perfect quality, i.e. with good surface uniformity, little or no image distortion, the smallest possible stellar images and thereby deep limiting magnitudes. Dr. André Muller of ESO has made a great personal contribution to this aim, and I am very thankful to him for information about the various improvements. Further details can be found in various papers; Muller (1979a, 1979b, 1980, 1983).

The ESO 1 m Schmidt telescope is basically extremely sturdy, and the optics is of reasonably good quality. Nevertheless, it was necessary to change and/or improve many features, the most important of which will be mentioned here:

Azimuth and Polar Setting

The telescope cradle is placed on four legs. Although this is a very stable system, it complicates azimuth and polar setting. The problem was overcome by introducing hydraulic jacks with spherical base and top, which take up the weight of the telescope during corrections, and which function like ball bearings allowing smooth corrections to within 1 arcsec. Azimuth and elevation are read from dial gauges and changes can be made through the night, especially after the installation of remotely controlled elevation setting.

Right Ascension Gear

An instability was present whenever the telescope passed the meridian. This was overcome by applying a small preload to the polar axis with a momentum of 32 kgm.

Oil System

The south end of the polar axis is supported by oil pads. Initially, the oil was heated during the night and the pads stuck due to change in viscosity. An oil cooling system was installed and the problem disappeared.

Declination Gear Wheel

In the original design of the declination gearing, springs were included to absorb earthquake shocks. However, this made the telescope very sensitive to wind, and oscillations frequently occurred. Therefore, a classical long arm declination control was installed allowing for declination corrections up to ± 1 arcmin. The telescope is now completely stable up to wind velocities through the slit of 10 metres per second.

Control System

The entire telescope control system for pointing and tracking was replaced by an ESO produced TCS system, similar to that used for the ESO 3.6 m telescope. This system is based on a Hewlett Packard 21MX computer. Pointing is possible with an accuracy of ± 3 arcsec and offset tracking with variable tracking rates etc. is also included.

Mirror Cell

Although the mirror cell is basically well designed, it was found that the positioning force was not succifient, causing collimation errors, especially when the telescope was placed in loading (horizontal) position. This was overcome by the installation of telescopic springs aligned with the center lines of the invar rods. They push the mirror with a constant force against the ends of the invar rods.

Plateholder

Of all problems, those connected to the plateholder in the center of the main tube were the most difficult. The longest possible exposure on medium grain emulsion (e.g. IIa-O) without serious image elongation was one hour, which was just sufficient for the production of the ESO (B) Atlas. However, longer exposures could not be satisfactorily achieved, and the main reason was finally located in the spring connection of the plateholder unit with the tube wall. This spring compensated for the weight of the plate carrier when it was lowered inside the tube to receive the plateholder. Consequently, a completely new plateholder unit was designed. It has no direct mechanical connection with the tube wall and is only supported by the focussing cylinder at the center of the spider. The plateholder is now placed directly by the observer in the center of the tube. To reach this area, he is lifted by a moving platform up through the loading hatch while the telescope is in loading position. The solution effectively eliminates the above mentioned problem and satisfactory exposures are now possible for up to 6 hours.

Guiding

While improving the plateholder stability, it became obvious that the mechanical connections of the two guiding tubes to the main telescope tube were not sufficiently stable. In order to benefit from the improved plateholder design, it became necessary to install a TV guiding system in the main tube, just off the plateholder. It is mounted on the plateholder unit to assure absolute stiffness between the photographic plate and the crosswire of the offset guider. As it is mounted inside the camera tube, it uses about 70% of the aperture and thus has an effective diameter of 84 cm. The guide probe and cross-wire are projected on a TV camera with an EBS tube and a window diameter of

40 mm. Guiding is done in the control room by remote control. In addition, the guider has a prism device which can be shifted into the field whenever objective prism plates are obtained, making it possible to reduce the spectra to star points and hence retain the guiding accuracy. Guiding down to magnitude 13 is now possible.

Electronic Cross

Since the guiding is no longer done at the center of the plate field, but approx. 3 degrees away, it is necessary to overcome the effect of differential refraction during longer exposures. An "electronic cross" device has therefore been installed which produces two crosses on the TV guide monitor. One cross is fixed and its position is calibrated with the optical cross of the guider which is also imaged on the monitor; these crosses are then solidly linked. A second cross is produced by computer software and can be moved with respect to the fixed cross. The correction includes the differential refraction effect and can also take into account offset guiding on moving objects. The device has proved extremely useful for recovering very faint comets by blind offset guiding. No fewer than six early recoveries were made during the recent years.

Dome Seeing

The Schmidt dome is well isolated and in view of the very small day-to-night temperature fluctuations on La Silla, good seeing is available at the very beginning of the observations every evening. However, it has been felt that there is still room for improvement, and it has been decided to further reduce dome introduced seeing by cooling the observing floor. The installation has not yet been terminated. Moreover, the control room with the observer has been moved from the dome to below.

Achromatic Corrector Plate

The original corrector plate of UBK7 glass was ordered with optimal correction at 4300 A. However, it turned out that a "redshift" had occurred and the best wavelength was 5000 A implying a loss of quality especially in the UV region. An achromatic corrector was manufactured by Grubb Parsons and installed in the summer of 1983. It is currently being tested.

Calibrators

In order to improve the usefulness of survey and other plates for photometric measurements, calibrators were designed (Muller, 1977) and installed in the telescope tube. Each projected 14 steps covering a total interval of $\Delta \log I = 1.4$. However, due to the necessity of blocking light from the sky in the area of the projected wedges, thereby unavoidably introducing a rather large area of vignetting, the calibrators were recently moved from the telescope to a table on the observing floor, and calibration is now made immediately following the telescope exposure.

Other Devices

A multi-diaphragm device has been constructed which allows the exposure on one plate of 36 different sky fields without overlapping. This is particularly useful for a supernova search, which is now carrried out by A. Muller and H.-E. Schuster and which has sofar resulted in the discovery of several supernovae. A Racine wedge is available for photometric calibration. A 4° objective prism gives dispersion 450 A/mm at 4340 A. A complete set of optical filters including interference filters to isolate astrophysically important emission lines is available.

3. PHOTOGRAPHIC MATERIALS AND THE HANDLING OF THESE

The times have passed when the astronomer just took a plate from the box and put it into the telescope to make the exposure. Nowadays, hypersensitization is necessary to take full advantage of the greatly increased capability of modern fine grain, high contrast astronomical emulsions. Various methods are available; cf. e.g. the review by Millikan and Sim (1978) and articles in the AAS Photobulletin.

At ESO, sensitization by baking in nitrogen has been used for many years, and the standard procedure is now nitrogen followed by forming gas. The entire procedure takes several hours and is the first, difficult step necessary to achieve satisfactory exposures. At various times, severe problems have been encountered, in particular in terms of large scale non-uniformity, which has made otherwise excellent plates useless for the sky surveys. Long series of tests have been necessary to arrive, in an empirical way, at the currently reasonably satisfactory procedures ensuring a uniform sensitization of the 30x30 cm plates. The problems were mainly connected to the gas flow pattern and small temperature gradients inside the hypersensitization box. Tests with hydrogen sensitization are underway, and it is likely that sensitization in pure hydrogen will be incorporated some time in the future. However, the need to ensure absolute safety has so far held back this project. It is believed that a maximum gain in speed of perhaps 30% over the current procedure is possible, e.g. for the IIIa-F plates for the ESO (R) Survey.

After exposure of edge marks, the plate is washed in Freon and exposed in the telescope. The distance between the optical filter and the emulsion is at least 6 mm allowing to avoid area desensitization, during exposure, a problem which has occurred in other telescopes, e.g. at the UK 48" Schmidt telescope. After exposure and calibration, the plates are processed to archival quality, following the latest ANSI and ISO guidelines. Accelerated aging tests have shown that the adopted method will, hopefully, assure plate survival for several hundred years if stored under controlled, optimal conditions. It includes the use of a large Palomar-type tray rocker (to ensure uniformity), a prolonged, agitated immersion in a stop bath, two agitated fixing baths, washing twice, Photo-Flo rinsing, swapping with cotton and vertical drying. The layout of the Schmidt dark room is a result of experience gained at other observatories and, in particular, at the ESO Sky Atlas Laboratory and encompasses several features which contribute to the safe handling of large astronomical plates in complete darkness (cf. West and Dumoulin 1974). Dust contamination is efficiently reduced by a high standard air conditioning system with adequate filters.

The entire process, from removing fresh plates from cold storage until they are put into the drying cabinet may last up to 10 hours, e.g. in the case of the plates for the ESO (R) Survey. This puts considerable strain upon the personnel. Without detailed and efficient preparations, it is not possible to maintain at an optimal level such a procedure. At the same time, it becomes quite clear that even minor technical problems may easily ruin a night's observations or at least incur significant loss of observing time and astronomical information.

4. PRODUCTION OF ATLASSES BASED ON SKY SURVEYS

Since the mid-1950's, several hundred copies of the Palomar Observatory National Geographical Society Sky Survey (POSS) have been distributed in the form of atlas prints on-paper and on-glass. In order to retain as much information as possible during the copying process, special reproduction methods were employed at the Graphic Arts Facilities of the California Institute of Technology, Pasadena (cf. Minkowski and Abell, 1963). However, since the POSS emulsions (103a-O and 103a-E) are rather coarse, it was not possible to use the same copying methods for the ESO (B) and ESO/SRC Surveys. Thus, when the ESO Sky Atlas Laboratory was set up in Geneva in 1972, a thorough investigation was made in order to optimize the reproduction techniques by making use of the most modern methods and materials. Many innovations are described in a report by by West and Dumoulin (1974), and further details may be found in the papers by West (1978) and West and Dumoulin (1980). In particular, it was decided to use film, rather than paper to allow quantitative measurement on copies, e.g. in transmission.

By making use of fine grain, high contrast materials and bringing down the contrast to approx. $\gamma = 1$, it has been possible to obtain a virtual grain to grain reproduction of the original plates. Major problem areas are now to assure complete physical contact between the originals and the copy plates during exposure without damaging the originals. Extensive use is made of drying copy plates in nitrogen and of Freon cleaning to avoid impurities, lipids, etc. Superior uniformity is assured by appropriate copying devices and processing techniques. Nevertheless, there is a decisive difference between a one time copying of a plate and the "industrial", large-scale production which is necessary in case of the ESO (B) and ESO/SRC Atlasses. The full attention which can be given to individual plates must be replaced by elaborate control methods. Extensive checks of Atlas prints at the ESO Sky Atlas Laboratory effectively assure that most faulty copies do not reach the buyers of the Atlas.

Special attention has also been given to the proper protection of Atlas copies by the use of Tyvek envelopes (glass copies) and two-layer transparent plastic envelopes (film copies). Special packing material has greatly reduced the risk of damage during transport to the subscribing institutes. An overview of the distribution of atlas copies throughout the world will be found in Figure 1; the total nos. are: ESO(B)-film, 42 copies; ESO(B)-glass, 20 copies; ESO/SRC-film, 178 copies; ESO/SRC-glass, 10 copies.

THE ESO SKY SURVEYS

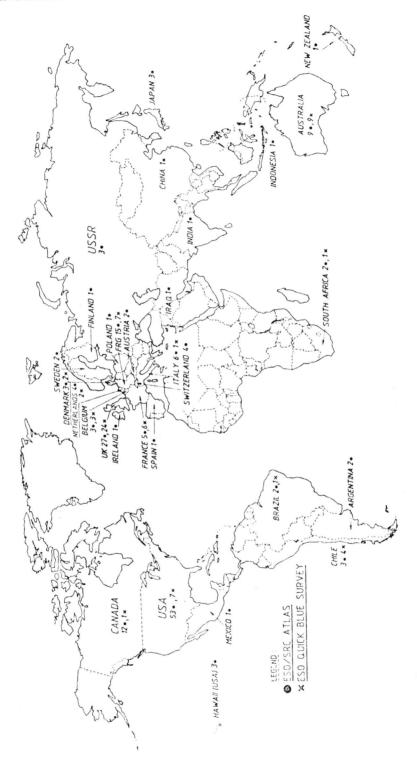

Figure 1

5. ASTRONOMICAL USE OF THE ESO (B) AND ESO/SRC SKY SURVEYS

The utility of the ESO and ESO/SRC Atlasses has been greatly enhanced by the availability of film copies of the survey plates. With the advent of fast measuring machines, many observatories are now able to perform quantitative measurements on their Atlas copies. A photometric transformation is necessarily introduced by the copying process. It has been found that in order to achieve the best possible calibration, it is preferable not to base the calibration curve on the imprinted step wedges only, but also, whenever possible, on calibrated stars and galaxies in the field. Two-dimensional digital devices like the CCD's now permit accurate calibration of objects and therefore the determination of zero points and calibration curves of entire survey plates. However, the danger of large-scale plate non-uniformities should not be underestimated and calibration by "local" objects, say within one degree is definitely recommended.

An example of the use of the ESO (B) Survey is the comprehensive inventory of objects in the southern sky known as the ESO/Uppsala Survey which was compiled by Lauberts (1982). It comprises more than 18.000 objects. Photometric measurement of most of these on the original ESO(R) plates is in progress (Lauberts, private communication) and secondary standards will therefore soon become available in virtually all survey fields.

Inspection of the ESO (R) half of the joint ESO/SRC Survey has already yielded a large number of red objects, many of which are hitherto unknown. Among these are Hα emission objects, red stars and distant clusters of galaxies. A future comparison of the SRC (J) and the ESO (R) plates will undoubtedly provide extensive finding lists of intrinsically red and/or reddened objects.

6. CONCLUSION

I have discussed some of the major technical problems connected with the production of high quality photographic atlasses, e.g. the ESO (B) and ESO/SRC Sky Surveys. It is, of course, not possible to exhaustively describe technical details in this short paper, but by calling attention to some of the most critical problem areas, it is my hope that current and future users of Schmidt-type and other wide-angle telescopes will realize that there is still much room for improvement. These telescopes are extremely powerful tools for many types of astronomical investigations, but in order to derive full benefits from the unique properties of the photographic emulsion as an unsurpassed panoramic detector, continued attention must be given to all those parameters which so easily deteriorate the performance.

REFERENCES

Lauberts, A., 1982, The ESO/Uppsala Survey of the ESO(B) Atlas, ESO

Millikan, A.L., Sim, M.E., 1978, in Modern Techniques in Astronomical Photography, eds. West and Heudier, ESO

Minkowski, R.L. and Abell, G.O., 1963, in Stars and Stellar Systems, Vol. III, ed. K.Aa. Strand, p. 481

Muller, A.B., 1977, ESO Messenger, 10, 10

Muller, A.B., 1979a, Abh. Hamburger Sternwarte, X, 2, 79

Muller, A.B., 1979b, ESO Messenger, 19, 29

Muller, A.B., 1980, ESO Messenger, 22, 18

Muller, A.B., 1983, ESO Messenger, 33, 26

West, R.M., 1972, ESO Bulletin, 10, p. 25

West, R.M. and Dumoulin, B., 1974, Photographic Reproduction of Large Astronomical Plates, Report from ESO Sky Atlas Laboratory

West, R.M., 1978, in Modern Techniques in Astronomical Photography, eds. West and Heudier, ESO

West, R.M. and Dumoulin, B., 1980, AAS Photobulletin, 23, 3

West, R.M., Schuster, H.-E., 1982, Astron. Astrophys. Suppl. Ser., 49, 577

DISCUSSION

J.A. Dawe: Why do you not put on your calibration at the same time as the plate is exposed in the telescope? And a possible relevant rider – What is the average humidity at La Silla?

R.M. West: To avoid the extensive vignetting that is introduced by the calibrators (sensitometers) in the telescope. With 5 degree centers for the survey, this would lead to parts of the sky not being covered by the Atlas. The average humidity is about 30%, thus, there may be less loss of sensitivity at La Silla than what is experienced in some humid places.

V.M. Blanco: I am delighted that you emphasized the high demand now being made on the mechanical stability of Schmidt telescopes. Martin McCarthy and I planned to talk next Friday about the same subject but, unfortunately, we will not be here. However, you have given our speech for us, and we thank you very much. I would like to add a comment to what you said. Namely, present-day large Schmidt telescopes were designed mostly prior or simultaneously with the development of low reciprocity failure emulsions and very large narrow bandpass interference filters. With these developments, observers now often need to make exposures as long as 3.4 or even 6 hours. Unfortunately, the telescopes were only too frequently designed for exposures of two hours or less. Users of the Curtis Schmidt at CTIO know this problem only too well. We have had to spend a tremendous effort in rebuilding parts of the Curtis telescope with some success, I am glad to say, but it has not been an easy task at all.

SKY SURVEYS WITH THE UK 1.2m SCHMIDT TELESCOPE

Russell D. Cannon
Royal Observatory, Blackford Hill, Edinburgh EH9 3HJ.

INTRODUCTION

On 1983 September 3 we will be celebrating the tenth anniversary of the formal hand-over of the UK 1.2m Schmidt Telescope (UKST) by the manufacturers, Messrs. Grubb Parsons of Newcastle-upon-Tyne, to the Science and Engineering Research Council (SERC). The completion of commissioning was marked in 1973 by the taking of the first completely successful sky-limited photograph on hypersensitised Eastman-Kodak IIIa-J emulsion, plate J149 of field 416. That plate itself led very quickly to exciting new scientific results, including the discovery of extraordinary and still apparently unique jets in the barred spiral galaxy NGC 1097 by Wolstencroft & Zealey (1975), a discussion of the clustering of 3000 faint galaxies (Dodd et al., 1975; 1976) and a suggestion that faint blue stellar objects are not randomly distributed (Hawkins & Reddish, 1975).

However, although that first plate had been obtained, it was to be another year before all the mysteries of plate hypersensitising, telescope guiding, focus control and plate processing were sufficiently well understood that good quality plates for the ESO/SERC Southern Sky Survey could be obtained routinely. Even now there are so many different faults which can spoil a survey plate that the success rate is only about fifty percent; the complicated quality control criteria are described by Cannon et al. (1978). Thus the early hopes of completing the survey in two or three years were not fulfilled, and in the end it has taken us just a decade to finish that job; perhaps this is one more instance where astronomical theory has had to be modified to fit new data!

Fortunately the 606 plates for the Southern Sky Survey do not represent anything like the entire ten-year output of the UKST; almost 9000 plates have been taken altogether and several other surveys are well underway, while special plates have been taken for several hundred research projects. In what follows I will describe the surveys and mention a few of the many discoveries which have been made, and say a little about future plans for the UK Schmidt.

THE UK SCHMIDT TELESCOPE

The UK Schmidt is situated on Siding Spring Mountain in New South Wales, Australia, close to the Anglo-Australian 3.9m Telescope (AAT). However the two telescopes are operated quite independently, since the UKST belongs to the SERC and is run by a team of astronomers from the Royal Observatory, Edinburgh (ROE). The best source of up-to-date technical information on the UKST is the Schmidt Telescope Handbook, which has recently been completely revised and is available on request from ROE. Briefly, the telescope has an aperture of 1.24m and a focal length of 3.07m, yielding a plate scale of 67.1 arcsec mm^{-1} and a focal ratio of f/2.5. The photographic plates are 356mm square and each one covers a field of $6\frac{1}{2}° \times 6\frac{1}{2}°$. All of these parameters are identical to those of the older Palomar 48-inch Schmidt, and indeed the UKST was originally conceived as a copy of that famous telescope. However technological advances, together with a desire to exploit the new Kodak IIIa-J fine grain emulsion, led to a significantly enhanced optical and mechanical specification for the UKST. The result is that the main UKST sky surveys penetrate about 1.5 magnitudes deeper on average than the original National Geographic Society - Palomar Sky Survey.

There are a few special features of the UKST which deserve mention. Primary pointing accuracy, using a hand calculator version of the AAT program written by P.T. Wallace to correct for flexure and atmospheric refraction, is now ± 6 arcsec r.m.s. over most of the sky (a tenfold improvement over the design specification); all plates are taken using an autoguider (Adam, 1971) mounted on an auxilliary telescope, without which the success rate for long-exposure survey plates would be much lower; the polar axis elevation adjustment has been motorised so that it can be set to the optimum value to minimise field rotation for every exposure (Wallace & Tritton, 1979); the Schmidt corrector plate is a cemented achromatic doublet (Wynne, 1981), believed to be the largest such in the world, which gives ~ 1 arcsec resolution through the full photographic wavelength range from the atmospheric ultraviolet limit at $3200Å$ to beyond $10,000Å$ in the infrared; there are two full-aperture narrow-angle objective prisms which can be mounted separately or together on the telescope to yield reciprocal dispersions from $2400Å\ mm^{-1}$ to $600Å\ mm^{-1}$ at $4300Å$; and there are several wide-field interference filters centred on astrophysically important lines such as H-alpha.

Innovations on the photographic side include automatic equipment for hypersensitising plates by a variety of techniques, the principal methods being soaking in nitrogen and then hydrogen gas at atmospheric pressure and at 20°C for IIIa-J and IIIa-F emulsions (Cannon et al., 1978), and bathing in dilute silver nitrate solution for IV-N emulsion (Hartley & Tritton, 1978); extensive sensitometric monitoring of every new batch of emulsion during hypering; stepwedge or spot calibrations which are put on to every plate during exposure in the telescope to facilitate the subsequent conversion of density or transmission measurements into intensities; and, most recently, the modification of the telescope plateholders to allow continuous flushing with dry nitrogen gas during exposure. This last feature prevents desensitisation of the hypered IIIa plates during exposure, and also results in better plate uniformity (Dawe & Metcalfe, 1982; and at this conference).

Making very high quality copies of original telescope plates on film or glass is essential for dissemination of Schmidt data in the form of sky atlases, and well-equipped laboratories have been established at ROE following the pioneering work of ESO in Geneva (now moved to Munich). The copying techniques have been described by Standen & Tritton (1979). There is a certain amount of friendly rivalry, and a lot of exchange of ideas, between the two establishments resulting from the perpetual struggle to maintain the highest standards. Some of the more specialised photographic techniques for image enhancement, such as high contrast printing and unsharp masking, will be described later by Malin and by Hadley.

Finally, and just as important for the successful exploitation of the UK Schmidt as a research telescope, we have set up plate libraries in Australia and Edinburgh which are well equipped with light tables, microscopes, Polaroid cameras and measuring machines. Astronomers are encouraged to come to Edinburgh in particular if they want to carry out research projects using UKST material. The system of the telescope together with the photographic laboratories and plate libraries is regarded as a 'national facility' for British university astronomers, but it is also very much an 'international' facility. By making originals and copies of plates freely available we are able to get much more astronomy out of the UK Schmidt Telescope than if each person simply kept his or her own plates, although individuals do have priority rights over plates taken especially for them for a limited period (up to two years), much as is the case for example with data from the IUE satellite. The operation of the telescope in Australia is in the hands of a team of half-a-dozen resident observers, and the telescope is run as a 'remote facility' even when not engaged in survey work. Thus astronomers are very rarely present when their plates are being taken. In these ways the telescope is able to support many more programmes for a much larger number of astronomers than if it was run as a 'common user' facility.

THE ESO/SERC SOUTHERN SKY SURVEY

The first major task of the UKST was to survey the southern sky from $-17°$ to $-90°$ declinations on Kodak IIIa-J emulsion. This set of 606 overlapping photographs comprises the SERC(J) half of the ESO/SERC Southern Sky Survey, covering the blue-green wavelength range from 3950Å to 5400Å. Complete sky coverage with good deep plates was achieved by 1980, although we are still trying to obtain top Atlas-quality plates for the last few fields. In fact a sort of equilibrium has now been reached, where we are having to replace the occasional survey plate because of accidental damage caused during copying or, more seriously, because of the problem of microspots or 'gold spot disease' (see Sim, this meeting).

Film copies of the ESO/SERC survey have been issued in instalments since 1976 to some 170 astronomical institutions; so far 507 fields have been issued. Glass copying has started more recently, also at the ESO headquarters in Munich. Ten sets are being made and up to now 180 fields have been issued. Added urgency has been given to this work by the requirements of the Guide Star Selection System at the Space Telescope Science Institute in Baltimore.

Numerous discoveries have been made on the SERC(J) Atlas photographs, and here I can give only a few examples. The most spectacular are very faint optical extensions discovered in nearby galaxies. For example, in NGC 1512 there is clear evidence for gravitational interaction with the companion galaxy NGC 1510 which has led to the formation of extensive but very faint spiral arms (Hawarden et al., 1979). In NGC 5291 an even larger scale phenomenon is seen: it seems that star formation, or rather dwarf galaxy formation, has recently been triggered over a region more than 100 kpc in extent (Longmore et al., 1979). The galaxy NGC 7531 (Cannon, in preparation) has a very low surface brightness companion. These, and the very peculiar galaxy NGC 1097 mentioned previously, are unique examples probably representing different physical phenomena. However, a whole new class of shell structure elliptical galaxies, possibly representing the aftermath of galaxy mergers, has been discovered by Malin & Carter (1983; Malin and Carter at this conference). Another fruitful type of object is represented by elliptical galaxies with dust lanes (Hawarden et al., 1981), the most spectacular example of which is the nearby radio galaxy NGC 5128 (Centaurus A). The latter has large scale optical extensions in approximately the same direction as the even larger radio lobes (Haynes, Cannon & Ekers, 1983). Turning to galaxies much closer to our own, the Carina dwarf spheroidal galaxy, which I discovered by chance on an SERC(J) survey plate (Cannon, Hawarden & Tritton, 1977), is the only new dwarf satellite of our own Galaxy to have been discovered in the last thirty years and brings the total number known up to seven.

I have concentrated in these examples on nearby galaxies, partly because this is work in which I have been involved, and partly because they make spectacular illustrations. However the SERC(J) survey has resulted in many new galactic discoveries such as the 'cometary globules' (Hawarden & Brand, 1976) and old planetary nebulae (Longmore, 1977; Longmore & Tritton, 1980). A systematic search for compact dust clouds has been carried out by Hartley, Hawarden, Manchester & Tritton (in preparation) and this has been followed up with radio studies of molecular absorption lines (Goss et al., 1980). On a much larger scale, very extended nearby filamentary material has been mapped out in the vicinity of the South Celestial Pole by King,

Taylor & Tritton (1979) and foreground dust has been detected in front of many important galaxies and star clusters. The 'J' survey has also been used for many systematic survey-type extragalactic projects, such as searches for peculiar galaxies (Arp & Madore, 1977), for Seyfert galaxies (Fairall, this meeting), the classification of galaxies (Corwin, de Vaucouleurs & de Vaucouleurs, 1982), cataloguing clusters of galaxies (Corwin & Abell, in preparation), and the identification of radio sources (Savage, this meeting). At the other end of the distance scale, mention must be made of the considerable number of comets and asteroids which have been discovered on survey plates.

All of the above work has been done simply by visual inspection of the sky survey photographs. A whole new phase of Schmidt telescope astronomy is now under way using machine measurements of plates, particularly using the very high speed machines COSMOS in Edinburgh and APM in Cambridge. The accepted original sky survey plates are kept in a safe in an air-conditioned room in Australia, but little information is lost in making the copies. More importantly, a large stock of rejected but still very good quality sky survey plates is held in the Plate Library in Edinburgh. One project which relies on these rejected survey plates is the search for very faint variable objects, which is turning up substantial numbers of quasars and distant RR Lyrae stars (Hawkins, 1980; 1981; this meeting). Another machine-based project using direct survey plates is the study of the clustering and superclustering of galaxies (Shanks et al., 1980; MacGillivray & Dodd, 1980).

THE UKST EQUATORIAL SURVEY

Many of the discoveries on the SERC(J) survey, such as old planetary nebulae and faint star clusters, were qualitatively predictable in that such objects had been found on the Palomar survey and it was to be expected that similar objects awaited discovery in the southern sky. What is more surprising and exciting has been the discovery of some qualitatively new classes of objects, typified by the elliptical galaxies surrounded by shells and the 'cometary globule' nebulae. Presumably similar objects exist in the north but have not yet been noticed. The reason for these new discoveries is that the combination of the high-performance UK Schmidt Telescope with Kodak IIIa-J emulsion, and the very dark Siding Spring site, has allowed us to carry out a systematic survey reaching unresolved objects about four times fainter than the limit of the Palomar northern survey (done on much coarser-grained 103a-O emulsion). More importantly, the higher contrast of IIIa-J emulsion makes faint extended nebulosity much easier to detect; the new features typically have surface brightnesses of around 26 mag arcsec^{-2}. Even more can be extracted from these fine-grain photographs by using various contrast-enhancing techniques, as described by Malin and demonstrated by Hadley at this meeting, and features as faint as 28 mag arcsec^{-2} (corresponding to less than one percent of the background sky brightness) have been recorded.

The upshot of these considerations is that it is obviously well worthwhile, now that the initial southern survey has been completed, to extend the IIIa-J survey northwards. Plans have been announced to refurbish the Palomar 48-inch Schmidt and repeat the northern survey. Since Palomar is at latitude 33°N while Siding Spring is at 31°S, it is very reasonable for the two new surveys to meet at the equator, and both telescopes will survey the 0° strip of sky to give a small area of overlap for comparison and cross-calibration. The UKST Equatorial Survey consists of 288 fields covering the zone from -18° up to +3° (i.e. plate centres at -15°, -10°, -5° and 0°) and is being done in two colours, using IIIa-J and IIIa-F emulsion. The optimum filter for the red (IIIa-F) survey is still under discussion. The original plan was to use the RG630 filter (cutting off at 630 nm), but this gives a rather narrow band since the emulsion cut-off is at 690 nm, and necessitates long exposures of up to two hours even on optimally hypersensitised plates. Tests are currently being carried out using an OG590 filter, giving a 70 percent increase in bandwidth. This should give almost as good an approximation to the standard photo-electric R band. The much narrower bandwidth had been chosen initially to give emphasis to H-alpha emission features, but this is no longer such an important consideration since the UKST is equipped with 80Å band interference filters centred on H-alpha (e.g. Elliott & Meaburn, 1976).

Originally the intention was to take pairs of blue and red photographs on the same night, as was done in the original Palomar survey, in order to minimise confusion between stars of extreme colour and variable stars. However, the much longer exposure times combined with the higher acceptance standards meant that the success rate for pairs of plates was unacceptably low. In any case, the Palomar survey already exists for this zone and will provide colour information on all but the faintest objects. Therefore the blue and red equatorial surveys are proceeding more or less independently, with the IIIa-J survey taking higher priority since these modern plates are again required for measuring Space Telescope guide star positions. Although the old Palomar Survey exists for this zone, the proper motions of potential guide stars are liable to be large enough that plates no more than ten years old are required to measure accurate offsets for Space Telescope targets. The current status of the Equatorial Survey is that Atlas-quality IIIa-J and IIIa-F plates have been obtained for 50 and 20 percent of the fields respectively; on-glass copying (four sets only) is already underway, and sets of Atlas film copies will be made in due course if there is sufficient demand from the astronomical community.

THE UKST NEAR-INFRARED SURVEY

The discovery that Kodak IV-N emulsion, although not designed as an 'a' series astronomical emulsion, could be hypersensitised by bathing in dilute silver nitrate solution (see Hartley & Tritton, 1978) led directly to its use on the UK Schmidt Telescope. It seemed particularly relevant for the UKST to carry out a near-infrared survey in view of the ROE's involvement in the UK 3.8m Infrared Telescope and British involvement in the IRAS satellite. Furthermore, such a survey could be carried in 'grey' time when the moon was up and so it did not conflict with the other major surveys. The survey involves taking pairs of plates, typically a 90 minute exposure on IV-N emulsion through an RG715 filter (covering the band 715 nm - 900 nm) and a matching 15 minute red exposure on IIIa-F emulsion through an RG630 filter. The first phase, now nearing completion, consists of 163 fields covering a band within ten degrees (plate centres) of the southern galactic plane plus the two Magellanic Clouds. Further details are given by Hartley & Dawe (1981). Some 70 sets of film copies of this Atlas are being made in the Photolabs at Edinburgh and distributed around the world; 140 fields have been issued so far. The Near Infrared Survey is now being extended to higher galactic latitudes. Rather little astronomical use seems to have been made of this survey so far, although Malkan et al. (1980) discovered one heavily reddened star cluster near the galactic centre. However, the latest reports from the IRAS group indicate that there are large numbers of new far-infrared sources to be identified and the 'I' survey may be particularly useful for this work.

OBJECTIVE PRISM WORK

A survey with the low dispersion (3/4°) prism (Nandy et al., 1977) was started several years ago as a low priority project. Most of the plates were taken in response to requests for specific fields. However there has been considerable discussion about the usefulness of such a survey. On the one hand there is no doubt that prism plates contain much more information than even a full set of broad-band direct plates, but only for stars or galaxies within a magnitude range of about two magnitudes and only to a plate limit some three magnitudes brighter (i.e. B ~ 19.5) than that of deep direct plates. It has also become apparent that, at least for visual inspection, the plates taken in the very best seeing are much more useful than the majority of the plates and that it would be difficult to carry out an all-sky survey to sufficiently high standards. When the new 2 1/4° prism became available (Cannon et al., 1982), it seemed desirable to compare its performance with that of the older prism before continuing with the survey. It is now clear that for some projects, such as identifying large samples of optical quasars, the low dispersion prism is greatly superior mainly because it reaches a fainter limiting magnitude. However perhaps the most effective argument against doing a large-scale survey is simply that it takes so long (several man-months) to scan a single plate, and that even then the results are best described as 'subjective' prism spectroscopy. A crucial new development has been the work of Clowes and his collaborators on the 'Automatic Quasar Detection' system (AQD) described by him at this conference. This uses COSMOS to scan prism plates and powerful computer software to analyse the data, which makes it possible to select large samples of quasars according to well-defined objective criteria; it also opens up the possibility of scanning larger numbers of plates and is causing a revival of interest in a large-scale prism survey.

Two important projects using the prism plates have been the selection of large samples of optical quasars, summarised for example by Smith (1983), and the confirmation of quasar candidates which had been selected as radio source identifications on direct survey plates by Savage (this meeting). The power of the Schmidt arises because it is not difficult to identify at least five quasars per square degree brighter than B magnitude 20, so that over 200 quasars can be found on each plate. This immediately makes it possible to find rare extraordinary quasars, and also permits some analysis of the clustering and possible alignment or association of quasars with other objects. However the powerful selection effects depending on magnitude, redshift and line strength have so far precluded the use of these visually selected samples for cosmological studies.

Quasars are of course not the only objects identifiable on the plates: the other $\sim 10^5$ images include interesting unusual types of star such as carbon stars (e.g. Mould et al., 1982), Wolf-Rayet stars and planetary nebulae (e.g. Morgan, 1983). Galaxies also give useful information; apart from the fairly obvious application to emission line galaxies, it turns out that even the low dispersion prism permits the determination of galaxy redshifts using the 4000Å feature to an accuracy of around 3000 km s^{-1}, for the study of super-clustering (Clowes, Cooke et al. and Parker et al., this meeting).

NON-SURVEY PROGRAMMES

In this review I have concentrated on the various sky surveys being done by the UK Schmidt Telescope. However no more than half of the usable telescope time is devoted to survey work. Frequently the seeing conditions are not good enough but still permit useful plates to be taken, especially for stellar photometry, or else the sky is bright so that only short exposure or interference filter plates can be taken. Even in 1973, when the SERC(J) southern survey had the highest possible priority, many special plates were taken for individual astronomers. Since then, and particularly since the high speed measuring machines became available, an increasing proportion of the best-seeing dark time has been used for non-survey work, so that recently about two-thirds of the plates taken each year have been for non-survey programmes. Altogether the UKST has supported some 600 programmes since 1973, with about 200 being 'active' at any time. Typically these programmes require only 2 or 3 plates each and requests are handled relatively informally on a 'first come, first served' basis by the UKST staff. Many such requests can now be dealt with by using existing plates from the ROE Plate Library or by providing copies made in the ROE Photolabs. However some programmes, such as the Manchester H-alpha survey of the Magellanic Clouds (Davies, Elliott & Meaburn, 1976), the University of California Asteroid Survey (Bus et al., 1982), searches for supernovae (Cawson, this meeting), faint variable stars searches, or studies of galactic structure, require larger numbers of plates and such programmes have to be approved by the SERC's Panel for the Allocation of Telescope Time.

Although our primary objective has to be the support of astronomical research in British Universities, we do our best to satisfy requests from astronomers in any institution in any country of the world, and at the most recent count about a third of our non-survey programmes involved foreign astronomers.

It would be unfair to describe any particular non-survey programmes among the hundreds which have been done; I hope that enough examples will be mentioned during the course of this conference to give an impression of the breadth and versatility of modern research which can be carried out with a large Schmidt telescope.

FUTURE PLANS

From what I have said above, a very simple arithmetical calculation will show that the present programme of the UKST will keep it more than fully occupied for another decade. However there is one particular exciting new development on the horizon, which may revolutionise our operation within the next few years. Preliminary experiments (described by Dawe here and Watson) are being carried out in Australia to use optical fibres in the telescope, in order to feed the light from many objects spread over the 40 square degree Schmidt field into a fast spectrograph with a modern electronic detector such as a CCD. This would free the Schmidt telescope from the limitations of the inefficient non-linear photographic plate, and make it directly competitive with much larger telescopes for many projects such as spectroscopy of clusters of galaxies or large samples of quasars. Indeed, it may make the concept of a really large new Schmidt-type telescope competitive with other proposals; a 2.5m Schmidt telescope with fibres could be as effective as a giant 7.5m telescope for many projects. However, no photoelectric detector comes close to matching the photographic plate in either resolution or number of picture elements, and so there will certainly be a long-term requirement for conventional photography as well.

Looking at the wealth of both 'expected' and unexpected results which have come from the UKST, both alone and in collaboration with the AAT, it is difficult to avoid the conclusion that a wide-field Space Schmidt telescope able to carry out a high resolution all-sky survey in the ultraviolet would be similarly successful, and would greatly enhance the effectiveness of the large Space Telescope. Such a proposal has been under discussion already for several years, starting with a NASA study followed by an unsuccessful ESA proposal.

Most recently, a group of astronomers at the University of Texas have been trying to set up an international project involving French, Italian, ESO and British collaboration. Unfortunately progress is slow at the moment, partly due to an international shortage of funds for astronomy and partly because of uncertainties over the best detector system to use. However there is an active IAU Working Group for Space Schmidt Surveys and further details can be obtained from its Chairman, K. Henize.

ACKNOWLEDGEMENTS

I have to thank Sue Tritton, Ann Savage and Dorothy Skedd, for help in preparing this review, and of course I must pay tribute to the pioneering work of Vincent Reddish who got the telescope built and to all past and present members of the UK Schmidt Telescope Unit, who have together contributed to its success.

REFERENCES

Adam, G.R.: 1971, Publ.R.Obs.Edinburgh, 8, 43.
Arp, H.C. & Madore, B.F.: 1977, Q.Jl.R.astr.Soc. 18, 234.
Bus, S.J., Helin, E.F., Dunbar, R.S., Shoemaker, E.M., Dawe J.A., Barrow J., Hartley, M., Morgan, D.H., Russell, K.S. & Savage, A.: 1982, NASA Technical Memorandum 85127 - Reports of Planetary Geology Program.
Cannon, R.D., Hawarden, T.G. & Tritton, S.B.: 1977, Mon.Not.R.astr.Soc. 180, 81P.
Cannon, R.D., Hawarden, T.G., Sim, M.E. & Tritton, S.B.: 1978, Occ.Rep.R.Obs.Edinburgh No. 4.
Cannon, R.D., Dawe, J.A., Morgan, D.H., Savage, A. & Smith, M.G.: 1982, Proc.astr.Soc.Australia 4, 468.
Corwin, H.G., de Vaucouleurs, A. & de Vaucouleurs, G.: 1982, Astron J. 87, 47.
Davies, R.D., Elliott, K.H. & Meaburn, J.: 1976, Mem.R.astr.Soc. 81, 89.
Dawe, J.A. & Metcalfe, N.: 1982, Proc.astr.Soc.Australia 4, 466.
Dodd, R.J., Morgan, D.H. Nandy, K., Reddish, V.C. & Seddon, H.: 1975, Mon.Not.R.astr.Soc. 171, 329.
Dodd, R.J., MacGillivray, H.T., Ellis, R.S., Fong, R. & Phillipps S.: 1976, Mon.Not.R.astr.Soc. 176, 33P.
Elliott, K.H. & Meaburn, J: 1976, Astrophys.Sp.Sci. 39, 437.
Goss, W.M., Manchester, R.N., Brooks, J.W., Sinclair, M.W., Manefield G.A., & Danziger I.J.: 1980, Mon.Not.R.astr.Soc. 191, 533.
Hartley, M. & Dawe, J.A.: 1981, Proc.astr.Soc.Australia 4, 251.
Hartley, M. & Tritton, K.P.: 1978, in "Modern Techniques in Astronomical Photography", eds. R.M. West & J-L. Heudier, pub. ESO, p.95.

Hawarden, T.G. & Brand, P.W.J.L.: 1976, Mon.Not.R.astr.Soc. 175, 19P.
Hawarden, T.G., van Woerden, H., Goss, W.M., Mebold, U. & Peterson B.A.: 1979, Astron.Astrophys. 76, 230.
Hawarden, T.G., Elson, R.A.W., Longmore, A.J., Tritton, S.B. & Corwin H.G.: 1981, Mon.Not.R.astr.Soc. 196, 747.
Hawkins, M.R.S.: 1980, Astrophys J. 237, 371.
Hawkins, M.R.S.: 1981, Nature 293, 116.
Hawkins, M.R.S. & Reddish, V.C.: 1975, Nature 257, 772.
Haynes, R.F., Cannon, R.D. & Ekers, R.D.: 1983, Proc.astr.Soc. Australia, in press.
King, D.J., Taylor, K.N.R. & Tritton, K.P.: 1979, Mon.Not.R.astr.Soc 188, 719.
Longmore, A.J.: 1977, Mon.Not.R.astr.Soc. 178, 251.
Longmore, A.J. & Tritton, S.B.: 1980, Mon.Not.R.astr.Soc. 193, 521.
Longmore, A.J., Hawarden, T.G., Cannon, R.D., Allen, D.A., Mebold U., Goss, W.M. & Reif, K.: 1979, Mon.Not.R.astr.Soc. 188, 285.
MacGillivray, H.T. & Dodd, R.J.: 1980, Mon.Not.R.astr.Soc. 193, 1.
Malin, D.F. & Carter, D.: 1983, Astrophys.J., in press.
Malkan, M., Kleinmann, D.E. & Apt, J.: 1980, Astrophys.J. 237, 432.
Morgan, D.H.: 1983, Mon.Not.R.astr.Soc., in press.
Mould, J.R., Cannon, R.D., Aaronson, M. & Frogel, J.A.: 1982, Astrophys.J. 254, 500.
Nandy, K., Reddish, V.C., Tritton, K.P., Cooke, J.A. & Emerson D.: 1977, Mon.Not.R.astr.Soc. 178, 63P.
Shanks, T., Fong, R., Ellis, R.S. & MacGillivray, H.T.: 1980, Mon.Not. R.astr.Soc. 192, 209.
Smith, M.G.: 1983, in Proc. 24th Liege Symposium, 'Quasars and Gravitational Lenses', in press.
Standen, P.R. & Tritton, K.P.: 1979, Occ.Rep.R.Obs.Edinburgh No.5.
Wallace, P.T. & Tritton, K.P.: 1979, Mon.Not.R.astr.Soc. 189, 115.
Wolstencroft, R.D. & Zealey, W.J.: 1975, Mon.Not.R.astr.Soc. 173, 51P.
Wynne, C.G.: 1981, Q.Jl.R.astr.Soc. 22, 146.

OBJECTIVE PRISM SURVEYS

Martin F. McCarthy S.J.
Vatican Observatory, Vatican City State, Europe

ABSTRACT

A survey of surveys made with objective prisms since the Hamburg Schmidt Conference in 1972 is presented. Ten outstanding achievements attained by these techniques are listed. Detailed accounts of particular surveys both general and specific follow. A table listing fundamental classical studies is given plus a list of the location and properties of the largest Schmidt-type cameras.

INTRODUCTION

The first coma free telescope was built by B. Schmidt in Hamburg in 1930. Thirteen years later the Morgan-Keenan-Kellman Atlas was published and opened up the HR diagram for large scale spectral classification. Today we consider the objective prism surveys, made for the most part with Schmidt-type telescopes in the years since 1972, when, under the gracious auspices of SRC, ESO and the Hamburger Sternwarte, we last met to talk together about the role of Schmidt telescopes in astronomy.

We should recall, as we attempt here to survey the objective prism surveys, just what kind of surveys have been carried out in this decade; next we can look at the reasons why these surveys were undertaken; finally, we can reflect on how such surveys have affected our science. Some surveys, begun in this period, turned out to be irrelevant or of secondary importance or could be done better by using other techniques (especially from the exciting new developments in multicolor photometry). Objective prism research in the past ten years has become more "problem oriented" than it was previously. Nonetheless, many, and these very important, surveys have been undertaken with objective prisms which continue to excite and illuminate our research efforts.

Some of the earlier papers which review progress and problems in objective prism survey work are listed in Table 1; I also commend to your attention the Symposia and Colloquia sponsored by the IAU et al. for example at Saltsjobaden (1964), Cordova (1971), Lausanne-Geneva (1975), Washington (1977), Vatican City (1978) and most recently in Toronto (1983). Also deserving of our attention are the Reports of IAU Commissions 45 and 28 at the General Assemblies: XV in 1973 at Sydney and in Poland, XVI in 1976 at Grenoble; XVII at Montreal in 1979 and XVIII at Patras in 1982.

Table 1.

Some Fundamental References for Objective Prism Work

Pickering, E. C., 1891, Harvard Annals, 26, vii.
Introduction to the Draper Catalogue

Lindblad, B. and Stenquist, E., 1934 Stockholm Ann. Bd. 11, No. 12.
On the Spectrophotometric Criteria of Stellar Luminosity

Ohman, Y., 1934, Astrophys. J., 80, 171.
Spectrographic Studies in the Red

Morgan, W. W., 1951, Pub. Obs. Univ. Michigan 10, 33.
Application of the Principle of Natural Groups to the
Classification of Stellar Spectra

Keenan, P. C., 1954, Astrophys. J., 120, 484.
Classification of S Type Stars

Nassau, J. J., 1958, Stellar Populations, Ric. Astron. Spec. Vaticana
5, 171 and 183.
M Type Stars and Red Variables; Carbon and S Type Stars

Bidelman, W. P., 1966, Vistas in Astronomy, 8, 53 (Ed. A. Beer).
Accurate Spectral Classification by Objective Prism Techniques

Stephenson, C. B., 1966, Vistas in Astronomy, 7, 59 (Ed. A. Beer).
Astrophysical Investigations Utilizing Objective Prisms

Blanco, V. M., 1965, Galactic Structure, (Ed. A. Blaauw and M. Schmidt) in Stars and Stellar Systems (Gen. Ed. G. P. Kuiper and B. Middlehurst)
Ch. 12, pp. 241-266
Distribution and Motions of Late Type Giants

McCuskey, S., 1965, Ibid., Ch. 1, pp. 1-26.
Distribution of the Common Stars in the Galactic Plane

Today we look at a mixed situation when we consider the Schmidt cameras in regular use for objective prism studies. Some Schmidt-type cameras have become inactive; some have been moved or are being moved to better sites. In the past decade we have all come to appreciate the importance of excellent seeing as a condition for superior quality research with objective prisms. Today we ask ourselves with more concern: "What will be the role for objective prism survey work in the final years of this century?"

The answer to this question comes rather immediately from a source heretofore little studied with objective prism techniques. I refer to the realm of the external galaxies as explored with objective prisms (mostly very small angled ones) attached to Schmidt-type telescopes. This story has been told and the steps of its early progress outlined by Malcolm Smith (1979) at the Secchi Conference. Since T. D. Kinman (1984) will be presenting his Invited Paper on Emission Line Galaxies, I shall not speak further of objective prism surveys of galaxies.

Ten Outstanding Achievements in Objective Prism Studies

I consider now some of the outstanding achievements and features of work attained through the objective prism surveys in the decade since the Hamburg Conference (Haug, 1972).

First. I begin by mentioning an object discovered on objective prism plates, listed as an object showing Hα in emission with suspected emission also at He I 6678Å. SS 433 indeed has peculiar emission features and its true nature still fascinates and eludes us. The complexities surrounding SS 433, listed in the emission star survey of Sanduleak and Stephenson (1977) as a 13.5 mag object, have stimulated much follow-up research and been the occasion for many new publications and conferences. We rejoice that many of the first successful observations of SS 433 and its fast shifting systems of spectral lines were made here at Asiago by Ciatti and Mammano (1981) and their colleagues. Sanduleak and Stephenson (1981) point out that neither the original nor follow-up plates would indicate that SS 433 was very peculiar. This fact may also serve as an encouragement for continuing the patient searches for emission objects. They are indeed a "mixed bag" of celestial objects but can, as we are seeing here, lead through subsequent research to new and most valuable results.

Second. The successful application of small angle prism methods to fainter (and so often to more distant) objects in Schmidt survey work. Besides the above mentioned research leading to the discovery of QSO and other emission line galaxies we note the successful application of small prism techniques to the study of stars of late type in nearby galaxies by Blanco and McCarthy (1975) and by Sanduleak and Philip (1977) and by our late colleague Bappu and Parthasarathy (1977). The first two mentioned studies were carried out with the thin prism at CTIO attached to the Curtis Schmidt. The Indian experiments were devoted to ultralow dispersion of stars and galaxies.

Third. The completion of the Michigan survey of spectral types in the southern sky and the publication of Volumes 1, 2 and 3 of the Catalogue (Houk and Cowley 1975; Houk 1978, 1982). This work is sponsored by the National Science Foundation.

Fourth. The introduction and successful adaptation of objective prism and gratings to Schmidt cameras of the largest size. We recall the interesting comments made at Hamburg by R. Minkowski (1972) on the reasons why no prism was made for the 1.2m Palomar Schmidt. Subsequently, paced by the successful introduction of the 1° prism objective combination for the 1.3m Tautenburg Schmidt, prisms of 0°8 and 0°2 have been installed and now combined on the 1.2m UK Schmidt while prisms of 2° and 4° have been attached to the Tokyo Observatory 1.05m Schmidt at Kiso, Japan. The success of these large diameter objective prisms have led to plans for installing prisms at the Palomar 1.2m Schmidt.

Fifth. The exploration of new regions of the spectrum for survey work with objective prisms has been the result of the new fine-grain emulsions developed by Eastman Kodak and of the new techniques introduced for hypersensitizing plates and finally, of the extended possibilities for making much longer exposures now available with new autoguiding techniques; to this should be added new methods for minimizing field rotation effects such as done with the UK Schmidt. A most useful aid in optimizing objective prism work was alluded to above: the successful transport of Schmidt cameras to regions where excellent seeing is normal.

Sixth. The establishment of new rapport among objective prism spectroscopists, slit spectroscopists (especially those with Reticon, Vidicon or CCD receivers at hand) and photometrists.

Seventh. The extension and improvement of objective prism techniques through the introduction of the combination grating-prism (the so-called 'grism') or other combination dispersion techniques (McCarthy and Blanco 1978).

Eighth. The successful application of objective prisms for use in determinations of radial velocities. Here the pioneering work of C. Fehrenbach (1978, 1983) with the GPO and the PPO has been crowned with success in both hemispheres and applied successfully to the discernment of members from nonmembers of our own nearby galaxies.

Ninth. The availability of very large interference and other specialized filters which make possible surveys for extremely faint objects. Examples: (a) searches for Planetary Nebulae with a 50Å bandpass filter centered at [O III] 5007Å and Hα; (b) investigations of SNRs with plates centered at Hα and again at SII 6317-30, and (c) discovery of low metallicity halo objects with filters limiting the observed spectrum to the H and K lines plus Hα (work of Preston and Schectman at CTIO).

Tenth. The progress of plans for Schmidt spectral surveys in space, especially with Space Telescope.

A word of caution may be inserted here immediately after our description of the outstanding achievements and improvements in objective prism spectroscopy in recent years. Many of the new observational techniques require very long exposures. These in turn lead to noticeable field rotation and flexure effects as the telescope tracks through large hour angles. It is therefore important not to overlook the fact that some of the best present day Schmidt telescopes were simply never designed for extremely long exposures.

Now we consider concrete instances of objective prism surveys. We shall be speaking here of prisms of all sizes: from the smallest apex prisms used chiefly for extragalactic objects [but also for special objects (PN, S and C stars) in our own galaxy] to the largest (10°, 12° and 15°) angled prisms, which serve to spread out the continuum in the spectrum and to enhance the contrast with emission features such as $H\alpha$. We shall first consider the largest (in area) prisms attached to the largest Schmidt telescopes, then we shall discuss in turn the kinds of objects detected in objective prism surveys.

The Largest Schmidt Cameras and Their Prisms

What is a large Schmidt telescope? Many definitions might be offered. I suggest this definition here: a large Schmidt is one whose correcting plate has a diameter of 1 meter or greater. Thus the

Table 2.

The Largest Schmidt Telescopes and Their Prisms

Telescope	Prism	Location / Dispersion
Tautenberg Schmidt 134/200/400 cm	Prism: 1°	East Germany 2500 Å/mm at $H\gamma$

By combining corrector and dispersor light losses in the 132 cm prismatic correcting plate are minimized and the weight is low. (Beck 1972)

Telescope	Prism	Location / Dispersion
United Kingdom Schmidt 120/180/cm	Prisms: 2°.2 0°.8	Siding Springs, Australia 830 Å/mm at $H\gamma$ 2440 Å/mm at $H\gamma$
Palomar Schmidt 122/183/307 cm	Prism: in planning stages	Palomar Mountain, California
Tokyo Observatory Schmidt 105/150/330 cm	Prism: 4° 2°	Kiso, Japan 170 Å/mm at $H\gamma$ 800 Å/mm at $H\gamma$
ESO Schmidt 100/162/305 cm	Prism: 4°	La Silla, Chile 450 Å/mm at $H\gamma$
Kristaberg Schmidt of Uppsala Obs. 100/135/300 cm	Prism: 7°	Kristaberg, Sweden 273 Å/mm at $H\gamma$

See Uppsala Ann. Vol. 5, No. 5

Hamburg Schmidt at Calar Alto would not be included since its corrector is 80 cm; similarly the Abastumani 70 cm meniscus telescope will not appear on this list, we note that both of these telescopes are larger than the many intermediate size Schmidts such as the Curtis, the Burrell, the Vatican, the Tonantzintla, the French-Liege, the Armagh-Dunsink-Harvard and the Torun Schmidts. In Table 2 we have listed the Largest Schmidt Telescopes along with the aperture in cm of the corrector plate, the spherical mirror and finally the focal distance; we also give the reciprocal dispersion at $H\gamma$ for the prisms used with the camera. It is worthwhile noting that none of the largest Schmidts has as yet completed a full sky objective prism survey. The reasons for this are that only recently has it become possible to construct and to mount such large pieces of glass and also that major efforts with large Schmidts have until now been concentrated and devoted to sky surveys in the <u>direct</u> mode.

OBJECTIVE PRISM SURVEYS

General Classification Surveys

General classification projects with objective prisms are usually carried out on blue sensitive plates and cover the range from 3800 to 4800Å at dispersions near 150 Å/mm to 250 Å/mm at $H\gamma$. Such surveys aim at the classification of all spectral types to a convenient limiting magnitude.

If the spotlight for the most exciting breakthroughs in objective prism work in this decade has centered on surveys of QSOs, emission line galaxies and on the 'thus far unique' object SS 433, the focus on the most thorough and extensive survey employing objective prism techniques must surely rest on Dr. Nancy Houk of Michigan. At Toronto she discussed the latest features of her current work and reviewed the present status and future prospects for this reclassification of the stars of the Henry Draper Catalogue in the MK system (Houk 1983). This work has begun with the 4° and 6° combined Michigan prisms on the Curtis Schmidt in Chile; it continues now with the new 10° prism attached to the Burrell Schmidt in Arizona. Three volumes of this work begun at Dec = -89° and reaching presently to Dec = -26° have now been published (Houk and Cowley 1975; Houk 1978, 1982). A generous overlap in declination zones is planned which will allow an intercomparison of the two telescopes, two different prism arrangements, and two different sites. The observational work for this project in regions north of Dec = -30° has been started at Kitt Peak under the direction of the Case astronomers. Bidelman will speak to us here of this work. This biggest objective prism survey is acknowledged and encouraged by us all. A feature of it has been the successful transfer of the two dimensional MK system of classification to this very large scale program of objective prism work. That this transfer could be made so serenely and successfully on plates covering the whole sky is a tribute to Houk and Cowley and to their predecessors Cannon, Maury and Mayall and of course, to Keenan, Morgan and Nassau, and in a most special way to S.

McCuskey (1965) who pioneered the first application of MK classification to objective prism projects for his monumental LF papers. We have with each new volume of the Michigan Catalogue thousands of new secondary and very localized standards for classification, all accomplished in a most uniform manner by most accomplished classifiers. It seems remarkable (yet most logical) that both the original HD work at Harvard and now the new MK classification at Michigan- Case have been made from objective prism plates.

A second general (i.e., covering all spectral types) project of classification is that due to the excellent work of our colleagues in the Soviet Union under the leadership of Professor E. Kharadze and Dr. R. Bartaya (1976, 1977, 1979, 1983). Their work is recognized and used more extensively each year. Since 1956 the 70 cm meniscus telescope at Abastumani with its objective prisms of 2°, 4° and 8° (yielding corresponding reciprocal dispersions of 1200, 660 and 160Å/mm at $H\gamma$) has been used to survey more than 100,000 stellar spectra in two dimensions. These stars are located in the Kapteyn Selected Areas and were observed through the 8° prism. This survey is complete for Areas 2 to 43; classifications have been made for stars to a limiting mag of m = 12. For Selected Areas 44 to 115 observational material has been obtained and the classification is in progress. We shall be speaking about other valued contributions from our Soviet colleagues when discussing specific surveys later on.

Another large scale general survey in progress is the study of stars at Intermediate Latitudes made by J. Stock and co-workers (1979, 1980) in Venezuela. This work includes besides spectral classification also the radial velocity measures plus the determination of positions and magnitudes for 10,000 stars in a region which extends from R:A: $11^h 47^m$ to $15^h 43^m$ and Dec -30° to -35°. We note once again that this large general survey is concerned with stars in the once very neglected Southern Hemisphere.

Other general classification surveys are less extensive but merit consideration here. Fehrenbach and Burnage (1981) give spectral types together with radial velocities, accurate positions and approximate magnitudes for 713 stars in four fields. Similarly, Amieux and Burnage (1981) present a catalogue for 169 stars for a field in NGC 3114. J. Drilling and A. Landolt (1979) have made MK classifications from low dispersion objective prism spectra for 608 of the 624 secondary UBV faint standards of Landolt. J. C. Doyle and C. J. Butler (1978) report on their classification of spectra taken with the ADH and its objective prisms. P. A. Krug, D. C. Morton and K. P. Tritton (1980) have made observations for the faint standards which are to be used with the objective prism plates taken with the UK Schmidt. E. Recillas Cruz (1982) has searched the deep spectral plates of the UK Schmidt for faint blue spectra of early type stars which may belong to the Magellanic Stream. Johannsen (1981) has classified 800 stars with the ADH.

Worthy of note, it seems to me, is the fact that certain of the large national plate archives containing low dispersion spectral data, especially those concerned with data acquisition from space, have recently become available to members of the astronomical community; perhaps more heed might be paid, especially when telescope time is very scarce and travel expenses are mounting, to this 'second hand' source material which may have for us, as for Ruth in the Old Testament, many golden items for our gleaning.

Specific Classification Surveys

Specific classification surveys limit themselves to a special group of objects. Such surveys usually employ preferentially lower dispersions; occasionally, however, the highest available dispersion is used along with strong filtering to cut down unwanted background radiation; this limits the range of the spectrum covered. So many conferences and publications have been dedicated to each of the main classes of objects detected in such specific objective prism surveys (QSOs, PN, OB and C Stars, etc.) that adequate coverage of these surveys here is not feasible. We shall cite with examples some of the more common types of objects surveyed with objective prisms.

Surveys of Emission Objects. The outstanding emission feature observed in stars, nebulae and galaxies is of course $H\alpha$. One of the features noted in the past decade since the Hamburg meeting has been the improvement of plate sensitivity in the red region as evidenced in the development of the III a F emulsion. Another significant advance has been the introduction of successful techniques for increasing the sensitization of red and near infrared sensitive plates. Such improvements in the photographic process and in the filtering mentioned above have helped to limit sky background contamination and to exclude night sky emission.

The objects detected in surveys for $H\alpha$ are a "most unnatural" group. They encompass many 'natural' groups. Certain of these true natural groups can be individualized only when more observations besides those which display the $H\alpha$ feature have been carried out; such further observations will most commonly involve slit spectroscopy at higher dispersion and resolution, multicolor photometry, or both. The objects which display $H\alpha$ prominently in emission include besides the Be stars, PN, H II objects, Wolf Rayet Stars, Novae, Of stars, Ke, Me, Ce and Se stars. Thus in searches for objects with $H\alpha$ in emission we will harvest many species of stellar (and nonstellar) objects. The harvest yields quite an array of different members of very different 'natural' groups which must be 'separated out' like fish after a big catch in a large net.

Several of the earlier published lists of early type stars showing $H\alpha$ in emission were collated by Bertiau and McCarthy (1969); a much more extensive list was published by Wackerling (1970), which included the important southern Be stars from Henize's survey (1967). The listing by Bidelman (1954) of the emission stars of late type remains today a fundamental source of information. Perek and Kohoutek (1967) have provided us with the General Catalogue of Planetary Nebulae; this has been supplemented by a listing of PN observed in the red spectral region by MacConnell (1982). The excellent Be Star Bulletins originated by Mme. Herman continue to contribute much to our picture of these early type emission stars so important for spiral arm tracing.

The important work of the Case-Hamburg Survey of High Luminosity Stars is extended now through the researches of many of the original observers and their students. From Case we note the studies of emission objects by Sanduleak and Stephenson (1977) which has given us SS 433 and its other not yet so famous companions. Other Hα emission stars have been published by Sanduleak and Bidelman (1980), by Stephenson (1979, 1981) by Bidelman (1981) and by Pesch (1984) who will be telling us here of some of the the faint emission stars observed at 1500 Å/mm at Hγ and at 18th mag. In his high-latitude red-region objective-prism survey, Stephenson (1983) has identified many emission line objects. This project is now two-thirds complete with 1000 of 1300 fields already photographed; it covers the sky north of Dec = $-25°$ and covers galactic latitudes beyond b = $-10°$. He uses a dispersion of 1000 Å/mm at Hγ and reports some 200 new Hα emission objects, noting that a few but by no means the majority are dMe stars. Among the interesting objects found, Stephenson mentions a faint C star with bright Hα. We may expect more surprises when higher dispersion spectra can be secured for these objects. Indeed studies of emission objects seem alive and well at Warner and Swasey Observatory. From Merida in Venezuela, Mac Connell (1981) reports the discovery of some 900 new Hα stars found on red-visual plates taken with the Curtis Schmidt. From Hamburg we will be hearing next from Kohoutek (1984) what the Hamburg Schmidt is doing with emission objects at its new site in Calar Alto, Spair. From Ohio State, another of the original Case-Hamburg researchers, A. Slettebak, reports continued work on emission objects.

Extensive work in surveying galactic fields for Hα emission objects continues with the Tonantzintla Schmidt under G. Haro and his coworkers (1982).

New combined efforts at objective prism survey work by the astronomers of Indonesia (Hidyat 1983) and Japan (Maehara 1983), constitute a most welcome manifestation of international cooperation in research. Samples of the first fruits of this collaboration are the researches of K. K. Hamajima et al. (1982) and by Maehara (1982). Both the Kiso Schmidt of Tokyo Observatory and the Bima Sakti Schmidt at the Bosscha Observatory in Java are used. The combined surveys have concentrated on regions of dark nebulosity and special attention is given to emission stars in the vicinity of Bok globules and to the discrimination of associated T Tau stars as distinct from the more distant Be stars, or from Mira variables or other Hα objects. These results are studied along with star counts, infrared studies, and surface photometry with narrow filters in order to trace the processes of stellar formation. Surveys have been completed for areas in Ara, Crux, and Monoceros and in the direction of the galactic center.

From Sweden, Welin's (1979) contributions to Hα emission studies are very important. At Castel Gandolfo, G. Coyne (1978, 1983) and his coworkers Cardon, DeGraeve, Lee, Mac Connell, and Wisniewski have published five lists with a sixth list now in press, in addition to this there is, also in press, a revised final listing prepared by Coyne and Mac Connell. There are no further plans to continue the search for Hα emission objects from Castel Gandolfo because of the deterioration of observing conditions there (bright lights from the Roman campagna).

Martinez, Muzzio and Waldhausen (1980) have listed approximate spectral types for 139 Hα emission objects in the Coalsack using the Curtis Schmidt at CTIO with a dispersion of 1350 Å/mm at Hγ. M. Kun (1982) has studied 110 Hα emission objects discovered in dark clouds along the galactic plane. Ogura (1983) has detected 140 new emission line stars in associations in Monoceros.

Each of us has enjoyed rediscovering the beauty of the emission features in Campbell's Hydrogen Envelope Star: HD 184738 (BWo + P) or made observations with objective prisms at the highest available dispersion of galactic Novae and appreciate their fascination. W. Seitter in her Atlases of Novae has given a splendid demonstration of the spectral features which are observable at objective prism dispersions especially when short exposures on unfiltered infrared IN plates reveal the full display of features from 9000 to 3500Å. However, there does not seem to be any formal survey work in progress directed at detecting Novae or peculiar emission objects. Like many other objects, they will be detected in Hα emission surveys and can then be subjected to more detailed exploration.

Surveys of Carbon Stars. Carbon stars have come into their own and are receiving much attention since we met at Hamburg in 1972. Among the pioneering giants here we note Secchi, C. D. Shane, Sanford, Merrill, Morgan, Keenan, Nassau and Blanco and others. One of these, Y. Fujita (1980) of Japan has presented his philosophy of spectral classification when he discussed the problems associated with Carbon Stars.

Fuenmayor (1981) in Venezuela surveyed regions near the galactic center and towards the galactic anticenter and confirms Blanco's (1965) observation concerning the concentration of C stars towards the anticenter and the absence of C stars in the direction fo the galactic center. Mac Connell (1978, 1981, 1982) has reported 34, 10 and 46 new C stars in surveys he has made along the Milky Way plane. Nandy, Smeriglio and Buonanno (1978) present the results of their C star surveys and Smeriglio and Nandy (1981) announce the discovery of new C stars in Cepheus. Kurtanidze et al. (1979, 1980) has found 39 new Carbon stars in a survey with the Abastumani meniscus telescope and he and West (1980) have found 10 new C stars. Alksnis et al. (1979)

reports the discovery of new Carbon stars on plates taken with the Latvian Schmidt and is studying the relation of observed IR colors with the C stars discovered. I. Platais (1979) also reports finding new C stars in several fields of the sky. Most of these observations were made along the galactic plane. Stephenson (1983) on the other hand has been looking at the distribution of R and N stars at higher latitudes and reports finding 100 new C stars. These 'out of the plane' Carbon stars should prove interesting for further photometric, kinematic and evolutionary studies.

Maehara (1983) at Tokyo has used red plates exposed through a GG 455 filter and the 4° prism on the Kiso Schmidt to survey 180 square degrees in Cassiopeia. He confirms 73 of the C stars already discovered by Stephenson and finds four new C stars and notes that for all C stars observed there is a scarcity of types earlier than C_4 on the Keenan-Morgan classification scheme. Similar studies at strategic points along the galactic plane and away from it might pay significant dividends for evolutionary studies.

Surveys of Other Late Type Stars. The classic Case researchers on objective prism spectra of M and S type stars in the near infrared are summarized by Nassau (1956, 1958). In the past decade, few new extensive surveys have been carried out to find galactic M stars. We mention here two unpublished studies of M stars in the direction of the galactic center. Blanco, Blanco and McCarthy (1978) report the discovery of 300 late M stars in the region of 0.12 square degrees in Baade's 'Clean' Window near NGC 6522, they also report one suspected C star. Recently these same investigators aided by DeGraeve and Meier have counted in an area of 0.12 square degrees some 750 late M stars plus a possible C and an S star; this latter investigation was carried out closer to the galactic center in Sgr I, another of the windows reported by Baade; we note that these surveys were not accomplished with Schmidt cameras but with a grism attached at the converging beam of the prime focus of the CTIO 4m telescope.

I believe that this 'drop off' in Schmidt M star surveys is due to the fact that there are so many other challenges in observational studies of evolution that the 'sheer plod' of M and S star surveys may have discouraged some possible investigators. It is as though we were all waiting for the day (which indeed may now be dawning) when automatic data processing and plate reduction techniques can accomplish classification tasks more efficiently than ever we could with low power microscopes used to examine very crowded fields with overlapping spectral images. Another factor, without doubt, has been the scarcity of giant M stars near the sun. This makes it difficult to determine the absolute luminosities of M giants. The grism technique used in the galactic center and in the Magellanic Clouds have given us for the first time the opportunity to obtain reliable luminosity measures.

Russian colleagues have been active in observing red stars and have used the Abastumani 70 cm meniscus camera to survey M and C stars in the northern part of the galaxy in a band 10° wide along the galactic plane. They have noted a tendency for a lesser concentration of high luminosity M stars towards the galactic plane than had been previously thought or expected (Kharadze 1983).

This work at Bosscha Observatory cited above includes the discovery of 2000 late type M giants and 59 supergiants in a region of 200 square degrees towards the galactic center and extending from l = 330° to 0° to 30° along the plane. One aspect of this study may be that we will have available for further study several new windows of transparency similar to those disclosed years ago by Baade. The Javanese and Japanese astronomers are examining their M star surveys in the light of observations of associated IR enhancement; such research may permit a close and more accurate glimpse of the regions near the galactic nucleus, and may complement the finds of the wonderful new IRAS satellite.

Mac Connell (1982) reports finding 328 M and MS stars on his Curtis Schmidt plates. Smiriglio and Nandy (1981) have been surveying a region in Cepheus for M stars brighter than $m_i = 13$.

Stephenson's above mentioned (1983) survey has yielded 100 new stars of type S. These plates also serve to identify late M stars which, as Stephenson notes, are, most usually, well known variable stars which often lack spectral types.

Pesch and Sanduleak (1978) have published a catalogue of probable dwarfs of type M3 or later. Alcaino and Pik Sin The (1982) report on M stars in the South Galactic Pole and Stephenson notes the presence of some 400 to 450 M dwarfs in his survey at high latitudes. We note that the unfinished portion of this survey will include the north galactic pole. The estimates and hypotheses about the numbers and densities of M dwarfs is a fascinating old 'ghost' which keeps 'rumbling' in the M star closet. New and shining nails for closing this case, in addition to those reported in 1976 by Faber and by King, were brought forth at the 1983 IAU Colloquium at Middletown, where the luminosity function was discussed. However, we all remember that it is difficult to bury 'ghosts'.

H. Bond (1980) by means of an objective prism survey has been searching for extremely metal-deficient stars, mostly of late type; this work has opened up the fascinating topic: 'Where have all the Population III Stars gone?' Such studies are of importance as we seek stars of very advanced age.

At Ohio State, McNeil and Schiller report surveys with the Curtis Schmidt of late type stars in the region of the South Galactic Pole. They use a dispersion of 580 Å/mm at $H\gamma$. McNeil (1981) found 2200 G, K and M stars in his survey which reached to 13.5 mag in an area of 81 square degrees. Schiller (1981) has surveyed a much larger area and finds 183 M giants to a limiting magnitude of 14.5 m_{pg}.

Surveys of Early Type Stars. The major work in discovering the intrinsically brighter OB stars is now completed through the successful Case-Hamburg studies. New searches and surveys extend to fainter limiting magnitudes in our own galaxy and to the intrinsically brighter OB stars in members of the local group. Sanduleak and Philip (1979) have observed many candidate OB stars in the Magellanic Clouds. Some of the greatest excitement in the OB star domain today concerns the IUE spectra but these are not a matter for our discussion here.

Orsatti and Muzzio (1980) and Forti and Orsatti (1981) have been searching for faint OB stars in the southern Milky Way and report the discovery of more than 200 OB stars on thin prism plates taken at CTIO.

At the Abastumani Observatory the 2° prism attached to the 70 cm meniscus telescope is used to search for O and B stars.

L. Erculiani Abati and H. Lorenz (1984) will be describing here some of the spectral studies they have been making at Asiago using the objective prism plates from the 1.2m UK Schmidt and the 1.3m Tautenberg Schmidt (plus our 'host' Schmidt here in Asiago). They will report first spectra on blue objects detected in the Asiago field, where UVX objects have been noted.

From Case-Western Reserve, Pesch (1984) will be reporting here on initial results of the low dispersion (1500 Å/mm at $H\gamma$) survey which can detect O and B stars to the 18th mag. This work is being done with the Burrell Schmidt in Arizona.

This survey of surveys with objective prisms must necessarily leave largely untouched the grand domain of specific studies of A-F-G-K stars. Many of the brighter stars of these types will, of course, be classified in the Michigan Catalogues; many have already been classified in the LF spectral surveys by McCuskey (1965) with the Case Schmidt in Cleveland; Kharadze and Bartaya and their colleagues will be giving us accurate types for fainter stars than the surveys with the Curtis and Burrell Schmidts record. As mentioned above, with the coming of automatic classification processes, both new and old objective prism plates (if they have some calibration keys available) can be surveyed and studied. Space density analyses according to spectral types may be made according to the methods of Schalen and the other methods of star counting cited by Bok. Such observations and analyses will prove most important for large scale evolutionary studies of sprial arm and interarm regions. In addition, we will be able to follow so much better than has been hitherto possible the changes in spectral type of so many variable stars found on Schmidt plates. This, in the words of W. W. Morgan, represents one of the last unexplored areas for spectral studies.

ACKNOWLEDGEMENTS

The author is grateful to colleagues for many observatories who supplied information on surveys made with objective prisms. Work has begun while the author was Visiting Astronomer at Kitt Peak National Observatory and completed at the Vatican Observatory at Castel Gandolfo. Special thanks are due to T. D. Kinman, V. M. Blanco, and C. Corbally for help and suggestions and to Helen Bluestein of the Steward Observatory staff who typed the manuscript.

REFERENCES

Alcaino, G. and The, Pik Sin.: 1982, Publ. Astron. Soc. Pacific, 94, 335.
Alksnis, A. and Alksne, Z.: 1979, Astron. Tsirk, No. 1081, 6; Sverolovsk, 106.
Amieux, G. and Burnage, R.: 1981, Astron. and Astrophys. Suppl. Ser., 44, 101.
Bappu, M. K. V. and Parthasarathy, M.: 1977, Kodaikanal Obs. Bull. Ser. A, 2.
Beck, H.: 1972, Role of the Schmidt Telescope in Astron. Hamburg, (Ed. U. Haug), p. 59.
Bertiau, F. C. and McCarthy, M. F.: 1969, Ric. Astron. Spec. Vaticana, 7, 523.
Bidelman, W. P.: 1954, Astrophys. J. Suppl. Ser., 1, 175.
Bidelman, W. P.: 1981, Publs. Astron. Soc. Pacific, 93, 129.
Blanco, B. M., Blanco, V. M., McCarthy, M. F.: 1978, Nature, 271, 638.
Blanco, V. M. and McCarthy, M. F.: 1975, Nature, 258, 407.
Bond, H. E.: 1980, Astrophys. J. Suppl. Ser., 44, 517.
Ciatti, F., Mammano, A., Bartolini, C., Guarnieri, A., Piccioni, A., Downes, A. J. B., Emerson, D. T. and Salter, C. J.: 1981, Astron. Astrophys., 95, 177.
Coyne, G. V.: 1983, Vat. Obs. Publ. 2, No. 3: Annual Report 1982.
Coyne, G. V., Wisniewski, W. and Otten, L. B.: 1978, Vat. Obs. Publ. 1, 275: VES Paper V.
Doyle, J. C. and Butler, C. J.: 1978, Irish Astron. J., 13, 229.
Drilling, J. S. and Landolt, A. U.: 1979, Astron. J., 84, 783.
ErculianiAbati, L. and Lorenz, H.: 1984, this volume, p. 475.
Fehrenbach, C.: 1983, private communication.
Fehrenbach, C. and Burnage, R.: 1978, C. R. Acad. Sci. Paris, 286, Ser. B, 289.
Fehrenbach, C. and Burnage, R.: 1981, Astron. and Astrophys. Suppl. Ser., 43, 297.
Fuenmayor, F. J.: 1981, Rev. Mex. Astron. Astrofis., 6, 83.
Fujita, Y.: 1980, Space Science Review, 25, 89.
Forti, J. and Orsatti, A.: 1981, Astron. J., 86, 209.
Hamajima, K., Ishida, K., Ichikawa, T., Hidayat, B., Roharto, R., Cont. Bosscha Obs. No. 74.
Haro, G., Chavira, E. and Gonzalez, G.: 1982, Bol. Inst. Tonantzintla, 3, 3.

Haug, U.: Editor, ESO, SRC, Hamburg Sternwarte Conference on The Role of the Schmidt Telescopes in Astronomy: Proceedings, 1972.
Henize, K.: 1967, Astrophys. J. Suppl. Ser., 14, 125.
Hidayat, B.: 1983, private communication.
Houk, N.: 1983, Conference on Criteria and Applications of MK Classification. Toronto, Ed. R. Garrison, in press.
Houk, N. and Cowley, A. P.: 1975, Univ. of Michigan Catalogue of Two Dimenisonal Spectral Types of HD stars, Vol. 1.
Houk, N.: 1978, Ibid., Vol. 2.
Houk, N.: 1982, Ibid., Vol. 3.
Johansson, K.: 1981, Astron. Astrophys. Suppl. Ser., 44, 127.
Kharadze, E.: 1983, private communication.
Kharadze, E. and Bartaya, R.: 1976, Proc. III European Astron. Meeting, IAU, Tiblisi, p. 17.
Kharadze, E. and Bartaya, R.: 1977, Astron. Nach. 298, No. 2, 111.
Kharadze, E.: 1979, Ric. Astron. Spec. Vat., 9, 127.
Kinman, T. D.: 1984, this volume, p. 409.
Kohoutek, L.: 1984, this volume, p. 311.
Kurtanidze, O. M., Natriashvili: 1979, Astron. Tsirk., No. 1036, 8.
Kurtanidze, O. M., Natriashvili, V. V., Natsvishvili, R. Sh.: 1980, Astrofisika, Tom 16, 190.
Kurtanidze, O. M. and West, R. M.: 1980, Astron. Astrophys. Suppl. Ser., 39, 35.
Krug, P. A., Morton, D. C. and Tritton, K. P.: 1980, Mon. Not. Royal Astron. Soc., 190, 237.
Kun, M.: 1982, Astrofisika, Tom 18, 63.
Maehara, H.: 1982, Cont. Bosscha Obs. 71.
Maehara, H.: 1983, private communication.
Maehara, H.: 1983, Publ. Astron. Soc. Japan, in press.
Mac Connell, J. D.: 1978, Astron. Astrophys. Suppl. Ser., 38, 335.
Mac Connell, J. D.: 1981, Astron. Astrophys. Suppl. Ser., 44, 387.
Mac Connell, D. J.: 1982, Astron. Astrophys. Suppl. Ser., 48, 355.
Mac Connell, D. J.: 1982, Bull. Amer. Astron. Soc., 14, 653.
Martinez, R. E., Muzzio, J. C., Waldhausen, S.: 1980, Astron. Astrophys. Suppl. Ser., 42, 179.
McCarthy, M. F. and Blanco, V. M.: 1978, Mem. Soc. Astron. Ital., 49, 287.
McCuskey, S.: 1965, Galactic Structure, Ed. A. Blaauw and M. Schmidt: Vol. 5 of Stars and Stellar Systems, Gen. Eds. G. Kuiper and B. Middlehurst, Ch. 1, 1.
McNeil, R.: 1981, Bull. Amer. Astron. Soc., 13, 357.
Minkowski, R.: 1972, in: Role of Schmidt Telescopes in Astronomy, Hamburg (Ed. U. Haug), p. 7.
Nandy, K. and Smiriglio, F. and Buonanno, R.: 1978, Publ. Royal Obs., Edinburgh, 9, 125.
Nassau, J. J.: 1956, Vistas in Astronomy, (Ed. A. Beer), Pergamon: London, 2, 1361.
Nassau, J. J.: 1958, Ric. Astron. Spec. Vaticana, 5, 171.
Ogura, K.: 1983, in press.
Orsatti, A. and Muzio, J.: 1980, Astron. J., 85, 265.

Perek, L. and Kohoutek, L.: 1967, Catalogue of Galactic Planetary Nebulae, Czechoslovak Acad. Sci.
Pesch, P.: 1984, this conference.
Pesch, P. and Sanduleak, N.: 1978, Astron. J., 83, 1090.
Platais, I.: 1979, Astron. Tsirk, No. 1043, 1.
Recillas Cruz, E.: 1982, Mon. Not. Royal Astron. Soc., 201, 473.
Sanduleak, N. and Stephenson, C. B.: 1977, Astrophys. J. Suppl. Ser., 33, 459.
Sanduleak, N. and Philip, A. G. D.: 1977, Publ. Warner Swasey Obs., 2, No. 5.
Sanduleak, N. and Philip, A. G. D.: 1979, Astron. Astrophys. Suppl. Ser., 35, 347.
Sanduleak, N. and Bidelman, W. P.: 1980, Publ. Astron. Soc. Pacific, 92, 72.
Sanduleak, N. and Stephenson, C. B.: 1981, Astrophys. J., 249, L19.
Savage, A.: 1983, private communication.
Schiller, S.: 1981, M.S., Thesis Ohio State University.
Smiriglio, F. and Nandy, K.: 1981, Occas. Rep. Royal Obs. Edinburgh, No. 7, 1.
Smith, M.: 1979, Ric. Astron. Spec. Vat., 9, 117.
Stephenson, C. B.: 1983, private communication.
Stephenson, C. B.: 1979, Bull. Amer. Astron. Soc., 11, 365.
Stephenson, C. B.: 1981, Bull. Amer. Astron. Soc., 13, 463.
Stock, J.: 1979, in First Latin American Regional Meeting IAU (Santiago, Chile, Ed. A. and H. Moreno), p. 160.
Stock, J.: 1980, Astron. J., 85, 1366.
Wackerling, L.: 1970, Mem. Royal Astron. Soc., 73, 153.
Welin, G.: 1979, Astrofisika, Tom. 15, 712.

DISCUSSION

P.S. BUNCLARK: Will the large amount of plate material used in the manual searches you have mentioned be readily available for use with the automated scanning facilities?

M. McCARTHY: The answer to your question depends on observatory policy and this differs markedly from one Institute to another. In national centers where space observations are deposited, these data (after a determined examination time assigned to the project scientists who actually directed and made the observations) do become available for open study by qualified persons. We know that traditionally photographic observational material obtained by individual guest astronomers remains the property of the host observatory - is used as long as required by the guest observers then should be returned to the institutes archives.

THE CASE LOW-DISPERSION NORTHERN SKY SURVEY

Peter Pesch and N. Sanduleak
Warner and Swasey Observatory
Case Western Reserve University, Cleveland, Ohio

ABSTRACT

The Burrell Schmidt-type telescope of the Warner and Swasey Observatory in its new location on Kitt Peak in Arizona is being used for a spectroscopic survey of the region $b > +30°$ and $\delta > +30°$. The plates, which cover $5° \times 5°$, are taken with the $1°.8$ prism which provides a dispersion of 1350 Å mm^{-1} at Hγ. Eastman Kodak IIIa-J plates, baked in forming gas, are used without filter to cover the spectral range 3300 to 5350 Å. The exposure times of 75 minutes reach a limiting blue magnitude of ~ 18.0 for threshold detection of an unwidened stellar continuum. The categories of objects which are being catalogued are blue and/or emission-line galaxies, probable HII regions, blue and/or emission-line stellar objects, known and probable blue stars, main-sequence late B and A-type stars, suspected field horizontal-branch stars of types A and F including RR Lyrae variables, suspected F and G-type subdwarfs showing a UV excess, faint carbon and late M suspected halo giants, and peculiar objects.

INTRODUCTION

When the Burrell Schmidt was moved to Kitt Peak in 1979, we began a large scale objective-prism survey of the northern sky ($\delta > +30°$, $b > +30°$) for a variety of galactic and extragalactic objects.

The Burrell Schmidt features a recently refigured 90 cm primary mirror, a new 61 cm corrector plate of Schott UBK7 glass a field of $5° \times 5°$ and a plate scale of $\sim 100''$ mm^{-1}. The prism used in our survey is made of Schott UK50 glass, has an apex angle of $1°.8$, and provides a dispersion of 1350 Å/mm at Hγ. Eastman Kodak IIIa-J plates, baked in forming gas, are used without filter to provide spectral coverage from 3300 to 5350 Å. Careful guiding produces unwidened spectra just under 0.03 mm in width. Exposures of 75 minutes reach a limiting magnitude of B $\sim$ 18.0 for threshold detection of a continuum of a stellar object.

GENERAL PROCEDURES

Each of the 19.6 cm x 19.6 cm photographic plates are scanned by both workers. Scanning is done by means of a binocular microscope or a converted aerial film scanner which projects an enlarged view of the plate onto a screen. Palomar Observatory Sky Survey prints are examined to eliminate overlaps and to aid in distinguishing between resolved and unresolved objects.

Equatorial coordinates are derived from measures of the spectra using Stephenson's analysis of the influences of the curved focal surface and the objective prism (Stephenson and Sanduleak 1977). The α measure is made by bisecting the narrow spectrum which is dispersed in the north-south direction, and has an accuracy of $\pm 3"$. The δ measure is made by setting on the long wavelength end of the spectrum. Since this cutoff depends on both color and brightness, the δ has an accuracy of $\pm 6"$. Crude estimates of m_B - to the nearest magnitude - are given to aid in identifying the objects. These are based on eye estimates of the continuum density near 4500 Å.

There have been many surveys, especially in the north galactic polar cap, for blue objects. Most of these have been multicolor rather than spectroscopic surveys. A comprehensive literature search is made to obtain additional information and previous designations of objects found in our survey. When finding charts are available, recovery of previously known objects is unambiguous. More commonly, only positions and magnitudes have been published and recovery is more difficult. In such cases, we prepare transparent overlays and search the region surrounding the published positions on our spectral plates.

DESCRIPTION OF THE CATEGORIES

A unique aspect of this survey is the wide variety of objects which are being catalogued. A list and description of our categories is as follows:

I. Blue and/or Emission-line Galaxies.

These are objects which appear nonstellar on our spectral plates and/or on the POSS prints and which have a bluer continuum than the average galaxy and/or which show emission. The emission most often seen falls near the long wavelength end of the spectrum and can be identified as the blended N_1+N_2 lines of [OIII]$\lambda\lambda$5007, 4959. Less frequently emission is seen near the short wavelength end of the spectrum; this identified with [OII]λ3727. Occasionally, an unresolved object will show both of these emission features. We consider such objects to be compact galaxies and include them in category I.

IA. Probable HII Regions.

When emission as described above is contained in knots distinct from the central or primary concentration of light in well-resolved and moderately bright galaxies, we consider these to be probable HII regions.

II. Blue and/or Emission-line Stellar Objects.

These are objects which are unresolved on our spectral plates and on both the E and O POSS prints. They generally show either a featureless, and relatively flat continuum extending well below the (absent) Balmer discontinuity or an emission feature or features located anywhere within the range 3300 to 5350 Å. Broad dips in the continuum are sometimes noted. Unless additional information is available, we do not attempt to identify a single emission feature, although work with the nearly identical Curtis Schmidt at CTIO (MacAlpine and Feldman 1982 and references therein) has shown that Lyα is most often the correct identification, in which case the object is a QSO with $1.7 < z < 3.3$. Until further observations are available, we do not know how many of the non-emission-line blue stellar objects are galactic stars and how many are extragalactic.

III. Known and Probable Blue Stars.

Objects in this category appear stellar on our plates and on the POSS prints. In most cases they are distinguished from category II objects by one or more of the following spectral characteristics: a) weak or broad Balmer lines, b) an incipient Balmer discontinuity, c) the continuum shortward of 3700 Å is more heavily exposed than the blue-green (4200-5350 Å) region. Also included in category III are objects with featureless flat spectra which might otherwise be put in category II except for the fact that they are known to have appreciable proper motion and thus must be galactic stars. Similarly, any blue object recovered by us which has been previously classified as a star on the basis of higher dispersion spectroscopy is placed in category III. When the Balmer lines are very broad and there is no Balmer discontinuity, the object is very likely an A-type white dwarf.

IV. Main Sequence Late B and A-type Stars.

These have spectra which exhibit a definite Balmer discontinuity and strong Balmer absorption lines.

V. Suspected Field Horizontal-branch Stars of Types A and F Including RR Lyrae Variables.

These have flat spectra with Balmer discontinuity but no apparent Balmer absorption lines.

VI. Suspected F and G-type Subdwarfs Showing a UV Excess.

Objects whose spectra have gradients in the blue-green like F and G stars but show unusually strong UV continua are placed in this category.

VII. Faint Carbon and Late M Suspected Halo Giants.

Carbon stars can be identified by the presence of the 4737 and/or 5165 C_2 bands. Late M stars are recognizable by the effects of the strong 4762, 4954 and 5167 TiO bands.

VIII. Peculiar Objects.

Spectra with emission and/or absorption features which do not fit into any of the above categories are placed in this category.

PRESENT STATUS OF THE SURVEY

Approximately 5000 square degrees (~ 225 fields or plates) are required to cover the northern sky to $\delta > +30°$, $b > +30°$. Two thirds of these plates have been taken; 25 have been scanned and measured. It is our intention to publish lists of objects with finding charts as we complete small areas. Particularly interesting objects will be announced as we encounter them. First results have been published in two papers (Sanduleak and Pesch 1982, Pesch and Sanduleak 1983). The completion of the converted aerial film scanner which makes it possible to combine the scanning and measuring is expected to speed up the project.

It is a pleasure to acknowledge the support of the NSF.

REFERENCES

MacAlpine, G.M. and Feldman, F.R.: 1982, Astrophys. J. 261 pp. 412-421.

Pesch, P. and Sanduleak, N.: 1983, Astrophys. J. Suppl. 51 pp. 171-182.

Sanduleak, N. and Pesch, P.: 1982, Astrophys. J. 258 pp. L11-L15.

Stephenson, C.B. and Sanduleak, N.: 1977, Publ. Warner and Swasey Obs. 2 pp. 73-99.

THE DETECTION OF FAINT IMAGES AGAINST THE SKY BACKGROUND

David F. Malin
Anglo-Australian Observatory

INTRODUCTION

This title was first used by William Baum (1955) in his contribution to an IAU Joint Discussion on photoelectric image tubes and their astronomical applications. The general tone of the ten or so papers collected in the 1955 <u>IAU Transactions</u> was optimistic. It seemed quite clear that the high quantum efficiency of the electronic devices existing at the time would soon be turned to astronomical advantage and that objects much fainter than those detectable photographically were well within reach.

    This period was a time when the concepts of signal-to-noise ratio and quantum efficiency were being extended to detectors of all kinds, mainly as a result of the work of Rose (1946). The photographic aspects were persued by Jones (1958) and Fellgett (1958) who emphasised that in a given image area a fine grained emulsion could count more photons than a coarse grained one, thus improving the signal-to-noise ratio of the image. The practical consequences of this were appreciated even earlier; in 1955 Baum was able to write:
'In principle we should be able ----- to place a fine-grain plate at the prime focus of the 200" telescope, expose it night after night on the same field, and eventually reach the same threshold of detection which an image tube system would yield ----. In practice, of course, reciprocity failure and the cost of telescope time would make such a procedure abortive'. Ten years later, Marchant and Millikan (1965) of the Eastman Kodak Company had evolved just such an emulsion and, initially at least, exposure times were inconveniently long, though not everyone was deterred by this. Of one of the first astronomical exposures with the new emulsion, Sandage and Miller (1966) say:
'The results were remarkable. A vast number of galaxies were visible over the entire plate - galaxies invisible on 103aO and 103aE plates exposed to the sky limit'. This fine grain, high contrast emulsion, now known as IIIaJ, was extremely slow but responded well to the hypersensitising techniques under investigation at the time. These techniques largely remove the bogey of reciprocity failure as well as giving a real

increase in sensitivity, thus allaying Baum's fears of prohibitively long exposure times. With careful hypersensitising the IIIa emission can be exposed to yield maximum output signal-to-noise (i.e., the sky-limited condition) in 60 minutes or so on an f/3 telescope, much shorter than the several nights anticipated by Baum in 1955.

His paper contains several ideas which were to be developed more fully in a better-known publication, his contribution to Hiltner's <u>Astronomical Techniques</u> (Baum, 1962). One idea however was dropped. That was something which he described as amplification and background subtraction and which to Baum (1955) 'does not seem photographically feasible'. This idea has in fact proved to be an extremely powerful method for detecting faint objects against the night sky, especially in combination with fine-grained emulsions exposed to the sky limit. The method is used to produce contact derivatives from original plates by means of a high contrast film in conjunction with a diffuse light source (Malin 1978a) and has been further developed to incorporate multiple-image superimposition, (Malin 1981) unsharp masking (Malin 1977, 1979) and as a means of making colour photographs of very faint objects (Malin 1980). Since the 1.2m Schmidt telescope (UKST) is a convenient source of sky-limited plates of excellent quality, it is natural that these photographic techniques should be applied to them. This paper briefly describes some of the results obtained, in the main from Schmidt material, but often followed up by plates from the 3.9m Anglo-Australian Telescope (AAT).

PHOTOGRAPHIC AMPLIFICATION

The purpose of photographic amplification is to enhance the visibility of faint photographic images. There are many ways to accomplish this of which the various methods of chemical intensification are perhaps the best known. Haist (1979) gives a thorough coverage of these possibilities. More complex methods of faint image enhancement by means of radioactive toners (Askins 1976) and 20 MeV electrons (Murray 1980) have also been tried. All these methods involve some chemical after-treatment of the processed emulsion but their main disadvantage is that they enhance background as well as image, which is why Baum (1955) dismissed this type of amplification for deep astronomical plates where the uniform density due to sky and chemical fog is already very high.

An alternative process, developed by Malin (1978a) is a non-destructive contact-copying procedure, which uses high contrast film. The copy exposure is adjusted so that only a very small range of densities around the sky background of the original plate appear on the copy positive. The sky background is not so much subtracted (to use Baum's terminology) as ignored and the increased contrast enhances the detection of faint objects superimposed upon the sky background density. There are other subtleties, mainly concerned with the use of a diffuse light source; these have been discussed elsewhere (Malin 1982). The high contrast but low density film positives produced by this process are printed on to contrasty bromide paper to produce the negative prints preferred for faint object detection. This very simple technique has revealed many new, low surface brightness structures, some of which are illustrated by the following examples.

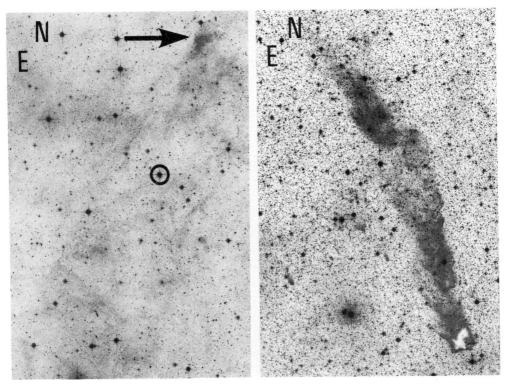

Fig 1. Extensive faint nebulosity has been found in several UKST plates centered near the SGP. The arrow indicates the only patch visible on the original. The circled star is SAO 166722, near 0^h 56.7 -25° 13 (1950).

Fig 2. About 7 faint cometary globules can be seen on this photographically amplified derivative from a deep IIIaJ Schmidt plate. Figs 1 and 2 are the same scale. The vertical height of each print is about 1.5°.

NEBULOSITY NEAR THE SOUTH GALACTIC POLE

During routine quality control inspections of plates at the UKST, Sue Tritton noticed that several plates of regions near the South Galactic Pole (SGP) showed evidence of very low contrast non-uniformities apparently covering small regions of the 6 x 6" fields. These effects were only visible to the experienced eye and would have passed unseen in regions of higher stellar density. When new plates of these fields were obtained, the non-uniformities remained. Photographic amplification of the plates showed that some fields were filled with extensive but very faint filamentary nebulosity. Although no detailed investigation of this nebulosity has been attempted, in view of its structure and distribution, it seems likely to be evidence of an extensive dust cloud which is reflecting light from the Galactic disc. Naturally, if such a cloud can reflect light from within the galaxy, it can attenuate light from beyond it, with obvious implications for those who study the luminosity, distribution and colour of faint objects in the direction of the SGP. The SGP itself seems clear of nebulosity, but the filaments seen in Fig. 1 can be traced to within 3° of the pole on the eastern side. Fig. 1 covers an area of sky about 1 x 1.5°.

The arrowed region of nebulosity, burned-out in this reproduction, is just visible to the unaided eye on the original plate. From previous quantitative work on faint features (Carter et al 1982) we estimate that the surface brightness of the indicated patch is about 26.5 mag arc sec^{-2} in B and that the faintest wisps visible in Fig 1 are about 1-1.5 magnitudes fainter.

COMETARY GLOBULES

Cometary globules appear to be a variety or perhaps a precursor of the well known dark clouds often identified as Bok globules. Cometary globules are distinguished by their long, faintly luminous tails extending from compact dusty heads, the whole object often spanning more than 1 degree of sky. They were first noted by Hawarden and Brand (1976) on deep IIIaJ plates taken on the UKST and have been studied by Zealey (1979) and Reipurth (1982).

Cometary globule (GG) number 22 in Zealey's list is quite easy to see on the original SRC IIIaJ plate, but the full extent of the tail is only apparent after photographic amplification (Fig 2). Also revealed by this process are several other, much smaller tails without any evidence of dark clouds at their origin. Like CG 22, these tails point in the general direction of the extremely luminous stars towards the centre of the Gum Nebula. At least seven such nebulous objects can be seen on the original print of Fig 2. Other examples of photographic amplification applied to Galactic objects appear in publications by Malin (1978a) Elliott and Malin (1979), Zealey et al (1980) and Gilmozzi et al (1983).

GALAXIES

The photographic amplification technique has proved to be particularly useful for determining the extent and structure of galaxies. Fig 3 indicates why this is so. Fig 3a is a normal contrast copy, representing the visual appearance of a deep IIIaJ plate taken on the UKST. Fig 3b is the result obtained by the photographic enhancement of this same plate. The galaxies are the well-known group in Leo which includes NGC 3379, the prominent E1 galaxy which has been adopted as a luminosity distribution standard.

This galaxy has been the subject of a detailed photometric study by de Vaucouleurs and Capaccioli (1979) who have obtained an expression for the luminosity profile of the galaxy to very faint limits. On the enhanced print, Fig 3b, the envelope of the galaxy can be detected out to at least 6.7 arc sec on either side of the nucleus in an EW direction, which corresponds to a surface brightness of 27.5 mag arc sec^{-2} on the basis of the de Vaucouleurs-Capaccioli profile (in B; the photometric correction from B to (IIIa)J is small and has been ignored here).

On the night the plate was taken, (1980 Jan 16) the UKST photoelectric sky photometer indicated that the sky in the direction of NGC 3379 was particularly bright, probably due to the combined effects of enhanced airglow near solar maximum and haze from bushfires burning in the nearby Pilliga Scrub (Dawe, private communication). The galaxy crossed the meridian during the exposure, culminating at a zenith distance of 44°N. The observing record shows that the night sky brightness varied from 22.16 (B, mag arc sec^{-2}) at the start of the 65 minute exposure to 22.04 at the end, which coincided with the start of twilight. If we take the mean night sky brightness as 22.1 during the exposure, then the faintest extremities of NGC 3379 as revealed on Fig 3b are 5.4 magnitudes fainter than the night sky. From this it seems evident that the photographic technique should be capable of detecting diffuse features as faint as 28.5 (B) mag arc sec^{-2} on plates taken under darker conditions, perhaps even fainter if there is an abrupt change in the luminosity profile of a faint object - a sharp-edged shell galaxy for example.

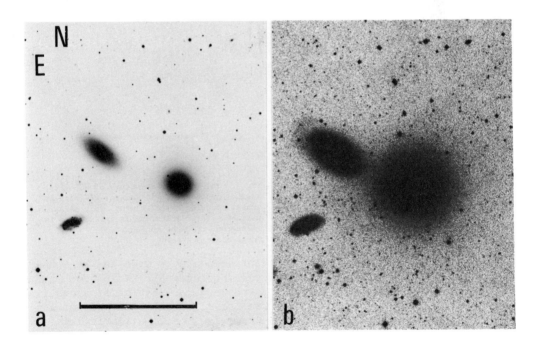

Fig 3. The NGC 3379 group of galaxies in Leo. Fig 3a represents the visual appearance of the original IIIaJ UKST plate while 3b shows the image derived from the same plate by the photographic amplification process. This shows features 5.4 magnitudes fainter than the night sky, including evidence of disturbance of the outer envelope of NGC 3384, the S0 galaxy N.E. of NGC 3379. Scale bar = 10 arc min.

Fig 3b confirms the overlap of the fainter isophotes of NGC 3379 with those of the S0 galaxy NGC 3384 to the NE of it as described by Barbon, Capaccioli and Terenghi (1975) who do not consider this to be the result of interaction between the two. It should be noted however that the NE end of NGC 3384 is disturbed, with a faint diffuse loop or extension clearly visible on the deeper photograph. Fig 3b also neatly demonstrates in a qualitative way the marked difference between the luminosity profiles of the three main galaxy types. (NGC 3389, E of NGC 3379 is an SAS 5 galaxy).

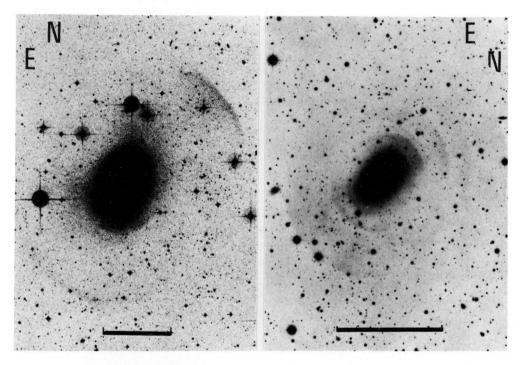

Fig 4. NGC 1344 was the first galaxy found to have a faint external shell. This photograph was made by combining the images of three UKST plates. The brightest part of the N.E. shell has a surface brightness of 26.5 mag arc sec^{-2} (B). Scale bar = 5 arc min.

Fig 5. The internal structure of NGC 3923, another prominent shell galaxy, revealed by unsharp masking an AAT plate. This galaxy has about 18 separate shells interleaved in radius out from the nucleus. Scale bar = 5 arc min.

The ready availability of excellent deep plates from the UK Schmidt, and our tendency to apply non destructive photographic techniques to them on an almost routine basis, has led to several interesting findings. Perhaps the most important of these was the discovery of a pair of low surface-brightness shells around the otherwise normal elliptical galaxy, NGC 1344 (Fig 4). This was soon followed by the detection of similar shells around several other elliptical galaxies by Malin and Carter (1980). Subsequent work by Carter et al (1982) has shown that in the case of NGC 1344, which seems to be the best example seen to date, the shell is composed of stars of colours consistent with spectral type G5-K4. In the same investigation, using optical and infra-red photometry, it was found that the brightest part of the NW shell of NGC 1344 had a surface brightness of 26.5 mag arc sec^{-2} in B. This shell was just visible to the educated eye on the original IIIaJ discovery plate.

The observation that the brighter shell-type galaxies could be found by direct inspection without photographic enhancement encouraged us to visually inspect each of the 606 6 x 6° fields of the SRC J sky survey south of -20° in the form of film copies, searching specifically for elliptical galaxies which showed a shell-like structure. 137 such galaxies were found and their positions and a brief description of each has been published in the form of a catalogue by Malin and Carter (1983).

The statistical properties of this sample of elliptical galaxies are of interest. About half of them appear to be isolated, while another third are members of small groups with 2-5 members. Only 5 examples (3.6% of the sample) are found in clusters or rich groups. This distribution is of course quite unlike that of normal ellipticals which are generally gregarious. Another significant finding is that only two members of the sample - Fornax A and Centaurus A - are powerful radio sources. The shells in Fornax A, NGC 1316, have been described in detail by Schweizer (1980) while those in Centaurus A (NGC 5128) are the subject of a recent paper by Malin et al (1983).

The nature and origin of the shells has been the subject of some speculation, but Quinn's (1982) thesis, which predicted that shells would be produced as a result of the merger of a disc galaxy and an elliptical, seems to fit the observed facts quite well. It would be expected, for instance, that shells at a considerable distance from the nucleus of the parent galaxy would be readily disrupted by tidal encounters with group members if it were in a cluster whereas the shell would be more stable if the galaxy was isolated. Quinn's theory also predicts that a series of concentric shells might be produced during the merger, each interleaved in radius outwards from the nucleus. Some examples of this have now been seen; NGC 3923 (Fig 5) is the best so far found, where about 18 shells can be counted. Fig 5 was prepared by copying an AAT plate through an unsharp mask (discussed later).

Some observational evidence for galaxy mergers has been presented by Schweizer (1983) who lists 32 galaxies he considers to be possible merger remnants on the basis of their morphology (internal ripples, tails, isophotal twists etc). 10 of these galaxies also appear in the Malin-Carter catalogue. Schweizer has obtained CTIO 4m plates of some of the galaxies in his list and one of these, IC 3370, shows crossed streamers forming an 'X' and 'ripples' in opposite quadrants of the X. The cruciform shape is clearly seen in Fig 6a but a much deeper photograph, made from high contrast derivatives from three deep IIIaJ plates (Fig 6b) shows the debris left over from what is almost certainly a merger. This kind of amorphous, low surface brightness material has been seen associated with several of the shell-type galaxies in the Malin-Carter catalogue.

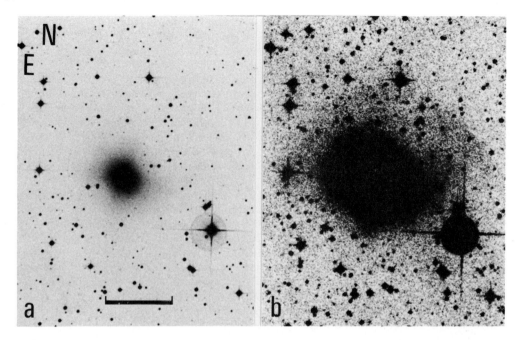

Fig 6. IC 3370 was identified by Schweizer (1983) as a merger remnant from the curious 'X' morphology of the galaxy seen in Fig 6a. A deeper image from three deep UKST plates reveals some of the debris of that merger. Scale bar = 2 arc min.

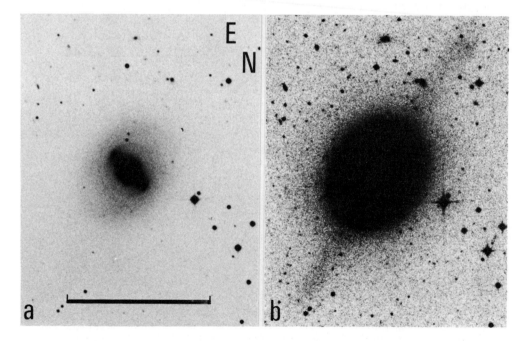

Fig 7. NGC 4643 appears to be a perfectly normal barred spiral on simple inspection of a deep UKST IIIaJ plate (Fig 7a). Only after photographic amplification are the remarkable extensions of the outer envelope visible (Fig 7b). Scale bar = 5 arc min.

While the photographic enhancement of images of elliptical galaxies
has been particularly fruitful (for other examples see Malin 1981a and
b) disc-type galaxies may also reveal unexpected features. The barred
spiral NGC 4643 appears to be a perfectly normal galaxy seen almost
face-on, with a very well developed bar running roughly NW-SE through
the nuclear bulge (Fig 7a). A print made from the same SRC IIIaJ
original, designed to reveal the faintest features (Figure 7b) shows
that the galaxy has a large diffuse envelope, which has a radial
luminosity profile more like that of an elliptical galaxy than a spiral
(compare Fig 3b). Even more surprising are the two faint ansae which
project from the diffuse envelope at right angles to the inner bar. We
are not aware of any other galaxy which shows features of this type,
which are quite unexpected. Note that the apparent angular extent of
the galaxy has increased from about 4 arc min in (a) to over 10 arc min
in (b). Although NGC 4643 seems to be a member of the southern extension
of the Virgo cluster, no other galaxy appears to be near enough to cause
the observed disturbance. Both the photographs in Fig 7 are printed
with the same scale and orientation.

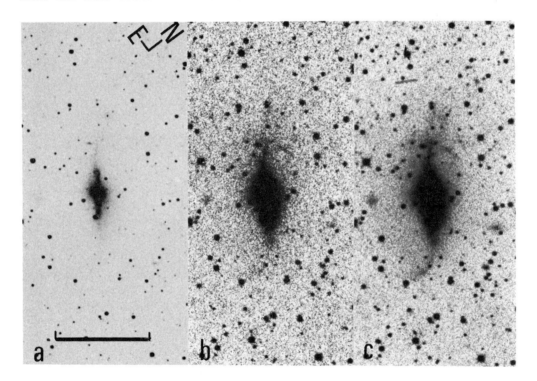

Fig 8. The effect of multiple-image superimposition on the signal-to-noise ratio is clearly
seen in this series of photographs of NGC 4672. Fig 8a is derived from a greatly enlarged
normal contrast copy of a deep UKST IIIaJ plate. The same plate produced the amplified
version seen in (b). Although deeper, much fine detail is hidden in the grain noise. The
effect of combining the amplified derivatives of four plates of similar quality is apparent
in (c) where the galaxy is seen to be highly disturbed. Note that there is no loss of
resolution in the image. The diameter of stars at the plate limit in (a) is the same as
those at the limit in (c), though these latter are much fainter.
Scale bar = 2mm on the original plate.

INTEGRATION PRINTING

The photographic amplification process described above enhances the apparent size of the silver grains which form the original image and the pictures produced always appear more grainy than those produced from normal copies. This effect is usually offset by a marked gain in the perception of faint images and is not obtrusive at modest enlargements. However, reduced granularity and improved perception of faint objects, particularly those with a scale length approaching the limit of resolution on Schmidt plates (e.g. stars) can be obtained by combining the images derived from two or more plates. This technique is always used to add together photographically amplified positives - there is little point in combining images which have not been enhanced to reveal the faintest information. The effect of this is seen in Fig 8. All three images are of the edge-on spiral NGC 4672 in the Centaurus cluster and are printed to the same scale, an enlargement of about 13x. Fig 8a is derived from a normal contrast copy of a deep IIIaJ plate while Fig 8b is the same image after photographic amplification. The effect of superimposing four high-contrast derivatives is shown in Fig 8c. Faint stars and low surface brightness features not seen in 8a and buried in the grain of 8b are seen clearly in Fig 8c. The improvement in signal-to-noise is evident and the disturbed outer structure of NGC 4672, only suspected from the appearance of Fig 8b, is amply confirmed in the four image superimposition. A simple device for combining multiple images has been described elsewhere (Malin 1980).

UNSHARP MASKING

In keeping with the title of this contribution, the previous sections have been concerned with the detection of faint images against the night sky. The photographic properties of fine grain and high contrast which are so useful in this endeavour inevitably have the disadvantage of limiting the dynamic range of the emulsion. In the case of IIIaJ plate exposed to the sky limit (i.e. minimum sky density of ~ 1.0 ANSI above fog) the available dynamic range is about 1 log exposure unit or 2.5 stellar magnitudes before the emulsion saturates at its maximum density of ~ 4.5. This compares with a dynamic range equivalent to 7.5 magnitudes and a maximum density of ~ 1.6 for a normal commercial camera-speed material. This limitation is a serious problem when information on the bright interior of a galaxy or HII region is required and the only image available is on a sky-limited IIIa plate.

This difficulty can be overcome in part by adapting a process called unsharp masking, long known in the graphic arts industry, to the special needs of astronomy. The process has been described in detail by Malin (1977) but in essence it is the reverse of Baum's (1955) background subtraction. An unsharp film positive of the plate is prepared by contact copying the original on to film through the glass support. The developed positive is then used as a mask to subtract the coarse detail from the original, leaving faint and small scale structure largely unaltered. The effect of this is clear in Fig 9, where a direct print

(a) from a deep UKST IIIaF plate of the Orion Nebula is compared with the unsharp masked version from the same plate (b). The technique has been particularly useful in exploring the dense images of elliptical galaxies, where low contrast structure is often hidden in the high density of the image of the elliptical envelope. (See Fig 5).

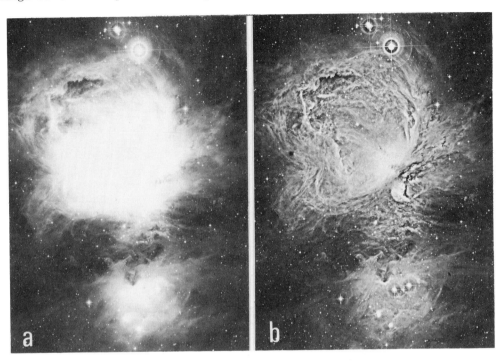

Fig 9. The effect of an unsharp mask on a deep UKST IIIaF plate of the Orion Nebula. Fig 9a is the best print which could be obtained by using the original plate as a negative in the enlarger. In Fig 9b much additional information appears. This details was hidden in the high density regions of the plate (D max 4.5) and was only revealed after printing through an unsharp mask in conjunction with a diffuse light source.

COLOUR PHOTOGRAPHY

The various processes of analogue image manipulation outlined above produce as a first derivative a positive film copy. Positive derivatives from separate plates taken in the standard photographic B,V and R passbands can be combined in the manner of James Clerk Maxwell to yield a 3-colour image representative of the true colours of the object. Such photographs show objects which are much too faint to be recorded by tri-pack colour films and reveal details of scientific as well as aesthetic interest. These pictures are now available in the form of 35mm slides (Malin 1983).

PLATE UNIFORMITY, THE ULTIMATE LIMITATION ?

The methodical record-keeping and quality control procedures which were soon established at the UK Schmidt quite quickly traced and eliminated many of the more obvious causes of plate non-uniformity. Generally these defects were seen on simple inspection and arose from one or more of the numerous hazards of shipment, storage, handling, hypersensitisation, exposure, processing and drying which plates must undergo. While the effects were always detected without the aid of photographic enhancement, their causes were often difficult to trace or when traced, not easily eliminated. The high standard of the SRC deep southern survey is a tribute to the skill and dedication of successive members of the UKST group which produced the original plates substantially free from visual defects.

This achievement was even more impressive when the photographic amplification technique began to be applied to UKST plates. The technique is not selective - any small density difference is enhanced, no matter what its origin. This property makes this simple process into a powerful diagnostic aid for revealing and tracing artefacts which ultimately limit the detection of faint objects. Fortunately, most of the faint features discovered turned out to be real objects, but one type of non-uniformity appeared to some extent on almost every hypersensitised IIIa plate and will be discussed here as an example of the subtle problems to be encountered in faint object photography.

In the UK Schmidt telescope, the plate is curved about a radius of 3.07m and is (usually) positioned close to a flat glass filter. At the corner of the plate the air gap between the emulsion surface and filter is $\sim$ 11 mm, which narrows to only $\sim$ 3 mm at the centre. The natural convective flow of air within this space is therefore increasingly restricted towards the plate centre. It is known that dried, hypersensitised plates lose speed if they are allowed to stand in room air before or during a long exposure to light (Malin 1978b). This is probably due to the re-establishment of low intensity reciprocity failure (dehypersensitisation) by absorption of traces of water vapour and will be more pronounced where air circulation is greatest, i.e. near the edge of the plate in the Schmidt. The result is a radial gradient in sensitivity, with its greatest effect at the plate edge (Dawe and Metcalfe 1982, Campbell 1982). On a IIIa plate exposed to the sky limit (density 1-1.3 above fog) the gradient can amount to a density difference of 0.1-0.15 from the centre to the edge of a 14x14 inch plate. While this is too small and gradual a change to be detected visually, its presence is obvious when a whole plate is photographically amplified and in our experience was, until recently the most obtrusive large-scale non-uniformity on hypersensitised IIIa plates from the UK Schmidt. This result is consistent with the work of Dawe, Coyte and Metcalfe (1983, in preparation) who found density differences from the plate centre to a radius of 107 mm ($\equiv$ 2°) equivalent to 0.05 magnitudes with small-scale point-to-point variation of $\pm$ $0\overset{m}{.}03$. Plateholders have recently been modified to overcome the problem by flowing nitrogen into the space between plate and filter during exposure. Preliminary results (Dawe, private communication) indicate a substantial improvement in both large and small scale plate uniformity together with a small increase in effective plate speed.

With the elimination of this last major user-induced problem, we are finally left with emulsion variations introduced in manufacture. Over the last 5 years or so, many plates from a variety of telescopes have been critically examined by photographic enhancement. Almost all the artefacts detected are originated by the user, very few seem to be the result of the manufacturing process. It should be emphasised that it is a remarkable achievement to manufacture and supply a detector 14 inches square which is able to detect variations in signal around the 1% level and which has a quantum efficiency of a few percent. And all this from an emulsion dewigned almost 20 years ago.

THE FUTURE

Over the last 100 years the development of astronomy and photography has been inextricably linked. Clear advances in astronomical understanding can be traced to the introduction of the dry gelatin plate in the 1880s, advances in dye sensitising in the 1920s and the introduction of Eastman Kodak spectroscopic plates in the 1930s, with special emulsions (type 'a') for long exposures appearing after the war. The last major improvement came with the introduction of the IIIa emulsion in 1965, though hydrogen hypersensitising (Babcock et al 1974) was necessary to make these products generally useful.

Major advances in emulsion making, as discussed in a recent Research Disclosure (Anon. 1983) indicate that substantial improvement in both the speed/granularity relationship and in the art of spectral sensitising have occurred in the last few years. These developments are appearing in improved colour negative and reversal materials now available from several manufacturers and stem largely from new methods of emulsion-making which give close control over the shape of the silver halide gain.

It is probable that some of the developments outlined in the Research Disclosure could be incorporated into a new generation of spectroscopic emulsion. With this possibility in mind, it might be useful to consider where improvements might best be realised, in granularity, speed or contrast, and what changes, (if any) in spectral sensitivity would best serve the astronomical community to the end of the century.

It is the personal view of this writer that the greatest gains will come from the photographic investigation of the faintest objects. For this, an emulsion with a (hypersensitised) speed and granularity similar to that of the current type IIIa, but with a much increased contrast level ($\gamma = 5-7$) is needed. Maximum output signal-to-noise might be achieved with less exposure if maximum contrast was attained at a lower density than is found in the present generation of IIIa products. Such an emulsion would have a very restricted useful dynamic range and an even smaller range over which output signal-to-noise is optimum, and users would be obliged to judge their exposures to within a few percent. This kind of control, which involves sensitometric tests on each batch of hypersensitised plates, has been the practice at both the UKST and the AAT for the past 6 years and has eliminated under- or over-exposure, with consequent saving of telescope time. An emulsion of increased contrast, optimally sensitised to the darkest part of the night sky and used in conjunction with enhancement techniques should reach 29.5 mag arc sec^{-2} at a dark site.

As a second option, a series of emulsions of much lower contrast, ($\gamma \sim 1.5-2$) sensitised in the photometric O, D and F bands and of much finer grain than the present IIa series would improve photographic photometry on small telescopes and be valuable for morphological studies of galaxies on Schmidt telescopes. Certain aspects of the art of spectral sensitising are specifically referred to as facilitated by the new emulsion technology described in the Research Disclosure. One area where such improvements might be of immediate benefit, particularly in objective prism work, is the design of an F sensitising with a much more uniform spectral response.

I emphasise that these are personal preferences and in no way reflect the views of the astronomical community as a whole. They are merely intended to stimulate discussion amongst those interested in furthering astronomical research with photographic detectors. There seems little doubt that with careful use, the present family of emulsions is equal and often greatly superior to any form of electronic detector for detecting faint images against the night sky background. The next generation of emulsions is keenly awaited.

REFERENCES

Askins, B.S. 1976. Applied Optics. 15, 2860-2864.
Babcock, et al. 1974. Astron. J. 79, 1479-1487.
Barbon, R., Capaccioli, M., Tarenghi, M. 1975. Astron. Astrophys. 38, 315-321.
Baum, W.A. 1955. in Trans. IAU. Vol. 9, 681-686.
Baum, W.A. 1962. Stars and Stellar System, Vol. 2., Astronomical Techniques, W.A. Hiltner, Ed. pp1-33. Univ. of Chicago Press.
Campbell, A.W. 1982. The Observatory. 102, 195-199.
Carter, D., Allen, D.A., Malin, D.F. 1982. Nature. 295, 126-128.
Dawe, J.A., Metcalfe, N. 1982. Proc. Astron. Soc. Australia. 4, 466-468.
de Vaucouleurs, G., Capaccioli, M. 1979. Astrophys. J. Suppl. Ser. 40, 699-731.
Elliott, K.H., Malin, D.F. 1979. M.N.R.A.S. 186, 45p-50p.
Fellgett, P.B. 1958. M.N.R.A.S. 118, 224-233.
Gilmozzi, R., Murdin, P.G., Clark, D.H. and Malin, D.F. 1983. M.N.R.A.S. 202, 927-934.

Haist, G.M. 1979. Modern Photographic Processing. Vol. 2 pp1-49.
  John Wiley and Sons, New York.
Hawarden, T.G., Brand, P.W.J.L. 1976, M.N.R.A.S. 175, 19p-21p.
Jones, R.C. 1958. Photog. Sci. Eng. 2, 57-65.
Malin, D.F. 1977. Amer. Astron. Soc. Photo. Bulletin. No. 16, 10-13.
Malin, D.F. 1978b. in Modern Techniques in Astronomical Photography.
  ESO Conference, Geneva. May 1978, 107-112.
Malin, D.F. 1978a. Nature. 276, 591-593.
Malin, D.F. 1979. Nature. 277, 279-280.
Malin, D.F. 1980. Vistas in Astronomy. 24, Pt. 3, 219-238.
Malin, D.F., Carter, D. 1980. Nature. 285, 643-645.
Malin, D.F. 1981a. Amer. Astron. Soc. Photo. Bulletin. No. 27, 4-9.
Malin, D.F. 1981b. J. Photogr. Sci. 29, 199-205.
Malin, D.F. 1982. J. Photogr. Sci. 30, 87-94.
Malin, D.F. 1983. Stars and Galaxies, 30-slide set available from
  Armagh Planetarium, Armagh, N. Ireland.
Malin, D.F., Carter, D. 1983. In press. Ap. J.
Malin, D.F., Quinn, P.J., Graham, J.R. 1983. Ap. J. Lett. In press.
Marchant, J.C., Millikan, A.G. 1965. J. Opt. Soc. Amer. 55, 907-911.
Murray, K.M. 1980. Photogr. Sci. and Eng. 24, 166-170.
Quinn, P.J. 1982. The Dynamics of Galaxy Mergers. Ph D. Thesis,
  Australian National University, Canberra.
Reipurth, B. 1983. Astron. Astrophys. 117, 183-198.
Anon, 1983. Research Disclosure. No. 1. Article 22534, 20-58.
Rose, A. 1946. J. Soc. Mot. Pic. Eng. 47, 273-295.
Sandage, A.R., Miller W.C., 1966, Ap.J. 144, 1238.
Schweizer, F. 1980. Ap. J. 237, 303-318.
Schweizer, F. 1983. IAU Symp. 100, Internal Kinematics and Dynamics of
  Galaxies, E. Athanassoula (ed). pp319-329.
Zealey, W.J. 1979. New Zealand J. of Sci. 22, 549-552.
Zealey, W.J., Dopita, M.A., Malin, D.F. 1980. M.N.R.A.S. 192, 731-743.

Comment by Russell Cannon during general discussion on Friday afternoon, relating to the discussion led by David Malin on new photographic emulsions.

    I support nearly all of the points made by David Malin, but would like to enlarge on two matters. (i) Regarding possible increases in emulsion speed, it is important that we have a real gain and not simply a decrease in the time it takes to reach a given photographic density. In other words, we want to retain all the properties of the excellent IIIa-J emulsion, and in particular there must be no loss of resolution; thus a halving of exposure time 'to achieve the same results' means we want a real increase in DQE by a factor of two. Such an increase would surely be welcomed by all Schmidt astronomers.

(ii) I think that Malin's plea for an increase in contrast is more of a specialist requirement, and to a certain extent will only produce yet more very faint features which are too faint to be studied further either photometrically or spectroscopically, as well as making a mess of more of our favourite galaxies, star clusters or whatever! More seriously, I suspect that when David says that he cannot get more out of the existing plates if he tries to push the copy contrast even higher because this simply brings up the grain, what he means is that his present techniques already extract all of the useful information from IIIa-J plates. Therefore what we need are not necessarily higher contrast original plates, but plates with a higher information content or better signal-to-noise; I think this must mean going to finer grain emulsions.

I would like to add my support to two other specific points: astronomers really would like to have a 'IIIa-O' emulsion for photometric work, and a flatter wavelength response version of the IIIa-F would be invaluable for objective prism spectroscopy.

ENHANCEMENT OF FAINT IMAGES FROM UK SCHMIDT TELESCOPE PLATES

B.W. Hadley, Royal Observatory, Blackford Hill,
Edinburgh EH9 3HJ.

INTRODUCTION

The UK Schmidt Telescope has taken several thousand photographs especially on fine grain Eastman Kodak IIIa-J and IIIa-F emulsions (see Cannon, 1984) which were hypersensitised before exposure in the telescope to improve their ability to record very faint signals.

Relatively little work has been done on further post-processing techniques to enhance photographically the faint signal from sources very near to the sky background level. Work is currently being undertaken by Malin (AAO) and others including staff at ROE to enhance faint photographic signals without a corresponding increase in noise, either by making high contrast enlargements from small areas of the original plate, or by making contact copies of the whole plate onto high contrast line material. The UK Schmidt photographs are 356mm x 356mm, covering a field $6\frac{1}{2}°$ square and over such an area there are several sources of large scale, small amplitude variations in the background density level. These include the vignetting function of the telescope (Dawe, 1984), non-uniform losses of sensitivity (Malin 1978a, Campbell 1982), limitations inherent in processing large plates uniformly, and the effects of ghost images (UKSTU Handbook 1980). These non-uniformities are enough to make direct high contrast copying of complete plates unsatisfactory.

MASKING

To compensate for the inherent non-uniformities, a photographic positive mask is made, higher in density where the sky background on the original plate is low and less dense where the backtround is high. In effect, the mask is a variable neutral density filter specifically matched to the original telescope plate calibrated so that the sum of the densities of the mask and plate will be the same at corresponding points in the field. In order to retain the small scale variations, the mask is made "unsharp" by separating the mask material from the original plate with a spacer. The spacer may be a sheet of glass, of thickness appropriate to the spacial frequencies required, or it may be convenient to place the mask material in contact with the back of the original plate, thereby using the plate itself as the spacer. The mask material, in ROE's case Kodak Aerographic Duplicating Film type 4421, is then exposed to a diffuse light source, through the original plate and spacer simultaneously, and processed to a gamma between 0.6 and 0.9. The mask thus follows the larger scale density variations of the original plate, but does not record the sharp detail from the original.

HIGH CONTRAST COPYING

After processing, the completed mask is relocated onto the back of the original plate and a positive copy is made by exposing through the pair onto the high contrast material which is in contact with the emulsion of the original plate. This exposure is also made to a diffuse light source. The significant point about this process (Malin, 1978b) is that the metallic silver grains in the original plate which are produced by the faint astronomical signal are located in the surface of the emulsion, but the fog or noise grains are distributed randomly throughout the emulsion thickness. By using a diffuse light source, the fog grains deep in the emulsion are not resolved on the positive copy, which records only those parts of the original at or very near the surface of the emulsion. Exposure and processing in this technique are very critical, because the noise and signal densities are very close to each other on or near the toe of the characteristic curve of the original plate and the high contrast copy. The objective is to copy the signal onto the straight line, high contrast portion of the characteristic curve, leaving the noise compressed in the low contrast toe of the curve. This is particularly difficult because the "elbow" between the toe and straight line portions of the characteristic curve is extremely sharp.

The high contrast positive is then printed onto a further high contrast film or paper where the fainter densities of the images are more easily distinguishable on a negative print.

Contrast enhance techniques have been particularly successful when applied to very deep UK Schmidt photographs of nearby galaxies. They make visible faint features not previously suspected, at brightness levels down to 28 magnitudes per square arcseconds. Since the average sky background, from atmospheric airglow, zodiacal light, and our own galaxy is about 23 magnitudes per square arcsecond in blue light on a dark night, this difference of 5 magnitudes, corresponding to a factor of 100 in brightness, means that we can detect features whose light output corresponds to only 1% of the overall night sky brightness.

ENHANCEMENT OF FAINT IMAGES FROM UK SCHMIDT TELESCOPE PLATES

Fig. 2. Copying the original plate by direct high contrast methods only makes the intrinsic non-uniformities more conspicuous.

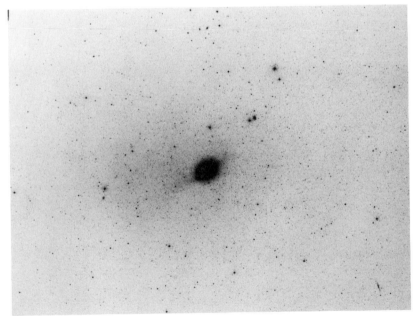

Fig. 1. The Original Plate

REFERENCES

Campbell, A.W.: 1982, Observatory 102, 195.
Cannon, R.D.: 1984, this volume, p. 25.
Dawe, J.A.: 1984, this volume, p. 193.
Malin, D.F.: 1977, ASS Photobulletin No. 16, p. 10.

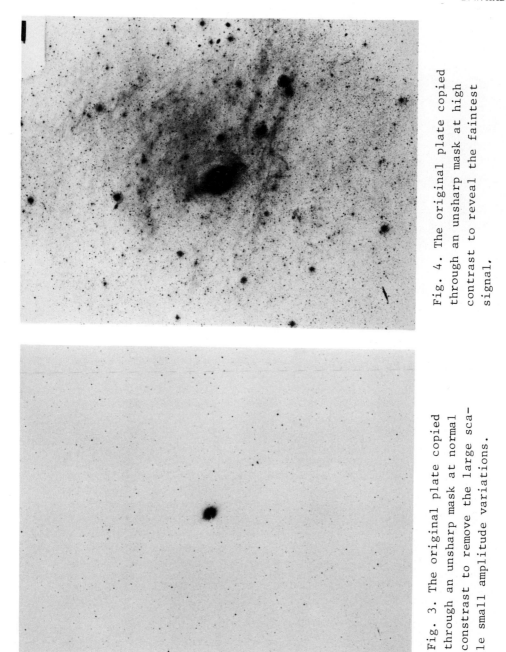

Fig. 4. The original plate copied through an unsharp mask at high contrast to reveal the faintest signal.

Fig. 3. The original plate copied through an unsharp mask at normal constrast to remove the large scale small amplitude variations.

Malin, D.F.: 1978a, 'Modern Techniques in Astronomical Photography', ed. West & Eudier, p. 107.
Malin, D.F.: 1984, This Volume.
UKSTU Handbook: 1980, p. 5.10.

STELLAR PHOTOMETRY WITH AN AUTOMATED MEASURING MACHINE

Gerard Gilmore
Royal Observatory, Blackford Hill, Edinburgh EH9 3HJ.

Visiting Associate, Mount Wilson and Las Campanas Observatories, Carnegie Institute of Washington.

ABSTRACT

The combination of wide field high quality telescopes, fine grain emulsions and fast automated measuring machines offers an unrivalled opportunity for progress in statistical astronomy. This review illustrates, with examples, the many steps which are necessary to realise this potential.

1. INTRODUCTION

The availability of large numbers of deep high quality photographic plates covering wide fields, and containing perhaps 100,000 usable images, presents new challenges in data reduction and data handling and new opportunities for statistical studies in astronomy. The existence of new fast automated measuring machines (COSMOS and APM in the UK) provides an unrivalled opportunity to exploit these data. Such exploitation however requires a careful understanding of the pitfalls as well as the advantages of automated photometry. This review outlines the practical factors involved in converting a Schmidt telescope plate into quantitative science, using as an example a survey of the stellar distribution in the Galactic spheroid. While most of the specific examples are relevant to plates from the UK Schmidt telescope measured with the COSMOS machine in Edinburgh, the principles are more generally valid. This paper discusses in turn the following topics: the operation of the measuring machine, and the significance of its parameters (section 2); the matching of several plates onto a single master reference set and image classification (section 3); photometric calibration (section 4); and some examples of the potential of fast measuring machines and Schmidt plates in statistical astronomy (section 5). Many other examples are presented in other contributions to this volume.

2) THE MEASURING MACHINE or WHAT HAS THAT MACHINE DONE TO MY PLATE?

The automated measuring machines may be used in either of two ways. In the first, they operate exactly as any other microdensitometer, recording the coordinate and density of every pixel in an area of plate. In the second, a real time decision is made by the operating system as to whether a pixel is above or below some prespecified threshold relative to the local sky background. Those pixels above this threshold are recorded, while those below are used to improve the current estimate of the local sky value. This procedure vastly reduces the number of pixels which need to be recorded, from the ~ 10^9 which are required to map all information on a single large Schmidt plate, to more manageable numbers. Apart from saving disk storage space, this greatly increases the speed of the machines, allowing an entire UK Schmidt plate to be mapped with 30μ pixels in less than half a day. After this measurement, the recorded pixel densities are converted to intensities, using an appropriate calibration curve, and processed by an algorithm which looks for connected pixels, joins them into an image, and deduces some coordinate, shape, structural, and total intensity parameters. It is these which are presented to the unsuspecting astronomer, and the derivation of these which must be understood to allow useful analysis of the results.

There are several parameters which must be specified for this stage of measurement which have a significant effect on the resulting data. These include the size of the measuring spot (the pixel size), the step size between consecutive measurements, the density to intensity conversion, the scale length for updating the local sky background value, the threshold above this level at which to accept pixels as 'signal', and the minimum number of joined pixels which constitute a 'real' image.

a) Pixel size: This is normally 16μ or 32μ with COSMOS and 7μ with the APM, although, as the intensity profile of the measuring spot shows extended wing structure, the physical significance of these numbers is unclear. In practice, the shape of the spot is a significant feature only in areas of steep density gradients.

b) Pixel spacing: This may usually be set at one-half the pixel size, following the Nyquist criterion. In practice, the optimum pixel size and spacing is dependent on the grain structure of the emulsion, the telescope plate scale, and the magnitude of the images of interest. A detailed analysis of these parameters on real plates has been carried out by Okamura et al. (1983), where details may be found.

c) Determination of sky: The primary disadvantage of fast measuring machines is their speed. This forces the use of relatively simple algorithms for sky estimation, and consequent high thresholds. The most reliable way to determine the sky background on a plate is obviously to scan the whole plate, fit some suitable polynomial to the data to estimate the background, and then rescan the plate. This however presupposes that the machine retains its coordinate and intensity system, and doubles the machine time per plate. A second method is to smoothly extrapolate the expected sky value from the values measured to date. A further complication is the usual choice of mean or mode as the best estimator of the true value. The practical consequence of these points is that typical thresholds above sky are 5%-10% - very high by the standards of those doing PDS photometry of galaxies - and that significant systematic errors in photometry, correlated with the local 'sky' value, can arise (cf section 4). The most important effect of these thresholds is to systematically distort the magnitude scale at faint magnitudes. This is discussed further in section 4.

A substantial improvement in this situation is possible, at the expense of considerable computer storage and time, by digitising and storing the entire plate. This then allows more sophisticated algorithms to be utilised to correct for variations in sky background, telescope vignetting and geometrical distortion, and so on. This technique is regularly applied to prime focus 4m plates, and allows thresholds of 1%-2% of sky to be reached, in spite of the extreme variations across these plates. See the COSMOS Users Guide for further details.

d) The density to intensity conversion: The choice of a calibration curve which is appropriate to a specific pixel at a specific position is an extremely complex problem. Factors which must be considered include the choice of wedge (continuously variable) or (discrete) spot calibration, the effect of exposing the calibration on sky or clear plate, the effects of time delays between image and calibration exposures and plate development, real variations in the true density-intensity relation with position and/or wavelength, and the systematic consequences of random errors in the adopted relation. These and other factors have been recently reviewed elsewhere (Gilmore 1983a) and so will not be repeated here. We note however that in practice it is not usually possible to do more than adopt a spline fit to the calibration provided on the plate. Projects which require a higher precision calibration must accept that very considerable effort is required to provide it.

e) Minimum image size: The minimum number of pixels in a 'real' image is a complex function of the seeing during the exposure, the telescope and emulsion scattering properties, the machine spot intensity profile, and the threshold set during measurement. In real cases, the minimum size is best set too low, so that most of the smallest (faintest) images are spurious. These may then be rejected by visual comparison with the original, or during the comparison with another plate (see below). While this is relatively straightforward, it is a very important consideration if faint object number counts are to be used as a test of completeness. It is not impossible to find a combination of too low a threshold and too small a minimum image size which mimics plausible number-magnitude counts much fainter than the true plate "limiting" magnitude. This problem is best eliminated, as are most others, by using several plates in each of several colours, and retaining images common to all (see below) and by remembering to look at the images on a plate which correspond to the machine output.

3)   INITIAL PROCESSING or WHAT DO I DO NOW?

After the machine processing outlined above, an astronomer will be presented with a set of magnetic tapes containing information on the position, intensity and shape of each of perhaps 10^5 groups of pixels identified by the algorithm as an 'image'. Many will be spurious. The next stage of processing is best illustrated assuming that serious photometry is intended. In this case, several plates in perhaps several colours on the same field centre have been measured. It is then necessary to transform them all to the same coordinate grid, match 'real' images, and classify them.

a)   Merging: This involves the transformation of the coordinate frame of each plate onto that of some master plate. In practice, this is achieved using a grid of stars common to all plates, which are identified (visually or by cross-correlation) and used to define the usual multi plate-constant transformation. For plates from the same telescope and with the same field centre such a transformation is usually accurate globally to a few microns, or ~ 0.1 pixels, allowing transformations to an astronomical reference frame with an accuracy of a few tenths of an arcsecond. The stars chosen for this initial transformation should be neither very faint (to minimise random errors in position) nor very bright (to minimise systematic errors due to the inclusion of asymetric image halo structure).

b)   Matching: When all image centroids are defined on the same reference frame, which may be an RA-DEC grid, it is straightforward to identify images which appear on more than one plate simply by position coincidence. This simple process however largely determines what science can be done with the resulting data, and so must be carefully considered. Some examples will illustrate this point. A common procedure is to take the 'deepest' plate available, and match all others onto it. This plate in practice is likely to be a sky limited IIIa-J plate, reaching to B ~ 22.5 or 23.0. The matching of an I band plate (limiting magnitude ~ 19.5) to this J plate will obviously retain every image on the I plate with B-I colour less than ~ 3^m. Thus, while the vast majority of images will be retained, a systematic rejection of late M dwarfs will result. This rejection is so efficient that the inverse procedure has been used with considerable success to identify extremely red stars (Gilmore and Hewett, 1984). Other classes of object which may be strongly biased against include variables and proper motion stars.

A more important effect for non-moving objects of neutral colour - just those usually required - is the effect of unresolved images. Deep Schmidt plates are already resolution limited over most of the sky (thereby precluding deeper surveys), so that a significant number of images are either unresolved or marginally resolved from other images. When adjacent images have different colours, or different plates are taken in different seeing, many images can be resolved on some and unresolved on other plates of a set. The reliable handling of these images is a very major task when processing large data sets, but must be carefully considered. At galactic latitude 50° below the Galactic bulge, 10% of images are affected by merging independent of apparent magnitude. These images must be reliably identified and handled in constructing the final data set. They are best found by comparison with deeper, larger scale plates. [The detailed analysis of crowded field photometry will not be discussed here. An excellent description is given by Newell (1983).] In the determination of the number of unresolved image pairs, it is striking how rapidly the surface density of stars may change across a Schmidt plate due to the variation in Galactic coordinates over the 6° field. Corrections for image crowding should be carried out in galactic rather than rectilinear plate coordinates. These corrections are themselves a useful scientific result, as they nicely map the gradient in the stellar surface density of the Galaxy. The determination of this is often the aim of the original project.

c) Image classification: When a final merged and matched set of images has been produced, it is usually necessary to decide what is a star and what is a galaxy. Images may be classified into one of four obvious categories:- definite star; definitely resolved; definite multiple image; and don't know. This classification can be carried out using surface brightness, image shape or extent, colour and other parameters. Examples are given by Kron (1980) and Reid and Gilmore (1982), and references in those papers. The details need not be discussed here. The important consequence for the intended scientific use is that no single classification scheme is optimised for all projects. A project which must include all stellar objects can do so only by accepting some galaxy contamination when classifying uncertain images, and vice versa. Provided that the consequences of this choice are known, and adequately calibrated this is not a problem.

An example of such a post-classification check is shown in Figure 1, which is the two-point correlation function for galaxies and faint K stars towards the south Galactic pole. The absence of any structure in the 'stellar' distribution on scales at which it occurs in the 'galaxy' distribution, at the same apparent magnitude and colour limits on the same plate, is a very powerful diagnostic of possible galaxy contamination of the stellar sample. [The lack of any significant structure in the stellar distribution is also a useful proof that no patchy obscuration is present in this field. cf Gilmore et al. 1984 for details.]

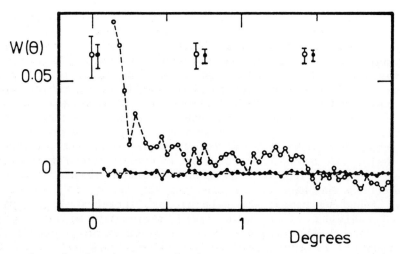

Figure 1 - the 2-point correlation function for galaxies (open points) and stars (solid points) at the south galactic pole. The absence of a galaxy feature in the stellar distribution precludes galaxian contamination of the stellar sample, and patchy interstellar reddening.

After these steps, the astronomer has a magnetic tape containing a coordinate, a shape, a label (star, resolved, etc.), and some intensity parameter for each of (possibly) several measures of every reliable image in some master list. It is now time to attempt photometry.

4) PHOTOMETRIC CALIBRATION or HOW BRIGHT IS THAT STAR?

Photographic photometric surveys divide naturally into those which require a reliable differential measure of luminosity or colour, and those which require 'true' magnitudes, free from zero point or colour scale errors.

a) Uncalibrated Photometry: An example of a project using purely differential photometry is a study of variable objects, where the aim is to find an image whose ranking, relative to a suitable set of nearby images, has changed significantly. The general problem of the detection of variability from uncalibrated data has been discussed by Gilmore (1979) and applied by Hawkins (this volume) to the particular case of faint quasars. A further discussion of this is given elsewhere (Gilmore, 1983a), and will not be repeated here.

b) Magnitude Scale Calibration: The basic philosophy underlying the calibration of magnitude scales using photometric standards which span the position, colours and magnitudes of interest is the application of consistent systematic errors. As the measurements of the standard are in error by exactly the same amount as the other images of interest, their use to determine the magnitude scale and zero point will automatically eliminate all these errors. A practical example of the application of this philosophy is described in Reid & Gilmore (1982), and the general problem is discussed by Gilmore (1983a). The following important points are discussed below: the meaning of a magnitude scale; the limits on systematic errors, the establishment of faint standards; and the accuracies attained in practice.

i) Types of magnitude: Apparent magnitudes for the relatively bright stars used as secondary standards in most photometry are usually obtained by photoelectric aperture photometry. Apparent magnitudes for faint standards are typically measured from sub-beam (Pickering-Racine) prism plates, or from CCD frames. These may produce magnitudes which are defined by either an aperture or a threshold or which may be total. There is no particular reason to expect these magnitude scales to agree either with each other, or with other observers. Considerable care must be taken when establishing faint standards that their magnitudes are defined in the same way as those of brighter stars, and in a way which is appropriate for their intended use. An illustration of the effect of thresholding on total magnitudes, of errors in the adopted local sky background, and of a real change in the local sky background (due to unresolved stars or gradients in the zodiacal light, for example) is shown in Table 1 (P.C. Hewett, pri.comm'n).

Table 1

Total Magnitude	10% Thresholded Magnitude	Sky error of +1%	Sky change of +3%
19.50	19.69	19.72	19.73
20.50	20.87	20.91	20.92
21.50	22.33	22.43	22.41
22.00	23.42	23.66	23.76
22.40	25.35	28.06	-

Calculated by P. Hewett using a sky brightness of 22.5 mag/arcsec^2, and a Moffat stellar profile with R = 2.5 and B = 3.0

---

ii) Establishing Faint Standards: Provided the points above are taken into consideration, reliable faint magnitudes can be established using sub-beam prism and CCD data. Recently Blanco (1982) has questioned the reliability of sub-beam prisms. Examples of the successful use of these devices, provided adequate care is taken, are presented by Reid & Gilmore (1982), Christian & Racine (1983) and Gilmore (1983a, esp. Figure 7). As always however, the only reliable test of a faint sequence is an independent attempt to derive a magnitude scale using several methods. For the study of the stellar distribution in the Galactic spheroid described by Gilmore (1983b), this is achieved by using photoelectric aperture photometry, sub-beam prism photographic photometry, and CCD photometry from two separate cameras in each field of interest. While this calibration then involves an enormous workload, and considerable overlap of observations, it does guarantee that the calibrating standards are themselves reliable - a sine qua non for useful photometry.

iii) Systematic errors: When a reliable consistent set of standards have been derived, they will usually be found inadequate to define any structure in the relation between instrumental and standard magnitudes. Some structure in the form of a change of slope of the relation is to be expected near the apparent magnitude at which the plate and measuring machine becomes saturated (m $\sim$ 19) and at the magnitude when the image halo rises above the threshold, and becomes included in the total image (m $\sim$ 12). An example of this is seen in Figure 2, which also illustrates the excellent agreement between photoelectric aperture and CCD magnitudes and those derived from sub-beam prism photometry. For stars with $V \leq 11^m$, a systematic error of up to $0.^m2$ may be present, while the magnitude scale is undefined for $V > 19.5$. As much published photometry is based on substantially fewer standard stars per magnitude interval than the example shown here, significant non-linearities in that data are possible.

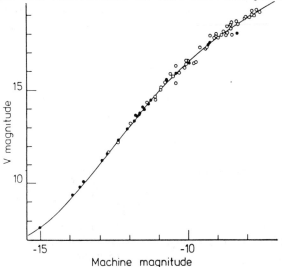

Figure 2 - A Schmidt plate calibration curve using standards from photoelectric and CCD observations (open points and all with V < 12) and sub-beam prism plates (solid points). The curvature is evident.

The second major class of systematic errors includes all those effects where a correlation exists between the residual magnitude from the calibration curve and a second parameter. The best known are those correlations with position ("field effects" - see Reid & Gilmore

(1982) for a discussion), with colour ("colour equations", cf Faber et al. (1976), and Blair and Gilmore (1982)) and with local sky background (cf Table 1). The existence of these effects must be carefully tested if reliable photometry is required.

iv) Random errors: If all the tests and calibrations discussed above have been carried out, a final object list with photometry is available. How good is it? A reliable estimate of the random errors can be derived by independently calibrating each plate in a set, and calculating the rms dispersion in the derived magnitudes for every star in the sample. Table 2 shows such results for a sample of ~ 20,000 stars with photometry from 5 V, 3 B and 3 I plates. This table shows that, with care and a lot of effort, reliable magnitudes and colours for large samples of stars can be derived with random errors of a few percent.

Magnitude	RMS(V)	RMS(B)	RMS(I)
10.5	0.09	0.07	0.07
11.5	0.06	0.09	0.08
12.5	0.05	0.06	0.10
13.5	0.05	0.06	0.10
14.5	0.05	0.06	0.10
15.5	0.06	0.07	0.09
16.5	0.06	0.07	0.08
17.5	0.08	0.07	0.11
18.5	0.15	0.09	0.21
19.5	0.26	0.19	-
20.5	0.42	0.34	-

5) SOME RESULTS or WHY DID I DO THAT?

What does one do with 20,000 accurate magnitudes and colours? This volume contains many examples of the type of astronomy which can be done only with large samples of carefully measured objects. Two others are shown in Figures 4 and 5 below.

Figure 3 shows the U-B/B-V diagram for a field in the southern hemisphere which is part of a survey for objects which do not have main sequence stellar colours. Some 250 square degrees have been surveyed as a pilot project, to prove the potential of this method for finding white dwarfs, HB stars, cataclysmic variables, and bright quasars. The high precision of the photometry and the very large number of objects measured allow a careful discussion of completeness and selection effects, and the detection of objects with less extreme colours than most other similar surveys to date. In particular, cooler white dwarfs and higher redshift quasars can be found. Further details are in Gilmore (1983c).

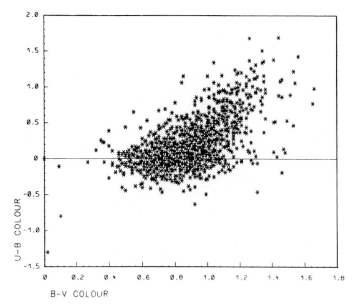

Figure 3 - A 2-colour diagram from a survey for stellar objects with non main-sequence colours.

Figure 4 shows the stellar number-magnitude-colour distribution towards the south Galactic pole. Similar data have been obtained in nine fields, and are being analysed to determine the structure and evolution of the Galactic spheroid, and the solar neighbourhood luminosity function. Further details may be found in Gilmore (1983b, 1983d, 1984) and Gilmore & Reid (1983).

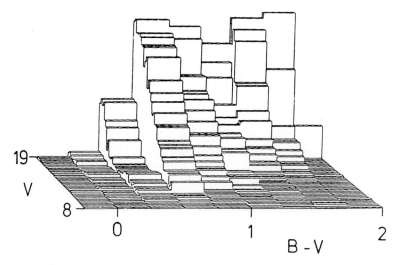

Figure 4 - The stellar number-magnitude-colour distribution of 22,000 stars near b = -90°. The prominent bimodality is evident for V > 17.

REFERENCES

Blair, M., and Gilmore, G.: 1982, PASP $\underline{94}$, 752.
Blanco, V.: 1982, PASP $\underline{94}$, 201.
Christian, C.A., and Racine, R.: 1983, PASP $\underline{95}$, 457.
Faber, S.M., Burstein, D., Tinsley, B.M., and King, I.R.: 1976, A.J. $\underline{81}$, 45.
Gilmore, G.: 1979, MNRAS $\underline{187}$, 389.
Gilmore, G.: 1983a, Proceedings of the Workshop on Astronomical Measuring Machines, eds. R. Stobie and B. McInnes. Occ.Repts. Royal Obs. Edinburgh, Number 10, p.259.
Gilmore, G.: 1983b, Proc. IAU Coll. 76, June 1983, in press.
Gilmore, G.: 1983c, Mem.dellaSoc.Ast. Italiana, in press.
Gilmore, G.: 1983d, Proc. IAU Symp. 106, May 1983, in press.
Gilmore, G.: 1984, MNRAS, in press.
Gilmore, G., and Hewett, P.C.: 1984, Nature, in press.
Gilmore, G., Reid, N., Hewett, P.C., and Morton, D.: 1984, in preparation.
Gilmore, G., and Reid, N.: 1983, MNRAS $\underline{202}$, 1025.
Kron, R.G.: 1980, Proc. ESO Workshop on Two Dimensional Photometry, eds. P. Crane and K. Kjär, p.349.
Newell, B.: 1983, Proc. Workshop on Astronomical Measuring Machines Op.cit. p.15.
Okamura, S., Davenhall, C., and MacGillivray, H.: 1983, Proc. of the Workshop on Astronomical Measuring Machines, Op.cit. p.95.
Reid, N., and Gilmore, G.: 1982, MNRAS $\underline{201}$, 73.

DISCUSSION

V.M. BLANCO: I must remark that the use of the Pickering prisms or so-called Pickering-Racine wedges can result in large errors in the extrapolation of photometric sequences. This led us to stop their use at Cerro Tololo. The main problem is that the beam produced by the small prism has a different f/ratio than the beam produced by the telescopes' full aperture. This results in appreciably different density structures for the primary and secondary images of a given star. So any of you who desire to use such a prism or wedge I recommend a careful reading of Racine's own paper wherein he reviewed this old idea. Also, you may want to read my paper in the PASP where are discussed the sizeable errors that can result from the use of these prisms.

ANALYSIS OF IMAGES WITH THE APM SYSTEM AT CAMBRIDGE

E. Kibblewhite, M. Bridgeland, P. Bunclark,
M. Cawson and M. Irwin
Institute of Astronomy
Madingley Road
Cambridge CB3 OHA, UK

The Automated Photographic Measuring (APM) system at Cambridge was started many years ago to analyse Schmidt plates, and its progress has been reported at a number of these conferences. Only brief details of its design will be given in this paper.

The system consists of a very accurate laser-beam scanning microdensitometer connected to a series of on-line computers. The microdensitometer (Figure 1) uses a 5500 kg x-y table to position the plate. The plate transmission is digitised to 12-bit accuracy at a rate of one sample every 4 microseconds, although the image processing hardware limits this speed in practice. Special-purpose hardware exists to convert the transmission measurements into density or intensity estimates, to smooth the spot-size digitally, to measure the background and to compute image parameters.

Previous papers have stressed the importance of accurate background measurements for this work (Kibblewhite et al. 1975, Kibblewhite et al. 1983). In the APM system the plate is scanned twice, once to measure the background using median estimators within 64 x 64 pixel areas (0.25 sq. mm), and then again to measure the images on the photograph. Special-purpose hardware interpolates between the smoothed background measurements so that a background estimate is derived for each pixel.

The machine now works in one of five modes, each of which is fully operational: (1) Background Mode as described above. (2) Image Analysis Mode in which the positions, shapes and sizes of images are measured. A plate can be measured overnight in this mode and software is available on the STARLINK VAX computer for star/galaxy separation and for the collation of data from a set of plates for automated detection of supernovae, variable stars or colour-excess objects. (3) Objective Prism Mode which produces an uncalibrated intensity versus wavelength data-set from each spectrum using a direct plate of the field to define the position of each image. These spectra are calibrated on the VAX and automated classification algorithms can pick up emission-line and other unusual objects (Hewitt, this conference). (4) Raster

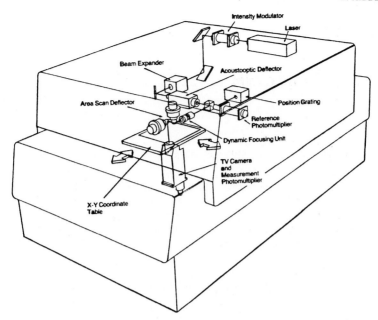

Figure 1. Schematic diagram of the APM microdensitometer.

Mode which digitises given areas of a plate at a speed determined by the magnetic tape-drive (50,000 samples per second). Software for a wide range of tasks is available. (5) Raster-90 Mode in which 90 x 90 raster-frames of hundreds of regions can be scanned automatically. Positions of the scans can be given in right ascension and declination and are usually obtained from selecting interesting images from scans made in either Image Analysis or Objective Prism modes.

The APM system is operated in a similar manner to most large telescopes. Astronomers come to the unit, learn how to use it, and can leave Cambridge knowing they have good data. The microdensitometer is controlled by a comprehensive software package which gives the astronomer great flexibility without the need for detailed knowledge of the workings of the machine. All plates can be aligned using standard reference-star catalogues stored on-line so that all measured image positions are given in right ascension and declination enabling data from a number of plates of the same area of sky to be efficiently collated on the STARLINK VAX. An interactive colour-display allows the user to have a 'quick look' at the data either to experiment with different scanning parameters to suit his objectives or to be certain that he has good data before leaving the facility.

The remainder of the paper will describe four projects currently under investigation which give an indication of the capabilities of the system. Intensity calibration is discussed by Pete Bunclark at this conference and objective prism capabilities are covered by Paul Hewitt.

Detection of low surface-brightness features

The background-scan of the plate contains important data for a variety of astronomical projects and complements the photographic techniques of Malin. In the Background Mode we scan the plate with an 8 micron spot and determine the median (or Maximum Likelihood) density level of the sky background within 64 x 64 pixel areas. The typical photographic noise per pixel is 0.05 density units so we are able in principle to estimate the density in these regions to 0.001 D. However, in practice quantisation errors increase this to about 0.002 D which for IIIaJ emulsion corresponds to a noise of 7 magnitudes below sky. Figure 2 shows a typical background map of a IIIaF UKSTU plate centred on NGC 3379 used to search for a low surface-brightness radio object (Schneider et al. 1983). No emission brighter than 27.2 magnitudes per sq. arc-second in R could be found which sets a lower limit on the mass-to-light ratio of the object at one.

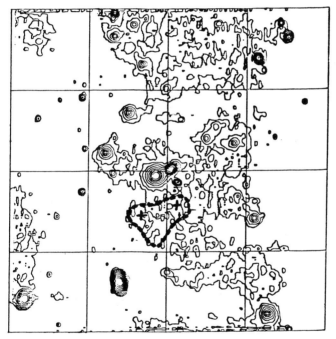

Figure 2. APM background map of an R plate centered on NGC 3379. Lowest contour corresponds to 27.2 mag/sq. arc-sec in R. Dotted line marks the position of the radio galaxy. The map covers 140 arc-min square.

A whole Schmidt plate can be scanned in this background mode in three hours and is extremely powerful for the automated detection of low surface-brightness galaxies. We have already catalogued these objects for Fornax (Cawson 1983) and are starting work on Virgo. At these low light levels the photographic plate is extremely linear, and accurate measurements can be made even in the presence of large-scale photographic non-uniformities.

Rotation of the Galaxy

One of the most interesting current projects is the determination of absolute proper motions of stars by using distant galaxies to define the reference-frame. Provided the centre of the galaxies are not saturated - which is the case for the majority of galaxies on a Schmidt plate - the positional accuracy of their measured centres turns out to be almost independent of their size and of the plate-scale.

Measurements of galaxy positions on Schmidt plates have an rms error of about one third arc-second so that the reference-frame can only be determined to $\approx 1/\sqrt{N}$ arc-seconds where N is the number of galaxies used for the plate-solution. Projects such as the Lick astrographic survey which have only a few dozen galaxies to define the co-ordinate system need a large number of plates to attain the required accuracy, and systematic effects due to the intrinsic differences in the shapes of galaxies, to photographic non-linearities and to colour-terms dominate. In the APM proper motion survey 30,000 galaxies are used to define the reference-rame and the field is measured in two colours. The galaxies

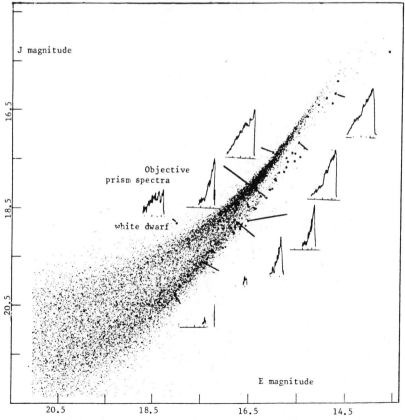

Figure 3a. Colour diagram of stars measured on the APM machine. Squares mark high proper motion (0.05 arc-sec/year) objects. Representative spectra are shown.

are distributed uniformly over the field and by using a 12-parameter plate-solution we can take out second-order differential refraction effects across the field of the Schmidt. 30,000 galaxies define the reference-frame to a few milli-arc-seconds and the importance of having large numbers of galaxies cannot be overstressed. Stars of approximately the same colour as the galaxies are chosen (K stars) by measuring a red and blue plate (Figure 3a), and the mean proper motion as a function of apparent magnitude is determined. This project has so far been carried out in only one field (8h 53, +17 34) at l=209 and we intend to determine the proper motion for a number of different galactic coordinates. Figure 3b shows the mean proper motion as a function of distance determined photometrically for groups of K stars of the same magnitude. This gives a formal value of (A cos 2 l + B) of 0.00041 arc-seconds per year. Corrections have to be made for the non-zero galactic latitude and better photometry bas to be obtained (so far the distance scale has only been determined from internal calibration of the data) but the experimental error of the proper motions is genuine and is less than one milli-arc-second per year per plate-pair. We expect to be able to reduce this to 0.2 milli-arc-second per year using E-copy plates and will measure Oort's constants to a few per cent. The accuracy of the technique is sufficiently high that the proper motions of the Magellanic Clouds and local galaxies out to Carina could be measured.

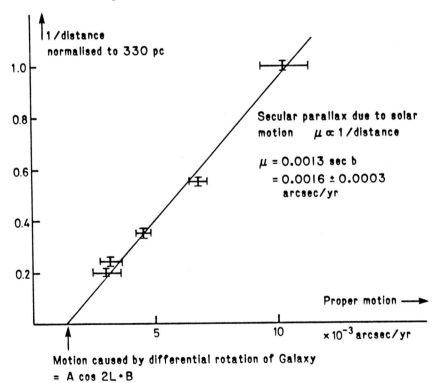

Figure 3b. Secular proper motion of K stars as a function of distance from the Sun. The slope is a measure of the solar motion.

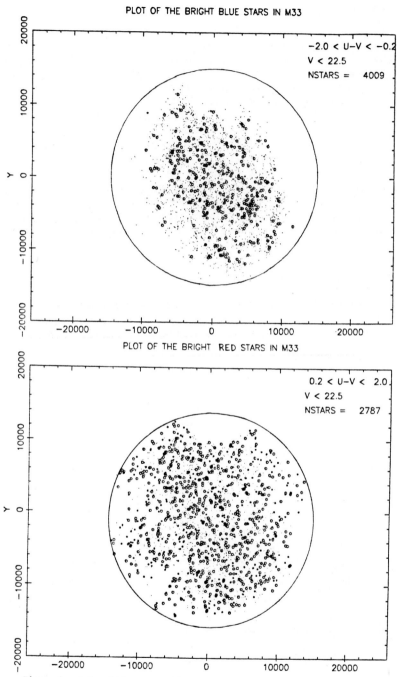

Figure 4. Red and blue stars discovered in M33 by APM machine. Circle defines the unvignetted field.

Bright Stars in Nearby Galaxies

The background correction of the APM system is sufficiently good that accurate photometry of bright stars within nearby galaxies can be routinely determined. Wendy Freedman and Barry Madore have measured and analysed M33, NGC 300, NGC 2403, M81 and M101 in 6 weeks this summer. Pete Bunclark will be talking about his calibration routines in this conference which use the measured areal profile of the stellar images and convert measured APM integrated intensities into a number directly proportional to magnitude over a range of at least ten magnitudes. A substantial fraction of the scatter is due to photo-electric photometry in the presence of the galaxy background.

UBV colours of 10,000 stars down to 22 magnitude were measured from a set of CFH prime-focus plates of M33. Figure 4 compares the distribution of the blue stars, which mainly form within the spiral arms, with the red stars which are either field stars or old red giants and are more uniformly distributed.

Supernova Search

The unique feature of the APM system in Raster-90 Mode - being able to scan a sequence of images from right ascension and declination positions - allows the system to be used for a fully automated supernova search.

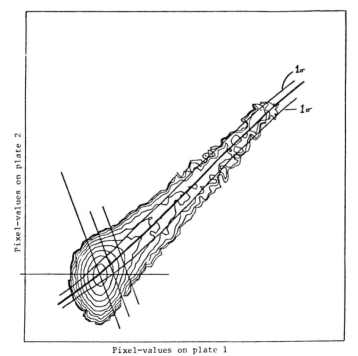

Figure 5. Calibration Mapping (see text).

Image Mode is first used to parameterise all the images within the search field from which the galaxies are chosen on the basis of surface brightness. The right ascensions and declinations of the galaxies are stored on disc and a 90 x 90 raster scan of each one is made. Subsequent plates are also scanned in Raster-90 Mode which operates at about 600 scans per hour. Early and late epoch scans are analysed off-line to search for supernovae. However, before scans can be subtracted they have to be calibrated and this is achieved by a statistical method of Calibration Mapping (Cawson 1983) which uses the data itself to remove possible differences in the photographic response of the two plates. Figure 5 shows a scattergram of pixel-values on one plate against corresponding values on the other plate. The curve of maximum density through the scattergram maps corresponding pixel-values onto the same scale enabling a flat difference-frame to be produced. This is analysed for groups of connected pixels with a significant difference and enables supernovae to be detected right to the centres of galaxies.

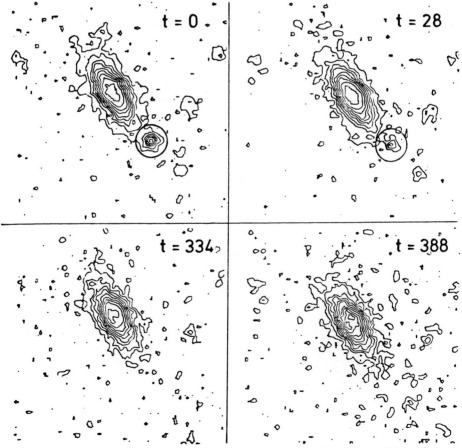

Figure 6. Contour plots of a supernova discovered automatically with the APM machine. Times are in days.

So far four supernovae have been found on a set of four UK Schmidt Telescope plates of area 406 (22h 48, -35). Figure 6 shows a contour plot of one of the supernovae on each of the four plates.

References

Cawson, M.G.M., 1983. Ph.D. thesis, Cambridge, England
Kibblewhite, E.J., Bridgeland, M.T., Hooley, T.A. & Horne, D., 1975. Image Processing Techniques in Astronomy, ed. de Jager 245-246.
Kibblewhite, E.J., Bridgeland, M.T., Bunclark, P. & Irwin, M.J., 1983. Proc. Astronomical Microdensitometry Conference, Washington (in press).
Schneider, S.E., Helon, G., Salpeter, E.E. & Terzain, Y., 1983. Ap.J. Lett., 273, L1 - L5.

REDUCTION TECHNIQUES - ESO FACILITIES

Philippe Crane
European Southern Observatory
Munich, West Germany

This paper describes the facilities available at the ESO headquarters in Munich for scanning photographic plates and for extracting information from these plates. ESO has an integrated system of three microdensitometers, a PDS 1010A, an Optronics S3000 and a Grant 800. Data analysis facilities include an HP 1000 series computer supporting an interactive data analysis system and two Vax 11/780 computers supporting general purpose computing and data analysis. Application areas currently active are: techniques for identifying objects in images and for classifying these objects, methods of analyzing extended images, and a package to reduce echelle spectra.

1. INTRODUCTION

ESO, the European Southern Observatory, has been operating a visitor oriented data analysis center since mid-1978. This was originally at CERN in Geneva and since October 1980 has been at the ESO headquarters in Munich. During 1982, there were 122 visitors coming to Munich specifically to use the ESO microdensitometers and data analysis facilities. Typical stays were of the order of 3 days, so that there was on average always a visitor using the facilities.

In Geneva, ESO supported an Optronics S3000 scanning microdensitometer, a Grant 800 single screw measuring engine, and a Hewlett-Packard HP 1000 system for data analysis. Just before the move to Munich, a PDS was acquired and this went into full operation soon after ESO's arrival in Munich. Also, at the time of the move to Munich, ESO acquired a Vax 11/780. Roughly one year later, a second VAX 11/780 was installed.

In the following sections, details of the physical installations and their capabilities are presented. Finally, some of the on-going research and development programs, both scientific and technical, are discussed.

2. MEASURING MACHINES

ESO has put some effort into building an integrated hardware and software system for the measuring machines (see Figure 1). Thus each measuring machine has its own control computer which handles the direct interaction with the particular measuring machine. The user then talks through a terminal to an HP 1000 system which in turn communicates with the control computer. This arrangement has many advantages. It allows all the capabilities of a real computer with disc files and so on. Thus, star catalogues, user catalogues of coordinates to be scanned, wavelength tables, etc. can be centrally stored, accessed, manipulated, and used to drive the various measuring machines. Figure 1 shows a schematic diagram of the overall layout of the system. Table 1 compares various aspects of the machine and also indicates the anticipated performance of "FIRST", the Optronics upgrade currently taking place.

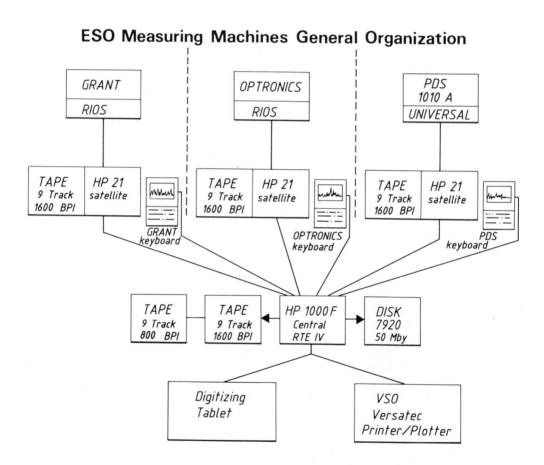

Figure 1: General organization of the measuring machine installation

Table 1: Summary of Measuring Machine Parameters

Machine	Scanning Area (cm)	Positional Accuracy (microns)	Typical Scanning Speed (mm/sec)	Density Range	Pixel Size (microns)
Optronics	35 × 35	0.5	10	2.5	10
PDS	25 × 25	1.5	10	4.0	10
Grant	40 × 10	1.0*	1	2.5	0-45
FIRST	35 × 35	0.5	10 256 pixels	2.5	10

* Relative accuracy but absolute accuracy determined by the lead screw temperature

An HP Digitizing Tablet is connected to the central HP 1000 computer. This can be used to determine the approximate position of objects to be measured or scanned.

The overall use of this installation is quite high. The PDS is used the most and has averaged about 150 hrs/month in 1983. The Optronics is next and is used an average of 100 hrs/month, while the Grant is used somewhat less than 50 hrs/month.

2.1. Optronics S3000

The Optronics S3000 is a granite based two dimensional scanning microdensitometer mounted on air bearings. The mechanical quality of this device is excellent and it has a repeatability of the order of 0.5 microns. It can accept plates up to 36cm by 36cm and scan areas up to 33cm by 33cm. Thus, it is ideally suited to doing astrometry on Schmidt plates. In the near future, an upgraded optical system will allow an increase in speed of about 250 using a reticon, and a micro-processor to preprocess the data will be installed. This project is called FIRST.

2.2. PDS 1010A

The PDS is the standard one used by almost all astronomical institutes. ESO, like most other institutes, has modified certain aspects of the electronics. In addition, we use a Hewlett-Packard computer to control the system. The PDS is the work horse of the measuring machine installation because it is better known, and easier to use than the Optronics.

A new high speed amplifier will be installed soon, and this should increase the scanning speed at densities up to 4 by a factor of two.

2.3. Grant Machine

The Grant 800 is a standard device which has been integrated into the measuring machine complex. It can be used for manual measurements or for one-dimensional scans. A Heidenhahn encoder is being installed on the Grant to improve the positional accuracy.

3. DATA ANALYSIS SYSTEMS

ESO supports two data analysis systems with different capabilities. The IHAP (Image Handling And Processing) system is based on Hewlett-Packard machines and has been in use for more than six years. A newer system, MIDAS, (Munich Image Data Analysis System) based on VAX computers has been available for about 1 year.

3.1. IHAP

The hardware resources of IHAP are shown in Figure 2. This configuration has been quite stable for a number of years. The IHAP represents the first attempt at ESO to build an interactive data analysis system. Major design aspects are: a special file system to enhance performance, command driven operation, a single very large program and a command macro facility. The IHAP system has proved very successful especially in supporting analysis of one-dimensional spectra.

The major application area of IHAP has been in spectral analysis. The Image Dissector Scanner data from La Silla is routinely analysed on IHAP. IUE spectra, and Image Photon Counting System data are also handled. In addition, many image analysis functions are available.

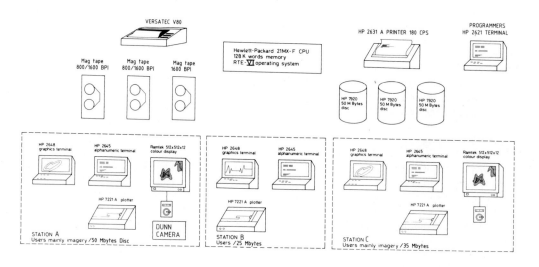

Figure 2: IHAP System Hardware components and user stations

3.2. MIDAS

The VAX based data analysis system was conceived to expand the resources and capabilities of the IHAP system. Three areas were to be enhanced. First, the total number of work stations on the VAX systems is six plus there are several terminals spread through the building. Second, the execution speed should be increased, and third, the ease of programming should induce astronomers to contribute application programs. In order to achieve these results, the hardware shown in Figure 3 was purchased.

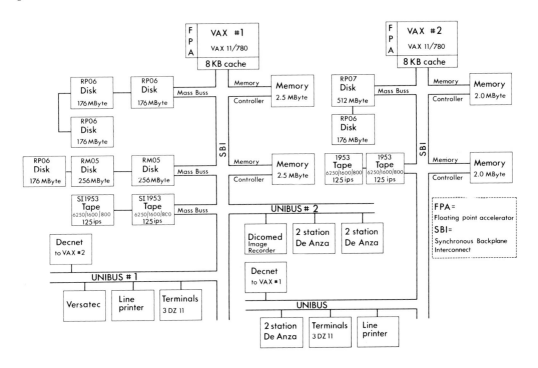

Figure 3: VAX computer installations for MIDAS

The MIDAS software developments were conceived to expand on the successes of IHAP and to allow easy integration of user programs. Thus MIDAS is command driven and makes use of extensive on-line help facilities. MIDAS includes a very extensive command macro facility, can be interfaced to a variety of graphics and display peripherals, and has been designed with very clean interfaces to the MIDAS data structures.

In addition to the usual image data structures, MIDAS supports named global variables for passing data between programs, table structures for storing and manipulating data in tabular form, and catalogue structures to aid in data management.

The concept of using well defined and simple interfaces to the MIDAS data structures has been quite successful. Figure 4 shows the logical structure of a MIDAS application program.

Structure of MIDAS Application Program

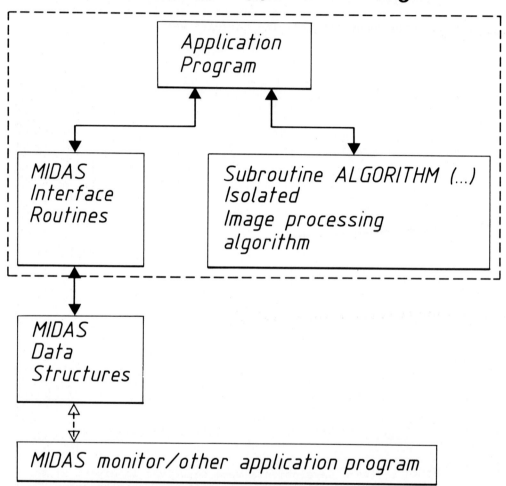

Figure 4: Structure of MIDAS Application Programs

Major MIDAS application areas at present include correction of CCD images, extraction of echelle spectra, full two-dimensional analysis of IPCS spectra, and image classification schemes.

4. SAMPLE APPLICATIONS

In addition to the day to day use, two major programs are being pursued at ESO using the facilities described above. First, the approximately 17,000 galaxies in the ESO/ Uppsala catalogue are being scanned and analysed for their major photometric properties. Figure 5 shows a typical galaxy from this sample and also shows the high quality of the Dicomed image recorder. The other project involves identifying and classifying faint objects. Figure 6a shows a CCD frame and Figure 6b shows the identified objects. The software developments from both of these projects are to be integrated into the general MIDAS system.

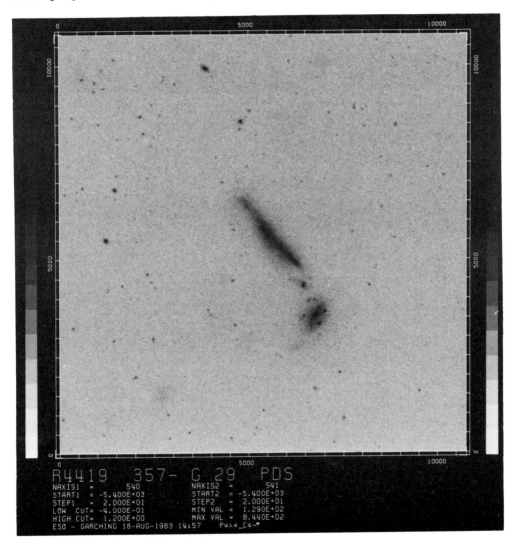

Figure 5: Discomed copy of a galaxy from the ESO/Uppsala survey being scanned and reduced at ESO.

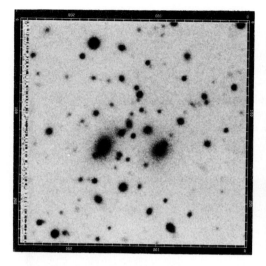 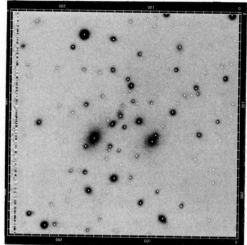

Figure 6a: CCD frame showing faint objects.

Figure 6b: Same as 6a but each automatically identified object has a dot in the middle.

5. CONCLUSIONS

ESO supports a major centre for the analysis of astronomical data. This is a growing centre which hopes to provide first class facilities for the ESO member states, and for the astronomical community at large.

ACKNOWLEDGEMENTS

I thank my colleagues in ESO Image Processing Group: K. Banse, P. Grosbøl, F. Middelburg, C. Ounnas and D. Ponz for their efforts in making these facilities possible. I would also like to thank Andris Lauberts for providing me with information on his work on the ESO/Uppsala catalogue, and Andje Kruszewski for providing me the details of his program for object recognition and classification.

REFERENCES

Banse, K., Crane, P., and Middelburg, F., 1980, 'Applications of Digital Image Processing to Astronomy', Proc. Soc. Photo-Opt. Instr. Eng. 264, pp. 66.
Crane, P., and Banse, K., 1982, Mem. Soc. Astron. Italiana 53, pp. 19.
Melnick, J., 1980, 'Two-Dimensional Photometry', ed. P. Crane and K. Kjar, European Southern Observatory, Munich, pp. 53.
Middelburg, F., and Crane, P., 1979, 'Image Processing in Astronomy', ed. G. Sedmak, M. Capaccioli and R.J. Allen, Osservatorio Astronomico di Trieste, pp. 25.

MACHINE-PROCESSING OF OBJECTIVE-PRISM PLATES AT THE ROYAL OBSERVATORY, EDINBURGH

Roger G Clowes
Royal Observatory, Blackford Hill, Edinburgh EH9 3HJ,
Scotland

ABSTRACT

This paper reviews the methods and early results of work in progress at the Royal Observatory, Edinburgh that involves the machine-processing of objective-prism plates. This work is in two categories: (i) semi-automated galaxy redshifts, and (ii) automated quasar detection. The galaxy redshifts are used, for example, to determine the radial velocities of clusters, to test cluster membership, to reveal superimposed clusters, and to reveal connecting bridges between clusters. Automated quasar detection uses selection criteria that are known, pre-defined and rigidly maintained to select complete samples of quasars. Unlike the earlier visual samples of optically-selected quasars these new samples are well suited to work in cosmology and the collective properties of quasars.

INTRODUCTION

At the Royal Observatory, Edinburgh (ROE) an extensive library of high-quality photographic plates has accumulated from its operation of the UK 1.2m Schmidt Telescope (UKST) near Coonabarabran in Australia. Supposing the plates to be digitised with 256 transmission levels and 16µm square pixels the library represents, at present, $\sim$ 2TByte of image information and sky background. In order to manage and analyse objectively even a small fraction of this quantity of data fast plate-measuring machines and excellent computing facilities are essential. In the UK there are two such measuring machines: COSMOS at ROE and APM at the Institute of Astronomy, Cambridge (IOA). Both ROE and IOA are nodes of the STARLINK network of computers dedicated to astronomical image and data processing.

In general, the machine processing of Schmidt plates divides into two classes, corresponding to direct plates and objective-prism plates. Normal requirements for direct plates are a set of parameters describing each image (e.g. magnitude, x-y coordinates, ellipticity and orientation) whereas normal requirements for objective-prism plates are pixel data for each spectrum. This paper reviews the work being done with objective-prism plates at ROE using COSMOS and STARLINK (the work with direct plates is reviewed at this Colloquium by Gilmore). This work is in two categories: (i) semi-automated galaxy redshifts, and (ii) automated quasar detection (AQD). Note that the techniques for obtaining galaxy redshifts could be adapted to provide a third category, stellar classification. The galaxy redshifts have been used, for example, to reveal the constituent clusters, connecting bridges, and diffuse components of superclusters, and to show that particular instances of rich clusters actually consist of two clusters aligned along the line of sight. AQD selects quasars from objective-prism plates according to selection criteria that are known, pre-defined and rigidly maintained. Unlike the earlier visual searches which it has replaced, AQD yields complete samples, and their selection effects are known.

UKST OBJECTIVE-PRISM PLATES

The UK Schmidt Telescope possesses two full-aperture objective-prisms. The low-dispersion prism (reciprocal dispersion: 2480Åmm^{-1} at 4340Å, resolution for 2 arcsec seeing: 74Å at 4340Å, apex angle: 44 arcmin) has been in use since 1975 and has made valuable contributions to surveys for quasars, intergalactic HII regions, Wolf-Rayet stars, and to stellar classification and the determination of galaxy redshifts. Probably the most well known contribution is that to the quasar surveys, in which many hundreds of new candidates have been discovered, including many of the instances of rare types. In general, the low-dispersion prism is used with Kodak emulsion IIIa-J, giving a wavelength coverage from the atmospheric cut-off at ~ 3200Å to the sharp emulsion cut-off at ~ 5400Å. At longer wavelengths the dispersion of this prism is too low to be very useful and so a high-dispersion prism (reciprocal dispersion: 830Åmm^{-1} at 4340Å and 2750Åmm^{-1} at 6560Å, resolution for 2 arcsec seeing: 25Å at 4340Å and 82Å at 6560Å, apex angle: 130.8 arcmin) was acquired in 1982 (Cannon et al. 1982). With Kodak emulsion IIIa-F the wavelength coverage can extend from ~ 3200Å to the sharp emulsion cut-off at ~ 6850Å, but it is usually restricted by a filter to $\sim 4950-6850$Å in order to obtain fainter limiting magnitudes by suppression of the sky background at short wavelengths and also to reduce the number of overlapped spectra.

For the galaxy redshifts and for AQD the typical plate requirements are deep (B>20), unwidened plates taken through the low-dispersion prism on emulsion IIIa-J. In addition, AQD uses second plates with the direction of dispersion rotated by 90° in order to minimise the number of overlapped spectra. Clowes (1983a) gives details of the variation with wavelength of the response of emulsion IIIa-J.

Each plate covers an area of 6.4°x6.4° at a scale of 67.12 arcsec mm^{-1}. Vignetting is $\Delta m \sim 0.1$ at a radius from the plate centre of 200mm (3.73°) (Dawe and Metcalfe 1982).

Most UK Schmidt plates (those with numbers exceeding 3148) were taken through the achromatic doublet corrector. This corrector introduces an image spread that is approximately constant with wavelength, at 1.50 arcsec, for wavelengths greater than ~ 3300Å. Earlier plates, however, including those used for Savage and Bolton's (1979) visual search for quasars, were taken through the singlet corrector. The singlet corrector introduces an image spread that descends steeply from 4.75 arcsec at 3200Å to a minimum at the corrected wavelength of 4200Å and then rises slowly to 2.75 arcsec at 5400Å and 4.15 arcsec at 6850Å; only at wavelengths close to 4200Å is the image spread of the singlet less than that of the achromat. Savage's (1983) Figure 1 plots manufacturers' data for the dependence of image spread on wavelength for both correctors. Only objective-prism plates taken through the achromat should be used for quantititive work.

Plates taken in excellent seeing (< 2 arcsec) with the low-dispersion prism have limiting continuum magnitudes $m_\lambda \sim 20.5$. With the high-dispersion prism (emulsion IIIa-F + GG495 filter), however, the higher dispersions and the brighter sky background cause a loss in limiting magnitude. No precise values of this loss are available at present but Savage (private communcation) estimates that quasars fainter than $m_\lambda \sim 18$, discovered using the low-dispersion prism, are not visible. One of the main reasons for having a high-dispersion prism is to discover quasars with redshifts greater than ~ 3.5; only $\sim$ 0-2 such quasars are expected per plate.

COSMOS

For the work on galaxy redshifts and AQD the COSMOS measuring machine is used in its mapping mode (MacGillivray 1981; Stobie 1982 - COSMOS user manual). It then becomes simply a fast densitometer which outputs the transmission values and, implicitly, the x-y coordinates of each pixel: it does not threshold or otherwise process the data. Digitisation is presently to eight bits (due to change to 14 bits in December 1983), which gives 256 transmission levels in the range 0-255. COSMOS, a flying spot scanner, has nominal spot-size options of 8,16,32µm (actual FWHMs = 13.9,25.5,31.1µm) and independent pixel-size options of 8,16,32µm. The "standard" selections for measurement of objective-prism plates are the nominal 16µm spot and 16µm pixels. Measurement of the maximum area that can be measured in one session - 287x287 mm^2 - then takes $\sim$ 6 hours and, with eight bits, the 306.8Mbyte of data occupy 12 magnetic tapes.

From $\sim$ March 1984 COSMOS should acquire a dedicated computer with a 1.2Gbyte capacity. Then, the mapping mode data with 14-bit digitisation for entire 287x287mm² areas will be written directly to the COSMOS disk, whereas at present the data are written to tape and then dumped in halves onto two 300Mbyte disks on the ROE STARLINK VAX 11/780. Ideally, the galaxy redshift software and the AQD software (both written on STARLINK) will be compatible with the COSMOS computer so that, because of the dedicated CPU time, processing in real time will be considerably faster.

STARLINK

The processing of objective-prism plates requires moderate processing power, large disk capacity, and excellent graphics facilities. These three requirements are provided at ROE by the STARLINK node, a VAX 11/780 and its peripherals.

STARLINK comprises six full-sized VAX 11/780 nodes and two smaller VAX 11/750 nodes, which are linked together by one or both of two networks. ROE has a full-sized node. The VAX 11/780 has 4Mbyte of memory, a floating-point accelerator (FPA), 1Gbyte disk capacity (of which 300Mbyte are non-STARLINK), graphics terminals, and high-resolution hard-copy devices. In particular, the node possesses two ARGS graphic monitors which can display 512x512-pixel images in colour or monochrome. For inspection, the appearance of objective-prism plates can be reconstructed (in monochrome, of course) on the ARGS monitors from the COSMOS data (see Figure 1).

The processing power of the VAX 11/780+FPA may be assessed using the Central Computer and Telecommunications Agency (CCTA) synthetic benchmarks for single-precision Fortran, which give VAX 11/780+FPA: 0.24, to be compared with, for example, PDP 11/45: 0.04, GEC 4080: 0.05, IBM 360/195: 1.00, and CRAY1 (scalar): 3.52.

THE TECHNIQUE OF SEMI-AUTOMATED GALAXY REDSHIFTS

Cooke et al. (1977) and Cooke (1980) began the work on galaxy redshifts from objective-prism plates using measurements with a Joyce-Loebl microdensitometer of Curtis Schmidt and UKST plates. The emulsion cut-off of emulsion IIIa-J was found to provide an adequate and convenient reference point for wavelengths (see Nandy et al. 1977). Redshifts could then be obtained by measuring the position with respect to the cut-off of the 3990Å continuum break - "the 4000Å feature". This continuum break corresponds to the sharp red edge of a general depression in the spectrum that begins with the strong CaII H & K absorption lines and extends shortwards, because of a multitude of other metal lines, to the atmospheric cut-off. The 4000Å feature is generally detectable in the UKST objective-prism spectra of ellipticals and nuclear bulges of spirals for $16 < B < 19$. Secondary absorption features might also be present. Using this method Cooke et al. (1981) found rms errors of $\sim 1800 \mathrm{Kms}^{-1}$ compared with slit spec-

MACHINE-PROCESSING OF OBJECTIVE-PRISM PLATES

Figure 1 is a photograph of a STARLINK ARGS graphic monitor displaying a small area (512x512 16μm pixels) of an image on disk that was formed from COSMOS measurements of an objective-prism plate. The image is of the cluster 2143-5732 in Indus (Corwin, 1981).

troscopy for the objective-prism redshifts of bright (B ~ 17) ellipticals in the cluster Abell 2670. Similar velocities were measured for the cluster Abell 140 but the wider dispersion suggested a superimposition of clusters. Cooke et al. (1981) concluded that the technique is useful for determining the radial velocities of clusters, for testing the cluster membership of ellipticals, and for resolving superimposed clusters.

During the early work the attempts to extract spectra from COSMOS mapping-mode data were essentially abandoned because the available processing power and disk capacity were too small. However, when ROE became a STARLINK node this approach was revived. Cooke et al. (1984) and Cooke et al. (1983) have now written software for the recognition and extraction of spectra from digital images on disk of the objective-prism plates. The software is not, of course, restricted to galaxy spectra, and it has already acquired another application in the AQD process.

The recognition and extraction of spectra begins with the formation of a compressed image on a 300Mbyte disk (one byte per pixel) of the measured area. Large areas required splitting over two disks. Figure 1 is a photograph of an ARGS graphics monitor displaying a 512x512-pixel area of such an image - the cluster 2143-5732 in Indus (Corwin, 1981). From this image an array representing the smoothed sky background is generated by smoothing with a median filter the weighted means from transmission histograms of 256x256-pixel areas (Cooke et al. 1984). Spectra are then recognised using the threshold criterion that N connected pixels should have transmissions that are at least T% brighter than the local sky background. The effects of particular values of N and T may be tested interactively on a small part of the image. Recognised spectra are extracted, and unrecognised spectra and empty areas of sky background are discarded. Each spectrum is extracted as a block of 8x128 pixels, located by the emulsion cut-off in the dispersion direction and by the centroid in the perpendicular direction. This block size is generally large enough to contain substantial areas of the local sky background. The recognition and extraction of spectra from the maximum COSMOS area of $287 \times 287 mm^2$ requires ~ 24 hours of CPU time using present software (but it is known that this time can be reduced).

There is an alterantive approach (Clowes et al. 1980) to the recognition and extraction of spectra. It uses coordinate transformations from direct plate to prism plate to locate on the prism plate a separate raster scan for each required object on the direct plate. The "little image mode" of COSMOS is not well suited to a large number of objects and so the same effect, if required, is then better obtained from software and mapping mode data. Work is in progress on this approach, which gives more accurate definition of the reference point for wavelength (the emulsion cut-off).

Beard et al. (1984) (see also Beard 1983, Cooke et al. 1983) have obtained galaxy redshifts in a segment of the Indus supercluster that was proposed by Corwin (1981). In the automated part of their technique Beard et al. (1984) used their recognition and extraction software to extract 50324 spectra of all types from an area of $255 \times 281 mm^2$, and converted them to one-dimensional approximate-intensity spectra. Then, in the interactive part of their technique, they used plots of these one-dimensional intensity spectra to separate galaxies from stars, and to obtain redshifts for 2294 of the galaxies by locating the emulsion cut-off and the 4000Å feature of each. Repeat measurements of the same plots gave rms errors of ~ 3000 kms^{-1}. The early work concentrated on bright (B $\sim$ 17) ellipticals whereas this recent work includes the types E-Sb (16 < B < 19), the extension to spirals and fainter magnitudes being responsible for the increase in random errors. An excess of "redshifts" in the range 0.10-0.12 was found to be caused by misclassification of stars with misleading features as galaxies. Figure 2 shows the approximate-intensity plots of a representative set of nine galaxies. In each case the 4000Å feature appears as a sharp discontinuity. The deduced redshifts are, reading from left to right and downwards, 0.051, 0.109, 0.056, 0.176, 0.058, 0.055, 0.085, 0.049, 0.041.

Direct plates will, in future work, be used to assist the separation of galaxies from stars, and to reduce magnitude-dependent errors in locating the emulsion cut-off of each galaxy by defining coordinate transformations from the direct plates to the objective-prism plates. Furthermore, new software being developed by Cooke, Kelly, Beard and Emerson (initial tests of it are reported in their poster paper presented at this Colloquium) should replace visual location of the 4000Å feature by automated pattern-matching of the galaxy spectra with standard references. They expect that this new software could be adapted to provide classification of stellar spectra.

Corwin found, using visual counts of galaxies on UKST direct plates, that the Indus supercluster comprised an approximately elliptical ring (major and minor axes of ten and eight degrees respectively) of clusters at a mean redshift of 0.076. Beard et al. (1984) found that the redshifts of 325 of their 2294 galaxies were within ± 0.01 of Corwin's mean redshift. They have shown that there is a 7h^{-1} Mpc bridge between two constituent clusters, 2151-5805 and 2143-5732, of the supercluster, and have shown that some galaxies which, with two-dimensional information only, apparently belong to 2151-5805 are distinctly members of a more distant, unrelated cluster.

Parker et al. (1983) used the technique to obtain approximate radial velocities for the rich clusters that were found in the deep (B < 22) galaxy survey of MacGillivray and Dodd (1983). Their results suggest a supercluster, at a mean redshift of ~ 0.12, that comprises not only clusters but also a large dispersed component of galaxies. Parker, MacGillivray, Dodd, Beard, Cooke and Kelly are presenting their latest results in a poster paper at this Colloquium.

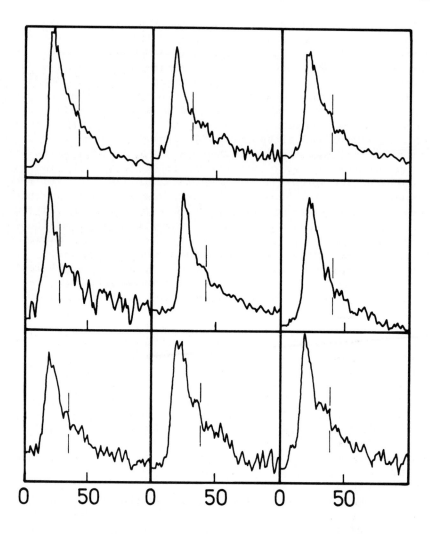

Figure 2 shows the approximate-intensity plots of a representative set of nine galaxies. The y-axis is relative approximate-intensity and the x-axis is pixel number (increasing pixel number corresponds to decreasing wavelength). In each case the 4000Å feature appears as a sharp discontinuity. The deduced redshifts are, reading from left to right and downwards, 0.051, 0.109, 0.056, 0.176, 0.058, 0.055, 0.085, 0.049, 0.041.

MULTI-OBJECT SPECTROSCOPY USING FIBRE-OPTICS SYSTEMS

Redshifts obtained with large telescopes would, of course, be preferred, and the recent development of multi-object spectroscopy using fibre-optics systems on telescopes such as the Anglo-Australian Telescope (AAT) provides the means, in principle (the field of the AAT system is soon to increase from 12x12 arcmin^2 to 40x40 arcmin^2). In practice, for galaxy redshifts, the large numbers of galaxies covering relatively large areas of sky together with their unavoidably non-stellar appearance cause multi-object spectroscopy to be impracticable at present, except for establishing wavelength references in small areas. For quasars, however, although the areas of sky remain relatively large the numbers are very much smaller and multi-object spectroscopy becomes more practicable: indeed, objective-prism redshifts, with random errors slightly larger than desired and $\sim$ 20% incorrect redshifts because of misidentified single lines, would be an unnecessary contaminant of the samples discovered by AQD.

THE TECHNIQUE OF AUTOMATED QUASAR DETECTION AQD

AQD was developed in order to combine the inherent sensitivity of objective-prism (and grism and grens) plates for detecting quasars with an objective detection procedure. Previously, this sensitivity was used inefficiently by visual searches because of their subjective selection criteria that varied systematically, and the resulting samples were properly useful only as sources of individually interesting objects and not for work on cosmology and the collective properties of quasars. With AQD, however, the selection criteria and consequently the selection effects are known, pre-defined and rigidly maintained; large, complete samples can be assembled from many plates. AQD is an extension of earlier work on objective-prism spectrophotometry (Clowes et al, 1980) and on the selection effects that operate in spectral searches for quasars (Clowes, 1981).

The AQD samples are well suited to studies of anisotropy on scales greater than $350h^{-1}$ Mpc ($q_o = 0$), self-clustering in two and three dimensions, cross-clustering with faint galaxies, the density contrast with sensitivity to values greater than 10% on scales of $150h^{-1}$ Mpc at $z \sim 2.3$ (but not its evolution), the luminosity function and its evolution.

AQD (Clowes et al. 1983a,b; see also Clowes, 1983b) uses the galaxy-redshift software (Cooke et al. 1984) for the recognition and extraction of spectra of all types from images on disk. At this recognition and extraction stage there is no testing of the quality of the spectra so the output blocks may contain, for example, overlapped, saturated and spurious spectra. Quality testing is performed by the AQD software. In future work quality testing will be assisted by the use of direct plates when coordinate transformations from direct plate to prism plate are used at the recognition and extraction stage. All spectra that survive quality testing, regardless of type, are converted to intensity (Clowes et al. 1980).

AQD can select quasars by emission lines (intended to be the primary method), absorption lines, spectral discontinuities, ultraviolet excess, and red excess. For selection by either emission or absorption lines a continuum spectrum is derived from the intensity spectrum and a noise spectrum is generated to represent the expected noise fluctuations of the intensity spectrum about the continuum spectrum. The criterion for recognising a spectral line is that its peak exceeds the specified signal-to-noise ratio with respect to the continuum. For selection by spectral discontinuities the ratio $(B-G)/G$ is formed for each element of the intensity spectrum, where B is the average of the next M elements to shorter wavelengths and G is the average of the next M elements to longer wavelengths. If this ratio is negative and it exceeds the specified limit then a discontinuity is recognised. For selection by ultraviolet or red excess two broad-band filters are defined for each case and corresponding magnitudes are derived from the intensity spectrum. Spectra are selected when the colours exceed the specified limits. Full details of the AQD process are given in Clowes et al. (1983b).

Figure 3 is a photograph of the ARGS monitor displaying the 8x128-pixel blocks of a representative set of AQD quasars selected by emission lines and spectral discontinuities from a field centred close to the South Galactic Pole. Using the same objective-prism plate as the visual search (Clowes and Savage 1983) of that field, AQD discovered (subject to confirmation) $\sim$ 50-100% more quasars brighter than B = 19.5 (Clowes et al. 1983b).

Figure 4 shows the intensity and superimposed continuum spectra of the top nine blocks of Figure 3. The y-axis is relative intensity and the x-axis is pixel number (increasing pixel number corresponds to decreasing wavelength). All except the fifth (counting from left to right and downwards) were selected by emission lines. The fifth was selected by a continuum discontinuity and it illustrates one algorithm by which very high redshift quasars might be selected when Ly-α is beyond the emulsion cut-off (however, high stellar contamination is expected).

The CPU time required for the maximum area is $\sim$ 24 hours for the recognition of extraction of spectra and $\sim$ 8 hours for selection of the quasars.

AQD, is intended for statistical work on defined samples of quasars, and, therefore, it requires calibration of the plate noise, the emulsion response, and the absolute flux scale. Such calibration is, as usual, difficult and time-consuming. For non-statistical work (eg the searching of many plates for close pairs) AQD can, of course, be used more crudely with only an invented calibration; the samples will then be similar to the undefined visual samples but will be spatially consistent on a single plate.

MACHINE-PROCESSING OF OBJECTIVE-PRISM PLATES 117

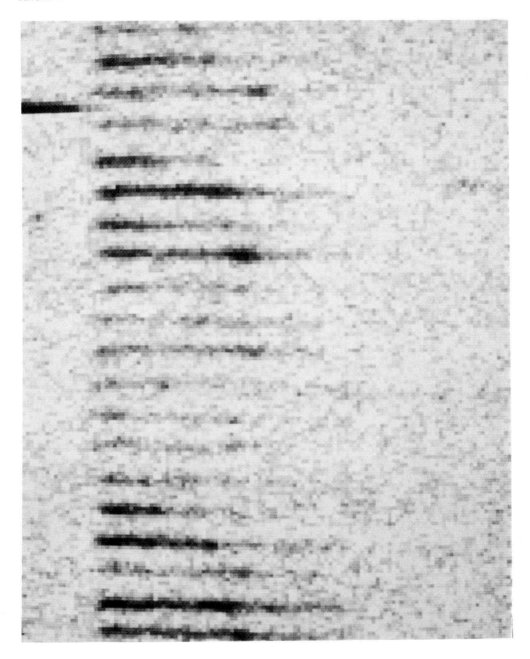

Figure 3 is a photograph of the ARGS monitor displaying the 8x128-pixel blocks of a representative set of AQD quasars. Wavelength increases to the left.

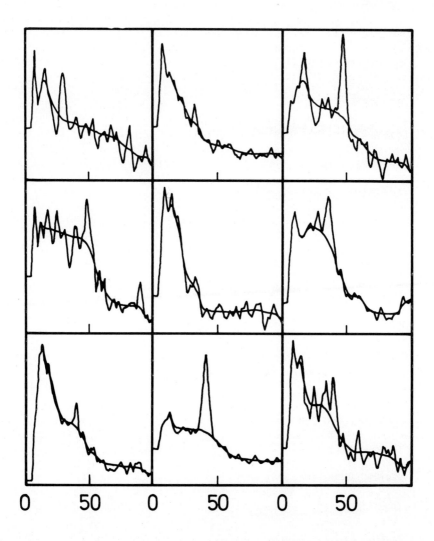

Figure 4 shows the intensity and superimposed continuum spectra of the top nine blocks of Figure 3. The y-axis is relative intensity and the x-axis is pixel number (increasing pixel number corresponds to decreasing wavelength). All except the fifth (counting from left to right and downwards) were selected by emission lines; the fifth was selected by a continuum discontinuity.

Unfortunately, AQD is presently restricted by the quality of the COSMOS data and of the calibration data. Noise in COSMOS data is correlated from pixel to pixel, and the degree of correlation is a function of transmission because the ratio of machine noise to plate noise is a function of transmission. However, the electronics of COSMOS have recently been rebuilt, and preliminary noise tests suggest that, in future measurements, the machine noise will be reduced to insignificance and that the degree of correlation will consequently be constant with transmission. The difficulty with the calibration data is that, for the emulsion responses at wavelengths shorter than $\sim 3950\text{\AA}$ (see Clowes 1983a), they do not exist because of attenuation caused by glass in the UKST calibraton spectrograph. Ideally, the emulsion response should be obtained from on-plate exposures rather than separate exposures in a calibration spectrograph.

REFERENCES

Beard,S.M. 1983. Occ.Rep.R.Obs.Edin., "Proceedings of the workshop on astronomical measuring machines held at Edinburgh, 28-30 September, 1982" p219, eds. R.S.Stobie, & B.McInnes.
Beard,S.M., Cooke,J.A., Emerson,D. & Kelly,B.D. 1984. Mon.Not.R.astr.Soc., submitted.
Cannon,R.D., Dawe,J.A., Morgan,D.H., Savage,A. & Smith,M.G. 1982. Proc.ASA, 4(4),468.
Clowes,R.G., Emerson,D., Smith,M.G., Wallace,P.T., Cannon,R.D., Savage,A. & Boksenberg,A. 1980. Mon.Not.R.astr.Soc., 193,415.
Clowes,R.G. 1981. Mon.Not.R.astr.Soc., 197,731.
Clowes,R.G. 1983a. AAS Photo-Bull., 32,14.
Clowes,R.G. 1983b. 24th Liege Astrophysical Colloquium: "Quasars and gravitational lenses" in press.
Clowes,R.G., Cooke,J.A. & Beard,S.M. 1983a, Occ.Rep.R.Obs.Edin., "Proceedings of the workshop on astronomical measuring machines held at Edinburgh 28-30 September 1982" p253, eds. R.S.Stobie, & B.McInnes.
Clowes,R.G., Cooke,J.A. & Beard,S.M. 1983b. Mon.Not.R.astr.Soc. in press.
Clowes,R.G. & Savage,A. 1983. Mon.Not.R.astr.Soc. 204,365.
Cooke,J.A., Emerson,D., Nandy,K., Reddish,V.C. & Smith,M.G. 1977. Mon.Not.R.astr.Soc., 178,687.
Cooke,J.A. 1980. Ph.D. Thesis, University of Edinburgh.
Cooke,J.A., Emerson,D., Kelly,B.D., MacGillivray,H.T. & Dodd,R.J. 1981. Mon.Not.R.astr.Soc., 196,397.
Cooke,J.A., Emerson,D., Beard,S.M. & Kelly,B.D. 1983. Occ.Rep.R.Obs.Edin., "Proceedings of the workshop on astronomical measuring machines held at Edinburgh 28-30 September 1982" p209, eds. R.S.Stobie & B.McInnes.
Cooke,J.A., Beard,S.M., Kelly,B.D., Emerson,D. & MacGillivray,H.T. 1984. Mon.Not.R.astr.Soc. submitted.
Corwin,H.G.Jr. 1981. Ph.D. Thesis, University of Edinburgh.

Dawe,J.A. & Metcalfe,N. 1982. ASA 4(4),466.
MacGillivray,H.T. 1981. In "Astronomical photography 1981", p277, eds. J.-L.Heudier, & M.E.Sim, CNRS/INAG.
MacGillivray,H.T. & Dodd,R.J. 1983. Occ.Rep.R.Obs.Edin., "Proceedings of the workshop on astronomical measuring machines held at Edinburgh 28-30 September 1982" p195, eds. R.S.Stobie, & B.McInnes.
Nandy,K., Reddish,V.C., Tritton,K.P., Cooke,J.A. & Emerson,D. 1977. Mon.Not.R.astr.Soc. 178,63P.
Parker,Q.A., MacGillivray,H.T., Dodd,R.J., Cooke,J.A., Beard,S.M., Emerson,D. & Kelly,B.D. 1983. Occ.Rep.R.Obs.Edin. "Proceedings of the workshop on astronomical measuring machines held at Edinburgh 28-30 September 1982", p233, eds. R.S.Stobie & B.McInnes.
Savage,A. & Bolton,J.G. 1979. Mon.Not.R.astr.Soc. 188,599.
Savage,A. 1983. Astron.Astrophys. 123,353.
Stobie,R.S. 1982. COSMOS user manual - available from ROE.

THE AUTOMATED DETECTION OF VARIABLE OBJECTS ON SCHMIDT PLATES

M.R.S. Hawkins
Royal Observatory, Blackford Hill, Edinburgh EH9 3HJ

This paper will review progress on the automated detection of variable objects on Schmidt plates, describe the various populations of variables which have been discovered, and summarise some of the most interesting results to date. So far, the searches have been based on sequences of a dozen or more sky limited UK 1.2m Schmidt plates taken on timescales from one day to several years. The plates were measured on the COSMOS measuring machine at the Royal Observatory, Edinburgh which typically detects between 100,000 and 200,000 images in the central 16 deg^2 of each plate. The measures for each field are combined, and calibrated using deep electronographic sequences, to give data sets of some 100,000 objects, complete in the magnitude range B = 13-21. The procedure is described in some detail by Hawkins (1983a) and results in a sequence of magnitudes accurate to about 0.1m for each object in the measured area. Field effects across the plates are allowed for, and an extensive set of diagnostics allows the distribution of errors as a function of magnitude, and on each plate, to be monitored.

The availability of a sequence of a dozen or more magnitudes for each object in the field enabled searches for variables to be undertaken according to various criteria. In the first instance a simple r.m.s. criterion was used, with checks designed to exclude spurious variation from plate flaws, proximity to bright stars and overlapping images. A number of variables were found in this way (Hawkins, 1981) and preliminary classifications made from an objective prism plate. In this early search it was clear that two types of variation were dominant - over a timescale of a day or less and over a year or more. This prompted the development of algorithms specifically designed to isolate the two populations.

By far the most numerous type of variable comprised those with timescales of a year or more. These objects were largely confined to the magnitude range B = 18-21 and spectra taken with the AAT showed them to consist of a population of quasars and active galaxies. A complete sample of these objects was obtained according to well-defined criteria, the most important of which was that the objects should vary by 0.3 mag over one year, but remain essentially constant within each year. The selection procedure is described in detail by Hawkins (1983a), and resulted in a complete sample of 77 objects, some 3% of all quasars. Detecting quasars by variability involves selection effects which are different from, and on the whole less serious than, those associated with other methods of compiling optical samples. In particular, the whole procedure may be carried out according to objective criteria, and systematic and random errors may be quantified.

The statistical analysis of several samples of variable quasars has now been carried out (Hawkins 1983a,d) in two high galactic latitude fields about 12° apart. The number/magnitude relation shows an increase by a factor of 5 per magnitude down to B = 21. There is as yet no definite indication of a change of slope in the range B = 18-21, but larger samples will soon be available which should clarify the situation. Since colour is not used either explicitly as in the case of UVX objects, or implicitly as for objective prism searches, in the definition of the sample, the colour distributions of the sample members can be investigated. This has revealed a reddening towards fainter magnitudes, clearly seen in the B/B-R relation but even more evident in the B/U-B relation (Hawkins, 1983d). It is however worth pointing out that even at the sample limit of B = 21, nearly all objects still have negative U-B colour. The reddening can in principle be accounted for either as a pure redshift at high z, or as a luminosity effect combined with a redshift cut-off. In this case the fainter end of the sample becomes dominated by low luminosity objects which are red due to the contaminating effect of an underlying red galaxy.

There are now several reasons for believing that the reddening is a luminosity effect. There is an increasing proportion of resolved objects in the faint red part of the colour-magnitude diagram implying a low luminosity regime. Furthermore when the colour of the faintest sample members is monitored as they vary, there is a tendency for the objects to become redder as they get fainter. This is not observed for bright objects in which the nucleus presumably dominates the galaxy light. There is also some evidence from redshifts obtained at random for faint red members of the sample that very high z objects do not dominate. On the contrary the redshifts measured tended to be low to moderate, but any definitive statement along these lines must await complete redshift coverage.

The availability of samples from two fields separated by about 12° has enabled the comparison of quasar members over this angular separation. The two samples were chosen according to essentially identical criteria, based on the requirement that they should vary by 0.3 mag over a 1 year and a 3 year baseline. The two samples contained 80 and 71 objects from Schmidt fields 287 and 401 respectively. These two numbers are clearly compatible with a uniform quasar distribution and form a first step to establishing the distribution on all angular scales.

In the magnitude range B = 13-18, the dominant type of variable was found to have a timescale of variation of 1 day or less. Follow up spectra on the AAT revealed that these objects are mainly RR Lyrae stars. A complete sample was obtained in field 287 at Galactic latitude -47° on the basis of well defined selection criteria, the main requirement being that the r.m.s. variation on 14 plates was greater than 0.20 mag. The sample has been used to examine the structure of the Galactic halo to a galactocentric distance of 60 kpc, about twice as far as any comparable method. A power law relation for the space density, with an index of -3, was found to hold out to 60 kpc. This is of great interest, as the globular cluster distribution which also has an index of -3 appears to cut off at about 30 kpc. The radial velocities of most of the sample were also obtained and used to provide an estimate of the mass of the halo, giving a value in excess of 1.4×10^{12} M_o. This is based partly on a lower limit set by an RR Lyrae star of unusually large velocity (Hawkins, 1983c) and partly on the velocity dispersion of the sample as a whole. It was also possible to obtain an estimate of the dynamical mass density as a function of radius, giving a power law relation with index -0.7.

A third category of variable objects found comprised cataclysmic variables. In the first instance they were included among the RR Lyrae stars, but their large amplitudes of 2-3 magnitudes and erratic behaviour clearly distinguished them (Hawkins, 1983b). A complete and well-defined sample of 3 objects was obtained, all of which were observed on the AAT, and each in its way turned out to be rather unusual. All were characterised by rather small amplitudes compared with prototype examples, and being very faint (B = 18-21) are apparently among the most distant of such objects known.

The present survey has answered the question of what populations of variable objects are present on deep Schmidt plates. There are undoubtedly rare objects of other types which will be discovered as new fields are examined, and the magnitude limit extended. This should also help to improve statistics for the currently identified populations. The technique has already found the faintest known Galactic RR Lyrae stars, and promises to be a powerful method of finding very high redshift (z > 3.5) quasars.

REFERENCES

Hawkins, M.R.S.: 1981, Nature, 293, 116.
Hawkins, M.R.S.: 1983a, Mon.Not.R.astr.Soc., 202, 571.
Hawkins, M.R.S.: 1983b, Nature, 301, 688.
Hawkins, M.R.S.: 1983c, Nature, 303, 406.
Hawkins, M.R.S.: 1983d, Proceedings of the 24th Liege Astrophysical Colloquium, p.31.

A FAINT GALAXY SURVEY FROM COSMOS MEASURES ON DEEP UKST PLATES

H.T. MacGillivray[1] and R.J. Dodd[2]

1. Royal Observatory, Blackford Hill, Edinburgh, Scotland, UK.
2. Carter Observatory, P.O. Box 2909, Wellington 1, New Zealand.

ABSTRACT

Deep photographs taken with the UK 1.2m Schmidt Telescope are being scanned with the COSMOS automatic plate-measuring machine in order to carry out a survey of galaxies over large areas of sky down to faint limits (B $\sim$ 21.5). Several fields have been examined, and the data are being used to investigate the properties and large-scale projected distribution of galaxies.

1. INTRODUCTION

Photographs taken with the UK 1.2m Schmidt Telescope (UKST) in Australia record the images of large numbers of faint galaxies. On any single plate are contained some 50000 - 60000 galaxy images down to B = 21.5 (a reasonable limit for most star/galaxy separation techniques), or $\sim$1300 galaxies/square degree. With the COSMOS plate-scanning machine at the Royal Observatory Edinburgh (ROE), we are attempting to extract this large quantity of information and make it available for statistical investigation. Large areas of the sky are being surveyed, producing a faint galaxy catalogue which will enable the properties and large-scale two-dimensional (2-D) distribution of galaxies to be examined from a completely objective sample. In this paper we describe the measurements and present an example for one area near the South Galactic Pole (SGP).

2. PLATE MEASUREMENT

The maximum square area of plate that can be scanned with COSMOS at present is an area of 287 x 287 mm² (equivalent to 5.35° x 5.35° on the sky for UKST plates). Since southern sky field centres are separated by 5° and the field of view of the UKST is $\sim$6.5°, then this means that there is a comfortable overlap of $\sim$1/3° on plates of adjacent fields. This is sufficient to ensure there are no gaps or missing "strips" in the sky coverage and that the magnitude system can be checked from plate to plate. The latter enables the calibration to be consistently maintained over the entire area under study.

At 16μm resolution, COSMOS carries out the scans on each plate in 6.5 hours. For image detection, a threshold at 7% of the night sky intensity level is usually applied (corresponding typically to the B = 25.6 mags/square arcsecond isophote). The information obtained for each image consists of both photometric (magnitudes) and geometric (positions, sizes, orientations and shapes) parameters. Star/galaxy separation is carried out on the data by means of surface brightness criteria (see MacGillivray and Dodd 1982b) for objects fainter than B $\sim$ 16.5, and by means of geometrical criteria for objects brighter than this value.

In other papers (MacGillivray and Dodd 1982a, b) we examined the usefulness of COSMOS for the photometry of faint galaxies. Depending upon the precise isophote used, we find that magnitudes for galaxies can be reliably determined at least in the magnitude range 13<=B<=23, the relationship between COSMOS magnitude and those from photoelectric or other photographic observations having a slope of 45°. We also find good agreement (with slope of 45°) between the magnitudes for the same objects in regions of overlap on different plates. Because of the high thresholds used in the present case, a distortion is introduced for the magnitudes of galaxies fainter than B $\sim$ 21 due to the fact that much of the light falls below the threshold level.

3. RESULTS AND DISCUSSION

Several areas of sky, including the SGP, are being scanned and complete coverages of large regions being constructed. For the purpose of illustration, figure 1 shows the result of combining the data for 2 survey fields near the SGP. Galaxies were counted in cells of dimensions 7' x 7' and isopleths drawn using a computer routine. Note that in the figure several high density regions can be seen, corresponding to the presence of rich clusters. Indeed, 4 clusters recorded in the Abell (1958) catalogue are particularly noticeable (viz. A118, A140, A141 and A155). Thus from this galaxy survey, criteria could be defined which would enable also an objective rich cluster catalogue to be obtained.

Data of this nature are suitable for objective studies of the 2-D distribution of galaxies using suitable algorithms (e.g. n-point correlation functions) and also for searches for non-random effects in the properties of galaxies (e.g. large-scale coherent alignment trends). Similar coverages in more than 1 passband are also planned and would help with understanding the present of galactic obscuration and its effect on the galaxy counts.

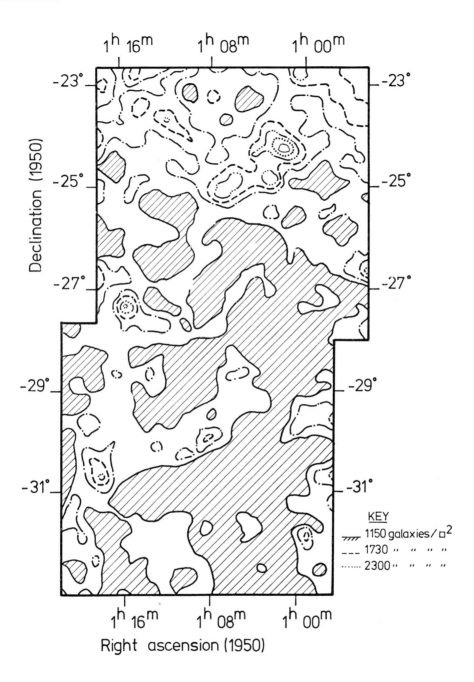

Figure 1 The distribution of galaxies detected down to B = 21.5 from COSMOS scans on the plates for the southern sky survey fields 412 and 475.

This large-scale survey is being coordinated with other surveys which are under way on objective prism plates, e.g. the automated quasar search (Clowes, 1984, these proceedings) and galaxy automatic redshift surveys (Cooke et al 1984 and Parker et al 1984, both papers included in these proceedings).

REFERENCES

Abell, G.O., 1958. Astrophys. J. Suppl., 3, 211.
Clowes, R.G., 1984, in "Astronomy with Schmidt-type telescopes", IAU Colloquium No. 78, ed. M. Capaccioli, Asiago, Italy, this volume, p. 107.
Cooke, J.A., Kelly, B.D., Beard, S.M. and Emerson, D., 1984. in "Astronomy with Schmidt-type Telescopes", IAU Colloquium No. 78, ed. M. Capaccioli, Asiago, Italy, this volume, p. 401.
MacGillivray, H.T. and Dodd, R.J., 1982a. Observatory, 102, 141.
MacGillivray, H.T. and Dodd, R.J., 1982b. in Proceedings of the Workshop on Astronomical Measuring Machines 1982, eds. R.S. Stobie and B. McInnes, Edinburgh, Scotland, p. 195.
Parker, Q.A., MacGillivray, H.T., Dodd, R.J., Cooke, J.A., Beard, S.M., Kelly, B.D. and Emerson, D., 1984. in "Astronomy with Schmidt-type Telescopes", IAU Colloquium No. 78, ed. M. Capaccioli, Asiago, Italy, this volume, p. 405.

VISUAL AND AUTOMATIC CLASSIFICATION OF GALAXY IMAGES

Allan Wirth
Harvard-Smithsonian Center for Astrophysics

ABSTRACT

The requirements for a classification system for galaxy images are discussed and the advantages and disadvantages of both visual and automatic classification techniques are assessed. The results of some preliminary experiments in the automatic classification of images are also presented.

1. INTRODUCTION

Of all quantities associated with an astronomical object, visual appearance is the most easily measured. The chief task of the morphologist is the creation of the scale upon which the quantity "appearance" is to be measured. Until very recently the primary technique used to construct such a scale for galaxies has been the visual inspection and ordering of galaxy images in what seems to be a reasonable way. In doing so we are utilizing the great power of the human eye-brain system to detect (or imagine) regularities of structure even in very low signal-to-noise situations. The hypothesis that such regularities represent the action of physical processes is basic to all morphological research. With the rapidly increasing capability of non-human image processing technology, it is becoming possible to attempt to construct a more objective classification system. In this paper, we will examine the advantages and limitations of the present classification techniques and briefly describe some preliminary work on the automatic classification of galaxies.

2. VISUAL CLASSIFICATION OF GALAXIES

The great advantage of visual classification over all other methods of studying the structure of galaxies is the rapidity with which it can be done and the huge number of galaxies within reach of present telescopes. On sky limited fine grain plates of scales between 100 and

60 "/mm taken in good seeing, it is possible to derive detailed classifications (i.e. full Hubble or Yerkes type) for galaxies larger than ~15" diameter, which is roughly equivalent to a magnitude of 17. Over the whole sky, there are ~500000 such galaxies, which, for an experienced classifier, represents less than half a man-year of work.

The primary limitations of visual classification of galaxies are twofold. The first stems from the difficulty the human mind has in perceiving relationships in a multidimensional classification space. This is, perhaps, the reason for the lack of success we have had in providing physical explanations for the observed patterns in galaxy classifications. Stellar spectral classification has provided very clear insight into the physics of stellar structure because nature has fortuitously allowed stars only a small number of independent physical parameters. If, for instance, galaxies form sequences in a six dimensional space, such sequences may be invisible to the human classifier. The second limitation is the difficulty of selecting the parameters to use in the classification. The eye may be drawn to the most striking structure in the image as a critical parameter, but it is not clear that such structure is physically important. As an example, the spiral structure seen in photographs taken in blue light may be classified in great detail but represents only a comparatively small physical variation relative to the underlying galactic disc. There may be other equally easily determined parameters that are unappealing to our esthetic views but of greater physical significance.

3. AUTOMATIC CLASSIFICATION

There are two ways in which automation may be introduced into the classification process, each being related to one of the deficiencies of visual techniques. First, there exist a large number of algorithms that may be used to search for correlations between the classification parameters of a set objects. These routines are in fairly wide use in the fields of medicine, ecology, geology and remote sensing (see, for example Sneath and Sokal 1973). Some very promising results have been obtained from the application of these methods to the available galaxy data (Whitmore 1983). The second automatic technique which may be applied to galaxy classification is the creation of machine generated classifiers. That is, to use objective algorithms to produce classification parameters directly from digital images of galaxies. As discussed elsewhere in these proceedings, this method has been used very successfully to sort faint images into galaxy or star bins. Its application to larger images of the sort of interest here has been hindered by the large number of numerical operations required to convert the image into a set of classification parameters. With the greatly increased processing power of computers now available, it becomes realistic to consider the wholesale classification of galaxies from such plate material as the new IIIaJ sky surveys.

Several experiments designed to explore the possibilities for automatic classification have been begun. Using standard techniques, extended images are located in PDS digitized CTIO schmidt plates. After subtracting the sky background, a number of image parameters are calculated. These include the first three moments of the intensity distribution, the isophotal axial ratio, and the slope of the intensity distribution at a given isophote. Some of these quantities are found to correlate with visual classification parameters, e.g., the ratio of second moment to total intensity relates to the Yerkes concentration class and axial ratio to the ellipticity parameter. However, the most interesting delineation, that between spiral and elliptical, cannot easily be made using only these quantities. In order to recognize spiral and irregular galaxies, it is necessary to be sensitive to higher frequency components in the image. One technique that allows a direct estimate of the relative importance of high frequency structure in an image is the production of a spatial power spectrum. Using an FPS-120B array processor on a VAX 11/780 host, it is possible to obtain the Fourier transform of a 256X256 image in somewhat less than one second. Such transforms have been produced for a number of galaxy images, using as input both the ordinary cartesian representation of the data and, because galaxies generally have axial symmetry, the same data mapped into polar coordinates. The transforms of the latter are more suited to easy interpretation by the human mind. The power at azimuthal frequencies in the range 2-6 cycles/2 radians is very strongly related to the apparent strength of the spiral structure in the image. It appears that a quite reliable discriminant between the early and late type galaxy bins may be constructed from this transform. One great advantage of working in polar Fourier space is that the power at various azimuthal frequencies does not depend upon the radial scale or position of the galaxy in the image. Thus, the same classifier may be used for both large and small galaxy images.

In order to make the fullest use of the objective classification power of machines, rather than attempt to interpret these derived parameters in terms of present classification systems or our current ideas of galaxy structure, it is better to combine both phases of automated classification. If, for a large sample of galaxies, a number of arbitrary parameters are generated and then the distribution of these points in the classification space is examined for clusterings and correlations, it is possible to allow the machine to select the most useful parameters. This process has been used to successfully create a device that rapidly searches images of blood samples for diseased cells. If a similar technique also proves successful for galaxy images, we may, for the first time, be able to obtain a truly objective classification system free of all prejudices, both physical and esthetic.

REFERENCES

Sneath,P.H.A., and Sokal,R.R.: 1973,"Numerical Taxonomy" (Freeman:San Fransisco)
Witmore,B.: 1983, Private communication.

A SYSTEM FOR OBJECT DETECTION AND IMAGE CLASSIFICATION ON PHOTOGRAPHIC PLATES

M.L. Malagnini[+,†], M. Pucillo[†], P. Santin[†], G. Sedmak[†], G.L. Sicuranza[+]
†Osservatorio Astronomico di Trieste
+Università degli Studi di Trieste

ABSTRACT

A system for detecting and classifying (faint) astronomical objects on photographic plates is presented. The system is planned to perform all the main operations, from the digitization of the plates with a PDS 1010A microdensitometer up to the final classification, under the control of the VAX-11/750 central processor. The aim of the project is to obtain an objective classification and reliable description of large amounts of objects, in a reasonable amount of time, for specific research projects. The detection and classification procedure is organized in modules, each of which is general enough to be applied to different kinds of analyses and to different classes of objects, according to astronomical requirements. An example of application to the problem of star/galaxy discrimination is given.

1. INTRODUCTION

The description of different interesting applications of systems for the automatic detection and classification of astronomical objects can be found in the literature (see, for instance, Jarvis and Tyson (1980), and references therein). Such systems are particularly important for the analysis of faint objects, whose dimensions are comparable with the seeing disk. Here we will describe the characteristics of the system under development at the Astronomical Observatory of Trieste, with reference to: (i) implementation of the hardware structure, (ii) development of suitable detection algorithms, for implementation both on-line with the acquisition process and off-line, and (iii) research on methodologies for image classification. The present configuration is planned for the classification of faint objects, with reference to the limit imposed by the S/N ratio of the images.

2. THE SYSTEM

The actual configuration of the hardware structure comprises a standard PDS 1010A, controlled by a PDP 11/45 computer. The main computer, VAX-11/750, is connected to graphic and pictorial terminals, a hard-copy device, and standard printer-plotters. Future implementations are fully described by Pucillo and Sedmak (1983). With reference to the actual structure, the operational scheme is illustrated. Starting from the digital image, the object extraction module provides a general catalogue containing, for each object detected, a set of suitable parameters. A new algorithm, based on information measures and transforms, has been applied (Santin, 1983). The second module is used for the selection of representative objects of the categories under study. For the definition of the training objects we make use of objective criteria derived from previous studies and of subjective criteria as well. Third, some smoothing is performed to clean up the images from spurious pixels. Next, the analysis of the training images is performed. We use a set of textural features extracted from the modified co-occurrence matrix proposed by Malagnini and Sicuranza (1980). The results are evaluated for different choices of the textural features, by using random samples of training images. Once the results appear to be stable, a final choice is made and a standard procedure is applied to all the objects of the general catalogue.

3. EXAMPLE

We present some partial results, referring to the analysis of a limited region extracted from an on-film copy, kindly provided to us by R. West, of the ESO Red plate No. 3183 (West and Kruszewski, 1981). Three regions of 4cmx4cm have been digitized with the PDS, using a 10 micron step and the slit of $20\mu \times 20\mu$. The test refers to an image of 780x780 pixels, corresponding to 9x9 arcmin^2. Our purpose is to discriminate faint images into the two classes of stars and galaxies. The detection algorithm produced a catalogue of 245 objects, with a mean area of 27 pixels. For the analysis, a window of 20x20 pixels, centered on the centroid of each object, has been chosen; objects too close to the borders of the test region have been discarded. Figure 1 shows the classification plot for the remaining 230 objects, with reference to two of the most significant parameters. These parameters have been determined according to the methodology described by Malagnini et al. (1983). The straight line represents a linear classifier, computed from the covariance matrix of the training sets; ellipses represent the equiprobability contours at the 95% and 99% confidence levels, computed from the statistical distribution of the parameters of

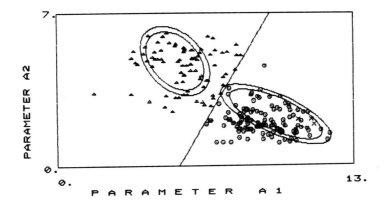

Figure 1. Classification plot for galaxies (open circles) and stars (triangles).

the training sets. The parameters A_1 and A_2 are measures of smoothness and contrast, respectively; therefore, galaxies have high A_1 values and low A_2 values, while stars have low A_1's and high A_2's. The plot appears rather scattered: only objects inside the ellipses can be reliably classified, while external ones cannot be classified as they are either multiple or very bright or extremely faint.
This kind of result gives an idea of the problems we intend to study further: (1) classification of individual objects of multiple images; (2) implementation of the procedure for taking saturation into account, and (3) a more precise quantitative definition of the limit of the classification.

Acknowledgements. We wish to thank F. Pasian for assistance during computer work. Partial support by the Italian MPI (Grant FD7) is acknowledged.

REFERENCES

Jarvis, J.F. and Tyson, J.A.: 1980, in *Application of Digital Image Processing to Astronomy*, SPIE Vol. 264.
Malagnini, M.L. and Sicuranza, G.L.: 1980, in ESO Workshop on *Two-Dimensional Photometry*, eds. P. Crane and K.Kjär
Malagnini, M.L., Pucillo, M. and Santin, P.:1983, in *Digital Image Analysis*, ed. S. Levialdi, Pitman Books.
Pucillo, M. and Sedmak, G.:1983, in NASA GSFC *Astronomical Microdensitometry Conference* (in press).
Santin, P.: 1983, preprint.
West, R.M. and Kruszewski, A.: 1981, Irish Astron. J., Vol. 15, No. 1.

AUTOMATIC ANALYSIS OF OBJECTIVE PRISM SPECTRA

P. Hewett, M. Irwin, P. Bunclark, M. Bridgeland,
E. Kibblewhite and R. McMahon
Institute of Astronomy, University of Cambridge

ABSTRACT

An automated system to measure and analyse large numbers of objective prism spectra from photographic plates using the Automated Plate Measuring (APM) facility at Cambridge is described. The system is being applied in a number of ways including automated quasar detection and subsequent clustering analyses, galaxy redshift surveys, and wide field searches for rare objects, such as carbon stars.

1. INTRODUCTION

The availability of large numbers (>50) of high quality low dispersion objective prism plates from the United Kingdom Schmidt Telescope (UKSTU) in Australia, together with increasing numbers of 4 metre telescope grism plates makes detailed spectral studies of significant areas of sky (hundreds of square degrees) a realistic possibility. The large number of spectra involved (typically >2000 per square degree to $m_J \sim 20$ at high galactic latitudes) means the use of high speed plate scanning facilities such as APM (Kibblewhite et al. 1983) or COSMOS (Stobie et al. 1979) is essential. The application of such plate scanning facilities can be regarded as an obvious development from the earlier human plate searches that have laid so much of the groundwork for the automated techniques now being developed. Crucial advantages in an automated, machine based approach include; (a) greatly improved homogeneity in selection procedures applied over large areas, (b) selection procedures are readily quantifiable and (c) large increases in speed are possible.

2. THE SPECTRUM MEASUREMENT SYSTEM

The APM Prism Reduction System (PRS) is based on a scan of a deep direct plate of the field to be studied. The APM control computer contains the complete SAO astrometric catalogue, and an initial alignment procedure provides coordinate transformations to convert machine

X-Y coordinates to right ascension and declination. The direct plate measurement provides complete lists of objects with positions accurate to 0.2 arcseconds, and a wide range of photometric and profile information. Magnitude limted samples may be derived from the data, and further subdivisions into object classes may be made - e.g. stellar, nonstellar or compact objects. A similar alignment procedure for the spectrum plate places the direct and spectrum images on the same celestial system, giving the exact position of all spectra corresponding to the object samples defined by the direct data. A global fit of the centroids of the direct images to the photographic emulsion cutoffs of the spectra allows the removal of second order geometrical distortions due to the prism, and establishes a precise wavelength zero point for all spectra, limited only by the object coordinate accuracy of 0.2 arcseconds. This procedure results in; (a) the virtual elimination of spurious images (due to noise) on the spectrum plate, (b) the ability to assign precise wavelength scales to objects (strong line quasars with no visible continuum for instance) which do not possess visible emulsion cutoffs, (c) the removal of all overlapping and confused spectra, and (d) reliable photometric and image classification data being available for all objects with measurable spectra.

An important feature of the PRS is the relatively complex measurement procedure, allowing the maximum possible signal to noise ratio to be obtained in the final one dimensional spectra. The mean profile shape of each spectrum is calculated by using the spectrum marginal sum perpendicular to the direction of prism dispersion. The calculation allows for any change in the intensity of the sky background over the extent of each spectrum. This mean spectrum shape is defined by the image profile and is effectively constant along the spectrum-saturation effects not withstanding. A smooth monontonic function is fitted to this marginal sum - thereby reducing problems due to nearby overlapping images - and used as a weighting function in the calculation of the intensity variation along the spectrum. The intensity at each wavelength is found by comparing the calculated mean spectrum profile shape with crossections of the data at each wavelength. Technically the determination of the object intensity scale factor k , at each wavelength is made by minimising the weighted sum of the squared error residuals between the data D_i and the spectrum profile P_i . For random noise this gives the most probable value for the intensity at each wavelength bin,

$$\text{minimize} \quad \sum_i (D_i - k \times P_i)^2 / \sigma_i^2 \quad \text{with respect to } k$$

where σ_i^2 is the noise variance at density D_i . The scale factor k that minimizes the sum is simply given by the weighted sum

$$k = \frac{\sum_i P_i \times D_i / \sigma_i^2}{\sum_i P_i^2 / \sigma_i^2}$$

The use of such a 'matched' estimator reduces the noise in the computed spectra by typically a factor two over straight forward integration. Spectra for brighter saturated images are obtained using the points in the sum that are unsaturated - effectively scaling just the wings of the profile onto the wings of the spectrum. The final spectra are output to magnetic tape and further processing takes place using the one dimensional intensity versus wavelength spectra - typically 50000 spectra are obtained from a high quality UKSTU IIIaJ plate.

3. SPECTRUM ANALYSIS PROCEDURE

The nature of subsequent processing depends on the type of project undertaken. Three main types of project can be distinguished; (a) sophisticated analysis of particular types of spectra - e.g. galaxy redshift determinations (Cooke et al. 1983 and this conference), (b) detection of specific types of rare objects with well defined, previously known spectra - e.g. metal poor stars and carbon stars, (c) general searches for complete samples of spectra that can not be unambiguously classified as main sequence stars or normal galaxies, in this case the type of spectra to be identified are not known beforehand. Subsets of such spectra are quasars and emission line galaxies. Examples of all three types of programme are already underway or are about to start.

The techniques for estimating galaxy redshifts are described by Cooke et al. (1983). In the case of specific object searches, template matching appears to be the most satisfactory approach and is being employed. Much of the work involved in this procedure is related to modelling the effects of the atmosphere, prism/telescope optics and the photographic process. Perhaps the most challenging projects however are of type (c) and considerable work has gone into developing a system that will identify all spectra that are not classified as main sequence stars or normal galaxies.

The PRS data are particularly suitable for this application and software is available to identify objects with specific features - emission lines, absorption lines, continuum breaks, strong uv excess - as well as more general techniques that examine all images on a plate and use cluster analysis techniques to identify anomalous objects in a wide range of parameter spaces. Objects identified by the latter technique include those with anomalous overall continuum slopes, peculiar colours, and featureless spectra.

The detection of objects exhibiting emission or absorption features provides an example of the techniques: an object continuum spectrum is defined using a combination of median and linear filtering techniques. The noise is determined globally as a function of wavelength and intensity from all the spectra, using the original spectra and the continuum fits. Then 'matched' filters are stepped along each spectrum to identify emission/absorption features. The filter technique is directly analogous to that used in the measurement of the spectra, although a Gaussian shape is used as little is gained from more sophisticated profile shapes. In order to detect resolved lines the procedure is repeated a number of times with the continuum-defining filter lengths increased, and the width of the matched filter enlarged. This 'matched' filter approach again offers large increases in the ability to reliably detect lines relative to the more common box filter technique. A catalogue of detected lines with associated detection probabilities can be derived in about an hour for a complete Schmidt plate data set of 50000 images.

4. SUMMARY

A fast automated spectral analysis system (based on the APM at Cambridge) applicable to a wide range of astronomical research is now in operation. Techniques for automated detection of quasars and other peculiar objects have been developed and surveys of large areas of sky are now underway.

5. REFERENCES

Cook, J.A., Emerson, D., Beard, S.M. and Kelly, B.D.: 1983. 'Proceedings of the Workshop on Astronomical Measuring Machines', Royal Observatory Edinburgh, Edinburgh.
Kibblewhite, E.J., Bridgeland, M.T., Bunclark, P.S. and Irwin, M.J.: 1983. 'Astronomical Microdensitometry Conference', Washington.
Stobie, R.S., Smith, G.M., Lutz, R.K. and Martin, R.: 1979. 'Image Processing in Astronomy', ed. Sedmak, G., Capaccioli, M. and Allen, R.J., Osservatorio Astronomico di Trieste.

REMARKS ON T-GRAIN TECHNOLOGY APPLIED TO ASTROPHOTOGRAPHY

Alain Maury
TE.S.CA., C.E.R.G.A., Caussols, Saint Vallier de Thiey (France)

If T-grain Technology can be used to manufacture emulsions of astronomical type, it will provide a better Detective Quantum Efficiency than any existing plate. As Dr. Millikan explains us, this increase in D.Q.E. can result in an increase in Signal to Noise ratio (i.e. increase in contrast or decrease in granularity) or in an increase of plate speed. As Kodak and we, their customers, don't have a lot of money to spend for three or more new emulsions, we have now to choose for a new emulsion, which might be a harder one (more contrast), a finer one (better granularity), or a faster one. Let us now examine these three possibilities:

HIGHER CONTRAST

As demanded by Dr. Malin, this could be an interesting solution for showing easily extensions in galaxies or in nebulosities, but my point of view is that it's the only advantage that higher contrast provide. It limits the useful photometric range, giving something like a direct isophote of the night sky (only one magnitude between plate fog and plate saturation!!). In these condition it's to be feared that some new problems of inhomogeneities occur ; so it might be an interesting solution for making some types of discoveries, but not for general use in Schmidt Astronomy.

FINER GRANULARITY

It seems that emulsion technology makes finer grain go with high contrast. This solution is nevertheless a better solution than the first one: high contrast amplification is feasible and photometric range might be conserved. But this solution can give a real improvement only during the ten best nights per year because of seeing effects and telescope quality. It might also give something very hard to measure with our actual measuring machines.

FASTER SPEED

I think this is the way in which this new technology will give the most important benefit in astronomical photography.
- Increase of plate quality in less time on high quality Schmidt telescopes due to a limitation of photometric change of the atmosphere during

the exposure, and also limitation of field rotation due to differential atmospheric diffraction as pointed out by Dr. Cannon.
- Short exposure also implies a greater probability of good exposure with telescope which suffers from flexure or other mechanical problems.
- If it's true that a two time increase in sensitivity, you can't take twice more plates, it's also true that in these conditions, you can take maybe between 40 to 80 percent more plates (depending on the operating facilities of your telescope). Two plates taken on the same field produce roughly a 40 percent increase in S/N ratio and a 99.9 percent increase in the chances of discovery of moving objects and discrimination of plate defaults.
- Faster speed also give new possibilities in the use of narrow band filters combination (U or B band photography with J type emulsion).
- This will also give more efficiency on some fields which are available during the short summer nights.
- Faster speed changes only the condition of exposures (Shorter exposure time or same exposure time with less hypering giving better signal to noise ratio). It doesn't change the condition of measuring: If astronomers have to change their detection system every decade, no reliable work can be produced.

CONCLUSION
      Let's wait the "T-a-J".

MICROSPOTS ON IIIa-J PLATES ("GOLD SPOT DISEASE")

M. Elizabeth Sim
UK Schmidt Telescope Unit, Royal Observatory
Edinburgh EH9 3HJ, UK

The UK 1.2 metre Schmidt Telescope at Siding Spring Observatory, Australia, has taken about 8500 plates since it was commissioned in 1973. A large fraction of these plates, including those for the southern and equatorial Sky Surveys, are on Eastman Kodak IIIa-J emulsion. A significant number of these processed IIIa-J plates have developed many small gold spots similar in size to faint star images. They are most likely to occur in areas of relatively high density, such as in bright star images or on the stepwedges, and are sometimes scattered around in non-image areas (fig. 1). An alarmingly high proportion of plates taken in 1974-75 (about 50 per cent) have developed gold spots, and the most recent affected plate found so far is J6517, taken in October 1980. Four early IIIa-F plates have also been found to have gold spots. The degree to which plates are affected may be slight, with spots appearing only on one or both stepwedges; moderate, with spots appearing along one, two, or three edges of the plates; or severe, when spots appear all round the edges or all over the plate. The affected plates are divided approximately equally between these three categories. It seems that the spots take about 3 years to develop, after the plate has been exposed and processed. This makes it very difficult to establish possible causes of the problem, which is almost certainly not curable. Unfortunately, gold spot formation involves changes in the structure of image silver and some silver migration. Even if it were possible to remove the gold spots, this could only be cosmetic since there is no way to restore the original structure of the image silver. It is therefore extremely important to find the causes of these spots as soon as possible, so that their occurrence on any more plates can be prevented.

The causes of microspot formation have still not been convincingly established, but the list of primary suspects include pick-up of contaminants in processing solutions, hydrogen peroxide fumes and fumes from fresh oil-based paints (1, 2). IIIa emulsions, unlike most others, are based on pure silver bromide, and there is some evidence to suggest that gold spots may be formed if the IIIa-J

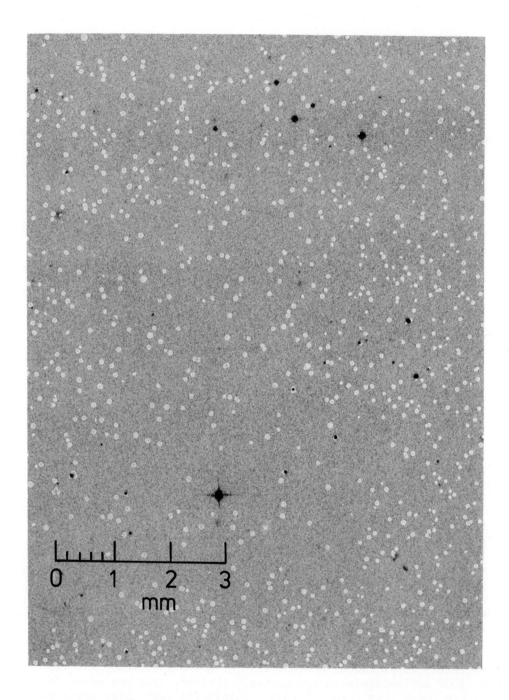

Figure 1. Microspots, which appear white in this print, may form around existing images or at random in the field.

picks up iodides or chlorides left in the processing solutions by other types of emulsion. To minimise this risk, the processing system at Siding Spring has been modified, along the lines of recent Kodak recommendations, to provide a separate, dedicated processing line for IIIa-J and IIIa-F plates so that they cannot pick up contaminants deposited into the solutions by other emulsions.

The soft foam packing that was used to protect plates during shipping is a potential source of hydrogen peroxide. Users are therefore asked to discard all such foam packing. Polystyrene sheets provide sufficient protection for the plates without any harmful chemical effects. Plates should never be stored in or near areas which have recently been painted, to avoid the risk of spot formation from contamination of paint fumes.

Investigations continue at ROE, Siding Spring, and Eastman Kodak into the occurrence and possible causes of gold spots on IIIa-J plates, and ways of preventing them. UKSTU has probably the largest, most homogeneous and well-documented collection of IIIa-J plates in the world. Andy Good in Edinburgh and John Dawe in Australia are systematically searching and monitoring all the available IIIa-J plates for signs of gold spots. So far 327 affected plates have been found, out of 1715 inspected. This includes accepted Sky Survey plates, about 40 of which are affected.

To protect plates from spot formation, Eastman Kodak have recommended treating plates for 3 minutes in a 1+19 solution of Rapid Selenium toner. This deposits a thin protective coat of selenium over the image silver. Further investigations of possible adverse effects of toning will be made before it is included in our processing system. If the affects of toning are acceptable, it may then be possible to treat unaffected and slightly affected plates to inhibit spot formation.

REFERENCES

1. Millikan, P.G., Miller, W.J., and Black, D.L. "Astronomical Photography 81", ed. Heudier & Sim, p. 153, 1981.

2. Feldman, L.H., Jnl Applied Photographic Engineering $\underline{7}$, 1, 1982.

INTERNAL CALIBRATION OF ASTRONONOMICAL PHOTOGRAPHS

P. S. Bunclark and M. J. Irwin,
SERC APM Facility,
Institute of Astronomy,
Madingley Road,
Cambridge CB3 OHA

ABSTRACT

By applying the principle that all stellar image profiles should be identical apart from a scale change, we can derive a relative calibration for a photographic plate without use of calibration spots or other indirect means.

1. INTRODUCTION

Despite the arrival in recent years of various forms of electronic two-dimensional detectors, photographic plates still have enormous advantages. For low cost they collect panoramic (~36 square degrees) data to faint limiting magnitude (~22). The serious drawbacks with photographs are their lack of dynamic range and a complex photometric response function. We describe here an improved version of the technique first presented in the Measuring Machines Workshop 1982 (Bunclark 1983) which partially solves the former problem and completely solves the latter. By requiring that the areal profile of a star be independent of its magnitude we can derive a "magnitude index" for stellar images which is linearly related to photoelectric magnitude over a range of up to 14 magnitudes. The linearity is reliable down to the plate limit and therefore also provides a means to extrapolate calibrating photoelectric sequences which typically end at around 17th magnitude. From the photoelectric sequence it is necessary to determine the slope and intercept of the calibration. We have developed the method specifically for use with SERC APM machine output, but it could be adapted to any data which provides profile information for large numbers of images in a field.

2. APM OUTPUT

The APM machine scans a given area of plate two times; the first pass divides it into 0.5mm x 0.5mm regions containing 64x64 pixels (or samples). A histogram is generated from these 4096 pixels and a modal estimate obtained which is taken to be the most likely background value. During the second pass, these background values spaced every 0.5 mm are interpolated to give a background estimate at every 7.5 micron sample position. A threshold above the local background is then subtracted from the sample value, and finally samples greater than zero are collated into images which are parameterised for output. The sixteen parameters are: integrated intensity, y-position, image number, x-position, 2nd x-moment, x-y cross moment, 2nd y moment, peak intensity, and an eight-level areal profile. Intensity/density output from APM has a range of 0-255. The profile of an image is sampled at eight levels set at threshold, t+1, t+2, t+4, t+8, t+16, t+32 and t+64. This logarithmic spacing measures the whole unsaturated part of bright images while providing sufficient sampling of faint ones.

3. PRINCIPLE OF THE INTERNAL CALIBRATION PROCEDURE

All stars are essentially point sources. The reason they appear as extended images is that they are smeared by atmospheric seeing, and the image suffers from scattering in the telescope optics and in the emulsion. These processes are independent of how bright the star is, so the final intensity distribution in the emulsion during exposure is identical for all point sources apart from a scale change. In actual fact since different parts of the field "see" a slightly different telescope geometry, there can be variations in image structure depending on position.

For a given image we have measures of the radius at some fixed levels above threshold. Suppose we had (i) a standard image whose profile was known at all points and (ii) the calibration curve so that we know the intensities of the levels at which the areal profiles are sampled. For a given image at a given level we know r and $I(r)$, the intensity at that radius. Then read off the standard profile $I_o(r)$, and the luminosity of our image is

$$L = \frac{L_o \, I(r)}{I_o(r)} \quad (1)$$

where L_o is the luminosity of the standard image. In fact, several profile measurements are always available, so several values of L can be computed, and averaged.

Further, if we know the brightness of all the stars and the standard profile we can derive the intensity of the levels (ie the calibration curve) by

$$I(r) = \frac{L \, I_o(r)}{L_o} \quad (2)$$

or finally, knowing the calibration and the brightness, we can derive

INTERNAL CALIBRATION OF ASTRONOMICAL PHOTOGRAPHS

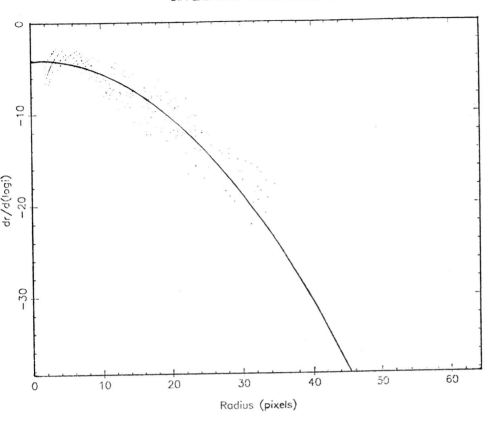

Figure 1. Inverse gradient of logarithmic profile used to generate initial profile estimate

the standard profile from

$$Io(r) = \frac{Lo\, I(r)}{L} \qquad (3)$$

Of course in practice we know none of the elements, and in fact would like to determine them. The procedure is to obtain a good first estimate of the standard profile and of the calibration. Then the brightnesses of the stars are computed from (1), and better estimates of calibration and profile from (2) and (3). The three elements are iterated cyclically (typically about ten times) until no further convergence occurs.

4. GENERATION OF FIRST GUESSES

The iteration converges fairly weakly, and if seriously incorrect starting points are chosen the process will diverge to nonsense. To form the first estimates of the profile and the calibration curve, only data near sky are used. This is because for a small range of intensity the relationship between incident intensity and measured density is fairly linear, so the calibration curve for this region can be taken as a straight line with 45 degree slope. Secondly, it has been found that when the inverse of the gradient of the logarithmic intensity profile is plotted against radius, the result is well represented by a parabola (figure 1).

$$dr/d(\log i) = a + br + cr^2 \qquad (4)$$

where a, b and c are empirically determined constants. Then the profile is given by

$$\log i(R) = \int_0^R \frac{1}{a + br + cr^2}\, dr$$

$$= \frac{2}{\sqrt{D}} \arctan\left(\frac{b + 2cR}{\sqrt{D}}\right) + K \qquad (5)$$

where K is the constant of integration, and $D = 4ac - b^2$ (D positive). As we are not able to determine the zero point at this stage, K may be taken to be zero. Now it is possible to derive the preliminary calibration of all eight areal profile levels. From the data, it is found at which value of threshold radius the areal profile at a level goes to zero. The difference in log I between the two levels is then simply the difference in log I between the peak of the profile and the

INTERNAL CALIBRATION OF ASTRONOMICAL PHOTOGRAPHS

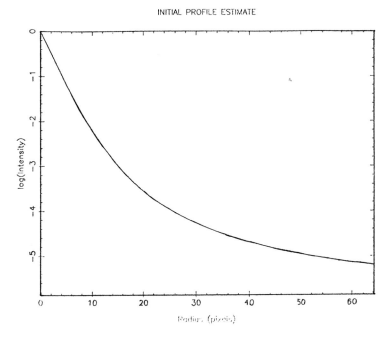

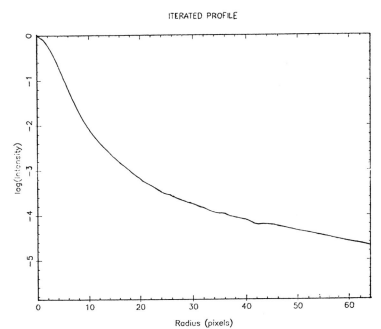

Figure 2. (Top) Initial profile estimate.
(Bottom) Final version of profile after iteration.

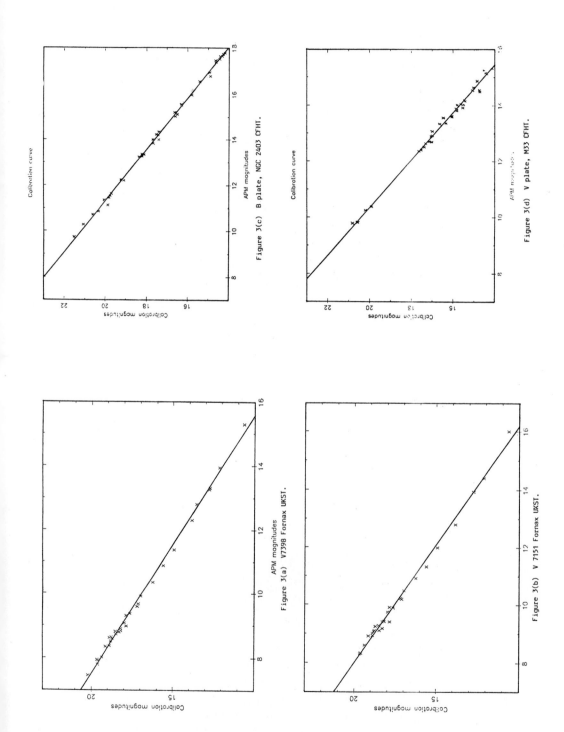

INTERNAL CALIBRATION OF ASTRONOMICAL PHOTOGRAPHS

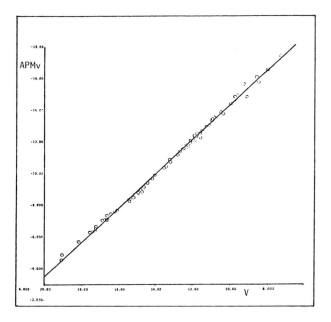

Figure 3(e) V plate, SGP UKST.

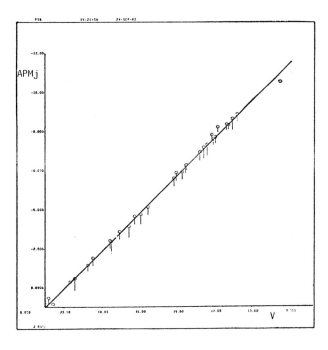

Figure 3(f) J plate, M31 PS.

value at the given threshold radius.

5. ITERATION

The iterative stage can now be entered, computing in turn "magnitudes" from profile and calibration, calibration from magnitudes and profile, then profile from magnitudes and calibration. The overall "shape" of the three sets of data are not much changed, but rather detail is added. Figure 2 shows a typical first estimate of profile from (5), and the final iterated profile. Finally the magnitude index is plotted against "APM magnitude", which is $-2.5\log(\Sigma D)$ for an image. This curve is used as the actual calibration, since the APM magnitude is less noisy than the magnitude index (the former is the sum of all points in an image, and the latter is computed from only a few samples of the image profile).

To achieve the final calibration, we must plot magnitude index against standard magnitudes for at least a few stars. This also serves to prove the method, and so several examples are shown in figure 3.

For stars, this calibration method is rigorous because stellar images define the solution and stellar sequences tie it to standard photometry. However, we also end with a calibration curve, and as all objects have an areal profile, it is possible to calculate isophotal magnitudes for extended objects by integration over their profile. A typical scan threshold puts the brightness of the outermost isophote at about 25 magnitudes per square arcsec. If desired, the integration may be stopped at a particular isophotal level (or extrapolated to one).

6. CONCLUSION

We have presented a method of calibrating astronomical photographs which does not require a deep sequence of standard stars. Further it does not use sensitometer spots which are notoriously unreliable. With the advent of measuring machines which routinely digitise whole plates, we feel this is the best possible way to carry out photographic calibration.

7. REFERENCE

Bunclark, P. S.: 1983, Proceedings of the Workshop on Astronomical Measuring Machines, Occasional Reports of the Royal Observatory, Edinburgh, No. 10, p149.

INTERNATIONAL HALLEY WATCH WIDE FIELD NETWORK FOR LARGE-SCALE PHENOMENA, CALIBRATION OF SCHMIDT PLATES USING STAR PROFILES

Daniel A. Klinglesmith III
Goddard Space Flight Center
Lab. for Ast. and Solar Physics
Greenbelt, MD 20771
USA

Stephen W. Rupp
Eleanor Roosevelt High School
Greenbelt, MD 20771
USA

1. INTRODUCTION

The International Halley Watch Wide Field Network for Large-scale Phenomena (Niedner, Rahe and Brandt, 1982) will be receiving several thousand original or first copy images of the large scale structure of comet Halley. These images will be coming from over 90 telescopes at 70 different observatories around the world. Some but not all of the photographic material will have associated sensitometry. It is anticipated that since this network is looking for changes in the plasma tail structure most, if not all, observations will be made with unfiltered Kodak IIaO emulsion. One of the major tasks of the network will be to study the plasma tail variations at the highest time resolution possible. In order to do this, it will be necessary to combine photographs taken at many observatories. This implies creating calibrated intensity images from a sequence of diverse photographic plates. We will need to provide a method for calibrating individual photographic plates, that is not only fast but also accurate using only information that can be obtained on a single Schmidt plate.

There exists in the current astronomical literature two methods that have been suggested to handle this problem. The first is due to Agnelli, Nanni, Pitella, Trevese and Vignato (1979). Their method assumes that the star profiles can be modeled with an analytical function and that this profile is circularly symmetric. Then only the magnitudes of a number of unsaturated stars on the plate in some standard magnitude system is needed to convert the density at a specific radius into an intensity transfer function. The second method (Zou, Chen and Peterson, 1981) eliminates the the assumption about the form of the stellar profile but requires that the intensity transfer function be approximated by a polynomial in opacitance.

This paper will show how well both methods provide surface brightness measurements on extended objects. The object used for this test is the Andromeda Galaxy, M31. The photographically calibrated results are compared to photoelectric scans by De Vaucouleurs (1958). Two plates have been obtained at the Joint Observatory for Cometary Research (Brandt, Colgate, Hobbs, Hume, Maran, Moore, Roosen, 1975) in Socorro, New Mexico.

They are 4 by 5 inch plates with approximately 8 by 10 degrees of the sky recorded. The exposures are long (16 and 40 minutes) so that the sky background is high. The density values for the sky are 1.0D and 1.5D respectively.

2. MODIFICATION OF THE AGNELLI METHOD

The Agnelli method assumes the total point spread function for the star plus the sky is given by

$$I(r) = I_o f(r) + I_s. \qquad (1)$$

We will assume that the sky background is constant for any one star profile, but need not be the same for all stars. The form of the analytical function that Agnelli et al. used was given by Moffat (1969) as

$$f(r) = [1 + (r/R)^2]^{-\beta}. \qquad (2)$$

The total flux from a star is given as the area integral of equation 1. If the flux is expressed as a magnitude, then the magnitude for one star can be expressed as

$$m_i = -2.5 \left\{ \log \frac{2\pi R}{2(\beta-1)} + \beta \log \left[1 + \frac{r_{ij}^2}{R} \right] + \log \left[I_i(r_j) - I_i(\text{sky}) \right] \right\}. \qquad (3)$$

This equation states that if we know the radius at which a particular density is reached for each star in a sequence of stars whose magnitudes are known on some standard magnitude scale; we can, using a non-linear least squares technique, solve a set of equation 3's for the three unknown quantities β, R and $\log[I_i(r_j) - I_i(\text{sky})]$. If this is done at "many" densities, with the constraint that the shape parameters are held constant, we would create the needed intensity transfer function.

3. THE DENSITY MOMENT SUMS METHOD

Zou, Chen and Peterson (1981) have proposed another method for doing the same type of problem. Their method assumes that the intensity transfer function can be approximated by a polynomial in opacitance,

$$I = \sum_{k=1}^{n} b_k (10^D - 1)^{pk} \qquad (4)$$

and that the magnitudes of a number of stars, with unsaturated density images are known. With these two assumptions the intensity for any one star can be written as

$$m_i = I_o 10^{-0.4 M_i} = \sum_{k=1}^{n} b_k \left[\sum_{j=1}^{m} (10^{D_{ji}} - 1)^{pk} - m(10^{D_{si}} - 1)^{pk} \right]. \qquad (5)$$

The only unknowns are the b_ks. Thus this set of equations, one for each known star, can be solved by least squares techniques to provide the required intensity transfer function (equation 4).

4. RESULTS

Two plates from the JOCR of M31 have been digitized and used as a test case for these two methods. A photoelectric sequence near M31 by Racine (1967) was used for the stellar magnitudes. Figure 1a and 1b show the extracted magnitudes of the 16 minute and 40 minute exposure respectiviely using the Agnelli et. al. method. The 16 minutes exposure does not reach the faint sky limit achieved by the 40 minute exposure. The values of "β" and "R" are given in table 1. The same digitized data was processed with the Zou, Chen and Peterson method with the results shown in figure 2. The values for the "b_ks" and the best value of "p" and "n" for each are shown in table 2.

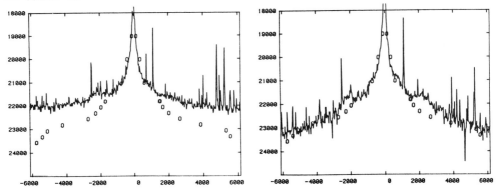

Figure 1a and 1b. Magnitude profile of major axis for M31 using Agnelli et. al. The lines are for the extracted magnitude and the circles are from De Vaucouleurs (1958). The left hand graph is for a 16 minute exposure and the right hand one is for a 40 minute expousre.

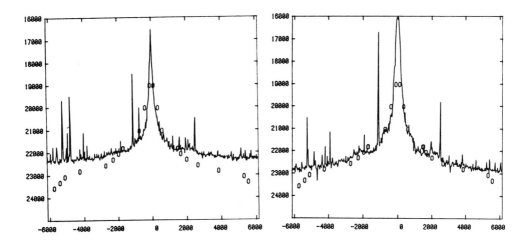

Figure 2a and 2b. Magnitude profile of major axis of M31 using Zou, Chen and Peterson method.

Both methods give similar results. For our application, cometary images where the stars are trailed due to the motion of the comet during the exposure, the advantage lies with the Zou et. al. method because it is not necessary to assume a shape for the star profile.

The availability of sensitometry would provide data on the nature of the intensity transfer function for each plate. However, it is not clear that we can use the same ITF curve all over a large Schmidt plate taken near the horizon and still expect to maintain 5 to 10% accuracy. With these methods and enough calibration stars, for example blue magnitudes from the Catalogue of Stellar Identification (Ochsenbein, 1976) we will be able to proivde an intensity transfer function varying with sky background. We will use what ever data we have to provide the best possible photographic photometry.

TABLE 1

PLATE	β	R
S 855	1.7	1.5
S1590	2.1	1.4

TABLE 2

	S 855	S1590
P	0.9	0.7
n	4	3
B1	1.233E-10	1.707E-09
B2	1.813E-13	1.101E-11
B3	-3.141E-17	-6.602E-15
B4	1.357E-21	

REFERENCES

Agnelli, G., Nanni, G., Pitella, G., Treverse, D., Vignato, A., 1979, Astron. Astrophys., 77, pp.45-52.
Brandt, J.C., Colgate, S.A., Hobbs, R.W., Hume, W., Maran, S.P., Moore, E.P., Roosen, R.G., 1975, Mercury, March/April, pp.12-13.
Moffat, A.F.J., 1969, Astron. Astrophys., 3, pp.455-461.
Niedner Jr., M.B., Rahe, J., Brandt, J.C., 1982, Proc. ESO workshop on "Need for Coordinated Ground-based Observations of Halley's Comet", Paris, pp.227-242.
Ochsenbein, F., Egret, D., Brshaft, M., 1976, Proc. of IAU Coll. #35, pp.31-36.
Racine, R., 1967, Astron. J., 72, pp.65.
De Vaucouleurs, G., 1958, Ap. J., 128, pp.465-488.
Zou, Z., Chen, J., Peterson, B.A., 1981, Chinese Astro. Astrophys., 5, pp.316-321.

THE 2020GM PDS MICRODENSITOMETER AT MUENSTER

Waltraut C. Seitter
Astronomisches Institut der Universität Münster, FRG

The 2020GM PDS microdensitometer of Perkin-Elmer, Optical Division, was installed at Muenster University during the summer of 1982. The present communication summarizes the greater part of work performed during its first year of operation and outlines plans for the future. The microdensitometer and the system configuration are shown in Figs. 1 and 2.

THE MONITOR

The Astronomical Data Analyzing Monitor ADAM, developed by D. Teuber, is part of the Astronomical Data Analyzing System ADAS at the Astronomical Institute of Muenster University. ADAM provides the software environment for application programmes and controls the execution of these programmes. The astronomer issues his commands in an application-related language. He is also able to shape the input sequence and the monitor response to resemble closely that of his home system.

TESTS

Mechanical and photometric tests of the microdensitometer were carried out by H.-J. Tucholke following the usual procedures. The orthogonality of the system is about 4", corresponding to a deviation of 10μ over the total travelling distance of 500mm. The mechanical stability of the machine lies within 0.2μ at all speeds (0-230mm/sec) and around 0.1μ at all but the smallest and the largest. Photometric stability is reached after 3 hours of warm-up time and is smooth for the sum of all components. The density readings show a maximum deviation of 2% relative to a standard wedge, generally it is much lower. For more detail see Teuber (1983) and Tucholke (1983a and b).

ASTROMETRY

Investigations of 160mm x 160mm plates by M. Geffert and H.-J. Tucholke

Fig. 1. The 2020GM PDS microdensitometer at Muenster

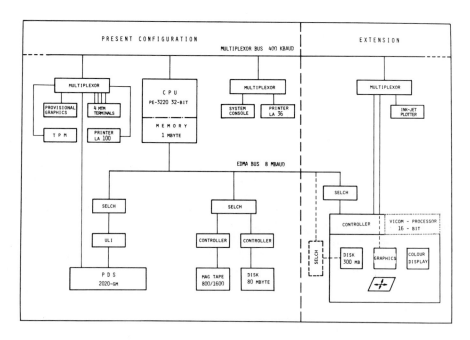

Fig. 2. The system configuration including the 2020GM PDS

show differences between measurements in position 0°+180° and position 90°+270° of 0.7μ in x and 0.5μ in y. Comparison with catalogue positions by Vasilevskis et al. (1979) gives deviations of 1.4μ in both x and y when linear reduction models are used, and 1.3μ in x, 1.1μ in y with quadratic models. The errors consist of four contributions: PDS measuring rods, fit routines of the programme, photographic plate and comparison catalogue.
The programme, based on work by R. Dettmar, includes the following steps: data sampling, reconstruction of stellar images from scan lines, transformation and comparison of two measurements of the same field scanned under different angles, iteration of parameters, comparison with catalogue star in the catalogue system.

STAR COUNTS AND PHOTOMETRY

The programme steps are: data acquisition with medium resolution, determination of pixels of maximum density under exclusion of multiple maxima due to saturation, determination of limits of stellar images centered on maximum pixels, intensity calibration (several methods in preparation).

The results are: stellar positions accurate to within scan step width, stellar luminosity functions and plots of star fields with symbols of different sizes representing different intensities.
The coarse data may serve as input for individual scans of stars leading to high positional and photometric accuracies.
The programme was developed and tested by W. Goerigk, T. Richtler and H.-J. Tucholke.

RADIAL VELOCITIES

The programme includes the following steps: data acquisition, search for comparison lines in given x-intervals, determination of line curvatures, derivation of the dispersion curve, rectification of the stellar continuum, identification of stellar lines and profile fittings, heliocentric corrections and Julian dates, determination of radial velocities for individual lines, mean values and the total radial velocity for the spectrum, error determinations.
The programme is fully automatic for normal stars. It was written and tested by R. Dettmar and R. Duemmler.

MORPHOLOGY OF GALAXIES

The programme sequence is: data acquisition, derivation of the characteristic curve using the Honeycutt-Chaldu method, determination of central coordinates of elliptical galaxies, sequential fits of ellipses to isodensity curves, determination of semimajor and semiminor axes a and b of ellipses, position angles Θ of major axes, ratios of major to minor axes, mean densities and mean intensities along ellipses.

The final results are shown as data tables and in the following plots:
$I(r)$ with $r = (ab)^{1/2}$, $\log I(r)$, $\Theta(r)$, $a/b(r)$ and $\log I(r^{1/4})$.

The program was written and is applied by H.-G. Scheuer with the assistance of E. Willerding. For more detail see Scheuer (1983a and b).

FUTURE WORK

Long-range aim is the reduction of entire Schmidt-plates.

The determination of BASIC DATA will include: stellar magnitudes in three (or more) colours, colour indices, colour excesses, stellar types from objective prism plates and/or colours, distances; isophotes of galaxies in different colours, radial velocities from objective prism plates, distances.

The FINAL DATA will be: distribution of stars in the galactic plane and perpendicular to it as a function of type, galactic isophotes; distribution of galaxies of different morphological types; positions, structures and physical properties of bright and dark clouds.

NUMERICAL EXAMPLE

350mm x 350mm Schmidt-plate, step width 15μ.

Measuring time 20 hours, total number of data points 1.1 Gigabyte. After arrival of 1000 scan lines on-line reduction commences. With an estimated number of 2 million stars and galaxies up to 21st magnitude on a typical plate the number of data points (total density and center position of each object) will reduce to 12 Megabyte. Another two Megabyte will give the values of a smooth sky background.

For the analysis of extended cloud regions digital storage of entire plates will be necessary.

PRELIMINARY TIME TABLE

September 1983	End of basic mechanical and photometric tests
1983	Morphology of southern elliptical galaxies from SRC-J atlas plates
after glass copy of POSS is received	Morphology of northern elliptical galaxies
1983	Radial velocities: stars and interstellar lines
1983/1984	Investigations of globular and galactic star clusters: membership, masses, variable stars
1983/84	Spectral classification from objective prism plates using line identifications, cross correlations, combination of procedures

1984	Reduction programmes using various image structures, photometric programmes of high accuracy
1984	Start collecting photometric sequences for plate calibrations
1984	Installation of image processing system and tests (assuming grant is provided)
1985	Tests of photometric properties of different wide-angle plates
1985/86	Reduction programmes for entire Schmidt-plates
1986	Test runs with three-colour Schmidt-plates and follow-up algorithms
1987	Commence long-range programme for the reduction of entire Schmidt-plates

For the 1983/84 programmes a total of about 650 direct and spectral plates is available, not counting the atlas plates.

The contents of this communication are the combined work of all full and temporary members of the Astronomy Department of Muenster University, including, besides the coworkers mentioned above, Dr. A. Bruch and Messrs. R. Budell, F. T. Lentes and C. C. Volkmer.

REFERENCES

Scheuer, H.-G. 1983a, Diploma thesis, Muenster University.
Scheuer, H.-G. 1983b, Astr. Astrophys. (in preparation).
Teuber, D. 1983, in "Astronomical Microdensitometry", D.A. Klinglesmith, ed., NASA (in press).
Tucholke, H.-J. 1983a, Diploma thesis, Muenster University.
Tucholke, H.-J. 1983b, Mitteil. Astr. Gesellsch. 60 (in press).
Vasilevskis, S., van Leeuwen, F., Nicholson, W., Murray, C.A. 1979, Astr. Astrophys. Suppl. 37, 333.

M.A.M.A. PROJECT: A NEW MEASURING MACHINE IN PARIS

Jean Guibert, Pierre Charvin, Patrick Stoclet,
Institut National d'Astronomie et de Géophysique,
77 avenue Denfert-Rochereau, F-75014 Paris

SUMMARY. A new photographic measuring machine is under construction at Paris Observatory. The amount of transmitted light is measured by a linear array of 1024 photodiodes. Carriage control, data acquisition and on line processing are performed by microprocessors, a S.E.L. 32/27 computer, and an AP 120-B Array Processor. It is expected that a Schmidt telescope plate of size 360 mm square will be scanned in about one hour with pixel size of ten microns.

I. THE MICRODENSITOMETER

The microdensitometer combines a X-Y movable carriage, a quasi monochromatic illumination source, and a multichannel photometer. High scanning speed and flexibility in machine control and data processing are expected from the design of hardware logic and choice of the computer system which includes a parallel processor and a real time oriented mini computer.

Mechanical design
The base is a granite block (1.8 x 1.1 x 0.3 m), plane to within 3 microns. The X-Y carriage consists of two superposed frames with independent perpendicular motions. Each frame is supported and guided by SCHNEEBERGER roller bearings travelling on rectangular ways, and is driven by a feedback loop controlled motor through a ball screw. The positional information is derived from incremental linear moiré fringe transducers (MINILID 300) from HEIDENHAIN. An overall repeatability of 1 micron is expected. The glass table accepts plates of size 550 x 550 mm with maximum thickness 6.35 mm. It can be rotated with respect to the upper carriage in view of some specific applications requiring proper orientation of the plate. The working area being 50 cm in diameter, a whole 360 x 360 mm Schmidt photograph can be measured with only one positioning.

Optical configuration

A regulated quartz-iodine lamp illuminates the slit image plane through a hot half-condenser, a broad-band interference filter, a set of neutral densities, and a cold half condenser. The light is quasi-monochromatic, with wavelength: 633 $\pm$ 25 mm. A beam splitter allows a small fraction of the light to be measured by a photodiode for the purpose of flux stability control. The illumination lens forms an image of the slit onto the photographic plate. Finally, the projection lens projects the plate area to be measured onto the 1024 element RETICON linear photodiode array. This array can be replaced by a T.V. camera in view of visual examination.

Two magnifications of 1.6 and 1/1.6 (obtained by rotating the projection lens) associate, to each photodiode, a plate element of 10 x 10 or 25 x 25 microns respectively. The field depth is 16 microns for a numerical aperture of 0.2. The system can be operated by remote control. Automatic or programmable focussing is also being considered to compensate for the variation in plate thickness.

II. MACHINE CONTROL AND DATA ACQUISITION AND PROCESSING

The configuration comprises two subsets :

- The host-computer, array processor, and standard peripherals (for data acquisition and processing).
- The microcomputer system (for management of the microdensitometer automatisms).

The host-computer is a S.E.L. 32/27, with 1 Mbyte memory, hardware floating point, input/ouput processor (IOP), two 80 Mbytes disk units, one 45 ips 800-1600 BPI tape unit, line printer, and consoles.

The AP- 120 B Array Processor (167 ns cycle), with 64 K 38 bits words memory is linked to the acquisition module by a GPIOP (General Purpose Input Output Processor) interface. Due to the parallel organization of the AP-120 B, floating point adds, floating point multiplies, control arithmetic operations, memory access, host/array processor and peripheral input/output data exchanges can all be overlapped in time.

The microcomputer system is essentially composed of :
. The acquisition module ;
. The 6809 master microprocessor which coordinates the different elements of the process (focussing, data acquisition...).
. two slave microprocessors in charge of the X and Y motions.

Data circulation

The output of the reticon array is fed into the analog to digital converter ; the data are then corrected for dark current and sensitivity differences from one photodiode to the other, for the inhomogeneities of illumination due to the optical system and for the fluctuations of the source of light.

After a transit through the GPIOP, the output of the acquisition module is compressed by the Array Processor which transmits it to the host computer.

III. OPERATING MODES

X as well as Y motions can be performed at velocities as high as 8 mm/sec in measurement mode, and 40 mm/sec in pointing mode. The acceleration will be limited to 0.5 m/sec^2, and the standard velocity taken equal to 3.6 mm/sec. With a 10 micron step, this corresponds for the whole set of 1024 photodiodes to a rate of 360,000 measurements per second. At this speed, a 35 x 35 cm plate is scanned in lanes of 10 mm in about one hour.

Each sample being coded with 10 bits, the raw data for the whole plate amount to about 10^{10} bits. Until storage devices of this class have become available and easy to operate, it is clear that one of the two following kinds of operating modes will have to be used :

- scanning of selected areas within the plate, either around successive positions, or along a curve.
- complete scanning of the plate (or of a significant part of the plate), with on-line data compression performed by the array processor.

We are currently investigating the algorithms best suited to these two kinds of data collection, as well as the processing techniques to be used downstream.

REFERENCE

T. Danguy, I.A.U. Working Group on Photographic Problems, meeting on "Astronomical Photography 1981", Nice, 1981, J.L. Heudier and E. Sim Eds.

IMAGE DETECTION SYSTEM FOR SCHMIDT PLATES

Hideo Maehara
Kiso Observatory
Tokyo Astron. Observatory

Tomohiko Yamagata
Department of Astronomy
University of Tokyo

1. INTRODUCTION

A 14-inch Schmidt plate contains 10^9 photographic grains and 10^5 to 10^6 images of stars and galaxies on it. Such a quantity of data is too large to be handled in a conventional way even for a big computer.

There is, in general, an alternative method to solve this problem; one is to store the data of all pixels on intermediate medium (e.g., magnetic tape), and reduce them into image parameters afterwards. The other method is to do all the processing simultaneously with the measurement. The latter is very useful for the automated detection of celestial images on large Schmidt plates.

In this short note, we describe our system operated at the Kiso Observatory. We have built a machine capable of the former method, and are extending and improving it toward a high-speed automatic processing of Schmidt plates. The previous stages of the system were reported in Maehara and Watanabe (1980) and Maehara (1981).

2. HARDWARE AND SOFTWARE

A measuring machine (isophotometer) at Kiso is equipped with a linear CCD (CCD121H) composed of 1728 elements with a 13-μm pitch. A plate is scanned perpendicular to the array of the detector with a 22-mm width. Transmission (T) values of all pixels are stored on a magnetic tape ranging from 0 to 1023.

The sampling speed attains up to 60 kHz, but it is usually limited to about 6 kHz due to the capability of the present mag. tape handler. The block diagram of the hardware is illustrated in Figure 1, where the measuring machine is to the left, the control computer is to the upper right, and the processing computer is to the lower right.

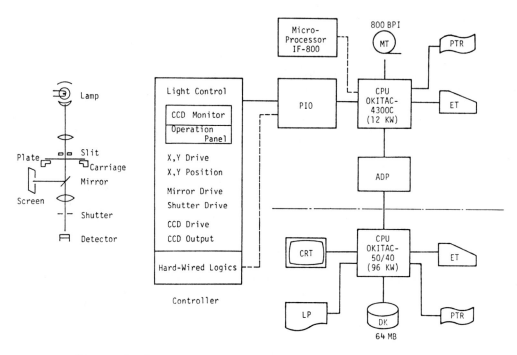

Figure 1. Block diagram of the hardware

The fundamental portion of the present line-up was built nearly ten years ago, and the successive extension of the capability has been followed. Recently, a module of hard-wired logic circuits has been attached to the system which corrects the non-uniformity of the sensitivity of detector elements, converts pixel data to intensity, and picks up image parameters along a scanning line. This module is being incorporated into the system by the software.

Two sets of processing programs are currently available to detect images from pixel data on a mag. tape. The former (Version 1) adopts a simple and straightforward algorithm suitable for the on-site minicomputers. It gives us several parameters including the central position, the fractional areas of some T ranges, and the magnitude parameter (Maehara 1981).

The other (Version 2) is a more sophisticated software for bigger computers which applys the moment analysis up to the second order of the pixel distribution (Stobie 1980). The discrimination of images among stars, galaxies, and overlapped objects is sufficiently achieved with the use of the moments and their combinations. Several kinds of image maps can be plotted in order to identify detected objects.

3. PERFORMANCE AND APPLICATION

The performance of the measuring machine was examined; the linearity and the reproducibility of pixel data are fairly good. Since the long-term stability of output values is affected by the variation of the source lamp, the calibration of the (absolute) sensitivity must be made every few hours. The spatial resolution of each detector element is not so high as its geometrical size. In this circumstance, the mechanism is being installed which is capable of putting narrow slits (0.1- to 0.6-mm width) in front of the plate.

The accuracy of image parameters is generally better for the software of Version 2. Its r.m.s. errors of the image position and the brightness are respectively about 0.2 arcsecond and 0.1 magnitude, though they depend on the plate quality. The reproducibility of these parameters are better than the above values.

This system has been used as an isophotometer to obtain isophotometric maps of nebulae and galaxies (e.g., Mizuno et al 1981, Watanabe et al 1982), though high-speed microdensitometers are frequently used for this purpose. In addition, it has been utilized to count stars and galaxies in selected areas of Schmidt plates.

Yamagata et al (1983) are working in the luminosity function of galaxies in poor clusters of galaxies. The determination of the cluster extension and the field correction is statistically carried out from the data in surrounding areas. The results thus obtained are discussed by comparison with rich clusters of galaxies. The scarcity of faint members characterizes these poor clusters.

A further improvement of this system will be made toward the fully automatic image detection for Schmidt plates.

The authors thank staff members of the Kiso Observatory for the assistance in setting up and examining the system.

REFERENCES

Maehara, H. 1981, American Astron. Soc. Photo-Bulletin No. 26.
Maehara, H., and Watanabe, M. 1980, in "Optical and Infrared Telescopes for 1990's", (ed. A.Hewitt), p677.
Mizuno, S., Sakka, K., Sasaki, T., and Kogure, T. 1981, Astrophys. Space Sci. 78, 235.
Stobie, R. S. 1980, in "Applications of Digital Image Processing to Astronomy", (ed. D.A.Elliott), p208.
Watanabe, M., Kodaira, K., and Okamura, S. 1982, Astrophys. J. Suppl. 50, 1.
Yamagata, T., Maehara, H., Okamura, S., and Takase, B. 1983, in preparation.

A NEW WIDE FIELD ELECTROGRAPHIC CAMERA AS AN OPTIMUM DETECTOR FOR SCHMIDT-TYPE TELESCOPES

P. J. Griboval and X. Z. Jia *
McDonald Observatory
The University of Texas
Austin, Texas U.S.A.

1. INTRODUCTION

The electrographic (E.G.) camera is one of the most powerful modern detectors having high sensitivity, extended spectral response, extreme spatial resolution, low background, no threshold, good linearity and large dynamic image. A combination of highly uniform electric and magnetic fields produces distortion-free and extremely sharp images which are recorded by electron sensitive emulsions [1] (Figure 1).

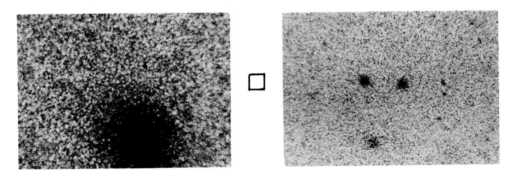

Figure 1. A microscope enlargement, with the same magnification, of a IIaO plate (left) and a medium-grained Eastman-Kodak electron sensitive film (right). At the same scale, a 15x15μm C.C.D. pixel element! (center)

It is essential, for optimum performance, to place the E.G. camera at the focus of a telescope whose optical resolution matches that of the camera. Since flux density varies in inverse ratio to the area of stellar images, a very high detection threshold can thus be achieved. Consequently, the most efficient use of an E.G. camera is at the focus of a Schmidt-type telescope. Unfortunately, there is, as yet, no E.G. camera having a field large enough to accomodate large telescopes of this type.

* Visiting scientist from Changchun Institute of Optics and Fine
  Mechanics, Changchun, China

2. THE McDONALD OBSERVATORY E.G. CAMERA

During the past 16 years the McDonald Observatory, with support from the National Science Foundation, has built and operated two magnetically-focussed, 5-cm E.G. cameras [2,3,4,5,6] . Years of intensive use, by astronomers from various countries and institutions, have proved the validity of the basic concept. Thousands of electrographs have demonstrated its capabilities for precise photometry [7] and astrometry [8]. Nevertheless, to build a large-field E.G. camera with the present method of loading the film at the electronic focus, through the back of the solenoid, would not be an elegant and efficient solution of this difficult problem for larger cameras. We had to develop a new approach to make the design, construction and operation of a wide-field E.G. camera simpler and more practical without loss of geometric or photometric accuracy.

Preliminary laboratory tests showed that a uniform magnetic field can be extended out of a solenoid by using a "correcting" coil placed on the same axis, at some distance from the main coil. An extensive computer analysis was made to determine the characteristics of such a coil pair, and to evaluate the field distribution in and out of axis, image distortion, rotation, curvature and modulation transfer function [9].
A coil pair built, according to these calculations for a 9-cm camera, gave a very uniform field (+/- 0.07%) inside the main coil and outside it, up to the correcting coil. Image distortion, rotation and curvature are negligible and the MTF is 25% at 200 lp/mm. A similar coil-pair for a 20-cm camera was also calculated and built with similar results.

3. A 9-cm E.G. CAMERA

The split solenoid solves elegantly the problem of building a large field E.G. camera (Figure 2). While the photocathode is within the main coil, the electronic focus can be placed between the two coils, permitting a direct, sideway access to it. Such a layout offers numerous advantages: (i) the film magazine design is much simplified as the film can travel straight from the supplying to the receiving spool. Since there is no need to roll the film on a roller of small radius, a thicker film base (7-mil Estar), which is easier to handle, may be used; (ii) the full area of the photocathode can be recorded; (iii) the design of the entrance and exit slit-like gate valves is much simplified, and they are much easier to seal; (iv) the film magazines can have large capacity; (v) a simple motor can drive the film loader under control by the observer at the telescope.

A 9-cm E.G. camera is under construction and should be ready for testing in the spring of 1984. It will serve as a pilot model for a larger, ~20-cm camera, of the same concept which - subject to adequate support - can be built for use with large-field telescopes. Such a camera would be the ideal detector for a diffraction-limited folded Schmidt-type space telescope.

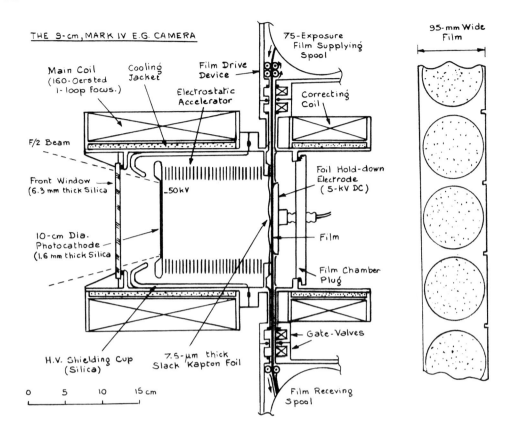

Figure 2. A sketch of the 9-cm, Mark IV, electrographic camera with a 75-exposure, remotely controlled, roll-film magazine. The film and photocathode chamber sputter ion-pump and H.V. feedthrough are at 90° with respect to the drawing plane.

REFERENCES

1. Griboval, P., Griboval, D., Marin, M., Martinez, J., Adv. Elect. Elect. Phys., 33A, pp. 67 (1972).
2. Griboval, P., Electrography and Astronomical Applications, Ed. G. L. Chincarini, P. Griboval, H. J. Smith, 55, Univ. Texas, Austin (1974).
3. Griboval, P., Adv. Elect. Elect. Phys., 40B, 613 (1976).
4. Griboval, P., Instrumentation in Astronomy III, S.P.I.E., 172, 348 (1979).
5. Griboval, P., Adv. Elect. Elect. Phys., 52, 305 (1979).
6. Griboval, P., Proc. I.A.U. Working Group on Photographic Problems (1981).
7. Opal, C., Bozyan, E., Griboval, P., Instrumentation in Astronomy IV, S.P.I.E., 331, 453 (1982).

8. Mulholland, D., Griboval, P., Publ. Univ. of Texas, Astronomy Dept. (1980).
9. Jia, X. Z., Griboval, P., Adv. Elect. Elect. Phys. (to be published).

DISCUSSION

D. CARTER: In the past, electrographic images have been ruined by the quality of the emulsion. What are your emulsions like now?

G. DE VAUCOULEURS: The Kodak emulsions are excellent. It would be catastrophic if their production were stopped for economic reasons. There is no alternative source. The astrometric quality and photometric uniformity of the films taken with the Griboval camera has been well documented by users (see references to above paper).

ASTRONOMICAL PERFORMANCES OF THE MEPSICRON, A NEW LARGE AREA
IMAGING PHOTON COUNTER

C. Firmani, L. Gutiérrez, E. Ruíz, L. Salas, G.F. Bisiacchi
Instituto de Astronomía, Universidad Nacional Autónoma de México
F. Paresce
Space Telescope Science Institute, Johns Hopkins University,
Baltimore, MD USA

The new detector MEPSICRON (microchannel electron position sensor with time resolution) is an image photomultiplier sensor for high spatial and time resolution, working in a photon counting regime. It has been especially designed for deep sky photometric pictures, for high resolution spectrophotometry with single or crossed dispersion spectrographs for long slit spectroscopic techniques, for high time resolution pictures and spectrophotometry especially related with speckles techniques and very fast varying sources as pulsars, and for Fabry-Pérot interferometry.

The first prototype of the MEPSICRON for astronomical use has been manufactured with a multialkaly photocathode of 25 mm diameter sensitive area (the design has been published in previous papers, Firmani et al. 1982 and 1983). The detector can be described as having three main parts: a proximity focused photocathode deposited on a quartz faceplate, an assembly of five high current 40:1 microchannel plates mounted in two stages with a V and Z configuration respectively, and a distortion free resistive anode.

A photoelectron produced by the incidence of a photon on the photocathode is accelerated toward the microchannel plates assembly where one electronic cascade with a gain of 10^8 electrons/count and very low statistical fluctuations is triggered. The electron cloud produced by the cascade is received by the distorsion free resistive anode that provides the partition of the total charge in four output pulses; the heights of these pulses are correlated with the position of the cloud centroid, i.e., the incidence point of the photon on the photocathode. The image processing system that recovers from the four output pulses of the detector the values of the x, y and t coordinates for each count, is based mainly on a pulse position analizer. The spatial resolution of 40μm and 52μm FWHM in the red and in the blue respectively, is uniform on the entire sensitive area. One array of 1024 by 1024 pixels provides a reasonable sampling of the image compatible with this resolution. The uncertainty of the photon arrival time is .2μs, however, the time coordinate is recorded with a precision of .1 ms which is sufficient for the majority of the purposes.

The average contribution of the dark current is 50 counts/sec on the entire sensitive area, or 1 count/pixel every 5 hours, with the detector cooled at -30°C. The low level of the noise of the dark count rate makes this detector very close to one which is photocathode quantum noise limited.

The maximum count rate for diffuse sources is 5×10^5 counts/sec; this limit is introduced by the electronic image processing system. The count rate for point images is limited by the recovery time of the microchannel plates; the statistical properties of the electronic cascade and consequently the spatial resolution begins to degradate above 50 counts/sec pixel. The geometrical stability of the pixels array has been carefully considered at the level of the electronic design and 1 pixel on several days represents an upper limit of the geometrical shift on the entire frame. Preliminary tests of the pixel to pixel sensitivity variations give results better than 5%; more accurate tests will be performed in the future. The dynamic range of the MEPSICRON is extremely high due to the high count rate for point images and the low contribution of the dark current; their ratio gives a rough estimate of the dynamic range greater than 10^6. During one exposure the image is integrated on a 2 M bytes digital memory with 16 bits/pixel introducing, in practice, the main limitation on the dynamic range. A real time imaging on a color screen is provided during the exposure, as well as a photon coordinate record on magnetic tape. A Nova 1200 computer interacts with the 2 M bytes memory and controls the input/output on the magnetic tape and disk units.

For astronomical use the MEPSICRON has been mostly coupled with the REOSC echelle spectrograph at the 2-meter telescope of the Mexico National Observatory at Baja California. The dispersion obtained is .14 A/pixel (blue) and .25 A/pixel (red) and the resolution .3 A FWHM (blue) and .4 A FWHM (red), compatible with a slit wideness less than 2 arc seconds. An exposure time of one hour is required to obtain a spectrum in the blue, with a signal to noise ratio $\simeq 10$ for a 16 magnitude star.

The spectrum of the galaxy MK35 obtained in the blue region with a 900 groves/mm echelette grating is shown in Figure 1; the exposure time was 10 minutes and the spectrograph entrance slit width 10 arc sec. The first strong emission line that appears from the bottom is Hγ, while the first emission feature from the top is the [O II]$\lambda\lambda$3726-9 doublet. A full resolution image of this doublet, in the same spectrum, is shown in Figure 2, where each pixel is visible as a small rectangle. Although the lines width is mainly due to the intrinsic velocity field of the galaxy, the two lines with 2.7 A of separation are completely resolved.

Fig. 1. The blue spectrum of the galaxy MK35

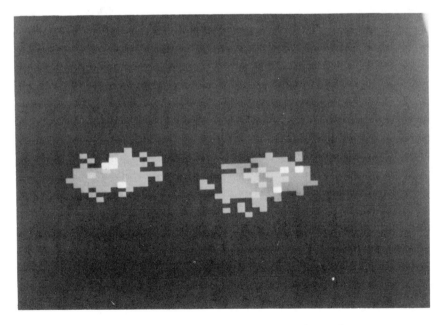

Fig. 2. Full resolution image of the $|O\ II|\lambda\lambda 3726-9$ doublet visible in Figure 1.

The spectrum of the 14.5 photographic magnitude, M4.5V star, Ross 368, is shown in Figure 3; the exposure time was 30 minutes. The spectrum has been obtained with a 200 groves/mm echelette grating working at the first order in the red (top) and at the second order in the blue (bottom).

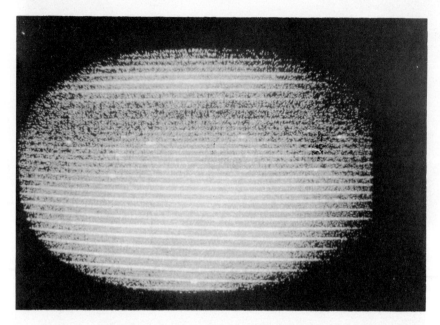

Fig. 3. The red (top) and blue (bottom) spectrum of the M4.5V star, Ross 368.

On the first order from the bottom, the [O I]λ5577 sky emission line is visible, while in the 17th and 18th order from the bottom, the K-CaII (left) and the H-CaII (right) emission lines are visible; this last line is near to and separated from Hϵ. The Balmer serie, in emission appears in the 7th (Hβ), 12th (Hγ), 15th (Hδ) order from the bottom. The strong emission feature in the red is Hα.

A large number of absorption lines have been identified (L. Carrasco and A. Serrano) showing the very complicated atmosphere structure of this type of stars.

REFERENCES

Firmani, C., Ruíz, E., Carlson, C., Lampton, M., and Paresce, F.: 1982, Rev. Sci. Instr. 53, 570.
Firmani, C., Gutiérrez, L., Ruíz, E., Bisiacchi, G.F., Salas, L., Paresce, F., Carlson, C., and Lampton, M.: 1983, Astron. Astrophys. in press.
Serrano, A. and Carrasco, L.: 1983, private communication.

THE APPLICATION OF OPTICAL FIBRE TECHNOLOGY TO SCHMIDT TELESCOPES

J.A. Dawe and F.G. Watson
U.K. Schmidt Telescope Unit, Royal Observatory Edinburgh

ABSTRACT: The potential of the Schmidt optical system, when combined with optical fibres and linear detectors, is assessed. Recent work on the use of optical fibres at the U.K. Schmidt Telescope is described together with anticipated developments. In conclusion, there is a speculative consideration of the construction of a large, alt-azimuth Schmidt telescope (LAST).

INTRODUCTION

The combination of the Schmidt telescope and the photographic plate is one of the most potent in the astronomer's armoury. But, because of its relative cheapness, the photographic plate is used in a rather indiscriminate fashion, only ~1% of the capacity of the emulsion actually containing data on astronomical images. Optical fibre technology permits us to be much more discriminating about the images we choose to record. It is now possible to transfer these selected images from the focal surface to the outside of the telescope for analysis. Such systems have already been tested on a number of conventional reflectors (e.g. Hill et al., 1980; Gray, 1983) but none, as yet, has been applied to the singularly appropriate case of a wide-angled Schmidt telescope.

CURRENT DEVELOPMENTS AT UKSTU

At the U.K. Schmidt Telescope, we are currently constructing a multi-object fibre-optics coupler (Watson and Dawe, 1983). This will transfer the light of several hundred images from the telescope to fast spectrographs on the dome floor. The fibres are supported by a thin 14x14 inch aperture plate which, in turn, is held in much the same manner as the photographic plate it replaces. The spherical mandrel behind the plate is perforated to permit the passage of fibres, and the plate itself may be rotated through several arcminutes. All-silica fibres are to be used, having a numerical aperture of 0.25

(f/2) and an attenuation of ~0.025 dB/metre at 500nm. With a core diameter of 50μm, they subtend 3.4 arcseconds on the sky. The aperture plates will be either metal, pre-drilled to accept fibres terminated in small ferrules (Gray, loc.cit.), or glass positive copy plates of the target field with the fibres cemented directly to them (Watson and Dawe, loc.cit.). The positional accuracy required will be ±25μm, and will be limited by (a) the precision of the fibre location in the curved focal surface, (b) differential refraction over the 6°.6 field (Watson, 1983), and (c) temperature effects.

Initial experiments, looking at a few tens of images, are to be performed using the ROE CCD Imaging Spectropolarimeter (ISP) (Mclean et al., 1980), with the polarization elements removed. Eventually, it is hoped to construct one or more optimized Schmidt-type spectrographs, when the efficiency of the whole optical train is likely to be ~14%. It should thus be possible to observe galaxies of surface brightness 18 magnitudes/square arcsecond, with a velocity resolution of 60km/s and a signal/noise of 50, in less than 20 minutes.

COMPARISON OF SCHMIDT VERSUS CONVENTIONAL REFLECTOR

Consider a telescope of aperture, A, having a field of view, θ, and equipped with n fibres. If the population of objects to be sampled has a number density on the sky, σ, greater than n/θ^2, all the fibres may be used simultaneously. A measure of the "effective aperture", a, will then be $\sqrt{n}$ A. If the number density of objects is so low that only one can be sampled at a time (i.e. $\sigma < 1/\theta^2$), the effective and physical apertures will coincide. For the intermediate range of number densities ($1/\theta^2 < \sigma < n/\theta^2$), $a = \theta\sqrt{\sigma}$ A, and we may write:

$$\log a = \log \theta A + 1/2 \log \sigma \qquad (1)$$

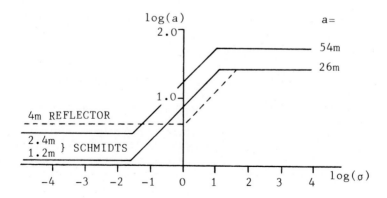

Figure 1. Effective aperture versus no. density.

These three cases are illustrated in Fig. 1 where we compare a conventional reflector ($A=4m$; $\theta=1°$; $n=50$) with two Schmidt telescopes ($\theta=6°.6$; $n=500$) of apertures $A=1.2m$ and $2.4m$, respectively. It is readily apparent that even the smaller Schmidt telescope is competitive with the 4m reflector. It may be argued that the values of n assigned to the telescopes are arbitrary (though we believe them to be realistic); even so, there will always be a range of σ over which the 4m reflector cannot compete, no matter how many fibres it has available.

The appropriateness of a fibre-equipped Schmidt for a given astronomical programme thus depends directly upon the number density of objects to be sampled. Such disparate entities as galaxies and Cepheids, quasars and M-type stars can all be found at sufficiently high densities to make such an approach attractive. This statement must be qualified by noting that, for objects fainter than the sky contribution (equivalent to 19.7 magnitudes in V over the fibre cross-section to be used at UKST), it becomes difficult to compensate for the use of a small aperture by increasing the integration time. We are thus led to consider further the larger Schmidt telescope, where the sky contribution amounts to 21.3 magnitudes over a 50μm fibre.

A LARGE ALT-AZIMUTH SCHMIDT TELESCOPE

In another paper (Dawe and Watson, 1983), we propose a large alt-azimuth Schmidt telescope (LAST) of 2.4m aperture, working with the well-proven Palomar/UKST combination of $6°.6$ field and f/2.5 speed. Novel features include:
(a) an alt-azimuth mount with co-rotating building. The focal-surface instrumentation is mounted on a ring girder, which rotates under laser control to compensate for field-rotation;
(b) a thin spherical mirror mounted on an active support similar to that used by the U.K. Infrared Telescope (Humphries, 1978);
(c) the separate mounting of various correctors and prisms on quadrant girders co-rotating with the main telescope (again under laser guidance);
(d) the use of a false focal cap, placed about 1/3 metre behind the true focus, onto which pallets of standard sizes can be mounted. Three types of detector would be accommodated in these pallets; viz., photographic plates, limited arrays of CCD's and large arrays of fibre optics.

The cost of the LAST would be about one-half that of a conventional 4m reflector. Unlike multi-mirror telescopes or arrays of small telescopes, it would truly exploit the statistical nature of observational astronomy, using a single optical train to look at many objects simultaneously.

REFERENCES

Dawe, J.A. and Watson, F.G., 1983, submitted to Optica Acta.
Gray, P.M., 1983, AAO preprint No. 182.
Hill, J.M., Angel, J.R.P., Scott, J.S., Lindley, D. and Hintzen, P., 1980, Astrophys.J., 242, L69.
Humphries, C.M., 1978, Sky and Telescope, 59, p.469.
Mclean, I.S., Cormack, W.A., Herd, J.T. and Aspin, C., 1980, Soc.Phot.Inst.Engng., 290, p.155.
Watson, F.G., 1983, Mon.Not.R.astr.Soc., in press.
Watson, F.G. and Dawe, J.A., 1983, submitted to Optica Acta.

DISCUSSION

T. SHANKS: What is the timescale for the completion of a practical fibre optics for the present UK Schmidt?

J. DAWE: We hope to undertake the initial positioning tests, photographically, towards the end of the year. Spectroscopic tests should take place in the first half of 1984, using the ISP, though these will be limited to only a small number of fibres.

IMAGES FROM LARGE SCHMIDT TELESCOPES

D.S. Brown
Grubb Parsons Ltd., Walkergate, Newcastle upon Tyne, U.K.
C.N. Dunlop and J.V. Major
Department of Physics, The University, Durham, U.K.

ABSTRACT: Calculations have been made of the intensity distribution of star images formed at the focal surface of large Schmidt telescopes. The calculations take account of atmospheric seeing, aperture diffraction, manufacturing errors, chromatic errors in the corrector plate, scattering in the emulsion, photographic response and grain structure. The results are expressed in terms of simulated microdensitometer scans across the processed emulsion.

Results are given for a system S1(1.24m, f/2.5) similar to the 48" U.K. Schmidt (with both singlet and achromat corrector plates) for several combinations of correctors and filters using IIIa-J emulsion for recording. The improvement in performance expected from the use of larger Schmidt telescopes is discussed and results are given for two designs with 2.5 metres aperture, S2(2.5m, f/2.5) and S3(2.5m, f/3).

1. INTRODUCTION

In a typical, large Schmidt starlight from an aperture > 1 metre is focussed at about f/2.5 - f/3.5 on a high contrast emulsion such as Kodak IIIa-J. The exposure is of such a duration that, after processing, the photographic density of the sky background is 1. The processed emulsion is scanned by an automatic microdensitometer with a pixel size of the scanning beam which is related to the angular structure of the images and the granularity of the emulsion. The detection of the image of the star depends upon its photographic density (the signal) relative to the fluctuation in the density of the background (the noise). Although it is expected that larger Schmidt telescopes will lead to an improvement in detection sensitivity through enhanced signal/noise ratios, the degree of improvement will depend upon the relative importance of the factors which contribute to the size and structure of stellar images. Several of these factors are examined here and the improvement in sensitivity of larger telescopes is determined in terms of the limiting apparent magnitude that will be detected.

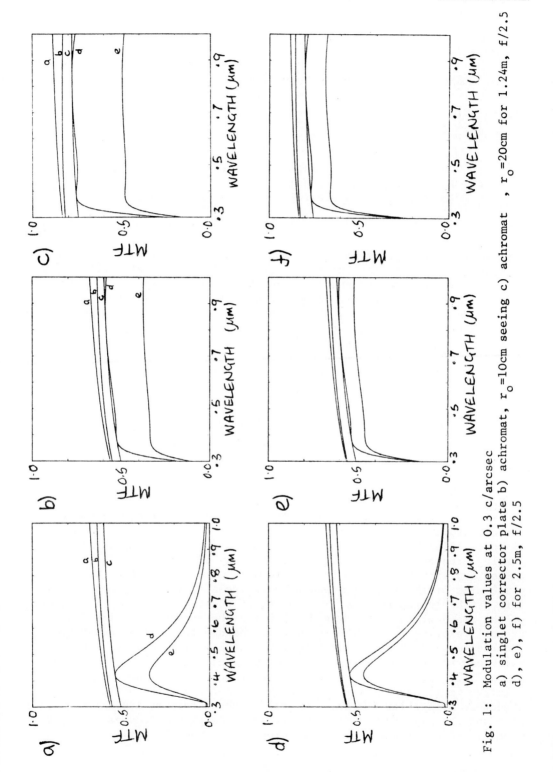

Fig. 1: Modulation values at 0.3 c/arcsec a) singlet corrector plate b) achromat, r_o=10cm seeing c) achromat, r_o=20cm for 1.24m, f/2.5 d), e), f) for 2.5m, f/2.5

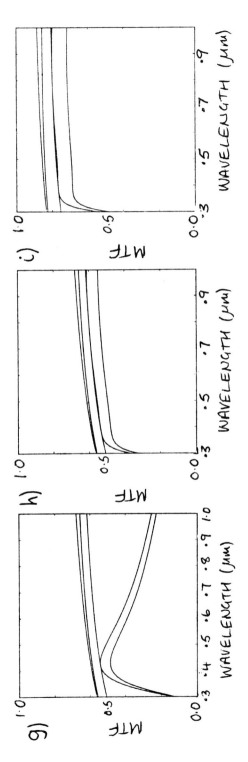

Fig. 1: Modulation values at 0.3 c/arcsec g), h) + i) for 2.5, f/3

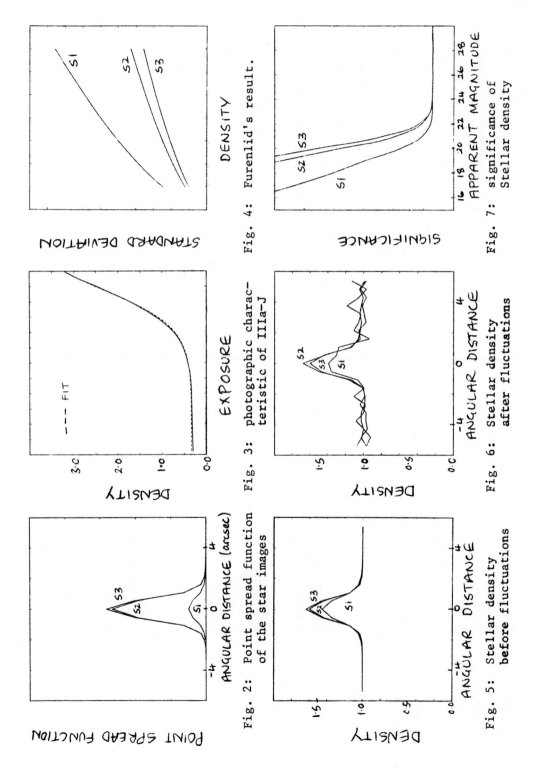

Fig. 2: Point spread function of the star images
Fig. 3: photographic characteristic of IIIa-J
Fig. 4: Furenlid's result.
Fig. 5: Stellar density before fluctuations
Fig. 6: Stellar density after fluctuations
Fig. 7: significance of Stellar density

2. BASIS AND RESULTS OF CALCULATIONS

The steps leading to the formation of the image are that the wavefront reaching the telescope is:
a) altered in intensity and phase by variations in the refractive index of the atmosphere (the "seeing", represented by r_o=10 cm for average seeing and r_o=20 cm for good seeing (ref. 1)),
b) diffracted at the aperture of the mirror,
c) changed in phase by the errors in the surfaces. This is akin to very good seeing with r_o=30 cm,
d) changed in phase by path length errors in the corrector plate. Two plates have been simulated, a singlet corrected at λ=0.42 µm and an achromat corrected at λ=0.38 and 1.0 µm,
e) scattered in the recording emulsion which is represented by the Kodak MTF data (ref.2).

A point source provides a wavefront characterised by unit modulations at all spatial frequencies, and each of the above processes is likely to reduce this modulation. The inverse Fourier transform of the modulated distribution leads to an image which necessarily has angular (or linear) breadth. The spatial frequency of each of the response processes has been determined in the usual way and the product of these forms the overall spatial frequency response or modulation transfer function, MTF, from which the image structure or point spread function has been found, PSF.

Representative results of the 5 processes are shown cumulatively in fig. 1 (a-c) for the S1 Schmidt. Modulation values at a single spatial frequency (0.3 c/arc sec) are shown as a function of wavelength. In all the figures the seeing reduces the MTF considerably although this improves slightly at longer wavelengths. Aperture diffraction and surface irregularities are of minor importance whereas the design of the corrector plates is very important. For the singlet (fig. 1a) the MTF is severely reduced away from λ=0.42 µm. However the achromat (fig. 1b) is very well corrected and contributes negligibly to the reduction of MTF except in the UV. Finally the effect of scattering in IIIa-J emulsion is seen to be quite important. Indeed in very good seeing (fig. 1c), the emulsion becomes the limiting feature.

The calculations have been repeated for 2 larger Schmidts S2 (in fig. 1d-f) and S3 (in fig. 1g-i). The importance of the focal length is seen here in reducing the residual sperical aberration (c.f. figs. 1a, d, g) and in reducing the effect of the emulsion scatter. So although telescope S2 will collect 4 times as much light as S1 and distribute it over a focal area 4 times larger for sky background, the improvement in MTF with longer focal length will lead to relatively narrower (and hence brighter) star images. This is apparent in fig. 2 where the point spread functions (PSF) of the star images are shown; these have been calculated in average seeing at λ=0.5 µm using the inverse Fourier transform of the MTF.

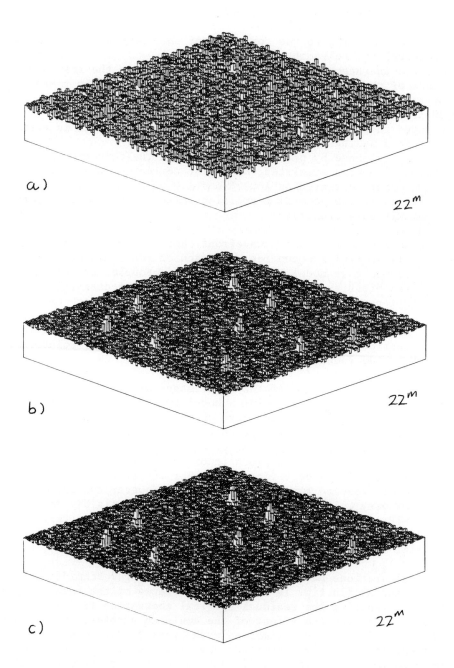

Fig. 8: a) Simulated densitometer trace of star field of Schmidt telescope, S1 (1.24m, f/2.5); b) S2 (2.5m, f/2.5); c) S3 (2.5m, f/3).

However the star image has to be detected by its photographic density relative to that of the sky background. The photographic characteristic of IIIa-J (ref. 2) has been fitted with a physical model (ref. 3) based on fixed grain size (~ 0.36 μm^2) and fixed quantum sensitivity of 2 photons/grain and this is shown in fig. 3. This gives a simple parameterization which is used to determine the exposure to give unit density for the sky background (22.5^m/arc sec^2) and this exposure is used to scale the exposures for stars of a variety of apparent magnitudes and hence to determine their photodensities which are then added to the background sky density.

The simulations are made in 2 dimensions in the photoplate. For S1 the densities were evaluated in pixels of 8 x 8 μm^2 where the point spread function of a stellar image covered a 21 x 21 matrix of such pixels (a total angular size $\sim$ 10 x 10 arc sec^2). A single pixel of the sky background covers about 400 developed grains and hence there will be random fluctuations of these from pixel to pixel giving the noise of the system. To simulate this the results of Furenlid (ref.4) have been scaled to pixels of the sizes used here. For the longer focal length telescopes it is the linear size, rather than the angular size, of the image which is increased. To give the same angular resolution the pixel size was increased from 8 μm to 16 μm for S2 and to 19.2 μm for S3. As these larger pixels now cover more grains the relative fluctuations will be reduced. This is seen in fig. 4 where Furenlid's results are shown for the 3 pixel sizes.

The 2-dimensional distributions of photodensity are then randomised by the standard deviation in density fluctuations corresponding to the density and pixel size. One dimensional simulated measurements of density through a star of 22^m are shown before fluctuations are added in fig. 5 and after fluctuations in fig. 6 for the 3 designs of telescope.

The limiting magnitude of stars that are just seen has been determined by quantifying a "significance" for the stellar density. For the field of a stellar image of 21 x 21 pixels the density is made up from starlight (S) and sky background (BG) and randomised according to the grain noise. A further 50 randomised sky background fields have been generated and a significance calculated where

$$\text{Sig} = \sum_{i=1}^{50} \left[(S+BG) - BG_i \right]^2 / 50$$

The significance is shown in fig. 7 for increasing stellar magnitude. When this is such that the star is "lost" in the noise, Sig approaches $\sqrt{2}$. This occurs at about 22^m for S1; at about 23^m for S2 and at about 23.5 for S3.

To demonstrate this improvement in sensitivity, 9 stars of 22^m have been generated in a stellar field of 121 x 121 pixels for the 3 telescopes. The fields are shown in fig. 8a, b and c.

3. SUMMARY AND CONCLUSIONS

Simulations have been made of the photographic densities of stellar images in 3 Schmidt telescopes (1.24 m, f/2.5; 2.5 m, f/2; 2.5 m, f/3.) The effects on performance of the main factors contributing to image size have been examined. The telescopes are usually limited by seeing but when this is good the 1.24 m, f/2.5 telescope would then be limited by the IIIa-J emulsion. This limitation can be removed by doubling the focal length with a corresponding increase in aperture diameter to compensate for loss in light density. When the focal length is increased further (with aperture constant) residual sperical aberration is reduced and the increase in image scale further improves signal/noise but at the expense of increased exposure time for sky limited plates. This is summarised in table 1 where the 3 telescopes were used to expose a background sky density of 1 in r_o=10 cm seeing at λ=0.5µm.

Table 1

	1.24m, f/2.5	2.5m, f/2.5	2.5m, f/3
Exposure	t	t	1.44 t
Signal/noise	S/n	2.93 S/n	4.34 S/n
Limiting sensitivity	22^m	23^m	23.5^m

4. REFERENCES

1. D.L. Fried, 1966, J. Opt. Soc. Am. 56, pp (1372-1379)
   J.C. Dainty and R.J. Scaddan, 1975, Mon. Not. R. astron. Soc. 170, pp(519-532)
2. Kodak publication p-315, 1973
3. J.C. Dainty and R. Shaw, 1974, Image Science (Academic Press)
4. I Furenlid, 1978, Modern Techniques in Astronomical Photography (ed Hendier and West pp(153-164).

THE DETERMINATION OF THE VIGNETTING FUNCTION OF A SCHMIDT TELESCOPE

J.A. Dawe
UK Schmidt Telescope Unit, Royal Observatory, Edinburgh.

ABSTRACT

Observational and geometric approaches to the problem of determining the vignetting function are discussed. A simple technique for producing a "flat-field" is described, and the various uses of such a flat-field to investigate the photometric accuracy and sensitivity of different emulsions are discussed.

INTRODUCTION

A necessary compromise in the construction of many wide-angle Schmidt telescopes is the presence of weak vignetting in the outermost portions of the field. However, in order for the full area of wide-angle plates to be of value for such problems as number counts of galaxies, photographic photometry to the plate edge, etc., it is essential that the effect of vignetting is known quantitatively.

There are two simple approaches to this problem. The first is theoretical, and involves ray-tracing through the appropriate geometry of the telescope. The second is empirical, and necessitates the imaging of a "flat-field", extended source onto the focal plane. A third method, which requires the extraction of the stellar background from real sky plates (Campbell, 1982; Dawe, 1983), is fraught with difficulties, but may be used as a check on the other two.

GEOMETRICAL APPROACH

Any parallel, on-axis beam of light entering the telescope corrector will encounter obstructions within the tube, e.g. the plate-holder support and spider, sensitometers, cables, etc. These restrict the amount of light intercepted by the mirror and place an upper limit on the effective aperture of the telescope.

An off-axis beam will encounter not only these obstructions but, if it exceeds a critical angle θ_c, will not be intercepted by the mirror over its entire cross-section. From a consideration of the geometry in Figure 1:

$$\theta_c = \arctan[(R-r)/2f] \quad (1)$$

(For the UKST, $\theta_c = 2°7$.) Thus, geometric optics predicts that the onset of vignetting is discontinous and that, for angles $\theta < \theta_c$, there should be no vignetting.

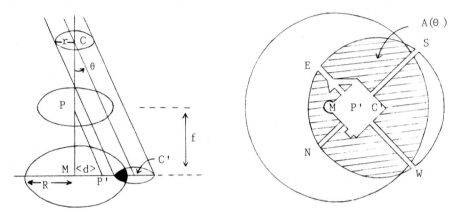

Figure 1. Vignetting geometry.   Figure 2. Mirror area intercepted.

A simple graphical method may be used to compute geometrical vignetting, making due allowance for internal obstructions and, in particular, asymmetric ones. On three separate pieces of tracing paper draw cross-sections of (1) the corrector, (2) the spider together with all associated equipment, and (3) the mirror, including the central hole if this exists. The vignetting, for a given angle θ, may now be derived by spacing the centres of the corrector and mirror sheets at equal and opposite distances, d, from the plateholder centre, P, where:

$$d = f \tan(\theta) \quad (2)$$

The line passing through the three centres, M, P' and C', will define the position angle over which the vignetting function is being sought. Figure 2 shows the result for $\theta = 3°8$ in the UKST, taken along a position angle of 45°. By definition, the equivalent magnitude change brought about by vignetting, is:

$$\Delta m(\theta) = -2.5 \log[A(\theta)/A(0°)] \quad (3)$$

where $A(\theta)$ = mirror area intercepted by a beam at angle θ. This function is illustrated in Figure 3.

EMPIRICAL APPROACH

A "flat-field" source may be provided for a Schmidt optical system by placing a back-illuminated, diffuse screen in contact with the corrector. In its simplest form, this can be a linen sheet stretched tightly across the entrance aperture of the telescope. The sheet should be indirectly illuminated by another diffuse screen, which in turn is symmetrically illuminated by a low-wattage lamp. The warp and weft of the linen define a large number of pin-holes at the centre of curvature of the mirror. These act as Lambertian scatterers of the incident radiation, providing a near-uniform illumination of the focal plane.

To determine the intrinsic vignetting function, a short exposure plate is taken of the flat-field source, together with appropriate sensitometry to determine the characteristic curve of the emulsion. Measurements of the density, at a grid of points on the plate, are made with a densitometer and converted into measurements of relative intensity, as a function of position in the field.

In the case of the UKST, departures from circular symmetry lie below the available accuracy of density measurement (i.e. <0.01 equivalent magnitudes). This is consistent with the predictions of the geometrical approach. The empirical mean vignetting curve, averaged over a number of concentric zones corresponding to various values of θ, is also shown in Figure 3.

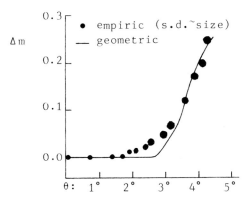

Figure 3. Vignetting function for UKST.

DISCUSSION

It is apparent, from Figure 3, that:
(a) the vignetting appears to commence at angles $<\theta_c (=2°.7)$,
(b) although the two vignetting functions agree at the plate centre

and edge, there is a zone, defined by $1°7<\theta<3°5$, where the geometric and empirical curves deviate from one another by up to 0.03 equivalent magnitudes.

Experiments with KODAK plates show that something like 25% of the light incident on a plate is reflected or scattered back. A simple analysis of this scattering problem is not possible here, but considerations of the fraction of this light, returned to the focal plane by the corrector and screen, lead us to suppose that it could account for a difference of up to $0.^m07$ between the two curves. That the observed difference is much less than this, and only occurs over a restricted range of θ, may arise from the complex scattering functions of the plate and corrector.

Unfortunately, because the presence of the diffuse screen is required for the "flat-field" approach and this also backscatters light, the resultant effective vignetting curve differs from that experienced by real-sky plates. However, the mean of the geometric and empirical curves is probably a fair approximation to the true effective vignetting curve. (For the UKST, it is known to be consistent with the vignetting, at the ($\theta=2°$) zone, derived from real-sky plates (Dawe et al., (1983)).

Once the vignetting function is known, it is then possible to investigate the intrinsic uniformity of photographic plates. This is particularly important for hypersensitized emulsions, which are known to lose sensitivity on exposure to moist air. Such studies have been carried out by Dawe and Metcalfe (1982), and Dawe et al. (loc.cit.), and have shown that point-to-point variations in the sensitivity of the IIIa-J emulsion lie below the range ±0.01 equivalent magnitudes, provided that it is kept in dry nitrogen. However, exposure to ambient air (even at a humidity as low as 45%) can increase this variation to ±0.04 equivalent magnitudes, as well as leading to larger-scale variations over substantial regions of the plate (typically 0.04 equivalent magnitudes over 2° on an UKST plate).

REFERENCES

Campbell, A.: 1982, Observatory, 102, p.195.
Dawe, J.A. and Metcalfe, N.: 1982, Proc.A.S.A., 4, p.466.
Dawe, J.A., Coyte, E. and Metcalfe, N.: 1983, submitted to A.A.S. Photo-Bulletin.

THE HIPPARCOS IMAGE

J-Y. Le Gall and M. Saisse
Laboratoire d'Astronomie Spatiale
Traverse du Siphon, Les Trois Lucs
13012, Marseille, France

Abstract :
One presents hereafter the HIPPARCOS satellite payload which is mainly constituted by a Schmidt telescope; a possible way to approximate the Schmidt mirror elliptic deformation profile is explained. Then, the signal expected from the optical chain is briefly described and one displays a residual chromatic effect which may introduce errors in the measure. To conclude, numerical values of this effect are given and one shows the necessity to take it into account in the data reduction process.

1. INTRODUCTION

The space Astrometry mission HIPPARCOS of the European Space Agency aims to build a catalog of the astrometric parameters (parallaxes, proper motion and positions) of 100,000 stars [1]. The Laboratoire d'Astronomie Spatiale is involved in this project as the responsible of optics analysis and develops, in close cooperation with Italian and Dutch Institutes, a mathematical model of the payload for the Data Reduction Consortium FAST [2]. One presents hereafter the results obtained during these last months and one shows the parameters which shall be studied very carefully in the future.

2. THE PAYLOAD

2. 1. Description

The sky is observed by a special telescope with two ways allowing to observe two fields of 0.9 degree by 0.9 degree separated by an angle of about 58 degrees [3]. An observed star is imaged on a modulation grid located on a field lens. Its image is transported by a relay optics and a detector determines the intensity then observed.

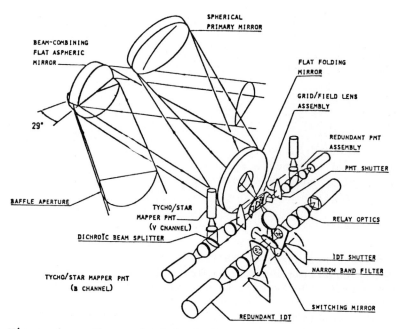

Figure 1. The payload including the special telescope with two fields separated by 58 degrees.

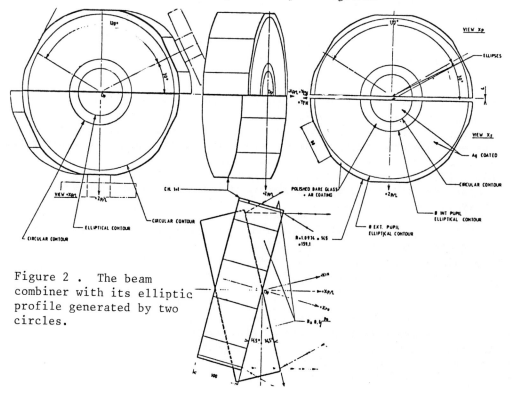

Figure 2. The beam combiner with its elliptic profile generated by two circles.

By knowing the separation angle of the two fields of view and the location of two stars on the grid, one can deduce the separation angle between the two stars. The final objective of the HIPPARCOS mission is to determine this angle with an accuracy of 2 milliarcseconds (after processing of the raw data expected from the satellite).

2. 2. The SCHMIDT telescope

The telescope which is an "All-Reflective" Schmidt system is formed by the Schmidt mirror which also insures the role of "beam combiner", so it works with a relatively strong incidence (14.5 degrees), a flat mirror devoted to fold back the system in order to adapt it to the satellite size constraints and the spheric mirror (figure 1). The Schmidt mirror working with a high incidence consequently it ought to have an elliptic deformation profile [4]. Such a profile is in fact very difficult to realize, regarding the optics dimensions and taking into account the manufacturing tolerances ($\lambda/60$ r.m.s on the wavefront, mirror diameter = 290 mm, telescope focal length = 1400 mm).
Incidentally, after the polishing, the system will have to be cut out in two equal parts (one by half pupil). In fact, the optical study showed that the utilization of the mirror having a revolution profile was acceptable, the elliptic form of the deformation profile being obtained in slightly decentring each half-pupil [5]. (figure 2)

2. 3. Grid, relay optics and detector

The grid located on the telescope focal surface is periodic, its period being equal to 1. 2 arcsecond on the sky which corresponds to a physical stepsize of 8 micrometers. So, the light falling down on the grid is modulated and, regarding the grid stepsize, diffracted. Then, the relay optics conjugates the grid surface with that of the detector ; The detector is an image dissector tube which counts photons. Its instantaneous field of view is about of 30 arcseconds of diameter and will be steered by the board computer. This kind of detector reduces the background noise and limits the confusion which would result from the surimposition of the light coming from different stars.

3. THE SIGNAL

In this paragraph we will assume the detector is perfect, the relay optics is without aberrations and collects the whole light diffracted by the grid and the system is monodimensional. If D is the signal due to the detector, T the distribution of intensity on the grid for a given wave-length and G the grid transmittance, it can be shown that D is the convolution of T by G : $D = T \otimes G$ [6]

3. 1. Analytical expression

We can then deduce FT (D) = FT (T). FT (G) where FT denotes the Fourier

Transform. As the grid has a periodic profile constituted with battlements, its Fourier Transform has only discrete terms corresponding to the frequencies $n \cdot \nu$ (ν is the grid frequency, $\nu = 125$ mm^{-1}). Incidentally, FT (T) is the telescope optical transfer function. Taking into account the pupil dimensions, the cut-frequency is between 2ν and 3ν over the interesting range of wavelength. Let a_0, a_1, a_2 be the values of FT (G) for the frequencies 0, ν and 2ν. Let $M_1 \cdot \exp(j\,v_1)$ and $M_2 \cdot \exp(j\,v_2)$ be the values of the normalized optical transfer function to the frequencies ν and 2ν. So the Fourier transform of D has only discrete terms, the values of which are a_0 for the continuum, $a_1 \cdot M_1 \cdot \exp(j\,v_1)$ for the frequency ν and $a_2 \cdot M_2 \cdot \exp(j\,v_2)$ for the frequency 2ν. So, the signal D can be written :

$$D = a_0 + a_1 \cdot M_1 \cdot \cos(2\pi\nu x + v_1) + a_2 \cdot M_2 \cdot \cos(4\pi\nu x + v_2)$$

introducing $x_1 = x + \dfrac{v_1}{2\pi\nu}$ and $x_2 = x + \dfrac{v_2}{4\pi\nu}$, it comes :

$$D = a_0 + a_1 \cdot M_1 \cdot \cos(2\pi\nu\, x_1) + a_2 \cdot M_2 \cdot \cos(4\pi\nu\, x_2)$$

This function depends :
* on x, that is to say on the angular coordinates of the considered star.
* on the starlight spectral distribution

3. 2. The CHROMATICITY

Let us consider now a geometrical point of the object field and take a blue star. The abscissa over the grid will be x_B and the phase corresponding to the first harmonic will be v_{1B}.
In these conditions :

$$x_{1B} = x_B + \frac{v_{1B}}{2\pi\nu}$$

Let us place now a red star in the same geometrical point. With the same conventions as before :

$$x_{1R} = x_R + \frac{v_{1R}}{2\pi\nu}$$

Now, let us calculate $x_{1B} - x_{1R}$:

$$x_{1B} - x_{1R} = x_B - x_R + \frac{1}{2\pi\nu}(v_{1B} - v_{1R})$$

The difference $C = x_B - x_R$ corresponds to the usually defined lateral chromatism ; incidentally, and the computation will verify it, v_{1B} and v_{1R} have no reason to be equal.

Let us put now : $X = \dfrac{1}{2\pi\nu}(v_{1B} - v_{1R})$

We find again a new parameter [7] introduced by the studies carried out at the ESTEC, which has been called CHROMATICITY. This parameter whose values depends on the spatial frequency induces, in the absence of classical chromatism, a residual chromatic effect in the images formation process; It is due to the repartition of the energy in the diffraction "rings" which depends on the wavelength.

3.3. Numerical results

The LASSO software has been especially developed to compute the CHROMATICITY. One made this software because the classical ones are not efficient enough to compute the CHROMATICITY with a sufficient accuracy (0.1 milliarcsecond). Moreover, a complete telescope analysis asks to modelize the thermo-mechanical effects and creates a lot of calculations which constraint the program to a very short execution time. The results obtained showed that the CHROMATICITY of the nominal instrument (theoretical values for the components profile and position) is of 1 milliarcsecond all over the field of view [8] .It has been established that the CHROMATICITY values depend on the payload thermo-mechanical conditions and least cases of distorted instrument have been evaluated to define the extreme conditions which will be endured by the payload.

4. CONCLUSION

This very quick study of the HIPPARCOS image showed the utilization of a Schmidt telescope in Space Astrometry. The results already established allowed the design of the telescope and displayed the CHROMATICITY. This effect appeared as a milestone in the data acquisition. Now, further investigations are undertaken to determine its real values and to propose solutions which might minimize its impact on the whole mission.

5. REFERENCES

1 : Bacchus, P. et Lacroute, P;, Prospects of Space Astrometry New problem of astrometry, p. 277-282, IAU Symposium N° 61, Perth, 1973.

2 : FAST CONSORTIUM; Proposal for Scientific Data Processing, ESA HIP 81/02, 1981

3 : Kovalevsky, J., The Satellite Hipparcos, p 15-20, ESA SP-177 Strasbourg, 1982

4 : Lemaitre, G. Optique astronomique et élastique, thèse d'Etat, Marseille 1974

5 : Lindegren, L., The All-Reflective Eccentric Schmidt (ARES), Lund, 1982

6 : Le Gall, J.Y, L'image de diffraction polychromatique de la mission d'astrométrie spatiale HIPPARCOS, Thèse de Docteur-Ingénieur Paris, 1983

7 : Vaghi, S., Astrometry Satellite. Apparent displacement of star photocentres. MAD Working Paper n° 95 (ESOC, May 1979)
8 : Saïsse, M. Preliminary results on ARS OTF with elliptical pupil on the complex mirror, Marseille, 1982.

THE SCHMIDT TELESCOPE ON CALAR ALTO

K. Birkle
Max-Planck-Institut für Astronomie,
Heidelberg, Federal Republic of Germany, and
Centro Astronómico Hispano-Alemán,
Almeria, Spain

During the last ten years the Max-Planck-Institut für Astronomie (MPIA) has installed four large telescopes at its observatory on Calar Alto, Spain, which is operated jointly with the Spanish National Commission for Astronomy as the German-Spanish Astronomical Center. Figure 1 shows the domes of the 1.2 m, 2.2 m, and 3.5 m telescopes of Ritchey-Chrétien type and the dome of the Schmidt Telescope. They all are Zeiss telescopes, the former three coming from Oberkochen, the latter one from Jena.

Originally, the Schmidt Telescope was built for the Hamburg Observatory and was in operation there since 1955. With the Calar Alto project of the MPIA entering into its stage of realization an agreement was also reached to move this instrument from Bergedorf to the much better observing conditions of Calar Alto. For this purpose, the MPIA erected a new dome building and bought a new fork mounting of Grubb-Parsons adapted to the latitude of the new site.

Regarding the properties of the telescope reference is made to the description of Heckmann (1955). Here, attention will be drawn only to its main characteristics: The free aperture of the correction plate is 80 cm, the spherical mirror has 120 cm diameter and 240 cm focal length corresponding to a field of $5^\circ.5 \times 5^\circ.5$ on the spherically curved photographic plates 24 x 24 cm in size. Figure 2 shows the instrument in its new housing. The telescope is equipped with two objective prisms whose essential parameters are given in Table 1. Like the correction plate they are made up of the Schott glass UBK 7 of high UV transparency.

In 1980 the Schmidt Telescope became operational on Calar Alto. In the meantime more than 500 plates have been taken, most of them by observers of the Hamburg Observatory.

Figure 1. The Calar Alto (West Long. +2°32'.7, Lat. +37°13'.4, Alt. 2168 m) with the domes of (from left to right) Schmidt, 1.2 m, 2.2 m, and 3.5 m telescope. The distance between the Schmidt and the 3.5 m telescope is about 350 m in a roughly north-southern alignment. At left the laboratory building.

Figure 2. The 80/120 cm f/3 Schmidt Telescope on Calar Alto

Up to now, however, it was not possible for us to operate the telescope in a normal routine way because of several serious technical failures, which led to restrictions and longer interrupts in the use of the telescope. We hope that we soon shall have overcome most of these surprises. Never-

TABLE 1

THE OBJECTIVE PRISMS OF THE CALAR ALTO SCHMIDT TELESCOPE

Refracting Angle	Free Aperture (cm)	Dispersion at H_γ (nm/mm)	H_α	Spectrum Length from H_γ to H_ϵ (mm)
1°.7	80	139	460	0.31
4°.0	80	59	195	0.73

theless, many of the direct and spectral plates taken hitherto are of very good quality as is demonstrated e.g. by the contributions to this Colloquium by H. Adorf and H.J. Röser and by L. Kohoutek. Investigations are in progress on the limiting magnitudes attainable with this telescope situated on Calar Alto, where seeing conditions with image diameters in the order of 1 arcsec are a well known phenomenon. So far estimates are showing that for instance with the $1°7$ prism and exposures of 1 - 2 hours on baked IIIaJ plates spectral information can still be obtained from starlike objects of apparent magnitude $B = 19^m - 20^m$.

REFERENCE

Heckmann, O.: 1955, Mitt. Astron. Gesellsch. 1954, p. 57.

THE 50/70-cm SCHMIDT TELESCOPE AT THE BULGARIAN NATIONAL ASTRONOMICAL OBSERVATORY

MILCHO K. TSVETKOV
Department of Astronomy and National
Astronomical Observatory, 72 Lenin Blvd.,
Sofia 1184, Bulgaria

The 50/70 Schmidt telescope at the National Astronomical Observatory (NAO), Bulgarian Academy of Sciences, is an example of the possibilities for a long-term operation and everlasting actuality of these types of telescopes. Originally, this telescope was constructed in the GDR in 1952 and mounted in the Potsdam Observatory, Academy of Sciences of the GDR (as such, it is on the list of West, 1974). In 1979, on the basis of a mutual agreement between the Bulgarian Academy of Sciences and the Academy of Sciences of the GDR, it was transferred to Rozhen, Bulgaria. The principal parameters of the instrument are as follows:

I. Main mirror: diameter: 70cm; material: Tempax; thickness in the centre: 10 cm; radius of curvature: 344 cm.
II. Schmidt-corrector: diameter: 50 cm; material: BK 7; thickness in the centre: 1.8 cm.
III. Objective prism: diameter: 50 cm; material: UBK 7; angle of refraction: $3.5°$; average dispersion in the field of $H_\gamma - H_\varepsilon$: 820 A/mm.
IV. Focal region: radius: 172 cm; size of plates: 13x13 cm; field: 16.4 sq.deg. (with a Piazzi-Smith corrector, size of plates: 16x16 cm; field: 25.4 sq.deg.); field free of vignetting: $\approx$ 9 cm ($3°$); focal ratio: 1:3.44.

Since June 1979, when the telescope began to function in Bulgaria, planned investigations have started both of the instrument itself and of the obtained particular scientific researches and observations. During the first three-year period about 1000 plates had been developed for 620 hours of effective observations. For investigating the telescope and establishing the methods of observations, approximately 200 plates had been developed, the basic part of the observations being carried out within the frames of the observational programs, i.e.
- Investigation of non-stable and flare stars in stellar aggregates;
- Photometric investigations of stellar clusters;

- Observations of small planets in the Solar system;
- Investigations of galaxies, etc.

Particular stress in the observational program was layed on observations of flare stars in stellar aggregates of different ages. During these observations at Rozhen (341.5 hours effective observational time) 26 new flare stars and 18 repeated flares in the area of Cygnus, Orion, the Pleiades, etc. were discovered (Tsvetkov, 1982).

The investigation of the telescope, including the Hartmann test of its system (Golev, etc., 1982), showed comparatively good characteristics - the technical constant of Hartmann T = 1.01 arcsec. The experiments on determining the limit star magnitude showed that in a 40-minute exposure of ORWO-ZU 21 plates within the photographic field of the spectrum, up to $19^m.5$ is obtained. The photometric researches showed that up to $1°.5$ from the field centre, photometric errors are insignificant.

The observations carried out and the observation material obtained thereof offered the possibility some statistical conclusions to be made, characterizing astroclimatic conditions and organization of observations. Similar to other observatories situated in a neighbouring latitude, the summer-autumn observational period was found to be considerably more effective for observations.

The data obtained from the observations offer the opportunity of making an attempt to calculate the efficiency of the observations and their organization as well. With regard to this, for convenience, the quantity k can be defined, referred to as relative efficiency of observations carried out:

$$k = n.m. \qquad (1)$$

where n denotes the number of the plates obtained, while m is the effective exposure time in minutes. In a practical case k is also a function of the parameter φ ($0 \leq \varphi \leq 1$) where φ characterizes above all the meteorological conditions under which observations are being carried out, etc.

Comparing the quantity k calculated for the particular observational periods during the recent three years, it turns out that on the average, observations in the summer-autumn period at Rozhen are twice more effective than in the winter-spring period. For calculating the efficiency of observations made in certain nights of favourable climatic conditions ($\varphi \approx 1$) it is convenient to apply the coefficient $\varkappa$ (kappa) where $\varkappa = k/K$. Here K denotes the absolute efficiency of observations equal to

$$K = T \cdot \frac{T}{E} = T^2/E \qquad (2)$$

where T indicates the entire possible observational time, while E is the optimal exposure of the given telescope, depending on its focal ratio and specified in compliance with Baum relation, 1968.

If $\varkappa$ is expressed in percentage, then we have

$$\varkappa = 100 \cdot n \cdot m \cdot E / T^2 \% \qquad (3)$$

The calculation of the coefficient $\varkappa$ for certain favourable observational nights offers the opportunity for an extra information on the observations to be obtained. In the given case the observations with the 50/70-cm Schmidt telescope at Rozhen showed that during particularly favourable observational nights the value of $\varkappa$ reaches approximately 40%. The calculations give above all an idea of the possibilities for effective work with the instrument, as well as estimation of the quality of organization and planning of observations.

References

Baum, W.A.: 1968, Astronomical Techniques, Ed. by Hiltner, W.A., 8.
Golev, V.K., Tsvetkov, M.K. and Vitrichenko, V.A.: 1982, Comptes rendus de 'l Academie bulgare des Scienc., 35, No.6, 729.
Tsvetkov, M.K.: 1982, Communications from the Konkoly Observatory, No.83, 206.
West, R.M.: 1974, Proc. Research Programme of the New Large Telescopes, Geneva, 321.

ON THE ASTRONOMICAL RESEARCH WITH THE TORUN 60/90 CM SCHMIDT TELESCOPE

A. Woszczyk
Institute of Astronomy
Nicolaus Copernicus University, Torun, Poland

ABSTRACT: The Torun Schmidt telescope and its scientific applications are described. A special attention is given to the Torun Objective Prism Sky Survey.

I. INTRODUCTION

The origin of the Torun Schmidt telescope goes back to the idea of a Polish National Observatory born about 60 years ago. In the fifties this idea took the form of the Central Astronomical Observatory of the Polish Academy of Sciences with a two meters reflector as principal telescope and a medium size Schmidt telescope as its companion instrument. The realization of this project started by placing an order for a smaller (cheaper) instrument in the Zeiss optical factory in Jena in 1958, and in 1962 the telescope was delivered.

Awaiting the construction of a Central Astronomical Observatory it was decided to install this Schmidt telescope in the Astronomical Observatory of the Nicolaus Copernicus University of Torun. Later, the idea of the big Polish National Observatory was abandoned and the smaller "companion instrument" of the original project became the largest optical telescope in Poland. This telescope is one of the four Zeiss medium size "universal" telescopes built more than 20 years ago for the Budapest, Jena, Peking and Torun Observatories.

II. TECHNICAL PARAMETERS OF THE TORUN SCHMIDT TELESCOPE

The Torun Schmidt telescope is in fact a Schmidt-Cassegrain telescope. Its basic technical parameters are the following:
1. in the Schmidt arrangement:
   - the main F/3 spherical mirror: glass BK7, diameter 900 mm,
   - the correcting plate: glass BK7, diameter 600 mm,
   - the focal distance: 1808 mm; the scale: 8.8 μm = 1",
   - field of view; circular, diameter $4°.5$.
In this arrangement two objectives prisms can be used: one in glass BK7 and another in glass F2. Both have a refractive angle of $5°$ and

give respectively a dispersion of 250 and 550 Å/mm near H-gamma.
2. in the Cassegrain arrangement the main mirror is used with full aperture; with a quasi hyperbolic secondary mirror an effective focal length of 13500 mm is provided. The focus can be obtained either directly through a hole in the main mirror or laterally by means of an additional plane mirror through the hollow axis of declination.
The telescope is mounted in fork and provided with an electric drive of Zeiss Gaber type.
The site of the telescope is the village of Piwnice near Torun, about 90 m above sea level, with an average number of 100-120 clear nights per year and a seeing ranging from 3 to 10 arcsec.

III. THE TORUN OBJECTIVE PRISM SKY SURVEY

The Schmidt telescope certainly has great advantage of high efficiency, specially for the galactic structure studies, through its high aperture ratio and through its large field. With its freedom from off-axis aberrations and its near perfect achromatism it is an ideal instrument for stellar spectroscopic exploration. The enormous amount of spectroscopic informations obtainable with Schmidt telescopes in a relatively small amount of observing time convinced Miss W. Iwanowska, former director of Torun Observatory, to initiate an Objective Prism Sky Survey (Iwanowska, W. 1963). In this project the accessible sky ought to be covered with photometrically calibrated objective prism plates. The collection of plates so produced could serve as a plate library on one hand, for further spectrophotometric investigations of selected regions or stars and, on the other hand, to secure a reference source for variable astronomical objects.

The survey is being carried basically on Kodak IIa-F and IIa-O plates with objective prism F2 (250 Å/mm). The spectra are generally widened to 0.2 mm. To avoid the overlapping of the spectra in the crowded regions, the Milky Way belt is also exposed with the prism BK7 (shorter spectra), or less widened, or taken in a shorter exposure time. The percentage of the overlapping spectra in the Milky Way region with the longest exposure time in good atmospheric conditions and with a widening to 0.2 mm goes from 45% for the emulsion O to 75% for the emulsion F. In the polar cap region the respective coverage goes from 5% to 20%, in both cases with the objective prism F2.

The plates are calibrated photometrically by means of a 12-spots tube sensitometer. The calibration plates are taken on a separate piece of plate for the respective emulsion and observing night.

Up to now, our collection of about 2000 plates covers 850 fields (4°.5 of diameter) on the sky (Fig. 1). The most complete coverage is for the region ±6° from the galactic equator, the galactic meridian ($\ell = 0°$ and $\ell = 180°$) and the galactic polar cap . The limiting magnitude of the different plates is from 8 to 12 mag.

Almost all the staff-members of the Torun Observatory participated in this project. The quality of this material for spectrophotometric studies was tested by several peoples (Smolinski J., Strobel A., Zaleski L., results published in the Torun Observatory Bulletin in the years 1969-

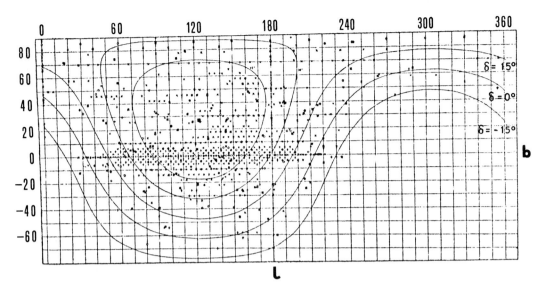

Fig. 1. Actual coverage of the sky (galactic coordinates l'' and b'') with the plates of the Toruń Objective Prism Sky Survey. + ÷ emulsion O, x ÷ emulsion F. Those symbols encircled mean that the given field was pictured at least two times with a given emulsion.

1976).

A system of quantitative three-dimensional classification of these spectra for F, G and K stars was elaborated by Strobel A. (1973, 1976). In this system the following stellar parameters: spectral type (with a precision of 0.1 of the type), luminosity class (with a precision of one class) and iron abundance [Fe/H] (with a precision of about 0.25 in [Me/H]) can be determined. Each determination of these parameters is based on several criteria of line intensities, spectral "jumps" and pseudo continuum levels. Another system of quantitative classification of our material, related to the narrow band photometric classification was proposed by Jeneralczuk J. (1981) for O-M stars.

IV. SPECTROPHOTOMETRICAL STUDIES OF INDIVIDUAL OBJECTS

Besides the sky survey the Toruń Schmidt telescope was basically used with objective prism for photographic spectrophotometry of individual objects.

Iwanowska W., strongly interested in the problem of stellar populations (see e.g. Iwanowska W., 1966), directed a series of paper in which spectroscopic characteristics of several stellar families (e.g. F, G, K stars, carbon stars, helium stars, RV Tauri variables, etc.) where investigated in conjunction with the position and movement of these stars in the galactic system.

Several emission lines variable objects were studied spectrophotometrically. Many Novae were observed and the evolution of their continuum as well as of their emission lines were followed over several months (among others, the papers of Glebocki, Krawczyk, Smolinski, Woszczyk for N Her 63, N Del 67, N Cyg 75, etc.). Symbiotic objects are also currently investigated (e.g. CI Cyg by Mikolajewski M.).

A special attention is given to the study of comets. The photographic photometry of the comets 1967n and 1968c in the CN and C2 bands were carried out by Typek J. and for several other comets the evolution of cometary emissions was investigated for a large range of heliocentric distances with the objective prism spectra.

Most of this objective prism researches were considerably extended with the aid of slit spectra of larger dispersion and resolution. In fact, the objective prism spectroscopy guided by the slit spectroscopy seems to us a quite efficient way of investigation.

V. OTHER APPLICATIONS OF THE TORUN SCHMIDT-CASSEGRAIN TELESCOPE

In its Schmidt optical arrangement the Torun telescope was tested by Swierkowska S. for the astrometric application. She concluded the good quality of this instrument for the determination of the astronomical objects position (precision better than 0".2 in both coordinates).

In the Cassegrain (or Nasmyth) optical arrangement the Torun telescope is used mainly with a small Cassegrain spectrograph (installed on one side of the fork) or with a photoelectric photometer (on the other side of the fork). Occasionally a photoelectric polarimeter was also used. The Torun Cassegrain spectrograph designed by Richardson H., from the D.A.O. in Victoria, gives several dispersions from 15 to 160 Å/mm and is used for the studies of the Ap stars, stars with a chromospheric activity and other peculiar objects.

The photoelectric photometer actually works in the fast photon-counting mode and is controlled by Polish made minicomputer Mera 305. Its maximal time resolution is less than 10 nanosec, the sampling time is 16 μsec and the dark current level 10-20 pulses per sec. This instrument is mainly used by its builder Wikierski B. and Dr. Turlo Z. to observe optical pulsars looking for the glitches.

REFERENCES

Iwanowska, W. 1963, Nauka Polska Vol. 11, p. 95
Iwanowska, W. 1966, in Vistas in Astronomy, Vol. 7, p. 133, A. Bear Ed.
Jeneralczuk, J. 1981, Master of Science Dissertation, Copernicus University
Strobel, A. 1973, PhD Dissertation, Copernicus University
Strobel, A. 1976, Bull. Obs. Torun, nr. 55, p. 9

JAPANESE ORBITING ULTRAVIOLET TELESCOPE PROJECT: UVSAT WORKING GROUP
REPORT

Masanori Iye
European Southern Observatory, 8046 Garching b. München, F.R.G.
Tokyo Astronomical Observatory, Mitaka, Tokyo 181, Japan

The ultraviolet-telescope satellite (UVSAT) project has been proposed to
the Institute of Space and Astronautical Science (ISAS/Japan) by the
UVSAT Working Group. The main telescope will be an F/4 Cassegrain with a
60cm primary mirror. A design study and some preliminary experiments of
an intensified CCD camera with an objective spectroscopic capability are
under way.

The main objectives of UVSAT would be to investigate i) the distribution and physical nature of UV sources in selected star clusters,
galaxies, and clusters of galaxies, ii) the structure of galactic
nebulae, and iii) the activities of stars and galactic nuclei. To these
ends, a fast optical system with a moderately large field of view is
desirable. The main UV telescope is at present proposed to be an F/4
Cassegrain system with a 60cm primary mirror. A number of focal plane
instruments are proposed for this telescope, e.g., a high resolution
spectrograph, a low resolution spectrograph, and a CCD camera (Kodaira,
1983).

The scale on the focal plane of the main telescope is 86 arcsec/mm.
An electrically cooled CCD with 500 x 500 pixels will cover a field of
view of up to 22 arcmin square with a spatial resolution of 2.6 arcsec. A
microchannel plate with a CsI photocathode and a phosphor rear window
will be mounted in front of the CCD in order to make it sensitive to UV
light. An exposure time of up to 30 minutes is planned and the attainment
of the required attitude control is endeavoured. There are three modes of
observation proposed for this CCD camera, namely, i) the direct imaging
mode through various filters, ii) the low resolution objective spectroscopic mode using a grism, and iii) the polarimetric mode using a double-
image prism. The direct imaging mode will provide photometric data of
point sources and extended objects. Expected limiting magnitude is
$m_v \sim 20$ mag for an A0 star. The polarimetric mode will help us to understand the physics of variability observed in many UV sources, e.g., cataclysmic variables and BL Lac objects.

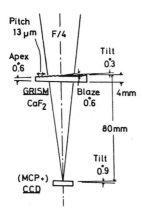

Figure 1. A tentative design of the grism camera.

Figure 2. A picture of the groove profile of an Si-grating produced by the ion-etching method. The blaze angle of this holographic grating is 3 degrees (Aritome and Namba, 1983).

Figure 1 shows a tentative design of a CaF_2 grism camera proposed for the objective spectroscopic mode. This grism camera will give a spectrum of each object for the wavelength range of 1300-2000 Å. The inverse dispersion will be 640 Å/mm at 1300 Å and 1460 Å/mm at 2000 Å. In order to make such a rather special grating of very shallow blaze angle, we are experimenting with an ion-etching method. Ion-etching is a method of carving the surface of a sample, on which a photoresist layer of an appropriate grating pattern should be coated, by a uniform flux of energetic ion beams. The necessary grating pattern of a photoresist layer is produced holographically and can be fixed chemically beforehand (Aoyagi and Namba, 1976). This method has proved to be very successful in making a Si or SiO_2 grating for use in optical region (see Figure 2). The preliminary experiment of making a CaF_2 grating by the ion-etching technique showed that this method is promising. However, further efforts are needed to establish the required surface accuracy of the groove profile.

Although UVSAT is not yet funded, it is hoped that UVSAT will be launched in early 1990's by an M3S-III rocket of ISAS in order to contribute UV astronomy in cooperation with other space missions, e.g., Space Telescope, Starlab, Space Schmidt, and so on.

The present report is based on the studies carried out by the UVSAT Working Group of ISAS which was organized by K. Kodaira of Tokyo Astronomical Observatory in 1980. The author is indebted to Drs. H. Ando, Y. Aoyagi, H. Aritome, H. Kawakami, K. Kodaira, S. Namba, K. Nishi, T. Onaka, W. Tanaka, and T. Watanabe among others in preparing this report.

REFERENCES

Kodaira, K.: 1983, Adv. Space Res., Vol. 2, No. 4, pp. 171-175.
Aoyagi, Y. and Namba, S.: 1976, Optica Acta, Vol. 23, No. 9, pp. 701-707.
Aritome, H. and Namba, S.: 1982, in Optical and Physical Characteristics of Diffraction Gratings (in Japanese), p. 75.

ASTROMETRY WITH SCHMIDT TELESCOPES

C A Murray
Royal Greenwich Observatory

1. INTRODUCTION

Photographic astrometry, including work with Schmidt telescopes, can be divided into two main fields, (i) the measurement of positions of objects relative to a reference frame of stars with known celestial coordinates, and (ii) the measurement of relative proper motions and trigonometric parallaxes from a series of plates taken on the same field. The former demands a knowledge of the absolute transformation between angles on the sky and measurements on a plate, whereas in the latter we are only interested in differential transformations from plate to plate. The potential value of Schmidt telescopes for both these fields of astrometry lies in the large area of sky and range of magnitude which can be imaged on a single plate. The former advantage is however, to some extent offset by the curvature of the focal surface which means that, in order to utilize the full field the plates must be constrained to the form of the focal surface during exposure.

Some ten years ago, a conference with a title very similar to that of this colloquium was held at Hamburg at a time when the two large Schmidt telescopes were being constructed at Siding Spring and La Silla. Since that time the surveys which have been and are being produced by these two telescopes in the southern hemisphere together with the earlier Palomar Survey (POSS) of the sky north of $-30°$ declination, have become essential sources of data for obtaining positions of celestial objects. Many astronomers now use these published atlases in the form of glass, film or even paper copies for the routine derivation of positions for identification purposes, although usually confining their measurements to a small area of a field. Hunstead (1974) has shown that a positional accuracy of $0\overset{"}{.}3$ can be achieved using paper prints of the POSS.

Smaller Schmidt telescopes, too, are are now regularly used for positional astrometry. In particular we may mention the work on Pluto here at Asiago (e.g. Barbieri et al. 1979) and optical positions of radio sources both here (Barbieri et al. 1972) and at Cambridge (Argue et al. 1979).

2. THE STELLAR REFERENCE FRAME FOR SCHMIDT PLATES

The main contribution to the error of a position derived from Schmidt plates at the present time arises from the inaccuracies in the catalogue positions of stars defining the reference frame. Since these are nearly always extrapolated from earlier observations, errors in catalogue proper motions produce a continual deterioration in the reference frame.

The AGK3 catalogue (Heckmann and Dieckvoss 1975) is now used almost universally as the source of reference stars in the northern hemisphere. The proper motions in AGK3 are based on observations made at about 1930 and 1959; the resulting error of a single star coordinate at the current epoch is about ± 0".4 and by the end of the century will amount to at least ± 0".5.

In the southern hemisphere the situation at the present time is much worse. The most readily available source of star positions is the SAO Catalogue (Smithsonian Institution 1966) which, in the southern hemisphere, is based primarily on photographic zone catalogues observed at Yale (Johannesburg) and at the Cape prior to 1950, with proper motions obtained from comparison with earlier meridian observations. The reference frame in the southern hemisphere will however soon be greatly improved when work on the Second Cape Photographic Catalogue (CPC2) is completed. The 5800 plates taken for this project, between 1962 and 1970 are being measured on the GALAXY automatic measuring machine at RGO. Although more than seventy five per cent of the plates have now been measured, completion of the project is now in jeopardy because of the recent decision of the Science and Engineering Research Council to withdraw support for GALAXY at the end of March 1984. A first instalment of CPC2, giving provisional positions at epoch 1962 with an average internal standard deviation of ± 0".1 in each coordinate for 50,000 stars in the zone $-38°$ to $-54°$, is now available from RGO and through the Strasbourg Stellar Data Centre. Proper motions for more than 20,000 of these stars which are in the SAO Catalogue have also been redetermined. Measurement is complete for the more southerly declinations and provisional positions for these should be available within a year or so. The final positions will be obtained by a block adjustment (de Vegt and Ebner 1974) of the measurements on all the plates covering the whole southern hemisphere.

A further dramatic improvement in the reference frame over the whole sky will becom available early in the next decade, with the completion of the HIPPARCOS mission which will give positions, annual proper motions and trigonometric parallaxes with average mean error of ± 0".002 for some 100,000 selected stars, and the associated TYCHO experiment which will give positions at mean epoch 1989 with accuracy of a few hundredths of an arc second for all stars brighter than about B = 11.

3. GLOBAL REDUCTION OF A SCHMIDT PLATE

The main distinction between Schmidt astrometry and what we may term 'flat field' astrometry, is that, in the former, distances measured along great circles over the spherical focal surface are proportional to the corresponding angular distances on the sky, whereas in the latter, linear distances in the focal plane measured from the tangential point are proportional to the tangents of the corresponding angles. To a high degree of approximation therefore, the difference between tangential coordinates and coordinates measured on a Schmidt plate can be described by a purely geometrical third order radial distortion with coefficient 1/3 if all angles are measured in radians. Although in principle, because of spherical symmetry, a Schmidt plate has no unique origin, corresponding to the tangential point in flat field astrometry, in practice the simplest model of elastic deformation during exposure will be symmetrical about some neutral point. Shepherd (1953) has shown that the effect of such deformations for a glass plate can be expressed as a linear scale change and a small third order radial distortion. Radial distortions and even more complicated transformations can in principle be derived from the adjustment of measurements of reference stars to their catalogue position, but this is generally impractical because of the errors of the positions obtained from the main source catalogues, AGK3 or SAO.

Special investigations of the astrometric performance of Schmidt telescopes are therefore carried out in particular fields in which relative positons with high positional accuracy are available, such as the region of α Persei (Heckmann et al. 1956), Praesepe (Russell 1983) and the Pleiades (van Leeuwen: to be published). The overlapping plate technique could also be used.

Anderson (1971) studied the performance of the Brorfelde Schmidt telescope using plates taken on the α Persei field for which the estimated mean error for one coordinate was ± 0".115. He showed that the radial distortion relative to tangential coordinates could be represented quite adequately by the geometrical term, with coefficient 1/3, over the field of 5°.3 diameter. Allowing for the estimated error of star positions he found the external mean error of one position measured on one plate to be about ± 0".20, compared with the internal error of ± 0".14; the difference between these two was attributed to residual plate distortions and emulsion shifts as well as unmodelled optical aberrations.

In order to examine plate effects Bru and Lacroute (1973) proposed an experiment in which a reseau of reference marks was photographed on to a Schmidt plate in its constrained (spherical) state. By intercomparing the measurements of these reference marks on a number of plates taken with the Palomar Schmidt, also on the α Persei field, Fresneau (1978) was able to show that the effects of plate errors were less than one micron, compared with his measurement error of 1.7μ. Fresneau further studied the error covariance function of the residuals from an empirical 20 parameter plate constant solution and found significant correlations between residuals extending to separations between stars of up to 5 cm (about 1°). He was able to improve the residuals by ± 1.7μ by applying a statistical smoothing process.

4. ATMOSPHERIC DISPERSION

A major potential error source in hight precision astrometry is the variation of atmospheric refractivity with star colour. In classical astrometry with photographic refractors this has never been too much of a problem since the objective itself filters out most of the ultraviolet light. In the case of a Schmidt, or for that matter any reflecting telescope, colour dependence of refraction on unfiltered plates can be large. Elsewhere (Murray and Corben 1979) I have shown plots of the average constant of refraction for various emulsions over a limited range of star colours; these were calculated by convolving the wavelength dependent constant of refraction for standard atmospheric conditions given by Allen (1973) with the

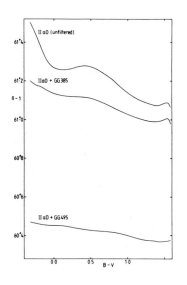

Fig. 1

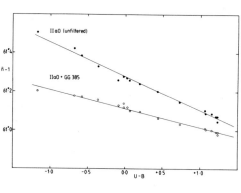

Fig. 2

transmissivity of the atmosphere and telescope, the emulsion sensitivity and stellar energy distributions. We have recently recomputed the refractivity in a slightly more rigorous fashion. Writing the refractive index for wavelength λ in the form

$$n(\lambda) = 1 + \beta_d(\lambda)\rho + (\beta_w(\lambda, - \beta_d(\lambda))\rho_w$$

where ρ is the total density of air and ρ_w the water vapour density, we have computed effective values $\overline{\beta_d}$, $\overline{\beta_w} - \overline{\beta_d}$ for various sensitivities and stellar energies for unreddened class V stars taken from the tables of Straizys and Sviderskiene (1973). Values of the effective refractivity $\overline{n} - 1$, expressed in arc seconds for unit air mass and the same standard atmospheric conditions as in Allen's table (loc. cit.) are shown as a function of B-V in Fig 1. The very marked dip in the curve for the IIaO emulsion corresponding to the Balmer discontinuity is obvious, and suggests that the refractivity in this case is much more conveniently expressed in terms of U-B. From Fig 2 we see that in fact to a high approximation the refractivity is linear with U-B. Least squares fits to the individual points which are shown in the figure are:-

IIaO (unfiltered) $\overline{n} - 1 = 61\overset{''}{.}281 - 0\overset{''}{.}176$ (U-B)

IIaO + GG 385 $\overline{n} - 1 = 61\overset{''}{.}111 - 0\overset{''}{.}098$ (U-B)

The IIIaJ emulsion shows similar characteristics.

It should be emphasized that these data refer to the refractivity corresponding to absorption in one air mass; this is not quite the same as the constant of refraction at zenith distance $\theta_o = 45°$. Of course the actual amount of refraction is approximately proportional to $\tan \theta_o$, so that relative displacements of up to one arc second between extreme red and blue stars at large zenith distance are quite possible.

5.   GALACTIC ASTROMETRY

I would like to conclude this review by discussing astrometric programmes for the measurement of proper motions and parallaxes.

By far the most extensive and successful programme of this nature which has yet been carried out with a Schmidt telescope is the survey for stars with large proper motion by Luyten, using plates taken with the Palomar 48-inch Schmidt. With the exception of about 160 very crowded Milky Way fields, Luyten has blinked, either with his automatic scanner or by eye, all the Palomar survey fields north of $-33°$ declination. Combining these new data with earlier surveys by himself and others he has now published a revised LHS Catalogue (Luyten 1979) containing all known stars with annual proper motions exceeding $0\overset{''}{.}5$ and three volumes of the NLTT catalogues (Luyten 1979-80) containing data for all stars with annual proper motions estimated to exceed $0\overset{''}{.}18$. The fourth volume of the NLTT Catalogue (Luyten 1980) covers the sky south of declination $-30°$. It contains roughly only half the number of stars in the corresponding zone of the northern hemisphere. Apart from a special search to faint limits with the Palomar Schmidt around the south galactic cap, this fourth volume is based largely on his earlier compilations, LTT (Luyten 1957) and the Bruce Proper Motion Survey (Luyten 1963). There is a rich field to be harvested here when the current southern hemisphere surveys are ripe for repetition; Luyten himself started his Palomar survey on red sensitive plates, in 1962, some fifteen years after the first POSS plates were taken. Maybe it is not too early now to plan reobservation of one or more of the southern hemisphere surveys; the first plates for the SRC 'J' and ESO 'B' surveys are already ten years old. Although for strict comparability with Luyten's work the extension to faint limits in the southern hemisphere should be based on the ESO red survey, this would delay proper motion measurements to near the end of the century.

A proper motion programme of rather a different character is being carried out at the Lohrmann Institute, Dresden with the aim of calibrating the proper motion systems of meridian circle catalogues (Böhme and Sandig 1979). Proper motions will be measured relative to extra galactic objects on plates taken with the 2m Universal Reflecting Telescope of the Karl-Schwarzschild Observatory, Tautenburg, in its Schmidt mode. Two exposures are obtained on each plate covering a $3°.1 \times 3°.1$ square field; Magnitude compensation is provided by means of a sub-beam prism with reduction of 3.9 magnitude.

While the detection of interesting stars with high proper motions may be described as the cream, there is much solid meat to be extracted from a study in depth of the large amount of astrometric and photometric data which can be obtained from plates taken on a single field.

A few years ago, the first results of an extensive study of a field near the south galactic cap were reported (Murray and Corben 1979). The plates for this programme were taken with the UK Schmidt telescope and measured on GALAXY at RGO. From an analysis of about fifty plates taken over three years we were able to obtain proper motions and parallaxes with standard errors of $\pm 0''.013$ and $\pm 0''.018$ respectively for stars brighter than $B = 14$. These parallaxes were no worse than those obtained in the routine parallax program of the least accurate of the long focus refractors. For the first time it has been possible to determine the statistical distribution of proper motions and parallaxes, as functions of colour, of a sample of stars limited only by apparent magnitude. For various reasons further progress has been delayed; however, more than 100 plates on this field have now been measured giving data for about 16000 stars brighter than $B = 18$. In addition nearly fifty plates have been obtained on two new fields, the Morton-Tritton field at $\alpha_{1950} = 22^h 03^m$, $\delta_{1950} = -18° 55'$ and Pickering's E8 region at $\alpha_{1950} = 20^h 05^m$, $\delta_{1950} = -44° 59'$

but future work on these fields is likely to be delayed because of the closure of GALAXY.

It is now nearly eighty years since Kapteyn proposed his programme of Selected Areas. This was a far sighted project from which most of our present ideas about the distribution of stars in the Galaxy has been derived. The main astrometric contributions to the programme have been the measurement of proper motions which have formed the basis for studies of statistical parallaxes and velocity distributions of stars out to several hundred parsecs from the Sun. Similar programmes to fainter magnitudes and larger samples using Schmidt plates will extend such studies to the population, space distribution and kinematics of stars to distances of several kiloparsecs.

One of Kapteyn's objectives was the systematic measurement of trigonometric parallaxes in each area, but for reasons which are obvious to us, nothing of value came from this aspect of the programme. With Schmidt astrometry we can now remedy this. It would for example be relatively straightforward to test the completeness of Gliese's catalogue of nearby stars, in which the average surface density is rather less than two per field of the UK or Palomar Schmidt telescopes. By analysing smaller parallaxes in a large number of fields it should be possible for the first time to obtain direct evidence for the true density and luminosity distribution in the solar neighbourhood, completely unbiased by the proper motion selection effects which have bedevilled earlier work in this field.

The availability of Schmidt telescopes, particularly the large ones, and facilities for accurate automatic plate measurement provide a great opportunity for astronomers today to pursue astrometric programmes which will make fundamental contributions to our knowledge of our Galaxy.

REFERENCES

Allen, C.W. 1973. Astrophysical Quantities 3rd edition. (London: Athlone press), p 125
Andersen, J. 1971. Astron. & Astrophys. 13, 40.
Argue, A.N., Clements, E.D., Harvey, G.M. and Murray, C.A. 1979 IAU Colloquium No 48, eds. F.V. Prochazka and R. H. Tucker (University of Vienna) p 156.
Barbieri, C., Capaccioli, M., Ganz, R. and Pinto, G. 1972 Astron. J., 77, 444.
Barbieri, C., Benacchio, L., Capaccioli, M. and Pinto, G. 1979 Astron. J., 84, 1890.
Böhme, D. and Sandig, H. U. 1979 IAU Colloquium No 48, eds. F.V. Prochazka and R. H. Tucker (University of Vienna) p 535.
Bru, P. and Lacroute, P. 1972. The role of Schmidt Telescopes in Astronomy, ed. U. Haug (Hamburg-Bergedorf) p 45.
de Vegt, Chr. and Ebner, H. 1974. Mon. Not. R. Astron. Soc. 167, 189.
Fresneau, A. 1978. Astron. J., 83, 406.

Heckmann, O., Dieckvoss, W. and Kox, H. 1956. Astr. Nachr. 283, 109.
Heckmann, O. and Dieckvoss, W. 1975. AGK3 Star Catalogue of Positions and Proper Motions North of $-2°.5$ Declination (Hamburg-Bergedorf)
Hunstead, R.W. 1974 IAU Symposium No. 61, eds W. Gliese, C.A. Murray, and R.H. Tucker (Dordrecht: Reidel) p 175.
Luyten, W.J. 1957. A Catalogue of 9867 Stars in the Southern Hemisphere with Proper Motions Exceeding 0".2 annually (Minneapolis, Minnesota: Lund Press).
Luyten, W.J. 1963. Bruce Proper Motion Survey - The General Catalogue (Minneapolis, Minnesota).
Luyten, W.J. 1979-80. New Luyten Two Tenths Catalogue (University of Minnesota).
Murray, C.A. and Corben, P.M. 1979. Mon. Not. R. Astron. Soc. 187, 723.
Russell, J.L. 1983. IAU Colloguium No. 76 (in press).
Shepherd, W.M., 1953 Mon. Not. R. Astron. Soc. 113, 450.
Straizys, V. and Sviderskiene, Z. 1973. Bull. Vilnius Astron. Obs. No. 35.

COMPILATION OF THE HIPPARCOS INPUT CATALOGUE -
AN EXTENSIVE USE OF SCHMIDT SKY SURVEYS

C. Turon-Lacarrieu
Observatoire de Paris-Meudon (France)

The ESA Astrometry satellite Hipparcos is due to be launched in early 1988. It will measure very precise positions, parallaxes and proper motions for about 100 000 stars. However, in order to be included in the Input Catalogue, the programme stars should have positions and magnitudes known in advance with respective accuracies of about 1" and 0.5 magnitude. This will require new astrometric and photometric measurements and observations. Sky Survey Schmidt plates will be extensively used, especially for astrometric measurements in the Southern hemisphere.

I - INTRODUCTION.

The European Space Agency Astrometry satellite Hipparcos is due to be launched in early 1988 by Ariane. It will be placed in a geostationary orbit for a nominal life-time of 2.5 years.

It will measure the positions, absolute parallaxes and proper motions of about 100 000 stars brighter than 13 (most of them brighter than 11) with an expected accuracy of 0.001 to 0.002 arc.sec. or arc.sec. by year.

The instrument is an all-reflexive Baker-Schmidt telescope which allows the superposition of two fields, 58° apart, in the focal plane. It is described in detail by J.Y. Le Gall and M. Saisse in this Colloquium.

The general organization and the various aspects of the scientific preparation of the mission have already been presented ("The Scientific Aspects of the Hipparcos Space Astrometry Mission" and references herein ; C. Turon-Lacarrieu 1983). Annual reports are issued by the "Project Scientist" in the "Bulletin d'Informations du Centre de Données Stellaires", Strasbourg Observatory (M.A.C. Perryman, 1981, 1982, 1983).

Let us only recall that ESA has charged four Consortia with the scientific aspects of the mission : the INCA Consortium for the Compilation of the Input Catalogue (led by C. Turon-Lacarrieu, Paris-Meudon Observatory, France) ; the FAST and NDAC Consortia for the Hipparcos Data Reduction (led respectively by J. Kovalevsky, Grasse, France, and by E. Høg, Copenhagen, Denmark) ; the TDAC Consortium for the Tycho Data Reduction (led by M. Grewing, Tübingen, F.R.G.).

II - COMPILATION OF THE INPUT CATALOGUE.

Due to the basic principles of the instrument : measurements of large angles (between stars in each field of view), continuous revolving scanning of the sky and photometric detection after modulation by a grid, it is necessary to compile and test in advance the entire catalogue of programme stars or "Input Catalogue". "Compile" means not only interrogate intensively the Strasbourg "Centre de Données Stellaires" (CDS) Data Base, but also perform new astrometric and photometric measurements and observations. Finally, each of the successive versions of the Input Catalogue will be tested by a complete simulation of the mission, in order to improve, from one version to the following, the uniformity of the programme stars spatial repartition, the adequacy of their magnitude distribution and to maximise the inclusion of high priority stars.

The different tasks and the structure of the INCA Consortium are described in C. Turon-Lacarrieu (1983).

205 observing propositions have been received by ESA, including about 800 000 stars (with many redundancies !). They are being merged, following the recommendations and priorities affected by the ESA Selection Committee. They are also being submitted to a careful study with respect to positions and magnitudes knowledge. Indeed a star, even with high priority, will not be kept in the final Input Catalogue if its a priori 1989 position is not known to about 1" and its magnitude to 0.5 magnitude.

These "technical" constraints, imposed by satellite operation and data reduction, will imply new measurements, either by the proposers or by the INCA Consortium, of a large part of the faintest ($m \gtrsim 9$) programme stars, and even of some brighter stars.

III - NEW GROUND-BASED OBSERVATIONS AND MEASUREMENTS.

III.1 - Astrometric observations and measurements.

Following the ESA requirements, 1989 positions should be known with an accuracy higher than 1".5 for all programme stars (r.m.s.) and higher than 1" for more than half of them, evenly distributed over the sky and brighter than $m_B = 10$. Conservative estimates of the number of stars to

be remeasured are the following :
- $\sim$ 4 000 Northern stars, mostly brighter than $\sim$ 9,
- $\sim$ 16 000 Southern "faint" stars (9-13),
- $\sim$ 15 000 Southern "bright" stars.

Northern observations will be mainly performed by the Bordeaux Transit Circle, Southern ones by <u>an extensive use of Sky Survey Schmidt plates</u> (Réquième, 1982).

Preliminary tests are being performed :
- comparison between the 3 involved measuring machines (Bordeaux, Hertsmonceux, Leiden) using copies of the same plate (ESO Quick Blue Survey) ;
- comparison and optimization of reduction programmes ;
- comparison between measurements performed with the same machine, using different copies of the same plate and the original plate (by courtesy of the Garching ESO staff).

III.2 - Photometric measurements.

The ESA requirement on magnitude accuracy (for integration time determination) is 0.5 magnitude. Moreover, the knowledge of a colour index is very desirable from a Data Reduction point of view.

It is very difficult to make even a conservative estimate of the number of stars to be remeasured as it strongly depends on the accepted proposals and on the proposers observing possibilities (Grenon, 1982).

Measurements will be made either by multicolour photoelectric photometry (already started) or by Schmidt plate measurements, but these are not well suited with the considered magnitude range. Another possibility being currently tested is the derivation of approximate magnitudes from prism-objective spectra taken for radial velocity measurements (Marseille and Haute-Provence Observatories ; cf. Fehrenbach and Burnage, this Colloquium), much better suited with stars in the 9-13 magnitude range.

REFERENCES

Grenon, M.:1982, "The Scientific aspects of the Hipparcos Space Astrometry mission", ed. M.A.C. Perryman and T.D. Guyenne, Strasbourg 22-23 Feb. 1982.
Perryman, M.A.C.: 1981, Bull. Inf. CDS 21, p.40.
Perryman, M.A.C.: 1982, Bull. Inf. CDS 22, p.87.
Perryman, M.A.C.: 1983, Bull. Inf. CDS 24, p.57.
Réquième, Y.: 1982, Strasbourg 22-23 Feb. 1982, op.cit.
Turon-Lacarrieu, C.: 1983, IAU Symp. N° 76 "Nearby Stars and the Stellar Luminosity Function", Middletown (USA) 13-16 June 1983, ed. A.G. Davis Philip, to be published.

SMALL BODIES OF THE SOLAR SYSTEM

Charles T. Kowal
Dept. of Astrophysics
California Institute of Technology
Pasadena, California, U.S.A.

Schmidt telescopes are ideal for many types of survey work, and this includes surveying the solar system. Schmidt cameras have the advantages of speed and wide fields for rapid coverage of large areas of the sky, and good image quality over the entire field. The small focal ratios also have the effect of enhancing low surface-brightness features. This is an advantage for studying cometary tails and comae.

Current work in the field of small solar system objects can be divided into three different areas: searching for new objects, recovering lost objects, and astrometry of known objects. Each of these areas requires its own techniques.

Searches for new objects generally concentrate on fast-moving asteroids, slow-moving asteroids, and planetary satellites. Again, each type of object requires a different technique for discovery. Of course, comets are also often found with Schmidt telescopes, but they are usually found during the course of other work, rather than during deliberate searches.

1. FAST-MOVING OBJECTS

Fast-moving objects include the Apollo, Amor, and Aten asteroids. These objects pass relatively close to the Earth, and thus have rapid angular motion across the sky. Motions of a degree or more per day are typical. For these objects, the wide field and speed of Schmidt telescopes are especially valuable. It is estimated that approximately 1000 Apollo asteroids with diameters greater than one kilometer exist in the solar system. Only about 60 of these have been discovered thus far. These objects are important in the study of the cratering history of the inner planets.

They may also be relevant to the study of comets, since it is suspected that some Apollo asteroids are "extinct" comets. Some of these asteroids are relatively easy to reach with space probes, and may become important sources of raw materials for manufacturing structures in space in the future.

Searches for fast-moving asteroids are being conducted by Eugene Shoemaker and Eleanor Helin, with the 0.46-meter Schmidt telescope at Palomar. These searches have resulted in the discovery of the first asteroid having a semi-major axis of less than 1 A.U., as well as some prime candidates for space missions.

Many Apollo-type asteroids have also been found in the course of other work, such as supernova searches, because these asteroids can appear anywhere in the sky, and are not concentrated toward the ecliptic.

Fast-moving objects are usually found by looking for relatively long trails on photographs. Shoemaker and Helin, however, use a stereoscopic technique in which two photographs of a region are examined simultaneously in a stereoscopic viewer. Any moving objects then appear to "stand out" from the stellar background. The two photographs are taken only about 30 minutes apart.

2. SLOW-MOVING OBJECTS

Slow-moving objects include all objects beyond the asteroid belt. These are the Trojan asteroids, and unusual objects like Chiron and Hidalgo. Trans-Plutonian planets could also be included in this category, as well as "Trojans" of the outer planets.

Searches for slow-moving objects usually require that two or more plates of a region be photographed one day apart. The plates are then examined under a blink microscope in order to find the moving objects. The photographs must be taken within 15 degrees of the opposition point (i.e. 180 degrees from the Sun). Near the opposition point, the apparent motion of an object is a function of its distance from the Earth and Sun. This allows us to discriminate between main-belt asteroids and the more distant objects. At greater distances from the opposition point, the main-belt asteroids move more slowly, and can mimic the motion of more remote objects.

A search for very distant objects has been conducted by Kowal, using the Palomar 1.2-meter telescope. About two-thirds of the ecliptic has been photographed during this survey, to a limiting magnitude of 20-21. As a result of this survey, the object 'Chiron' was discovered in 1977, near the orbit of Uranus. As by-products of the search, one Apollo asteroid and two comets were discovered. The asteroid Adonis and Comet Taylor, (1916 I), were also recovered, after having been lost for many decades. An expanded survey of this type, but using automated blinking techniques, would greatly increase our knowledge of the population of small bodies throughout the solar system. It is already clear that the solar system does not consist of only the known planets, comets, and a well-defined asteroid belt.

3. SATELLITES

Although planetary satellites are usually studied with long-focus instruments, the satellite systems of Jupiter and Saturn cover such a large angular extent that wide-angle Schmidt telescopes can be used to great advantage. This was demonstrated in 1974, when Kowal discovered J XIII (Leda). The entire satellite system of Jupiter spreads across five degrees of the sky.

Because of internal reflections and scattered light from these bright planets, it is necessary to block the planet's light with an opaque mask. Satellites close to the planet are covered by this mask, or buried in the scattered light, so only the outer satellites can be photographed.

4. OTHER WORK

Wide-field telescopes are of obvious benefit in the study of comet tails, which can be many degrees long. Since the features within these tails can change on a timescale of hours, it is advantageous to photograph a comet at several observatories around the world for continuous time coverage. Such a co-operative effort will be conducted during the International Halley Watch in 1985-86.

The recovery of lost asteroids and comets is another area in which Schmidt telescopes have been used very successfully. The original observations of lost objects are re-analyzed, and new orbits and ephemerides are computed. Then, one or more plates are taken of the area around the ephemeris position. In this way, many objects that had been missing for decades have been recovered with a moderate expenditure of telescope time, (but with a considerable expenditure of computer time)!

5. FUTURE WORK

The demise of astronomical photography is often predicted, but the future of Schmidt telecopes seems secure for the rest of this century. One photographic plate contains a vast amount of information, and great progress could be made by the use of automated retrieval of this information. Research on small bodies in the solar system could well be revolutionized by the use of automated plate scanning, position measurement, and "blinking" of photographic plates. The use of such devices is not yet widespread, because of the great expense involved.

References:

Kowal,C.T., Aksnes,K., Marsden,B.G., and Roemer,E.: 1975, Astron. J. 80. pp. 460-464.

Kowal,C.T., Liller,W., and Marsden, B.G.: 1979, "Dynamics of the Solar System", I.A.U. Symposium No. 81. pp. 245-250.

Shoemaker,E.M., and Helin, E.F.: 1978, "Asteroids: an Exploration Assessment", NASA Conference Publication 2053.

WIDE-FIELD IMAGING OF HALLEY'S COMET DURING 1985-1986 USING SCHMIDT-TYPE TELESCOPES

J. C. Brandt, D. A. Klinglesmith III, and M. B. Niedner, Jr.,
Laboratory for Astronomy and Solar Physics,
NASA/Goddard Space Flight Center, Greenbelt, MD 20771, USA

J. Rahe
Remeis Sternwarte, Universität Erlangen-Nürnberg, Bamberg, FRG

ABSTRACT

Photographic imaging of the plasma- and dust-tails of bright comets requires fast ($f \leqslant 4.0$), wide-field ($FOV \geqslant 5°$) optics for the proper recording of these large, low surface brightness features. Schmidts and astrographs are well-suited to this task and a large number of these instruments around the world will be turned toward Halley's Comet in 1985-1986 in support of the Large-Scale Phenomena Discipline of the International Halley Watch (IHW). This "worldwide network" should provide imagery with a time resolution never before realized in the study of a comet, and major breakthroughs in the understanding of highly-variable, elusive plasma processes in comets are expected. The imagery will also provide support for the GIOTTO, VEGA, AND PLANET-A deep space probes to the comet.

1. THE NEED FOR A WIDE-FIELD NETWORK TO STUDY HALLEY'S COMET

The rapid variability of cometary plasma tails, and the associated need for high-time resolution imagery, have been known since the regular application of photography to the study of these tail systems began in the 1890's. A full review of the subject is not possible here, but it is important to note that E. E. Barnard in 1905 (Barnard 1905) advocated that wide-field photographs of bright comets be taken as often as possible--preferably every half hour--in order that the rapid changes at times of high cometary activity could be studied effectively. Such time resolution has not, unfortunately, come close to being realized in any past comet, due almost entirely to the short observation times at individual sites (1-2 hours typically) and to the lack of coordination among observers around the world.

Many of the tail disturbances noticed by Barnard and his contemporaries are now known to be disconnection events, or DE's, in which the entire plasma tail uproots itself from the head of the comet, drifts away in the anti-solar direction, and is replaced by a new plasma

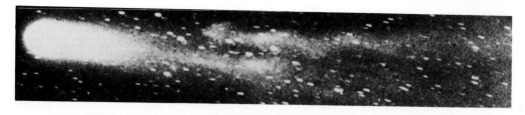

Fig. 1: Yerkes Observatory photograph of Halley's Comet on 1910 June 6 showing a Disconnection Event in the plasma tail.

tail. An example of a DE in Halley's Comet in 1910 is shown in Fig. 1. Another class of plasma-tail transient is helical wave structure, an example of which is shown in comet Kohoutek 1973XII in Fig. 2. The physical causes of DE's and helices are thought to be magnetic reconnection at interplanetary sector boundary crossings (Niedner and Brandt 1978, 1979; Niedner, Ionson, and Brandt 1981) and the Kelvin-Helmholtz instability (Ershkovich 1979), respectively.

The time-scales associated with DE's and helical waves are short. Although the persistance times of detached tails can be several days, recession speeds from the cometary head are large enough--50-200 km s^{-1} (Niedner 1981)--that an accurate kinematical description might require 30 or more images spanning each DE (never before achieved); hence, the need for near-hourly images. Moreover, high-time-resolution imagery is necessary to ascertain the onset time of the DE, a parameter of importance when attempting to find a correlation with solar-wind events. DE's typically occur every 1-2 weeks. The actual growth of helical waves has never been observed in a comet tail, and the inference is that the growth time must be $\ll$ 1 hr. As pointed out by many workers, knowledge of this time-scale is critical to our understanding of these waves. Despite the knowledge gained in recent years from a re-examination of historical data, much remains to be done on a bright comet which exhibits the full array of large-scale phenomena and whose arrival time is known. Only Halley's Comet meets these requirements.

2. THE LARGE-SCALE PHENOMENA DISCIPLINE OF THE IHW

The International Halley Watch (IHW) is well-known to most astronomers around the world. It has formally been endorsed by the IAU

Fig. 2: Helical waves in comet Kohoutek's plasma tail (JOCR)

at the 1982 meeting (in Patras, Greece), and it will almost certainly be the major ground-based effort directed at Halley's Comet. The purpose of the IHW is to encourage the observation of Halley around the world, to help guide the observations when necessary (or when asked), and to collect as much of the data as possible for inclusion in a permanent Halley Archive. Large-Scale Phenomena, one of seven IHW Disciplines, is administered by Discipline Specialist John C. Brandt. The goals of this Discipline are to assemble a uniquely complete set of imagery to advance the study of rapid plasma-tail disturbances, and to support the deep space comet missions by providing data on the state of the "entire comet" at the times of the encounters. The photometric effort of the Discipline is described in detail in this volume by D. A. Klinglesmith.

As a result of several calls for worldwide support, the Large-Scale Phenomena Network presently consists of ~75 facilities around the world, the majority of which are Schmidt cameras and astrographs, but also included are patrol cameras, Celestron-class Schmidts, and 35mm cameras. Given the basic goal of hourly coverage, the present network is in good condition although additional facilities are always vital (and welcome) as contingencies against poor weather, etc. A source of large concern is the understandable paucity of observatories in the Southern hemisphere: in 1986 March, the time of greatest scientific interest, Halley will primarily be a S. hemisphere object. Funds permitting, expeditions equipped with Celestron Schmidts to remote islands are not out of the question.

Wide-field observational techniques--including exposure times, emulsions, and filters--have been discussed elsewhere (Niedner, Rahe, and Brandt 1982). They are meant to be guidelines and suggestions, and are by no means rigid "instructions". The Large-Scale Phenomena Discipline Specialist Team sends out circulars several times each year and one of the issues planned for 1984 will discuss observational techniques in more detail. Individuals not already in the network or on our mailing list, but who wish to be, should contact J. C. Brandt or M. B. Niedner (Code 680, NASA/GSFC, Greenbelt, MD 20771, USA), or J. Rahe (Remeis Sternwarte, Universität Erlangen-Nürnberg, Bamberg, FRG).

REFERENCES

Barnard, E. E.: 1905, Astrophys. J., 22, pp.249-254.
Ershkovich, A. I.: 1979, Planet. Space Sci., 27, pp.1239-1245.
Klinglesmith, D. A.: 1983, this volume.
Niedner, M. B.: 1981, Astrophys. J. (Suppl.), 46, pp.141-157.
Niedner, M. B., and Brandt, J. C.: 1978, Astrophys. J., 223, pp.655-670.
Niedner, M. B., and Brandt, J. C.: 1979, Astrophys. J., 234, pp.723-732.
Niedner, M. B., Ionson, J. A., and Brandt, J. C.: 1981, Astrophys. J., 245, pp. 1159-1169.
Niedner, M. B., Rahe, J., and Brandt, J. C.: 1982, Proceedings of ESO Workshop "The Need for Coordinated Ground-based Observations of Halley's Comet", Paris, 29-30 April 1982, pp. 227-242.

OBJECTIVE PRISM SPECTROSCOPY OF THE TAIL OF COMET AUSTIN 1982 G

K. Jockers
Max-Planck-Institut für Aeronomie
D-3411 Katlenburg-Lindau, FRG
Visiting Astronomer at Observatorium Hoher List, Daun, FRG

L.G. Balázs
Konkoly Observatorium, Budapest, Hungary

The main emissions of the cometary plasma tail are due to CO^+ (400/402 and 425/427 nm) and H_2O^+ (aroud 650 nm). Emissions of CO_2^+ at 367 nm have also been observed (Swings and Page 1950). These ions relate much more directly to the presumed mother molecules of the cometary nucleus (like H_2O, CO_2, CO) than most strong emissions of the neutral cometary coma (Huebner and Giguere, 1980, see also Jockers, 1982). Therefore it is important to determine column densities of cometary ions. So far only very few measurements of ion column densities are available (Wyckhoff and Wehinger, 1976a, b; Arpigny, 1965).

In principle, the determination of ion column densities should be best possible from objective prism spectra. Since the wavelengths of the emissions are widely separated, the low dispersion of a prism may be sufficient. The reduced dispersion of a prism in the red spectral region is even favourable because the red H_2O^+ emissions are more extended in wavelength than the blue CO^+ emissions. A slitless spectrograph will always record the ion tail even if it deviates from the projected antisolar direction because of solar wind influence. The large field of a Schmidt telescope is also of advantage. If, during the exposure, the telescope is guided to follow the proper motion of the comet the spectra of the field stars will be widened and provide a means for an absolute calibration of the plate. Objective prism spectroscopy has been applied to comets before. As far as we know, however, no red sensitive emulsions were used so that the H_2O^+ emission was not recorded.

Objective prism spectra of Comet Austin 1982 g were obtained with the Schmidt telescopes of the Observatorium Hoher List (34 cm/138 cm) and of Konkoly Observatory (60 cm/180 cm). Reproductions of some of the plates are shown in Fig. 1. 103 a-E and 098-02 emulsions were used. A short dust tail and a narrow plasma tail are visible. To our surprise the plasma tail was strongest in the red spectral region (H_2O^+). Traces of CO^+, however, seem to be present on all plates. Reduction of the inclined spectra, which are close to plate background, turned out to be more difficult than expected so that we are not able to present ion

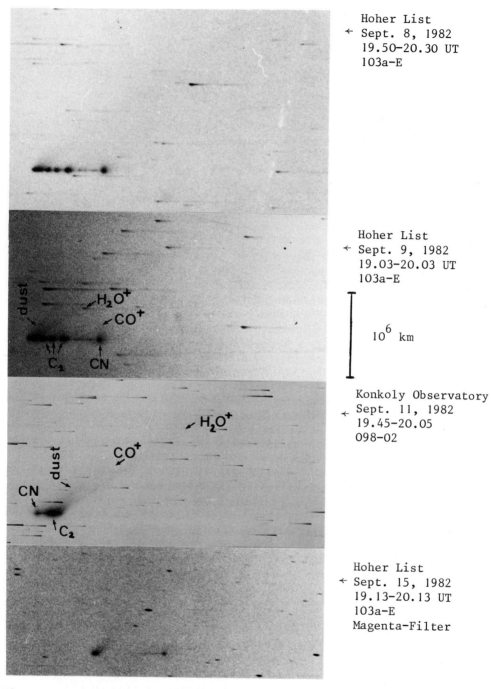

Fig. 1: Photographic reproductions of the objective prism spectra of Comet Austin 1982g. Note that on the Konkoly spectrum the dispersion is reversed.

column densities yet.

Two results which are apparent from Figure 1 may be worth mentioning in view of possible future observations.

1) For stellar applications Schmidt prisms usually are designed for a certain <u>linear</u> dispersion. For an extended object like a comet the <u>angular</u> dispersion is the figure of merit. Therefore Schmidt telescopes with a shorter focal length (and a larger refracting angle of the prism) are more useful. For large Schmidt telescopes the use of colour film (Lamy and Koutchmy, 1982) or separate exposures in a red or blue wavelength band defined by coloured glas filters and appropriate emulsions may be more useful. In this case, however, no widened spectra of background stars will be available for the calibration.

2) Since comets usually have small elongations from the sun the background sky is a serious problem. The spectrum of Sept. 15, obtained during a period of poor sky transparency, was taken through a magenta gelatine filter (T = 55 % at 425 nm, T = 84 % at 650 nm, T < 1 % between 470 and 605 nm) to reduce the contribution of the background sky. More experience with this filter is necessary.

K. Jockers thanks Prof. H. Schmidt and Prof. E.H. Geyer for the possibility to observe at the Observatorium Hoher List and A. Hänel, B. Nelles and F.T. Lentes for assistance at the telescope.

REFERENCES

Arpigny, C.: 1965, Mem. Acad. R. Belgique, Coll. 8 $\underline{35}$, Fasc. 5.
Huebner, W.F., and Giguere, P.T.: 1980, Astrophys. J. $\underline{238}$, 753.
Jockers, K.: 1982, ESO Workshop on Comet Halley (ed. P. Veron, M. Festou, K. Kjär), p. 149.
Lamy, Ph.L., and Koutchmy, S.: 1982, ESO Workshop on Comet Halley (ed. P. Veron, M. Festou, K. Kjär), p. 242.
Swings, P., and Page, Th,: 1950, Astrophys. J. $\underline{111}$, 530.
Wyckhoff, S., and Wehinger, P.A.: 1976a, Astrophys. J. $\underline{204}$, 604.
Wyckhoff, S., and Wehinger, P.A.: 1976b, Astrophys. J. $\underline{204}$, 616.

MISSING MATTER IN THE VICINITY OF THE SUN

John N. Bahcall

The Institute for Advanced Study, Princeton, New Jersey 08540

ABSTRACT

The combined Poisson-Boltzman equation for the gravitational potential is solved numerically for a detailed Galaxy model. The main result - obtained by comparing the calculated densities with observations of F dwarfs and K giants - is that about half of the mass density in the vicinity of the Sun has not yet been observed.

I have solved the combined Poisson-Boltzman equation for the gravitational potential of Galaxy models consisting of realistically large numbers of individual isothermal disk components in the presence of a massive unseen halo. The calculations were carried out with different assumptions about the unseen matter and the predicted number densities of F dwarfs versus height above the plane were compared with the observed distribution given by Hill, Hilditch, and Barnes (1979). I have also calculated the expected distribution of K giants and have compared these results with the observations described by Oort (1960).

The basic result I obtain is that the amount of *unobserved* material in the disk is at least as large as 50% of the observed material for all of the models that are discussed. In the best-estimate model, the amount of unobserved material is approximately equal to the amount of observed matter in gas, dust, and stars. This unseen disk material may be different from the missing mass inferred to be in extended galactic halos. Also, the disk material must be dissipational.

The special features of the work described here are: (1). the solutions are self-consistent (the star densities are determined by the common potential they create); (2). the Galaxy models contain realistically large numbers of disk components (from 10 to 23); (3). a massive halo is included; and (4). quantitative estimates of the uncertainties are determined. The details are given in Bahcall (1984) [Paper I]; only the results are summarized here. Earlier work is reviewed by Oort (1965).

Table 1 describes a Standard Galaxy model for the observed mass components. The disk luminosity function and the z-velocity dispersions are taken from Wielen (1974). The decomposition of the faint disk stars is based on the grouping according to H and K intensities described by Wielen (1974). The other components are taken from the standard Bahcall and Soneira (1980, 1984) model (hereafter, referred to as the B&S Galaxy model). The mass fractions are defined in terms of the total *observed* mass density (in stars, gas, and dust), i. e.,

$$A_i = \frac{\rho_i(0)}{\rho_{obs}(0)}. \tag{1}$$

Table 1. The Galaxy Model for Observed Components

Component	B & S Mass Fraction (A_i)	$\langle v_z^2 \rangle^{1/2}$ (km s^{-1})
Main Sequence Stars		
$M_V < 2.5^m$	0.021	4
$2.5^m \leq M_V \leq 3.2^m$	0.015	8
$3.2^m \leq M_V \leq 4.2^m$	0.031	11
$4.2^m \leq M_V \leq 5.1^m$	0.035	21
$5.1^m \leq M_V \leq 5.7^m$	0.025	20
$5.7^m \leq M_V \leq 6.8^m$	0.037	17
	0.0358	8
	0.0626	13
$M_V \geq 6.8^m$	0.0536	15
	0.0626	20
	0.0834	24
Giants	0.016	~20
White dwarfs	0.052	21
Atomic H and He and Molecular H and dust	0.469	4
Spheroid	0.001	~100 km/sec
Total	0.0958 $M_\odot pc^{-3}$	

The basic equation used is the combined Poisson-Boltzman equation for the potential. This equation describes how the gravitational potential at a given height above the plane can be calculated from the mass densities and velocity dispersions that are specified in the plane of the disk for any number of isothermal disk components - some observed as stars, dust, or gas and some unobserved - plus a halo mass density (constant, to first approximation, with height above the plane). The *dimensionless* form of the combined Poisson-Boltzmann equation is:

$$\frac{d^2\varphi}{dx^2} = 2\left[\sum_{i=1}^{N_{obs}} A_i e^{-\alpha_i \varphi} + \sum_{j=1}^{N_{unobs}} B_j e^{-\beta_j \varphi} + \varepsilon\right], \quad (2)$$

with $\varphi(0) = \left[\dfrac{d\varphi}{dx}\right]_0 = 0$. The gravitational potential has been divided by the square of a velocity dispersion which is taken here to be $(10 \text{ km s}^{-1})^2$ for numerical convenience. The quantities $\alpha_i = \left[(10 \text{ km s}^{-1})^2 / \langle v_z^2 \rangle_i\right]$, with a similar definition for the unobserved β_j. The height z above the plane is taken to be $z = z_0 x$, where the unit of length is $z_0 = \left[(10 \text{ km s}^{-1})^2 / 2\pi G \rho_{obs}(0)\right]^{1/2}$. The quantity N_{obs} is the total number of observed mass components (15 for the standard case considered here, see Table 1) and N_{unobs} is the number of unobserved mass components. For the B&S Galaxy model, $z_0 = 196.6$ pc. The unobserved mass fractions B_j are defined, by analogy with equation (1), as the ratio of the mass density in component j to the total observed mass density. Finally ε is defined as the ratio of $\rho_{halo}^{eff}(0)$ to $\rho_{obs}(0)$. The effective halo mass density is equal to the total halo mass density for a constant rotation curve but is slightly different if the rotation curve is not exactly flat.

The isothermal approximation adopted here requires that the absolute value of the logarithmic derivative of the velocity dispersion be much less than the absolute value of the logarithmic derivative of the density, i.e., $\left|\dfrac{\Delta(\langle v_z^2 \rangle)}{\langle v_z^2 \rangle}\right| << \left|\dfrac{\Delta \rho(z)}{\rho}\right|$.

The fractional change in the velocity dispersion of the F stars is less than or of order 0.1 over the first 200 pc in z (cf. the first four rows in Table 6 of Hill et al. 1979), while the density changes by a factor of 3. Thus the isothermal approximation appears to be well satisfied for the Hill et al. (1979) sample of F stars. Radford (1976) found that the velocity dispersion of G and K giants is constant to an accuracy of about 10% for z less than 400 pc. Hartkopf and Yoss (1982) obtained a similar result for the separate velocity dispersions of metal-poor and normal composition giants (cf. their Figure 8). Eggen (1969) found an approximate constancy of the velocity dispersion of A stars to about 300 pc. It is well known that at moderate and large values of z each separate disk component can be described by an exponential density profile (see the many references to the original data that are given in the caption of Figure 2 of B&S), which suggests that the isothermal approximation is also reasonable for the other disk stars.

I have neglected also the cross terms involving $\langle v_z v_R \rangle$. Assuming Oort's (1965) hypothesis of a tilted velocity ellipsoid pointing in the direction of the Galactic center, the ration of omitted to include terms is of order $(z z_s / h R)$, where z_s is the scale height of the stars and h is the scale length of the disk. At the solar position, the correction due to the cross terms is less than 0.01 for all $z \lesssim 1$ kpc.

The results obtained with the Galaxy model described in Table 1 are $\rho_{Total}(0) = 0.188 \pm 0.02\, M_\odot \text{pc}^{-3}$, σ_{Disk} (to infinity) = $64 \pm 5\, M_\odot \text{pc}^{-2}$, and $(M/L)_{Disk} = 2.7 \pm 0.4$ solar units. For this calculation, the unobserved matter density was assumed proportional to the observed matter density everywhere. The best-estimate volume density given above is 4% larger than the best-estimate when all of the disk stars fainter than $M_V \geq 6.8$ mag are combined (the best-estimate column density is 2% smaller in the present case).

I have solved (cf. Paper I) numerically the differential equation (2) for a number of assumed distributions of the unseen matter and, in each case, for many values of the parameter characterizing the amount of unobserved material. The results of the theoretical models were fit to the observed spatial densities of F stars tabulated by Hill et al. (1979). The total number of mass components that were included varied between 11 and 31, depending on the assumption made about the unobserved matter. In Paper I, all of the faint disk stars ($M_V \geq 6.8$ mag) were combined.

Table 2 summarizes the requirements on the missing mass in the disk assuming that it is distributed like the observed interstellar matter, faint M-dwarfs, white dwarfs, or young massive stars.

Table 2. Some Candidates for the Missing Mass in the Disk

Candidate for the Missing Mass	Observed Mass Density ($M_\odot pc^{-3}$)	Unobserved Mass Density ($M_\odot pc^{-3}$)	$\frac{\rho_{unobs.}}{\rho_{obs.}}$
Interstellar Matter	0.045	0.144	3.2
Faint M-dwarfs ($M_V \geq 12.5^m$)	0.0093[a]	0.07	7.4[a]
White dwarfs	0.005	0.07	14
Young massive stars ($M > 1.6 M_\odot$)	0.002	0.144	35

[a] Assumes that the number of stars per absolute magnitude is constant for $16.5^m \geq M_V \geq 13$ and is equal to the value given by Wielen (1974) at $M_V = 12.5^m$.

The total volume density is better determined than the total column density. The extreme range of volume densities that were found in Paper I corresponds to a ratio of (maximum allowed/minimum allowed) = 1.5, versus 2.9 for the corresponding ratio of extreme column densities. Moreover, the theoretical uncertainties (isothermal approximation, separable potential) affect the calculated column density more strongly than the volume density.

Oort (1960) gives an often reproduced curve (smoothed) distribution of K-giants as a function of height above the plane for z=0 to 3 kpc. I have begun a reinvestigation of the K-giant distribution using this data set.

The amount of disk matter can be inferred in a simple way only if the tracer stars are essentially uncontaminated by spheroid stars.

In the B&S Galaxy model, the ratio (spheroid K-giants)/(disk K-giants) is: 0.00 at z=0, 0.02 at z=0.6, 0.09 at z=0.9 kpc and reaches unity at 1.2 kpc. For larger values of z, the spheroid K-giants are much more numerous than the disk K-giants, the ratio reaching 72 at 3 kpc. It is not known what fraction of the spheroid K-giants were included in Oort's data set, which was based on the low dispersion HD and BSD catalogs. In fact, the fraction of spheroid K giants that are listed as K giants in these catalogs must depend upon the - as yet unknown - metallicity gradient of the spheroid. I have, therefore, come to the conclusion, reluctantly, that Oort's data cannot be used beyond a maximum distance which is somewhere between 0.6 to 0.9 kpc.

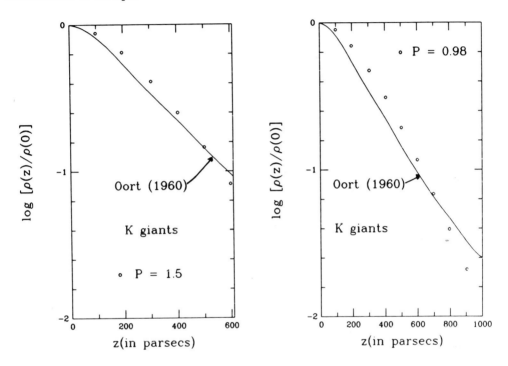

Figure 1 shows the comparison between a best-fitting theoretical model and Oort's data for two-values of the maximum distance. For the models shown in Figure 1, I have assumed a velocity dispersion for the K-giants of 20 km/s [Radford 1976, Hartkopf and Yoss 1982] and a halo mass density corresponding to $\varepsilon=0.1$. The unseen disk material was assumed proportional to the observed disk material.

The results for the K-giant sample are consistent with the analysis, described above, of the F dwarfs. If Oort's curve is used to 0.6 kpc, the best-fitting total mass density is $\rho_{Total}(0) = 0.24 \, M_\odot \, pc^{-3}$ and the column density is $\sigma(\infty) = 65 \, M_\odot \, pc^{-2}$. If the analysis is extended to 0.9 kpc, the best-fitting values are $\rho_{Total}(0) = 0.20 \, M_\odot \, pc^{-3}$ and $\sigma(\infty) = 69 \, M_\odot \, pc^{-2}$. Unfortunately, the results are rather sensitive to the maximum distance above the plane that is accepted in the calculation and to the not very well known velocity dispersion of the K-giants $(\rho_{Total} \sim <v_z^2>_{K-giants}^{3/2})$.

The missing matter at the solar position must be in a disk since if it were in a spheroidal component the calculated rotation velocity at the solar position would be much larger than the observed value (see Section 5b of Bahcall and Soneira 1980). The reason that so much spheroidal material would be required is that a given amount of matter is much less efficient at producing the needed z-acceleration if it is placed in a spheroidal rather than a disk configuration, approximately in the ratio of the scale heights (3 kpc to 0.3 kpc).

The slope of the faint end of the disk luminosity function that is required in order to hide all of the missing matter in stars fainter than 0.1 $M_\odot$ can be calculated easily. Suppose that the disk luminosity function has the form: $\Phi(M_V) \propto 10^{\gamma M_V}$ stars per absolute visual magnitude, for faint ($M_V \geq 13$ mag) stars. Then the minimum γ that is required varies from about 0.01 to about 0.05, depending mostly on the assumed mass - visual luminosity relation for faint dwarfs.

The observationally important implication is that all of the missing mass in the disk could be in faint stars *if* the slope of the disk luminosity function has a small *positive* value for large absolute visual magnitudes ($M_V > 13$ mag). This work was supported by the National Science Foundation grant no. PHY-8217352 and NASA grant no. NAS8-32902.

REFERENCES

Bahcall, J. N. 1984, *Ap. J.* **276** (Paper I).
Bahcall, J. N. and Soneira, R. M. 1980, *Ap. J. Suppl.* **44**, 73-110.
Bahcall, J. N. and Soneira, R. M. 1984, *Comparisons of a Standard Galaxy Model With Stellar Observations In Five Fields*, (submitted to Ap. J.)
Eggen, O. J. 1969, *P. A. S. P.* **81**, 741.
Hartkopf, W. I. and Yoss, K. M. 1982, *Astron. J.* **87**, 1679.
Hill, G. Hilditch, R. W., and Barnes, J. V. 1979, *M.N.R.A.S.* **186**, 813.
Oort, J. 1960, *Bull. Astr. Inst. Netherlands* **15**, 45.
Oort, J. 1965, in *Galactic Structure*, ed. A. Blaauw and M. Schmidt (Chicago: University of Chicago Press), p.455.
Radford, G. A. 1976, PhD Dissertation (esp. Table 4, page 26), Cambridge University.
Wielen, R. 1974, *Highlights of Astronomy*, **Vol. 3**, 395, ed. Contopoulos, G., (Dordrecht, Holland: D. Reidel).

GALACTIC RESEARCH WITH SCHMIDT-TELESCOPES

Wilhelm Becker
Astronomical Institute of the University of Basle

The use of Schmidt Telescopes for the determination of space density gradients is discussed. Space densities can be determined by a three colour photometry for main sequence stars with M< 8.0 and later type giants of population I and for metal poor stars in higher galactic latitudes with M< 7.0. A general synthesis of the present available results is not yet possible. But the determination of a density gradient for later type giants along a galactic radius, the distribution of metal poor stars in a galactic meridian and around the galactic centre seems to be possible with the existing observational material.

In order to obtain a correct idea about the Schmidt-telescope, it is useful to consider the situation before its invention and its actual and possible impact upon galactic research.

Up to this event around 1931, galactic research was dominated by the astrograph and by the parabolic mirror, the former allowing photometry as well as spectral classification of stars in larger fields, but with brighter limiting magnitude. As typical limits for an astrograph with four lenses of a diameter of, say, 40 cm, one may assume about $V = 16^m$ for photometry and $V = 13^m$ to 14^m for spectral classification. Photometry of fainter stars with a large parabolic mirror, however, may have been limited then, at about $V = 20^m$.

Already the first description of the Schmidt-telescope in 1931 let it appear as the ideal successor of the astrograph in galactic research: the error-free field size could be considerably enlarged and, even more important, the optics were no longer restricted by the diameter limit of about 40 cm given for multiple-lens astrographs.

It should, however, not be disregarded that the new frontiers for galactic research were opened up not only by the invention of the Schmidt-telescope but also by two other, approximately contemporary, discoveries bringing about the access to larger fields and to fainter stars, as well as to the near ultraviolet spectrum barred to the astrograph, i.e.: UV-transparent glass and aluminization instead of silverplating of mirrors. All three accomplishments together opened up the great new possibilities for galactic research.

Those astronomers who continued to work with Schmidt-telescopes along Kapteyn's line were left to their own initiative with respect to their choice of galactic fields. The first such investigations stem from the Cleveland Observatory where a large program of nine galactic fields was carried out by Mc Cuskey and Nassau.

The situation was rather drastically changed by the first opportunity to realize a new photometric method at the Schmidt-telescope, i.e.: three-colour photometry, proposed already in 1938 in the form of the RGU-system with the purpose of replacing spectral classifications in pushing the limiting magnitudes by several unities practically to the limit of an optical telescope. Without spectral types, it brought about the determination of absolute magnitudes as well as of interstellar reddening. The reason for that is - roughly speaking - the first order deviation of stellar radiation from black body radiation being a function of luminosity, and its not being influenced by interstellar extinction.

The most informative among the three spectral regions of three-colour photometries is the ultraviolet one around about 3700 Å which became realizable only owing to the mentioned inventions of UV-transparent glass and mirror aluminization.

The method of three-colour photometry was applied first with the help of the Schmidt-telescopes of the Cleveland- Ann Arbor- and the Asiago-observatories, and later, from 1961 on, with the 48 inch Palomar Schmidt. The limiting magnitude reached about 19^m5 in G. The possibilities were enormously enhanced, thereby, leaving far behind the praxis, however.

I should like to talk just about this praxis and some of its results, i.e.: about the investigation of star fields in the Milky Way and in higher galactic latitudes. The ultimate goal of the photometry in these fields is the evaluation of density functions, possible for main-sequence stars not less luminous than about $M(G) = +8^m$. These functions are the basis for a synthetic view of the density distribution. Since it is hardly foreseeable if such a synthesis can be achieved, it is appropriate to choose the position of the fields such that they let expect at least partial synthesisses restricted to certain facts.

Therefore, the fields within the Milky Way were mainly chosen along the directions close to the galactic centre and anticentre, and the ones in higher galactic latitudes lie mostly close to the galactic meridian

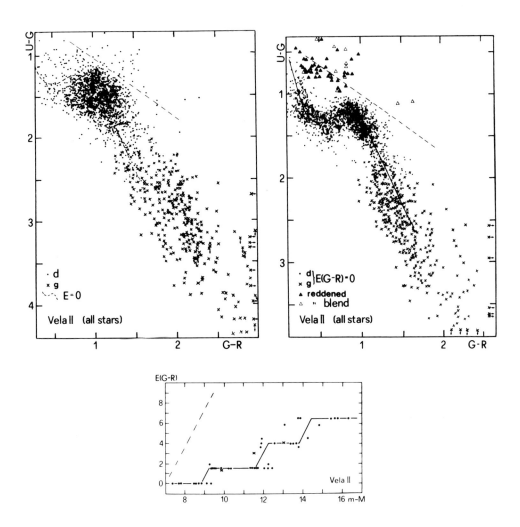

Figure 1: The distribution of all stars in the field, down to the limiting magnitude in the two-colour diagram; left: observed values of the colour indices, right: colour indices corrected for interstellar reddening (below).

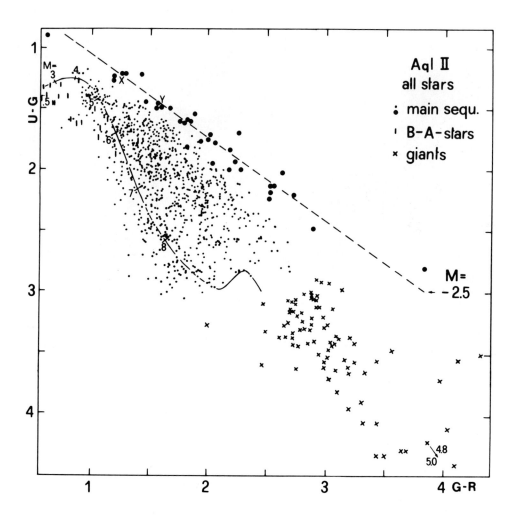

Figure 2: The distribution of all stars in this star-poor field in the two-colour diagram, being the result of an interstellar reddening which increases with increasing distance.

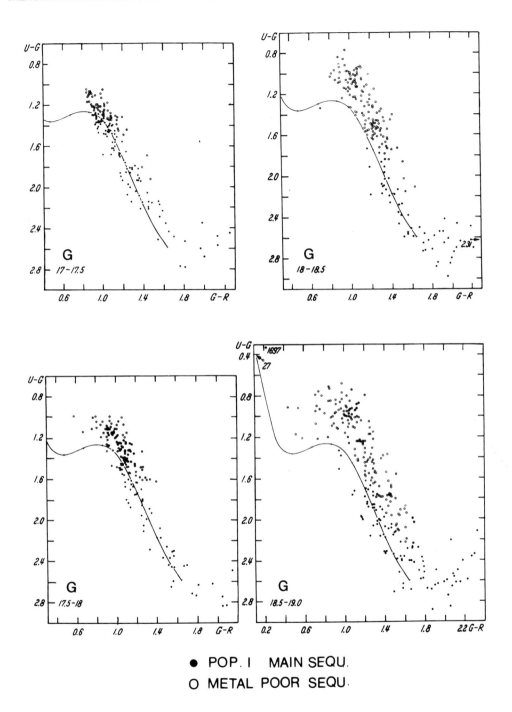

Figure 3: The metal-poor main-sequence stars shifted along the blanketing lines (up and to the left).

(perpendicular to the galactic plane and containing the sun and the galactic centre).

In all these fields, between 500 and 2500 stars should be measured down to the limiting magnitude, leading to field sizes of between 0.1 and 2.5 square degrees for star-rich and star-poor regions, respectively. These numbers can easily be managed without automatic reduction devices, even for at least five plates per colour, necessary for keeping the errors reasonably small. It is upon these errors, of course, that depends the accuracy of the absolute magnitudes and of the colour excesses and, therefore, the reliability of the density functions, too. Also the identification of the late-type giants and of the metal-poor main-sequence stars is sensibly influenced by the accuracy of the colour indices. The mean errors of the magnitudes lie predominantly between $+0^m02$ and $+0^m03$ and they increase to about twice these values for the faintest stars.

The interdependent discussion of the two-colour diagrams plotted for consecutive intervals in apparent G-magnitude presents a good control for the successful total elimination of systematic errors in the catalogue magnitudes of a given field. If the stars of a field, after correction for interstellar reddening, do not scatter evenly around the main-sequence and the late-type giant branch, but show, instead, a systematic deviation, it is due to a rest-error of up to several hundredth of a magnitude from previous experience, which can usually be identified and eliminated according to the specifications of the deviation.

Up to now, nine volumes, containing the magnitudes and colour indices of 7000 to 10 000 stars each, from totally 48 fields have appeared as Publications of the Basle Astronomical Institute.

Three typical examples shall give an idea about the vastly varying behaviour of two-colour diagrams in different star fields: the first gives the distribution of all stars in a star-rich Milky Way field with low extinction, the second the one in a galactic strong-absorption field and the last the distribution pattern in a field of higher galactic latitude. The first figure shows most prominently the abundance of stars in the middle spectral classes and their separation from the late-type giants. The discussion leads at first to the interstellar reddening function as seen in the figure. After the elimination of the reddening effect the stars are distributed in the two colour diagram as seen at the right side diagram in figure 1. The second figure is typical for a complete mixing of stars from almost all spectral classes, but also for the existence of a large amount of late-type giants. The discussion leads to an interstellar reddening which increases with increasing distance up to about 2.5 magn. The resulting density gradients are not as good as in the case before. The last figure reveals the deviating position of metal-poor main-sequence stars.

For the determination of the density-functions one does not use these overall diagrams, containing all the stars of a field, but rather the already mentioned partial diagrams, fractioned according to consecutive intervals in apparent magnitude, yielding much more detailed information. Unlike the methodic tool of the two-colour diagrams, the density functions for the different fields are directly comparable and form so the basis for a synthesis. This term, however, - as said before - should not awake exaggerated expectations. Even with a considerably increased number of fields, the region covered with our density functions will remain relatively limited. Due to the given limits for Schmidt-telescopes, the density functions will be complete within a few hundreds of parsecs for the absolutely faintest and up to about 10 000 pc for the absolutely brightest stars. Much more favourable is the situation if the synthesis aims at more specific points of view. As said before, the fields have actually been selected correspondingly. I should like to point to three such results here: First, to a density gradient for late-type giants along a galactic radius, from centre to anticentre, second, to the density distribution of metal-poor stars within the accessible part of the galactic meridian, and third, to the behaviour of the metal-poor population in the neighbourhood of the galactic centre.

For the density functions of the late-type giants in the directions to the galactic centre and anticentre, we had at our disposition four fields in each case, covering up to 3 and 11 kpc, respectively, depending on the limiting magnitudes and on the extinction situations. The density functions of all these fields (fig. 4) can be interpreted homogeneously: Assuming the galactic centre at a distance of about 8 kpc from the sun, a density maximum is reached at about 6 kpc towards the sun, where from, the density decreases steadily, passing the solar neighbourhood, out to a distance of about 19 kpc from the galactic centre by a factor of 200.

Within the distance interval from 2 to 5 kpc from the galactic centre, the density remains at a constant level about 2.5 times below the maximum. Because of the strong extinction, it is not possible to approach closer distances from the centre.

The density distribution of the metal-poor main-sequence stars north to the disk can be evaluated from the results in seven fields, two of them pointing to the anticentre side, only. Figure 5 shows the isodensity lines in the galactic meridian for the three consecutive intervals in absolute magnitude M(G): ($4^m 5^m$), ($5^m, 6^m$) and ($6^m, 7^m$). On the whole, their behaviour is about similar, with the exception that the gradient for the absolutely brightest group ($4^m, 5^m$) is somewhat too steep for SA 54 and a bit too flat for SA 57. Apart from that, the systematic behaviour consists in the fact that the isodensity lines approach

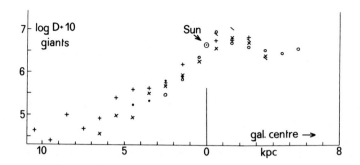

Figure 4: The density function for late-type giants obtained from four fields in the direction to the galactic centre and from four other fields in the one to the anticentre, reduced to z = 0 by use of the function D(z) for the solar neighbourhood, because of the different galactic latitudes (+2.°5 to -3.°6).

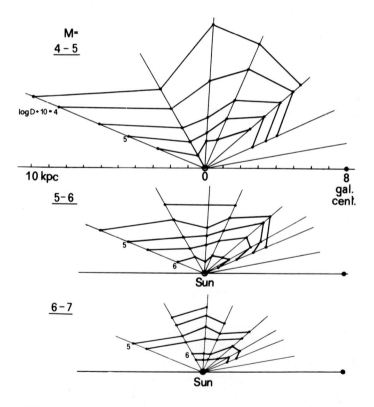

Figure 5: Isodensity lines for three consecutive intervals in M(G) in the galactic meridian (perpendicular to the galactic disk and containing the galactic centre and the sun).

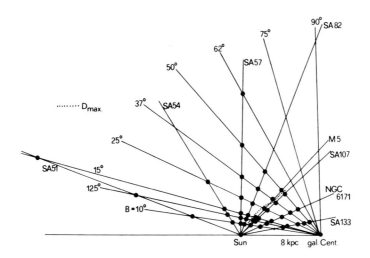

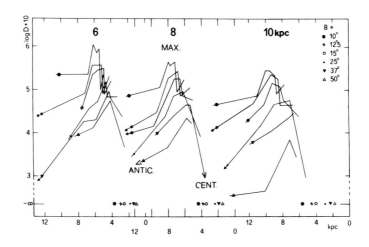

Figure 6: Upper part: Transformation of the heliocentric density gradients for the metal-poor main-sequence stars within the galactic meridian of figure 5 into galactocentric gradients. Lower part: Galactocentric density distribution for different galactocentric latitudes and different solar distances from the galactic centre (6, 8 and 10 kpc). The maximum densities follow the dashed line in the upper part.

continuously the galactic plane with increasing angular distance from the galactic centre, starting at about $40°$ to $65°$ (SA 107 to SA 82, respectively). At smaller angular distances, however, the isodensity lines behave completely different: approaching the galactic centre, the densities decrease strongly, and at about $10°$ from the centre the metal-poor stars are almost completely missing in the corresponding fields. The details of this drastic density drop, however, are not yet fully revealed by the available data.

One can visualize this density decrease towards the galactic centre more perceptually by transforming the observed (solar-centered) density gradients into such which have their origin in the galactic centre (fig. 6). In all their directions, the densities first increase from an undeterminably small value near the centre steeply to a maximum, where from, they decrease again but slowly. These maxima lie rather exactly on an ellipse centered in the galactic centre and with its major axis in the galactic plane. They are reached at distances between about 7.5 and 5.0 kpc from the galactic centre at $10°$ and $50°$ galactocentric latitude, respectively.

Further partial synthesisses might consider: the density distribution in the farther solar neighbourhood, the possible connection between densities and spiral structure, the density behaviour perpendicular to the galactic plane and the distribution of the interstellar absorption clouds whose distances follow as a by-product from the discussion.

All these results, of course, will obtain considerably more reliability if the number of fields will be increased, and they could penetrate into deeper space if the limiting magnitudes of the photometric measurements could be pushed to fainter values. To increase the number of fields presents no serious problem, more critical is, however, the supply of photoelectric scales down to sufficiently faint magnitudes. The spatial extension encounters principal difficulties, as far as Schmidt-telescopes are concerned. They stem from the fact that the exposure times for U-plates at the big Schmidt-telescopes (about 40 minutes at the 48" Palomar Schmidt) cannot be extended much more without darkening the plate background to a degree where the measurement of faint stars at the iris-photometer becomes too inaccurate. I suppose that only the use of large parabolic mirrors might help to overcome this threshold.

SPACE DISTRIBUTION OF RED GIANTS AND THE GALACTIC STRUCTURE

K. Ishida
Tokyo Astronomical Observatory
Mitaka, Tokyo, Japan

ABSTRACT

Stellar content contributing to near IR radiation do not show radial differentiation in the Galaxy. Late-type giants and supergiants supply about 70% of the total volume emissivity at the K band, in the solar vicinity within 1 kpc, and also at the distance of several kpc in the Scutum region.

1. STELLAR CONTENT IN THE SOLAR VICINITY

Space number density of stars in the immediate solar vicinity and its dispersion from the galactic plane were derived for every spectral-luminosity groups by Ishida & Mikami (1982) by simulating the cumulative number of apparent magnitude for the stars in the catalogue of Two-Micron Sky Survey done by Neugebauer & Leighton (1969). Stellar content thus derived makes possible to predict volume emissivity at every broad color bands in the visual and near IR wavelength. At the K band, late-type giants and supergiants supply about 70% of the total volume emissivity in the solar vicinity within 1 kpc.

Accordingly red giant is a most useful probe to investigate over-all space distribution of stars in the Galaxy, because 1) they have bright absolute magnitude in near IR wavelength where interstellar extinction is small (Ishida & Mikami 1978), 2) show clear spectral features to be discriminated from early type stars (Iijima & Ishida 1978), 3) belong to the disk population (Mikami & Ishida 1981), and 4) are a major contributor to the volume emissivity of near IR radiation of the Galaxy (Ishida & Mikami 1982).

2. OBSERVATIONAL MATERIAL

We have three different kinds of observational material available to specify a model of over-all space distribution of stars in the Galaxy;

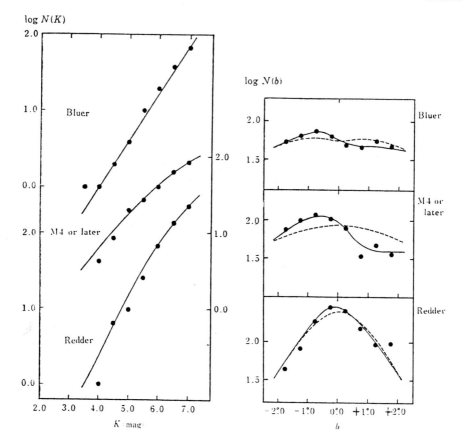

Fig. 1. Cumulative number function of point sources in 1.5 sq deg. of the Scutum region.

Fig. 2. Galactic lat. distribution of point sources brighter than K=7 in the Scutum.

viz. 1) objective prism plates (62.6"/mm and 1100Å/mm at the atmospheric A band) taken with the Schmidt telescope at the Kiso Observatory of the Tokyo Astronomical Observatory, which are used to know spectral type of stars out to a few kpc from the sun, 2) survey of two-micron point sources with broad band photometer at the Agematsu IR Observatory of Kyoto Univ., which presents information of deep inner part of the Galaxy, 3) contour map of two-micron surface brightness obtained by the small balloon-born telescopes (Hayakawa et al 1978, Okuda 1981), which is an edge-on view of the inner part of our Galaxy.

3. DETECTION

Nearly 900 point sources brighter than about 8 mag in the K band were detected in 1.5 sq deg. of the Scutum region ($l=26°-27°, -2° \leq b \leq +2°$) with the 1-m telescope (Kawara et al 1982). They were observed by broad band photometer for the H and K bands, giving K mag and H-K color. Cumulative number function N(K) of apparent mag K=4.0-7.0 is compiled for 469 point sources, and about 400 sources fainter than K=7.0 are rejected.

4. CLASSIFICATION

The near IR point sources were identified to stellar images on the photographic plates, and then to spectral images on the objective prism plates taken on IN emulsion through RG 695 filter. Spectral type of "M4 or later" is assigned to 115 stars, of which mean color is H-K=0.68. The rest of stars are divided into "bluer" group (H-K<0.68) of 93 stars and "redder" one of 261 stars. Cumulative number function N(K) is drawn for each of the three groups in figure 1 and galactic latitude distribution in figure 2. The "bluer" group is expected to be K-type and early M-type giants, the "M4 or later" is self-explanatory, and the "redder" group is supposed to be red supergiants at large distance contaminated with extremely cool giants, respectively. Mean absolute magnitudes and dispersions from the mean appropriate for each group are assumed for luminosity function.

5. MODEL FITTING

The luminosity function is integrated with weight of space distribution of number density of stars on each line of sight to simulate the cumulative number function and galactic latitude distribution. The model of space distribution contains main structural features of the exponential disk (scale radius 2.3 kpc), the 5 kpc ring, and the bulge of the Galaxy. The model of space distribution of interstellar absorbing matter also consists of three components. See Mikami et al (1982) for detailed discussions.

6. STELLAR CONTENT IN THE SCUTUM REGION

The cumulative number function and the galactic latitude distribution of the "bluer", "M4 or later", and "redder" groups are simulated by integrating with the space distribution of the assumed spectral-luminosity group. The numerical parameters in the space distribution include dispersions of the space distribution from the galactic plane, and the space number density in the immediate solar vicinity. The simulation is successful for the cumulative number function of the two-micron point sources with the assumed stellarcontents. The observed asymmetric distribution to galactic latitude is due to heavy interstellar extinction deviated toward north from the plane to a few kpc from the sun (see figure 2).

The over-all feature of the surface brightness contour map is also reproduced by the model if we assume that 30% of volume emissivity is radiated by objects which do not show up in the cumulative number function of the present survey observations. It is concluded that late-type giants and supergiants occupy about 70% of the total volume emissivity at the K band. This is nearly the same proportion as that obtained in the solar vicinity within 1 kpc. It is indicated that stellar content contributing to volume emissivity of near IR radiation do not show radial differentiation in the Galaxy.

Some local deviations from the proposed model are interpreted as a galactic window in a case (Hamajima et al 1981, Ichikawa et al 1982), and a local concentration of supergiants in another case (Mikami et al 1982).

REFERENCES

Hamajima, K., Ichikawa, T., Ishida, K., Hidayat, B., and Raharto, M.: 1981, Publ. Astron. Soc. Japan 33, 591.
Hayakawa, S., Matsumoto, T., Murakami, H., and Uyama, K.: 1978, Publ. Astron. Soc. Japan 30, 369.
Ichikawa, T., Hamajima, K., Ishida, K., Hidayat, B., and Raharto, M.: 1982, Publ. Astron. Soc. Japan 34, 231.
Iijima, T., and Ishida, K.: 1978, Publ. Astron. Soc. Japan 30, 657.
Ishida, K., and Mikami, T.: 1978, in A.G. Davis Philip and D.S. Hayes (eds.), the HR Diagram, IAU Sym. No. 80, Reidel, p. 429.
Ishida, K., and Mikami, T.: 1982, Publ. Astron. Soc. Japan 34, 89.
Kawara, K., Kozasa, T., Sato, S., Okuda,H., Kobayashi, Y., and Jugaku, J. : 1982, Publ. Astron. Soc. Japan 34, 389.
Mikami, T., and Ishida, K.: 1981, Publ. Astron. Soc. Japan 33, 135.
Mikami, T., Ishida, K., Hamajima, K., and Kawara, K.: 1982, Publ. Astron. Soc. Japan 34, 223.
Neugebauer, G., and Leighton, R.B.: 1969, Two-Micron Sky Survey, NASA SP-3047, Washington, D.C. (IRC).
Okuda, H.: 1981, in C.G. Wynn-Williams and D.P. Cruikshank (eds.), Infrared Astronomy, IAU Sym. No. 96, Reidel, p. 247.

OBJECTIVE PRISM SURVEY OF THE OUTER GALACTIC HALO

Kavan Ratnatunga and K.C. Freeman
Mount Stromlo and Siding Spring Observatories
Research School of Physical Sciences
The Australian National University

The aim of this survey is to locate samples of very distant field halo stars, to study the kinematics and metal abundance distribution in the outer regions of the galactic halo. Field halo K giants were chosen for the study, as they are intrinsically bright stars whose evolution is well understood. Halo K giants, near the tip of the giant branch and between 10 and 40 kpc from the sun, will have apparent magnitudes in the range $13 < V < 18$, and colours $(B-V) > 0.9$.

The distant halo K giants in this range of colour and apparent magnitude are greatly outnumbered by the nearby disk K dwarfs. Therefore a very efficient procedure is required to separate the two populations. The Mg b and MgH feature near $\lambda 5100$ A is a good luminosity discriminant in the colour range $0.9 < (B-V) < 1.4$. It can easily be identified on medium resolution objective prism spectra. Our objective prism plates were therefore taken on hypersensitized IIIa-J emulsion with a Schott GG475 filter, to select the wavelength range from 4700 A (filter cutoff) to 5400 A (emulsion cutoff). This limited spectral range allowed longer sky limited exposures, to reach fainter limiting magnitudes. The short image length also reduces problems of image crowding, which is an important consideration when exposing to faint magnitude limits, even in high latitude fields.

Stars of the required colour range are first selected by photographic photometry, using direct IIIa-J + GG395 and IIIa-F + RG630 Schmidt plates. Locations, magnitudes and colours of all stars in a region are determined by scanning the full Schmidt plate on the Mount Stromlo PDS microdensitometer. Automated software has been developed to locate and estimate magnitudes and colours of all stars from these PDS scans. Although for practical reasons the digitising is done with a 50 micron square aperture, the images are centered to an accuracy of 7 microns and the standard errors of the measured magnitudes and colours are 0.07 and 0.10 magnitudes, respectively.

For automated analysis of the spectra, the individual objective prism spectral images of all stars located in our range of colour and

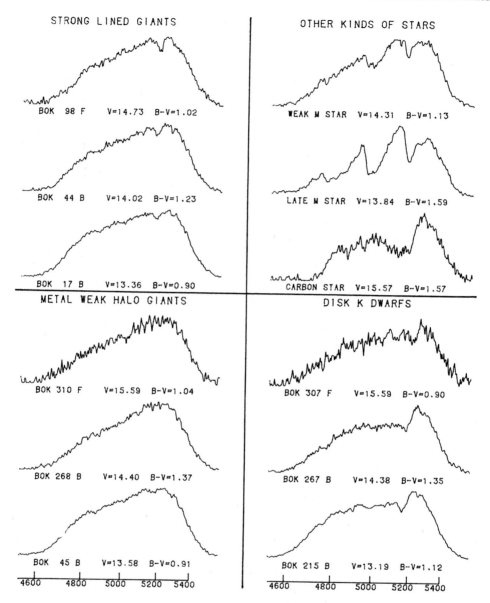

Figure 1. Plots of objective prism spectra, after image analysis. The vertical axis is intensity, on the instrumental system (arbitrary units). The nonlinear wavelength scale (Å) is shown. The panels give three examples each of strong-lined giants, metal-weak halo giants, disk K dwarfs, and late-type stars, over the range of magnitude and colour as given. The differences between these stellar types are clearly visible, and form the basis for the classification procedure used in this survey. The spectra are from an ESO objective prism plate, 450 Å/mm at Hγ, using IIIa-J emulsion and a GG475 filter.

apparent magnitude are scanned on the PDS as 2D arrays. The image profiles perpendicular to the dispersion, for a few photoelectric standards in the field, are used to bootstrap a wavelength-dependent intensity calibration. The mean cross-sectional profile of these standards is also used as a template in the analysis, for integrating the intensity at each wavelength. This procedure is able to resolve any partly merged spectra, and also corrects for saturation in the brighter images. The software uses the maximum amount of useful information in the image, and therefore gives spectra with the best possible resolution. Spectral features which are not visible in the faint images, even at high magnification, are seen on the digital spectra after image analysis. Since the analysis of the PDS scan of the image is automated, it may be done online, and only the evaluated spectrum need be recorded for classification.

Figure 1 shows examples of spectra determined by this procedure. Qualitative classification from the digital plots turned out to be very efficient in identifying sufficiently large samples of halo K giants, so quantitative methods with automated spectral classification have not yet been fully developed.

Most of the objective prism plates used for this survey were taken with the ESO 1-m Schmidt in Chile (4-degree prism, dispersion 450 A/mm at $H\gamma$). All the direct plates were taken with the UKSTU 1.2-m Schmidt in Australia, and also some objective prism plates (3-degree prism, dispersion 830 A/mm at $H\gamma$). Plates of the northern fields have been taken for us with the 1-m Schmidt of the Tokyo Observatory at Kiso, Japan (2-degree prism, dispersion 800 A/mm at $H\gamma$). The success of this survey is entirely dependent on the very kind cooperation of these Schmidt observatories, and we wish to take this opportunity to thank them.

We have completed surveys of about 20 square degrees in each of three fields (SA141, 189, 127) which can be easily observed from the southern hemisphere. Locations, magnitudes and colours of about 80,000 stars have been determined, complete to a limiting magnitude of V = 18. Objective prism spectra of about 6000 stars, with colours (B-V) > 0.9 and magnitudes in the range 13 < V < 16, have been measured and classified. Two hundred metal-weak halo giants, between 10 and 40 kpc from the sun, have been positively identified in this survey, and are now being studied with slit spectroscopy, to measure their radial velocities and chemical abundances. About 400 giants with metal abundances similar to 47 Tuc have also been discovered. Two faint carbon stars with V = 15.5 were also found.

These automated survey procedures have been developed and used successfully to locate large samples of distant field halo K giants. The same procedures could be applied to make any other similar objective prism survey to faint magnitudes a practical and quick process.

A SURVEY FOR O-B STARS IN THE PUPPIS WINDOW

David J. Westpfahl, Jr.
Physics Department, Montana State University,
Bozeman, MT 59717   U.S.A.
(visiting astronomer, CTIO, supported by NSF under
contract AST 78-27879)

I. INTRODUCTION

In the 1950's Haro (Haro and Luyten 1962) developed a survey method in which three images, roughly corresponding to U,B, and V, are exposed on one plate. The images are separated by several arc seconds by moving the telescope between exposures. Exposure times are chosen so an unreddened star of spectral type A0 to A5 shows three images roughly equal in size and density. The plates are examined by eye to find stars by their colors. A red star shows a strong central image flanked by two weaker images, while a blue star shows a weak central image flanked by two stronger ones. This method allows surveys for stars and galaxies with very red or very blue colors to be carried out with any telescope, and in particular with the Schmidt telescopes at Tonantzintla and at Palomar which were not equipped with objective prisms.

II. THE SURVEY METHOD

This method has several advantages over the standard objective prism technique. 1. The method is useful over the range 12<V<18 on one plate (Haro and Luyten 1962), or about two magnitudes wider than objective prism surveys. 2. Crowding is less of a problem in dense Galactic fields because the three images cover less area on the plate than even an unwidened spectrum. 3. The method remains useful even during poor seeing when objective prism images lose their detail. 4. It works equally well for stars and extended objects, such as galaxies and gaseous nebulae.

There are also some disadvantages to the method. 1. It gives incomplete spectral information, so cannot be used to identify objects with emission lines. 2. It is selective on the basis of color only, so a reddened B5 star may have the same appearance as an unreddened A0 star. This means that desirable stars may be missed in a blue-star survey, but also assures that the survey will be over-selective, missing some late B-type stars, instead of being under-selective and including undesirable A and F-type stars. 3. The method is not sensitive to small differences in color near spectral type A0 because the colors of the stars combined with the effects of interstellar reddening give small changes in the relative appearances of U,B, and V images.

In 1981 the author decided to undertake a survey for faint blue stars in the Puppis window, where absorption in the plane of the Milky Way is low enough that it should be possible to see stars at the edge of the Galaxy's disk (Westpfahl and Christian 1979). A survey method was needed which could be used in crowded fields near the Galactic plane and which could allow identification of stars as faint as eighteenth magnitude in V to assure that the edge of the galactic disk could be observed. Following the suggestions of Chromey (1981) it was decided to use an updated version of Haro's method with a new choice of emulsion and filters in order to increase the discrimination of the method near spectral type A0 and allow for better plate resolution. To get better discrimination, ultraviolet, blue, and red images (U,B,R) were chosen instead of U,B,V, because the B-R color changes faster than the B-V color near spectral type A0. This required an emulsion with extended red sensitivity. IIIa-F plates were chosen for the spectral sensitivity of the F-Class emulsion and the finer grain of IIIa emulsions.

The determination of required exposure times and the survey for blue stars in the Puppis window were undertaken with the Curtis-Schmidt telescope at Cerro Tololo Inter-American Observatory during 19-27 December, 1981. Exposure times were calibrated using the spectral types and magnitudes of the standard stars in Selected Area 98 from Landolt (1973) and using the spectral types and colors of two clusters in the Puppis window, NGC 2362 (Johnson 1950, Johnson and Morgan 1953) and NGC 2483 (FitzGerald and Moffat 1975). The final exposures on IIIa-F plates were 50 minutes through a UG-1 filter for the ultraviolet image, 20 minutes through an RG-610 filter for the red image, and 25 minutes through a Wratten 47A filter for the blue image. The images were displaced twenty arc seconds from each other using the micrometer eyepiece of the guiding telescope. This resulted in three uniformly-spaced images of roughly equal size and density for stars of spectral type A0-A2. Following each three-image plate a conventional single-image plate was taken of the same field to provide finding charts and help eliminate confusion from crowding.

III. THE SURVEY AND PRELIMINARY RESULTS

The survey covers 13 fields corresponding to $235 < l < 245$, $-17.5 < b < 17.5$. These fields were chosen to give coverage of the window of low absorption near $l=240$, $b=0$ to study the extent and rotation of the Galactic disk, and to cover nearby fields out of the plane to look for possible warping of thickening of the disk. This will also allow comparison of the nature of blue objects at low latitude $|b|<5$, with those at intermediate latitude, $10<|b|<20$.

Each plate has been examined by eye to search for blue stars, which show a faint central red image flanked by brighter blue and ultraviolet images. The blue stars have been divided into three natural groups, blue, very blue, and extremely blue, according to the relative sizes of the three images. Two of the fields overlap fields observed in detail by FitzGerald and Moffat (1975) and Nordstrom (1975). Comparison of the blue stars in those surveys with stars in the present survey has allowed rough determinations of the spectral types falling in each natural group. Stars classified as blue have all three images roughly equal, and are unreddened A2-B7 stars or reddened B8-B4 stars. Those classified as very blue have the U and B images roughly equal and both larger than the R image. They are unreddened B7-B4 stars and reddened B5-O9 stars. Those classified as extremely blue have the U image larger than the B image, which is larger than the R image. All are B3 or earlier. Altogether 2035 blue stars have been found, of which 416 are very blue or extremely blue. These 416 stars are distributed about equally for $|b|<5$, but for $5<|b|<10$ there are more stars below the plane than above, and for $10<|b|<17.5$ there are more stars above the plane than below. Any conclusion about warping or thickening of the plane should be postponed until after the photometry and spectroscopy are complete. It is interesting to note that it has long been suspected that blue stars were more common below the plane than above in this region. The large number of stars with $|b|>10$ is, however, a surprise.

Photometry of the very blue and extremely blue stars was started in January, 1983, using the 1.5m and 0.9m telescopes at CTIO. All of the 50 stars observed have B-V colors of slightly reddened O-B stars, and all fall in the magnitude range $13.0<V<18.6$. This confirms that the stars are within the desired range of colors and magnitudes for galactic structure studies. The UBV photometric observations will be continued and slit spectroscopy started in January, 1984.

REFERENCES

Chromey, F.R., 1981, private communication
FitzGerald, M.P., and Moffat, A.F.J., 1975, Astron. Astrophys. Supp.20, 289.
Haro, G., and Luyten, W.J., 1962, Bol. Obs. Tonantzintla Tacubaya 22, 37.
Johnson, H.L., 1950 Astrophys. J. 112, 240.
Johnson, H.L., and Morgan, W.W., 1953, Astrophys. J. 117, 313.
Landolt, A.U., 1973, Astron. J. 78, 959.
Nordstrom, B., 1975, Astron. Astrophys. Suppl. 21, 193.
Westpfahl, D.J., and Christian, C.A., 1979, Bull. Am. Astron. Soc. 11, 415.

DISCUSSION

M.SANTANGELO: Have you searched for systematic errors due to the variations of the atmospheric conditions during the long exposures? Have you also searched for an abnormal extinction law in this region?

D.WESTPHAL: I have searched the regions where neighbouring plates overlap, and find very good agreement, suggesting that such systematic errors are not present. I have not searched for abnormal extinction, but extensive observing in the Puppis window by FitzGerald and coworkers shows only normal extinction.

STATISTICS OF A-TYPE STARS AS POSSIBLE INDICATOR OF STAR FORMATION

L.G. Balázs
Konkoly Observatory, Budapest, Hungary

ABSTRACT

Space distribution and kinematics of stars are determined by the initial distribution, i.e. the distribution of spatial positions and kinematical data just after the birth, and dynamical effects the stars experienced during their life. Evidence is presented that the space distribution of A type stars perpendicular to the galactic plane might be a result of periodic star formation having a characteristic time of $10^8 < \tau < 6 \times 10^8$ years.

INTRODUCTION

The space distribution of stars as derived from observations is a result of two main physical processes. The first important process is the star formation which produces an initial distribution of stars, characterized by their spatial, kinematical and physical data. The second main process is the dynamical evolution by means of regular and irregular forces which act on the stars and give them their present distribution. The distribution observed is a superposition of subsystems of different ages having probably different dynamical histories and initial conditions.

In the case of OB stars the dynamical evolution is unimportant since their lifetime is too short in comparison to evolutionary time scales. In the case of A type stars, however, the life times can be some 10^9 years and the dynamical evolution can change the initial distribution drastically.

SPACE DISTRIBUTION OF A TYPE STARS

Space distribution of A type stars differs significantly from space distribution of OB stars. Unlike OB stars they do not display spiral arms, they distribute more smootnly but some concentrations can be observed (McCuskey, 1965). There is also a strong difference in the distributions perpendicular to the galactic plane. The OB type stars concentrate strongly in the galactic plane. Removing the members of the Gould belt the scale height of these stars is less than 100 pc (Stothers and Frogel, 1974). The space distribution of B type stars can be well represented by the following simple formula in the direction perpendicular to the galactic plane:

$$D(z) = D(o) \exp(-\Phi(z)/\sigma_w^2) \qquad (1)$$

where $D(z)$, $\Phi(z)$ and σ_w^2 are the number density, the gravitational potential and the velocity dispersion in the z direction, respectively. This expression does not fit the respective space distribution of A type stars. A satisfactory fit can be achieved in the (z)<600pc domain, however, if one fits the actual distributions superposing two components each having the form of the right side of equation (1) but the velocity dispersions have a ratio of about $\sigma_{1w}:\sigma_{2w}=1:2$ (Wolley and Stewart, 1967)

$$D(z) = D_1(o)\exp(-\Phi(z)/\sigma_{1w}^2) + D_2(o)\exp(-\Phi(z)/\sigma_{2w}^2) \qquad (2)$$

The ratio $D_2(o)/D(o)$ measures the percentage of the component of higher velocity dispersion compared to the total density in the galactic plane. This quantity is independent of the possible distance scale errors, therefore it is very appropriate for comparing the results of different authors.

Using the $D(z)$ distributions of stars of different spectral types obtained by different authors I found a dependency on spectral type in the ratio defined above (Balázs, 1975). At spectral types earlier than A0 $\log(D_2(o)/D(o))<-2$, i.e. $D_2(z)$ makes little contribution to the total density in the galactic plane. Passing towards later spectral types, however, the logarithmic ratio jumps up to about -1 at A0 and gradually increases afterwards to about -0.5 at F8. The jump means that $D_2(z)$ suddenly becomes more prominent in the total density distribution at A0 spectral type.

The jump at A0 on the $\log(D_2(o)/D(o))$ v.s. spectral type plot can be easily accounted for the discontinuous star formation. Namely, velocity dispersions, i.e. σ_w^2, are increasing with increasing stellar ages (Wielen, 1974). This increase can be adequately described by assuming a diffusion proces which causes an increase of the variances of z co-ordinates as well (Wielen, 1977). The observed $D(z)$ of a given spectral type is composed of two subsystems having different velocity dispersions. Combining the increasing σ_w^2 with uniform distribution of stellar ages the presence of two characteristic velocity dispersion would be difficult to explain. Supposing discontinuous age distribution, however, the presence of two subsystem is a natural consequence. If the characteristic life time of stars of a given spectral type is smaller than the time required for increasing the velocity dispersion up to σ_{2w}, only the small dispersion component, i.e. σ_{1w}, is significant. As a consequence of discontinuous star formation the larger dispersion component appears if the lifetime of a star is greater than the characteristic time between two consecutive star formation activity. The sudden increase of the prominence of $D_2(z)$ results in a jump in the $\log(D_2(o)/D(o))$ v.s spectral type diagram. In our case this jump appeared near A0. The lifetime of these stars and the accuracy of classification on small scale spectra gives an estimate for the period of star formation of

$$10^8 \text{yrs} < \tau < 6 \times 10^8 \text{yrs}$$

AGE DISTRIBUTIONS

The interpretation of the space distribution of A type stars in terms of discontinuous star formation and diffusion processes is based on many speculative elements. Age distributions of these stars revealing the characteristic period expected above would give important support for proving the scenario outlined in the preceding paragraph. Any method which is capable of determining the positions of stars on the HR diagram can give the age of the stars if one uses appropriate theoretical isochrones. Studying the age distribution of A type stars brighter than 6.5 mag Balázs and Tóth (1981) found a probable periodicity of 4.5×10^8 years.

The study of age distributions requires accurate location of stars on HR diagram and can be done on limited samples only. Determining space distributions, however, is possible on much more extended samples and Schmidt telescopes are well suited to this purpose. The combination of results obtained from these big samples with work yielding age distributions can give insight into the star formation processes going on in our Galaxy.

REFERENCES

Balázs,L.G., 1975, Mitt.Sternwarte Ung.Ak.Wiss. No. 68.
Balázs,L.G.,and Tóth,I., 1981, AG. Mitt. Nr. 55, 119
McCuskey,S.W., 1965, Distribution of Common Stars in the
    Galactic Plane in "Galactic Structure" ed. A.Blauw and
    M.Schmidt, Univ. Chicago Press, p.1.
Stothers,R. and Frogel,J.A., 1974, Astron.J. __79__, 456
Wielen,R., 1974, Highlights of Astronomy __3__, 395, ed.
    G. Contopulos, Reidel Publ.Co. Dordrecht
Wielen,R., 1977, Astron.Astrophys. __60__, 263
Woolley,R. and Stewart,J.M., 1967, Monthly Notices Roy.
    Astron. Soc. __136__, 329

DISCOVERY OF NEW BRIGHT PECULIAR STARS OF THE NORTHERN SKY

William P. Bidelman
Warner and Swasey Observatory
Case Western Reserve University
Cleveland, Ohio 44106

ABSTRACT

   I am now engaged in an objective-prism program directed toward the discovery of new peculiar and interesting stars in most of the northern sky. The plates utilized are 20-min. blue exposures taken with the Warner & Swasey Observatory's Burrell Schmidt telescope, now located on Kitt Peak. The dispersion of the spectra is 108 A/mm at $H\gamma$, the limiting magnitude about m_{pg} = 10.5, and the plate quality generally excellent. This paper discusses the history and status of the project and gives examples of the variety of objects being found.

   When the 24-36-inch Curtis Schmidt was transferred from its Michigan site to the far better skies of Cerro Tololo in 1966, it was agreed that the University of Michigan would have one-third of the time of the telescope. Since we were not used to having so much clear sky available, this necessitated consideration of the question of exactly what to do with the telescope. Luckily, shortly before this I had had occasion to inspect a number of the blue-region plates taken with the telescope's combined $4°$ and $6°$ prisms--dispersion 108 A/mm at $H\gamma$--and had been extremely impressed (Bidelman 1966). I had even gone so far as to suggest that "such a large project as the reclassification of the stars of the Henry Draper Catalogue appears to be quite feasible and highly desirable.." (After all, Curtis' initials were H.D.!). So it did not require much effort for me to decide that the major work of the telescope would be a southern hemisphere all-sky survey, and, with the aid of the National Science Foundation and especially my associate Darrell J. MacConnell, this in fact was the case for a considerable number of years.

   Even though I was on record as having thought it "feasible" to reclassify all of the Henry Draper Catalogue stars, I myself had no intention of doing this, nor had Dr. MacConnell. But what we did do, to begin utilization of the beautiful plates that kept coming, was to start a socalled "early-result" program in which we, assisted by several gifted graduate students, scanned the plates for all peculiar stars, supergiants, and late-type dwarfs. The previously unrecognized objects were then published. The results of this work appeared in the papers by Bidelman and MacConnell (1973) and MacConnell and Bidelman (1976), in a number of short reports, and in papers by MacConnell, Frye, and Upgren (1972), and Upgren, Grossenbacher, Penhallow, MacConnell, and Frye (1972). Two later papers of similar nature, from inspection of southern plates taken subsequent to the original program, are by Bidelman (1981) and Bidelman and MacConnell (1982).

I left Michigan in the fall of 1969, to come to C.W.R.U. a year later, Quite unbeknownst to me, Dr. Nancy Houk, who had heard me talk about this project when a graduate student at Case, was greatly interested in actually carrying out the "grand design" of reclassifying the HD stars, and eventually persuaded the U. of Michigan to give her access to all of the plates taken at Cerro Tololo. The results of this happy circumstance are almost unbelievable: she has by now published MK classifications for some 94,000 stars south of $\delta = -26°$. She is presently working on the zone from $\delta = -26°$ to $-12°$, which she will finish sometime in 1985. In view of her far more extensive experience and greater use of spectral standards, her classifications for stars noted in the "early-result" program should in general clearly supersede the earlier results, though there are some cases in which the classifications were done on different plates (she is using shorter exposures for the brighter stars).

With the move of the Warner & Swasey Observatory's Burrell Schmidt to its Kitt Peak Station in June 1979 and with the acquisition of a new single $10°$ prism for it, it became practicable to extend the all-sky survey to the northern sky, and fortunately Dr. Houk has expressed her willingness to continue her classifications to $\delta = +90°$. The new prism gives spectra of exactly the same dispersion as the southern plates; the only difference is a substantially greater extension of the northern spectra to the ultraviolet. To allow for this, new plates of spectral standards are being obtained, and some fields will be classified using plates taken with both telescopes. We are responsible for obtaining the plates with the Burrell; they are being taken by Resident Observer Rik Hill. The status of the project is as follows: all plate-taking from Cerro Tololo has now ceased, after the entire southern sky has been covered and even quite a number of plates have been taken as far north as $\delta = +35°$. The total number of plates to be taken with the Burrell is 679 20-min. and 659 4-and 1-min. plates, of which we now have accumulated 191 and 220 acceptable plates respectively. There are thus a substantial number of plates yet to be taken, and since there are several concurrent W&S programs only about one-third of the time available to us is presently being devoted to the $10°$ survey.

Now that these northern plates are becoming available, I have again begun an early-result program. There are two reasons for doing this. The first is that the plates go to approximately $m_{pg} = 10.5$, i.e., appreciable fainter than the HD. Thus if interesting new objects are below the HD limit they will not generally be noted unless some such search is made for them. And second, some HD objects are of sufficient interest that they deserve rapid publication. A first paper, listing 175 objects found on the first 100 fields to be covered from Kitt Peak, is in a recent issue of the Astronomical Journal.

I will now show a number of slides illustrating the types of things that show up on the plates. Some are taken from the Atlas of Objective-prism Spectra published by Houk, Irvine, and Rosenbush (1974). The first slide shows the O-type stars. HD 102567 is V801 Cen, the x-ray source. The second shows some B5 stars, and a nice B-type shell, HD 33599. The next shows the stars near A5, with a good A-type shell, HD 104237, which is probably a companion to ε Cha, a metallic-line star and a peculiar A. The shell star is a 6th mag. object which has bright Hα but nothing much else seems to be known about it. The bottom star, HD 8783, is a striking Sr-Eu-Cr star. It is 7.5 mag., and yet is not classified as peculiar in the HD. Thus it is evident that the HD missed practically all of the fainter Ap stars. On our plates, the silicon stars are the easiest to detect; the manganese and λ Boo stars are almost never detected. One phenomenon occasionally seen in the Sr stars is a marked shallowness of the K line. The behavior of the Sr $\lambda 4077$ line is rather odd: it is surprisingly strong in many Am stars and I have also noted it to be anomalously strong in some stars that otherwise appear to be quite normal F's. Incidentally, at first we did not publish the new metallic-line stars, as we thought them too common to be of interest. I have changed my mind on this point. Distinguishing between Am and δ Delphini stars is very subtle and uncertain; some stars classified as either may be normal objects. Also it is very difficult, if not impossible, to distinguish between true Am's and composites consisting of late A or early F giants and earlier type main sequence A stars. If the late-type component shows a G band the distinction is clear, as in the case shown on the next slide. There is no doubt that this is a composite spectrum, with the late-type star being a G or K giant.

Turning to later types, we have the Ba II stars and the stars that should but do not show the G band. The former are rather easily confused with normal supergiants or even some giants with strong CH and CN features. Stars can be erroneously classified as no-G-band stars if they are actually too late in type to be expected to show the G band, and one must be quite careful about this. Another notable thing in late-type stars is the occasional presence of strong Ca II emission, as in the next star HD 81410 = IL Hya. This is an extreme case, but weaker cases are fairly common. These stars are without exception spectroscopic binaries related to if not actually RS CVn stars. The T Tauri stars, which might be confused with them, do not show up appreciably in my survey. However, one does have to take some care in the discovery of emission objects. The next slide shows a spectrum that seems to have nice hydrogen and Ca II emission, but it is actually two stars a few seconds apart lined up in a north-south direction! At the latest types we note the carbon and S stars and the M-type Miras with their characteristic abnormal Balmer decrement. Dwarfs of types K0 and later are easy to classify and I am publishing these if they are not already known. And, of course, one occasionally finds a symbiotic, with strong He II $\lambda 4686$, or a flare star with a normal Balmer emission decrement.

There are even more exciting objects: one occasionally sees subdwarf O's or B's, though these are generally already known. The B stars with strong helium lines, like HD 168785 shown in the next slide, are of great interest to theoreticians. Now lest you jump to the conclusion that everything is easy, let me show you one of our mistakes. MacConnell, Frye, and Bidelman (1972) published this star, CoD $-37°9248$, as a new hydrogen-deficient F-type star; however, several people have subsequently shown that it is markedly variable in light and is in fact a very weak-lined high-velocity star---persumably a new RV Tauri or semiregular variable that we just happened to catch at the phase when the hydrogen emission just balanced the absorption. I herewith apologize.

Finally I must mention the type of object that is perhaps the most satisfying of all to discover--a star of very low metal abundance. We have found many new ones, of course; in fact most of the known field weaklined giants have been found on objective-prism plates through the efforts of Howard Bond (see Bond 1980)and ourselves. The other day I found a low-metal star that really made my day: it is shown on the last slide. The object is the 9th magnitude star BD $+44°493$, which has a proper motion of about one-tenth of a second per year. This star has been classified twice as a B star, but it is clearly an extremely weak-lined G, and probably a giant as well. This discovery is a convincing illustration of the value of the early-result program, for the star, not being in the HD, would not have been classified by Nancy Houk and may well have remained unnoted for years. I suspect that it will turn out to be one of the most metal-deficient objects ever found, and I trust that it will not long remain in the obscurity from which it has finally partially emerged.

REFERENCES

Bidelman, W.P.: 1966, in Vistas in Astr. 8, 53.
Bidelman, W.P.: 1981, Astron. J. 86, 553.
Bidelman, W.P. and MacConnell, D.J.: 1973, Astron. J. 78, 687.
Bidelman, W.P. and MacConnell, D.J.: 1982, Astron. J. 87, 792.
Bond, H.E.: 1980, Astrophys. J. Suppl. 44, 517.
Houk, N., Irvine, N.J., and Rosenbush, D.: 1974, Dept. of Astronomy,
    U. of Michigan (Ann Arbor, Mich.)
MacConnell, D.J. and Bidelman, W.P.: 1976, Astron. J. 81, 225.
MacConnell, D.J., Frye, R.L., and Bidelman, W.P.: 1972, Publ. Astron.
    Soc. Pacific 84, 388.
MacConnell, D.J., Frye, R.L., and Upgren, A.R.: 1972, Astron. J. 77, 384.
Upgren, A.R., Grossenbacher, R., Penhallow, W.S., MacConnell, D.J., and
    Frye, R.L.: 1972, Astron. J. 77, 486.

DISCUSSION

L.O. LODEN: Can you give some numerical value of the completeness of the discovery of chemically peculiar stars, particularly with respect to the type of peculiarity? (A significant systematic tendency may impose a selection effect upon subsequent frequency statistics.)

P. BIDELMAN: I'm afraid not. But for the more conspicuously abnormal stars we should be very nearly complete, in the relevant magnitude range. For the marginally abnormal the incompleteness will be serious. I should add that all of the objects noted on the objective-prism plates should be subjected to further spectroscopic and/or photometric study before they are accepted as completely certain. We are preparing only finding, not definitive, lists. Also we are only publishing objects not previously known.

USE OF THE UK SCHMIDT FOR A PHOTOMETRIC INVESTIGATION
OF RED STARS IN THE ORION NEBULA MOLECULAR COMPLEX

A.David Andrews and Brian McGee
Armagh Observatory
Northern Ireland
United Kingdom

As a region of recent star formation and intense stellar activity the Orion Complex extending over several square degrees is of paramount importance in the study of early evolution and activity in the stellar chromospheres of low mass stars. Although red stars form a rather inhomogeneous group, including foreground K-M dwarfs, distant K-M giants, carbon stars and a variety of reddened objects, in principle these can be separated from a combination of photometric, spectroscopic and kinematic data. The proximity of the Orion Complex and its relatively high galactic latitude mean that future astrometric and radial velocity work will yield useful kinematical information on all member stars, including the enigmatic pre-main sequence stars on the lower part of the HR diagram. For these stars, differences in spectroscopic properties, photometric behaviour, kinematics and the interstellar environment contain the key to some of the problems, presently insurmountable, in the early evolution of the entire Orion Complex.

A recent 'Photometric Atlas of the Orion Nebula'(Andrews 1981), based on UBVI UK Schmidt plates, has been searched for possible K-M type stars, objects with V-I $\geq$ 1.4, to limiting magnitudes, V = 16.2 and I = 12.6. Over 1300 such objects were found within 21 sq.degs centred on the Orion Trapezium. Preliminary results show large differences across the region, particularly in the CM diagrams of V-I versus V. This survey will be combined with previous photometric and objective-prism work, as well as flare-star detections of K-M objects, to investigate variations in the properties of red stars across the region. The total density distribution of over 16000 stars from the 'Photometric Atlas' is shown in Fig 1, which may be compared with Fig 2 which shows the general anti-correlation with the 2.6mm CO emission-strength given in the Columbia Survey (Kutner et al.,1977) delineating the molecular clouds. Red stars alone are plotted in Fig 3. The V-I versus V diagram (Fig 4) has been re-plotted for red stars in 45 rectangular zones at 2 min and 1 degree intervals (Fig 5). Zonal differences are clearly seen to contribute to the overall scatter in Fig 4. The red stars are fully listed in the 'Photometric Atlas', and will be the subject of further research at Armagh into the early evolution of low mass stars in Orion and other young clusters.

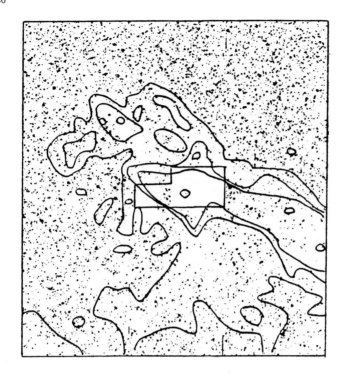

FIG 2

Computer-simulated Star Field to V = 16.2, for same region as Fig 1, showing contours of CO emission strength at 2.6mm wavelength. See Kutner et al., 1977.

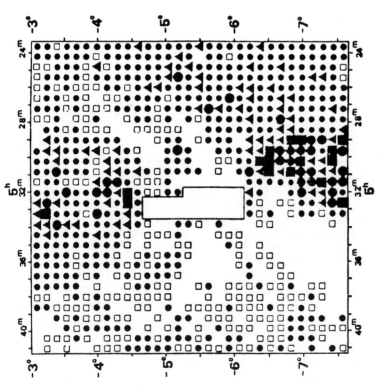

FIG 1

Stellar Density Distribution to V = 16.2 from GALAXY measures of UK Schmidt plate. See for details in 'Photometric Atlas of the Orion Nebula' (Andrews 1981).

INVESTIGATION OF RED STARS IN THE ORION NEBULA MOLECULAR COMPLEX

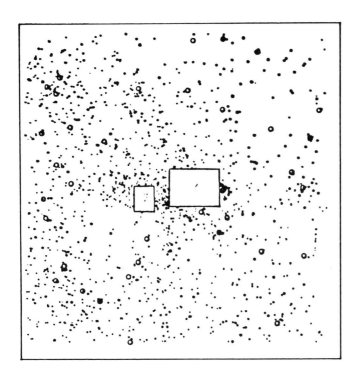

FIG 3

Distribution of Red Stars (V-I ≥ 1.4) in same region as Figs 1 & 2, with symbol size indicating V magnitude. See 'Photometric Atlas'. Large circles are reflection halos on the I plate. N.B. Transparent area to south-west.

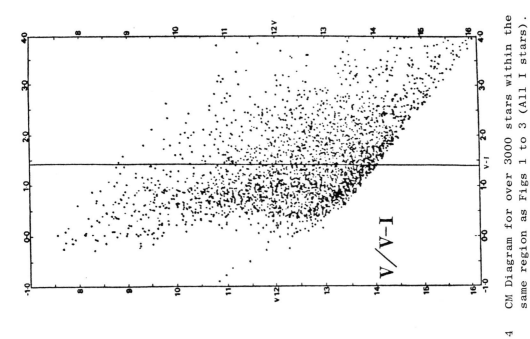

FIG 4 CM Diagram for over 3000 stars within the same region as Figs 1 to 3 (All I stars).

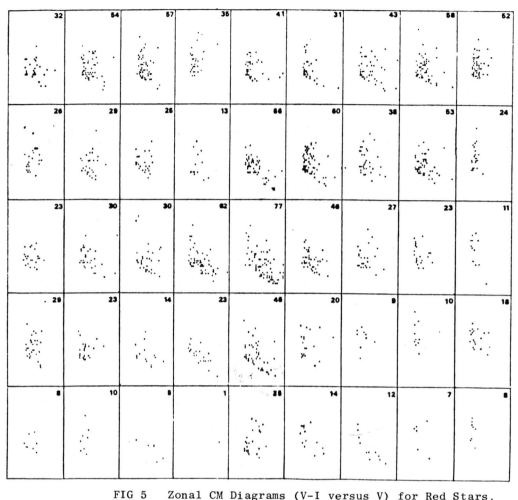

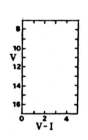

FIG 5 Zonal CM Diagrams (V-I versus V) for Red Stars, showing individual diagrams for 45 zones across the total area of Figs 1 to 3, and numbers of objects. The inset to the left gives the scale. Large differences are noted in various regions of the Orion Molecular Complex

REFERENCES

Andrews,A.D., 1981 'A Photometric Atlas of the Orion Nebula', published at Armagh Observatory.
Kutner,M.L.,Tucker,K.D.,Chin,G. and Thaddeus,P., 1977 Astrophys.Journ.215,pp.521-528.

A SURVEY OF NORTHERN BOK GLOBULES AND THE MON OB1/R1 ASSOCIATION
FOR H-ALPHA EMISSION STARS

Katsuo Ogura
Kokugakuin University, Higashi, Shibuya-ku, Tokyo 150

Tatsuhiko Hasegawa
Astronomical Institute, Tohoku University, Sendai 980

1. INTRODUCTION

The most direct approach to the suggested relationship of Bok globules to star formation is to search for actual spots of star birth in or around them. But no protostellar objects have hitherto been found by infrared and/or radio observations. Optically relevant to this are the discoveries of some Herbig-Haro objects and pre-main sequence stars associated with cometary globules inside the Gum Nebula (Reipurth 1983).

Young objects which are found most often in regions of star formation are emission-line stars of T Tau and related types. We have made a systematic survey of Bok globules in search for such stars. Objects are taken from the catalogue by Bok and Cordwell(1973). Among them the globule in NGC 2264 is rather peculiar; it is located in a well known site of recent and active star formation. So we searched for new emission-line stars in a larger area, i.e., in the whole association Monoceros OB1 and R1.

2. OBSERVATIONS AND REDUCTIONS

Deep objective-prism plates in the red were taken with the 105/150-cm Schmidt telescope at Kiso Observatory. A 4° objective prism was employed in conjunction with a Schott RG610 or RG645 filter and Kodak 103aE emulsion (hypersensitized), giving a dispersion of 700 Å/mm at Hα-line. Exposure time was 90-120 min, and no widening was applied. The limiting magnitude of the detected emission-line stars is estimated as $m_r \sim 15.4$ for the globules and $m_r \sim 17.0$ for Mon OB1 and R1.

We also took 4° objective-prism plates in the blue-violet. Kodak IIaO emulsion (hypersensitized) and a Fuji SP-4 filter were used. This combination isolates the spectral interval 3700-4500 Å which includes CaII H and K lines, at a dispersion of 120 Å/mm, as well as higher Balmer lines. The limiting magnitude is, however, rather bright: $m_b \sim 12.5$.

The summary of our results is presented in the table. Approximate red

Objects Name	Surveyed Area	Detected Stars	Known Stars	Suspected Stars	Comparison Fields
Barnard 335	20'	0	0	0	0
Barnard 145	30'x105' E-W	5	1	1	0
Barnard 343	40'	5	1	5	1
Barnard 361	68'	6	4	2	0
Barnard 362	50'	7	0	3	3
Barnard 157	20'	1	0	1	0
Barnard 161	15'	2	0	0	2
Barnard 163	20'	2	0	0	1
Barnard 367	15'	1	0	0	0
Barnard 164	30'x48' NW-SE	1	0	0	1
Lynds 1225	30'	4	0	0	1
Barnard 5	45'x66' NE-SW	3	2	1	0
Barnard 34	80'	3	0	2	2
Lynds 1622	90'	16	4	2	0
Barnard 227	50'	4	0	0	2
Mon OB1 and R1	7.11 □°	137	–	–	–

and blue magnitudes of the stars were estimated from their image diameters on the Palomar Sky Survey prints. We determined their celestial coordinates from the positions on the PSS prints as well.

3. DISCUSSION

3.1. Globules

Out of the 60 emission-line stars detected near the 15 northern globules, 80% are newly discovered stars (cf. the fourth column). We are concerned mainly about how many of the stars in the table are associated with the globules and are of T Tau type. Six of them (two in B5 and four in L1622) are already known as T Tau stars. Spectral classification with our prism plates in the blue-violet can be applied only to 11 brighter stars, most of which are found to be of early type. But one star in L1622 shows H and K lines in emission, and probably a new T Tau star. As to the rest of the stars, we have no information on their nature at present.

Simple statistics show about half of these objects can be young stars associated with globules. We have set up arbitrarily a comparison field of 30'-diameter circle for each globule. A total of 13 emission-line stars are found in the 15 comparison fields (the last column). This means 4.4 stars/square degree. On the other hands the average density near the globules amounts up to 7.6 stars/square degree. Therefore, the surface density of emission-line stars in the vicinities of globules is higher than that in ordinary fields by a factor near 2, or more if we take the depression of star numbers in globules into consideration.

The present survey has mainly picked up the candidates for young stars possibly associated with Bok globules. But it is now clear that some globules do have T Tau stars in their vicinities. At least these are B5, L1622 and the globule in NGC 2264. We can infer the evolutionaly status of these globules either that they have already formed some young stars, or that they are also the products of star-forming processes themselves.

These three, however, are not typical globules on account of their incomplete isolation. On the other hand, the most typical globule B335 has no emission-line stars in its vicinity. This may suggest that Bok globules have some varieties in regard to their origin and/or evolutionaly status, and that we may have a higher possibility to find young objects near globules subjected to some external forces rather than near isolated ones.

3.2. Monoceros OB1 and R1

In addition to 84 $H\alpha$-emission stars found by Herbig (1954), we discovered even larger number (about 100) of new stars concentrating on NGC 2264. Indeed a considerable part of the stars in the cluster show $H\alpha$ emission.

The distribution of emission-line stars in Mon R1 is rather sparce, particularly in its northern portion; star formation in this region seems to be well separated from and less active than that in Mon OB1. Here, however, we notice a very interesting feature, a horseshoe-shaped concentration of emission-line stars in the southwestern part. This is just coincident with the structure of the CO cloud revealed by Kutner et al (1979). As discussed by them and by Herbst (1981), star formation in Mon R1 is thought to have triggered by some energitic event in its southwestern part, which is now associated with an incomplete ring of a CO cloud and an expanding HI shell.

Further details of these works can be found in Ogura and Hasegawa (1983) and Ogura (1983). We have just finished the observations to extend the survey to the southern globules in the catalogue by Bok and Cordwell (1973) with the Schmidt telescope at Bosscha Observatory.

REFERENCES

Bok, B.J., and Cordwell, C.S. 1973, in "Molecules in the Galactic Environments," ed. M.A. Gordon and L.E. Snyder (John Wiley and Sons, New York), p.53.
Herbig, G.H. 1954, Astrophys. J., 119, p.483.
Herbst, W. 1981, in "IAU Symp. No. 85, Star Clusters," ed. J.E. Hesser (Reidel, Dordrecht), p.33.
Kutner, M.L., Dickman, R.L., Tucker, K.D., and Machnik, D.E. 1979, Astrophys. J., 232, p.724.
Ogura, K. 1983, submitted to Publ. Astron. Soc. Japan.
Ogura, K., and Hasegawa, T. 1983, Publ. Astron. Soc. Japan, 35, p.299.
Reipurth, B. 1983, Astron. Astrophys., 117, p.183.

DISCUSSION

M. McCARTHY: I too congratulate Dr. Ogura and colleagues for the fine exploration of Bok globules. This spring I was able to obtain slit spectra with the Reticon attached to the Steward Observatory 1.3 m reflector at Kitt Peak and I can confirm the reality of Hα features in 15 of these stars and note additional spectral features in several of these. Our reductions are not yet fully done. It is a new exciting research project.

PRESENT STATE OF THE WORK ON AUTOMATIC SPECTRAL CLASSIFICATION
AT TARTU

V. Malyuto
W. Struve Tartu Astrophysical Observatory
Toravere, Estonia, USSR

An automatic spectral classification technique for objective-prism spectra is being developed. The procedure of selecting standard stars is described. The applicability of a linear polynomial regression model for choosing and calibrating the spectral criteria for T_{eff}, M_v, [Fe/H] is demonstrated.

1. INTRODUCTION

Using objective-prism spectra obtained with the 70-cm meniscus telescope at Abastumani Astrophysical Observatory (D=166 A/mm at H_γ) and facilities for registration and processing of astronomical images available at Tartu Astophysical Observatory, we are developing a technique for automatic spectral classification of stars of spectral range F-K. The work has been reported by Malyuto and Pelt (1981, 1982). In the following we describe the current state of the work on automatic spectral classification. Our logical outline is similar to that proposed by West (1973).

2. CHOICE OF STANDARD STARS AND DATA SOURCES FOR THEM

Choosing standard stars we excluded known double and variable stars and tried to avoid unrecognized ones by requiring that the estimates of T_{eff}, M_v, [Fe/H] obtained by means of different methods were in mutual agreement. Main data sources were the catalogues of Strasbourg Stellar Data Center (CDS) 1010, 2007, 2057, 3042, 3052, 3054 and 5074.

Preference was given to stars with highly accurate photoelectric estimates of [Fe/H] in extensive lists by Nissen (1981), Campbell (1978) and Christensen (1976). These stars amount to 50 per cent of the sample. Another main source of the adopted [Fe/H] was Cayrel's catalog (CDS 3054) where results of high dispersion analysis were presented. The adopted T_{eff} values were drawn mainly from two sets

of homogeneous data with high internal accuracy: the photoelectric
β-indices for F stars and the V-K or R-I color-indices for G-K stars.
The adopted M_V values were mainly obtained either from most accurate
trigonometric parallaxes or taken from the extensive list of Wilson
(1976).

Judging from the mean errors of the used data the internal
accuracy of the adopted T_{eff}, M_V, [Fe/H] values for our standard
stars is expected to be about ± 80 K, ± $0^m.7$ and ± 0.15, respectively.
To increase the accuracy of M_V for standard stars within 200 pc, we
proposed a list for observations of trigonometric parallaxes with the
HIPPARCOS Satellite.

The final list of standard stars contains about 150 F-K stars of
all luminosity classes and with the [Fe/H] between +0.3 and -2.0.
The observations have been performed at Abastumani Observatory in
collaboration with Dr G. Jimsheleishvili. Observational material for
some stars obtained by earlier investigators, above all by Dr M.
Shiukashvili and Dr R. West, was also used. The spectra (totalling
about 400) have been digitally recorded on magnetic tape in Tartu. In
future we intend to use the PDS machine installed at Tartu Observatory.

3. TRANSFORMATION OF SPECTRAL DATA

The procedure of transformation of spectral data has been
described by Malyuto and Pelt (1981,1982). The transformation results
in the so called equivalent areas (confined by the continuum, spectral
energy distribution and integration limits). Each equivalent area
includes one spectral line or a tight group of lines. It has been
shown that these quantities and, in particular, their ratios are rather
insensitive to the spectral resolution and to the graininess of
photographic emulsion.

4. SELECTION OF CRITERIA

We have adopted equivalent area ratios as spectral classification
criteria. The number of possible ratios exceeds 1000 per each spectra.
Therefore one needs a simple and effective algorithm for selecting
the ratios, most sensitive to different physical parameters in order to
exploit them subsequently as classification criteria. We have tried
to use models of linear polynomial regression for this purpose. The
values of T_{eff}, M_V, [Fe/H], their powers and products were treated
as predictors, and the measured ratios as response variables.

Because our measurements of equivalent areas are not yet
completed, we analysed 10 central depth ratios, measured by Malyuto
(1977) in the same spectra for 72 F2-G8 stars, to try applicability of
linear regression models. One may expect that in general central depth
ratios behave similarly to area ratios. This assumption is supported

by comparisions made by Malyuto and Pelt (1981).

TABLE

No.	Identification	R	Partial derivatives with respect to T_{eff}	M_v	[Fe/H]	Type of sensitivity
1	4300/H_γ	0.94	-9.84	0.04	1.17	T_{eff}, [Fe/H]
2	4272/H_γ	0.95	-4.10	-0.08	0.60	T_{eff}
3	4251/H_γ	0.92	-4.46	0.02	0.28	T_{eff}
4	4226/H_γ	0.95	-3.71	-0.01	0.40	T_{eff}
5	4200/H_ζ	0.94	-3.75	0.31	0.38	T_{eff}, M_v
6	4172/H_ζ	0.93	-3.74	0.30	0.41	T_{eff}, M_v
7	4128/H_ζ	0.92	-2.73	0.24	0.14	T_{eff}, M_v
8	3883/H_8	0.90	-3.64	0.14	0.73	T_{eff}, M_v, [Fe/H]
9	(3871+3860)/2H_8	0.92	-5.65	0.32	0.95	T_{eff}, M_v, [Fe/H]
10	(3826+3815)/2H_{10}	0.90	-2.45	-0.12	0.63	T_{eff}, [Fe/H]

Inspired by the visual appearance of graphs of corresponding relationships, we started with the following model:

$$Y = \beta_0 + \beta_1 X_1 + \beta_2 X_2 + \beta_3 X_3 + \beta_{11} X_1^2 + \beta_{22} X_2^2 + \beta_{33} X_3^2 +$$
$$\beta_{111} X_1^3 + \beta_{222} X_2^3 + \beta_{333} X_3^3 + \beta_{12} X_1 X_2 + \beta_{13} X_1 X_3 + \beta_{23} X_2 X_3. \quad (1)$$

Here Y is the analysed ratio and X_1, X_2, X_3 stand for T_{eff}, M_v, [Fe/H], respectively. Next we tried to simplify the model by gradually excluding supposedly redundant terms beginning with higher degrees. Multiple correlation coefficients R and partial derivatives with respect to T_{eff}, M_v, [Fe/H] at the average values of these parameters were calculated at each step. In Table we present results for the model with 10 predictors (the $\beta_{333} X_3^3$ and $\beta_{23} X_2 X_3$ terms of (1) were excluded). The results are very similar to those for the model (1). In the last column we indicate the type of sensitivity of criteria. The physical parameters to which a criterion is most sensitive, according to Malyuto (1977), are indicated. We see that this model is capable for separating those ratios which are sensitive to different sets of physical parameters (the criteria of differing type were chosen on the basis of the corresponding derivatives). Multiple correlation coefficients differ slightly.

5. CLASSIFICATION

If the chosen model of linear polynomial regression is adequate in detail, it may be used not only for selection of criteria, but also for classification. Sets of relations of type (1) for each star may be considered as systems of non-linear equations with respect to T_{eff}, M_V, [Fe/H]. Using the above model (1) without the last term and the criteria listed in the Table, we composed and solved these systems of equations for our 72 standard stars. The method of steepest descent was used to actually find the solutions. The results turned out to be encouraging. From comparision of program results with standard T_{eff}, M_V, [Fe/H] values and taking into account their mean errors indicated in Section 1, we found that our model gives the T_{eff}, M_V, [Fe/H] values with the mean errors ±170 K (±140 K if stars with calculated T_{eff} > 6200 K are excluded), $±0^m.8$ ($±0^m.6$) and ±0.31(±0.18), respectively. Lower accuracy for hotter (F type) stars is due to lower sensitivity of criteria and, possibly, to the inadequacy of the used model in this temperature region.

The detailed version of this paper will be published in a report to be issued by W. Struve Tartu Astrophysical Observatory.

REFERENCES

Campbell, B.: 1978, Astron. J. 83, pp.1430-1437.
Christensen, C.: 1978, Astron. J. 83, pp.244-265.
Malyuto, V.: 1977, Tartu Astrofuus. Obs. Publ. 45, pp.150-172.
Malyuto, V., Pelt, J.: 1981, Preprint of the Academy of Sciences of the Estonian SSR, A-1, pp.3-15.
Malyuto, V., Pelt, J.: 1982, CDS Inf. Bull., No.23, pp.39-40.
Nissen, P.: 1981, Astron. Astrophys. 97, pp.145-175.
West, R.: 1973, in C.Fehrenbach and B.E. Westerlund (eds.), "Spectral Classification and Multicolor Photometry", IAU Symp. 50, pp.109-124.
Wilson, O.C.: 1976, Astroph. J. 205, 823.

LA MESURE DES VITESSES RADIALES AVEC UN PRISME OBJECTIF ASSOCIE A UN TELESCOPE DE SCHMIDT

Ch. Fehrenbach - R. Burnage
Observatoire de Haute Provence

ABSTRACT

Radial velocity measurements with an objective-prism mounted on a Schmidt telescope.

A 62 cm diameter objective-prism is mounted on the CNRS-University of Liège Schmidt telescope at the Haute Provence Observatory. The field is $4 \times 4°$ and the limiting magnitude is 12.5 on IIIaJ hypersensitised plates. The dispersion is 200 A mm^{-1} at 4220 A. The plates are measured with a special machine and data are reduced by means of a computer with a correlation method. Stars of all spectral types are measured. The probable error is of some 4 km sec^{-1} over a mean of at least 3 plates. Already several lists of radial velocities of stars belonging to field situated at $-30°$ of galactic latitude have been published. We have also started radial velocity observations for the Hipparcos Program.

RESUME

Un Prisme Objectif de Fehrenbach, de 62 cm de diamètre, est monté devant le télescope de Schmidt CNRS-Université de Liège à l'Observatoire de Haute Provence.

Les clichés de $4 \times 4°$ permettent d'atteindre la magnitude de 12.5 sur plaque IIIaJ hypersensibilisées. La dispersion à 4220 A est de 200 A mm^{-1}. Les clichés sont mesurés par un instrument spécial et réduits par une méthode par corrélation informatisée. Une erreur probable de 4 km sec^{-1} sur une moyenne de 3 clichés est obtenue pour les étoiles de tous types spectraux. Plusieurs listes d'étoiles situées à $-30°$ de latitude galactique ont été publiées. L'équipe est maintenant engagée dans le programme Hipparcos.

LE PRISME OBJECTIF A CHAMP NORMAL

Nous indiquons les propriétés principales de ce prisme. On se reportera pour tous les détails à une publication (Fehrenbach, 1966) et pour le prisme décrit ici à Fehrenbach et Burnage (1975).

La propriété essentielle est la déviation nulle, dans tout le champ, pour une longueur d'onde $\lambda_c \sim$ 4220 A. Un prisme en Flint à haute dispersion est accolé à un prisme de Crown Ba de même indice, mais moins dispersif. Les prismes ont le même angle et forment une lame à faces parallèles.

Il est possible, par retournement du prisme de 180° autour de l'axe optique du télescope, de juxtaposer un spectre direct (D) à un spectre renversé (R). La position relative D,R est une fonction de la vitesse radiale de l'étoile.

Des prismes de ce type sont installés à l'Observatoire de Haute Provence et à l'Observatoire Européen Austral, mais le remplacement de l'objectif dioptrique par un télescope de Schmidt constitue un progrès essentiel, dû à la qualité de la combinaison de Schmidt, de la lame et du P.O.

La dispersion des spectres varie dans un champ de $\pm 2°$ de $\pm 0,5\%$, mais si le spectre D s'allonge, le spectre R se raccourcit, de sorte que la distance de la même raie dans les spectres D et R ne varie pas dans le champ.

La longueur d'onde λ_o de retournement varie un peu avec la température; si donc pour un cliché donné, le prisme objectif est un vrai étalon de longueur d'onde, il varie un peu d'une nuit à l'autre et nous devons connaître en principe au moins la VR d'une étoile du champ. En fait, nous avons l'habitude de rattacher l'ensemble des mesures à quelques étoiles mesurées spécialement avec le spectrophotomètre à corrélation (Coravel).

Nous n'indiquons pas les précautions nécessaires à prendre pour obtenir des clichés de grande qualité et sans erreurs de champ.

Le Prisme Objectif SPO, associé au télescope de Schmidt-Liège-Observatoire de Haute Provence, a les caractéristiques du tableau :

<u>Schmidt</u> lame	62 cm
Distance focale	208 cm
Champ couvert	4 x 4°
<u>Prisme objectif</u>	angle de 12°,5
Prisme en Flint et en Crown Ba	
Dispersion	200 A mm^{-1} pour λ_o = 4225 A
Déplacement relatif R,D	1μm 7,20 km sec^{-1}
Magnitude limite	plaque IIIaJ traitée 2x1^h magnitude 12
Programme Hipparcos	2 x 15 min magnitude 10

MESURE DES CLICHES

La méthode de mesure est décrite en détail dans la publication technique O.H.P. et dans la thèse de Burnage (1983). Nous ne donnons ici que le principe.
Les spectres D, R, sont mesurés simultanément en 1375 points espacés de 4 µm avec une fente de 10 µm de largeur. Les transmissions du cliché sont enregistrées sur une bande magnétique et traitées ensuite sur l'ordinateur. Celui-ci contient en mémoire les fichiers d'un certain nombre de spectres étalons(D',R'). On calcule le coefficient de corrélation pour le spectre D avec le spectre D'. Si nous déplaçons D par rapport à D', nous déterminons un pic de corrélation entre les fichiers D et D' dont le maximum correspond à la superposition exacte des profils spectraux D, D'. Le même calcul, appliqué à R, R' nous permettra, par différence, de déterminer la position relative D, R par rapport à D', R' et donc la VR de l'étoile par rapport à l'étoile étalon.
Naturellement, la corrélation n'est possible que si le spectre mesuré et l'étalon sont de types voisins. L'ordinateur contient en mémoire un certain nombre de spectres étalons de B à M dont on connaît bien les différences de VR.
Les VR déterminées avec ce système d'étalon sont donc toutes raccordées entre elles. Le rattachement au système standard sera fait à l'aide de quelques étoiles, de préférence G à K, de VR connues (catalogue, ou mieux mesures spectrophotométriques Coravel).
Nous ne donnons pas ici les détails des calculs, explicités dans le travail de Burnage. Indiquons toutefois qu'il est important de filtrer les spectres de la courbe du fond continu, car la corrélation doit se faire sur les raies spectrales et non sur les pentes du fond continu.

RESULTATS

Indiquons qu'en plus de la VR, l'enregistrement donne simultanément α et δ à 2" près, le type spectral selon des critères de Morgan-Keenan par un programme mis au point par Simien (1982), la magnitude photographique à 0.1 près.
Nous allons présenter quelques résultats pour 1400 étoiles dans 8 champs (Fehrenbach, Burnage, 1981). L'étude de la précision montre que la VR est connue à 4 km sec^{-1} pour une moyenne de 3 à 4 clichés, quelque soit le type spectral. Le déplacement relatif correspondant est de 0,3 µm. Cette précision s'explique par le principe même de la technique où les spectres D et R sont obtenus dans les mêmes conditions, à l'inverse de ce qui se passe pour les spectrographes classiques.
Elle s'explique aussi par la qualité de l'optique (Lame et Prisme), la possibilité d'utiliser des plaques à grains fins (IIIaJ) et la méthode de

mesure et de calcul.

Nous sommes actuellement engagés dans la mesure des étoiles du programme Hipparcos, travail commencé à titre expérimental avant même de connaître les étoiles à mesurer, en collaboration avec l'Observatoire de Marseille (M.Duflot).

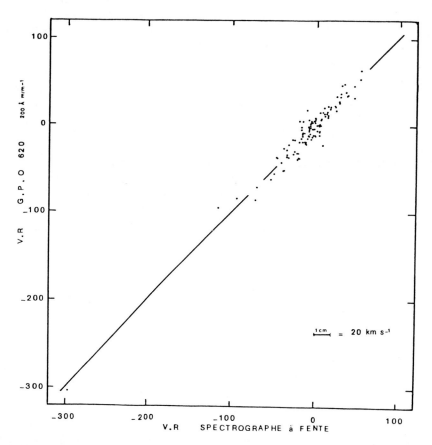

REFERENCES

Burnage,R.:1983, Thèse Université de Marseille I

Fehrenbach,Ch.:1966,Objective Prisms and Measurement of Radial Velocities in Advances in Astronomy vol.4

Fehrenbach,Ch., Burnage,R.:1975,Compt.Rend.Acad.Sci.Paris ser.B 281,481.

Fehrenbach,Ch., Burnage,R.:1981,Astron.Astrophys.Suppl.Ser. 43,297.

Fehrenbach,Ch., Burnage,R.:1982,Astron.Astrophys.Suppl.Ser. 49,483.

Simien,F.:1982, Thèse Université de Marseille I.

THE YOUNG OPEN CLUSTER NGC 2384

S.M. HASSAN
King Saud University
Kingdom of Saudi Arabia

ABSTRACT

Combined photoelectric – photographic UBV-photometry for NGC 2384 is discussed. The results achieved so far are summarized and indicate that this cluster is young enough to be used as spiral arm tracer out to a distance of 3.27 kpc. The colour excess $E_{(B-V)}$ is 0.31, the apparent distance modulus is $13^m.50$ and its earliest spectral type is B0.

INTRODUCTION

NGC 2384 is an irregular chain like object of the type IV3p (Ruprecht, 1966) lying in Canis Major at the coordinates:
R.A. $07^h\ 20^m.7$ $l^{II} = 235°.4$
Dec. $-20°\ 50'$ (1950.0) $b^{II} = -2°.4$

This cluster has been previously investigated by few authors. Distance determinations of 2600, 4550, 2400 and 2200 pc have been quoted by Trumpler (1930), Collinder (1931), Zug (1933) and Barkhatova (1950) respectively. Angular diameters of 4.5', 2.8', and 2.5' as well as linear diameters of 3.4, 3.7, 3.2 and 4.5 pc have been estimated for NGC 2384 by the same authors. Hayford (1932) assigned a B0-B1 spectral type while Zug (1933) noted the same spectral type for the cluster under consideration together with 0.31 as an estimate for the colour excess in front of it.

Besides, it has been concluded by Isserstedt & Schmidt-Kaler (1964) and Neckel (1967) that interstellar absorption, in this section of the Milky Way, limited by galactic longitudes 231° and 256°, is small. This gives a higher probability of reaching the outer spiral arms in this region in which, until now, no open clusters have ever been observed. That is why NGC 2384, together with some other objects, has been scheduled for observations in our program. Unfortunately, at the time when observations have been secured, it was impossible to reduce the photographic plates because of the inavailability of a photoelectric standard sequence. However, in 1972, Vogt and Moffat published the first photometric

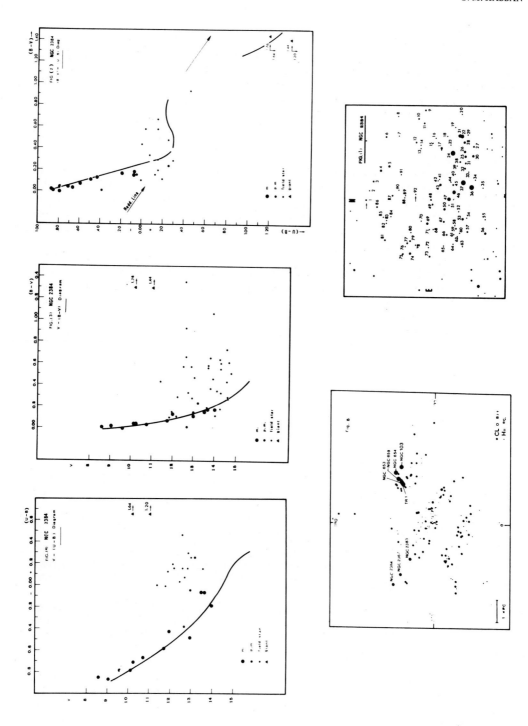

results of the cluster based on 15 stars – for two of which H_β photometry is available – observed photoelectrically. They could reach a distance of 3.28 kpc to the cluster as well as an estimated total visual absorption (A_V) of $0\overset{m}{.}95$ along the direction towards it. Also, an earliest spectral type B0 has been assigned to NGC 2384 according to their investigation.

The plate material secured in the Newtonian focus of the 74" Reflector at KOTTAMIA, which consists of:

   4 Plates : 103 aD + GG 14
   4 Plates : 103 aO + GG 13
   4 Plates : 103 aO + UG 2

has been measured using the Askania Astrophotometer available at Helwan. The 15 photoelectrically observed stars by Vögt & Moffat (1972), when combined with their corresponding iris readings, enable smooth V, B and U calibration curves to be attained. It is to note that neither magnitude nor colour equations could be detected, while controlling the calibration curves used in the reductions.

In the overall, 92 stars have been measured for the present study. They are identified, with the same numbers as those given in the catalogue enclosed at the end of this paper, in Fig. (1). The limits of completness of our photometry are 15.00, 15.30 and 14.00 in the V, B and U spectral bands respectively.

The mean error of the V-magnitudes and colours for the program stars given in the catalogue at the end of the manuscript are as follows:

   V   : $\pm$ 0.02 magn.
   (B-V) : $\pm$ 0.02   "
   (U-B) : $\pm$ 0.03   "

COLOUR-COLOUR AND COLOUR-MAGNITUDE DIAGRAMS

The visual magnitudes (V) and the (B-V) & (U-B) colour indices, for the stars which could be measured and which are tabulated in the catalogue enclosed, have been utilized in constructing the CCD presented in Fig. (2) and the CMDS illustrated in Figs. (3) and (4).

The fitting of the Standard Zero Age Main Sequence (Schmidt-Kaler, 1965), to our apparent diagrams, applying Becker's method (1954) could yield the following results for NGC 2384:

   (m - M) $13\overset{m}{.}50$ Distance   3.27 kpc
   E_{B-V} 0.31   No. of probable Physical Members 12
   E_{U-B} 0.22
   A_V $0\overset{m}{.}93$ No. of possible Members 3
   $(m - M)_o$ $12\overset{m}{.}57$ Spectral type B0

It is worthy to point out that the sliding method along the reddening line, oftenly used by Johnson (1958), for reddening determination, has led to values of 0.31 and 0.22 for the excesses E_{B-V} and E_{U-B} exactly the same as those achieved using the CMDS illustrated in Figs (3) and (4). Moreover, the value of E_{B-V} achieved in this paper is in a good agreement with that given by Zug (1933). Generally, our results are comparable to those noted by Vögt & Moffat (1972).

NGC 2384
Catalogue

Star No.	V	(B-V)	(U-B)	Star No.	V	(B-V)	(U-B)
1	13.51	0.15	-0.07 m	41	15.03	-	-
2	15.06	-	-	43	12.98	0.15	-0.05 p.m.
3	13.04	0.47	0.25	45	13.78	0.37	-
4	11.40	0.43	0.00	46	11.73	0.07	-0.59 m
5	11.93	0.15	0.12	47	10.75	0.03	-0.67 m
6	13.78	0.65	-	48	13.68	0.56	-
9	12.60	1.35	-	49	14.00	0.12	-
10	13.92	0.46	-	50	10.22	1.76	1.64
11	12.62	0.92	0.46	53	12.68	0.01	-0.39 p.m.
12	12.74	0.33	0.07	55	14.22	-	-
15	14.88	-	-	56	12.64	0.66	0.15
16	15.01	-	-	57	14.60	0.29	-
17	12.22	0.29	0.15	58	14.35	-	-
19	14.61	0.27	-	59	14.24	0.18	-
20	14.27	0.60	-	62	14.67	-	-
21	9.59pe	0.00pe	-0.79 m	63	14.67	0.61	-
22	0.05pe	0.02pe	-0.87 m	64	14.87	-	-
24	10.26	0.04	-0.71 m	66	12.88	0.57	0.15
25	12.99	0.11	-0.49 m	67	13.61	0.19	0.15
26	12.50	0.57	0.03	68	15.02	-	-
27	13.22	0.23	0.25pm	69	12.17	0.11	0.19
28	11.82	0.10	-0.01	73	14.27	0.33	-
29	15.06	-	-	74	11.03	1.44	1.20
30	11.98	0.13	-0.43 m	75	14.21	0.64	-
31	14.54	0.43	-	76	14.54	0.73	-
32	13.98	0.17	-0.19 m	77	14.09	0.15	-
33	15.05	-	-	78	14.95	-	-
35	14.66	-	-	79	13.68	0.58	-
36	10.14	0.04	-0.79 m	81	13.92	1.06	-
37	8.58	0.01	-0.85 m	85	14.77	0.51	-
38	13.64	0.18	-0.07 m	86	12.92	0.28	0.29
39	15.02	-	-	87	14.27	0.55	-

The three stars No. 26, 28 and 69 (7,3,15 in Vögt list) are generally confirmed as probable field stars although the last two are almost lying on the main sequence curve in the V-(B-V) diagram illustrated in Fig. (3). According to the author's view stars No. 50 and 74 may be considered as member super giants in the cluster. Besides, star No. 27 lies above the main sequence in Fig. (3) and if it is considered as a star in the phase of gravitational contraction (pre - main sequence contraction stage) due to the fact that its reddening is compatible with its membership as pointed out by Vögt & Moffat (1972), the star No. 43 in the author's notations can be considered to represent the same situation as star No. 27 (No. 4 in Vögt numbers) and consequently, the author considers it as a possible member belonging to NGC 2384. In this way, 12 stars could be discriminated as probable physical members in the region of the cluster under investigation together with 3 possible members belonging to it.

The large spread of the stars in the region of NGC 2384 and its rather irregular shape has made it difficult to determine an exact angular diameter and consequently a linear one for it. However, if one considers the area in which most of the probable physical members are populated, an approximate angular diameter of 4'.2 as well as a linear one of 3 pc could be reached for the cluster under investigation.

Moreover, it is important to point out that the distance of NGC 2384 (3.27 kpc) - determined in the present investigation is large when compared with that distance of 1.97 kpc found out by Vögt & Moffat (1972) for NGC 2383, although these two objects are lying in the same direction (almost the same galactic longitudes and latitudes) and having almost the same interstellar reddenings. Also, a spectral type B0 could be assigned to the cluster under study here in this paper.

The spectral type B 3 for NGC 2383 as given by Vogt & Moffat (1972) and the spectral type B0 for NGC 2384 (the present work) together with the type B 1 achieved by the author for NGC 2367 (published elsewhere, 1975) indicates that these three stellar systems are young enough to be used as spiral arm tracers. They are located, as shown in Fig. (5), somewhat outside the extension of the local arm (Becker & Fenkart, 1970).

REFERENCES:

Barkhatova, K.A., 1950, Azh., 27, 182.
Becker, W. and Stock, J., 1954, Z.Astrophys., 34,1.
Becker, W. and Fenkart, R., 1970, IAU Symposium No. 38, 205.
Collinder, P., 1931, Ld An, 2Lk.
            , 1931, Ld An, 2Cr.
Hassan, S.M. and Marei, M., 1975, Helwan Obs. Bull. No. 116.
Hayford, P., 1932, Lick Obs. Bull. 16, 53.
Isserstedt, J. and Schmidt-Kaler, Th., 1964, Z. Astrophys., 59,182.
Johnson, H.L., 1958, Ap.J., 126,126.
Neckel, Th., 1967, Veröff. Landessternwarte Heidelberg-Königstuhl 19.
Ruprecht, J., 1966, Bull. Astron.Czechosl.
Schmidt-Kaler, Th., 1965, in H.H.Voigt (ed.),Group VI/1,Springer-Verlag (Berlin), p.284
Trumpler, R.J., 1930, Lick Obs. Bull. 14,171.
Vögt, N. and Moffat, A.F.J., 1972, Astron. & Astrophys. Suppl.Ser.,7,2, 134.
Zug, R., 1933, Lick Obs. Bull. 16,132.

ASIAGO SCHMIDT SURVEYS OF VARIABLE STARS AND SUPERNOVAE

L.Rosino
Astrophysical Observatory of Asiago of the University of Padova

Survey methods used at Asiago for finding supernovae and new variable stars are reviewed. Some by-products of the supernova searching are indicated. It is shown that the use of infrared emulsions is particularly effective for the discovery of new Mira-stars in Milky Way fields.

1. INTRODUCTION

Large fields telescopes are particularly suited for statistical researches over wide sky areas. I shall give here briefly some notices on the researches carried out at Asiago with the Schmidt telescopes of the Astrophysical Observatory on variable stars and supernovae.
The following Schmidt telescopes are available at Asiago:
a) Schmidt telescope 40-50-100 cm. Mean limiting magnitude B $\sim$ 17.5. The telescope has been recently transferred to the Mount Ekar Station, 1350 m on the sea level, 5 km from Asiago, far from the city illumination. With this telescope we have obtained since 1958 more than 15,000 photographs, each photograph covering an area of about 50 sq.degrees. In addition, nearly 2000 photographs have been obtained through an UBK7 objective prism of 12°, giving a dispersion of 45 nm/mm at H_γ.
b) Schmidt telescope 67-92-215 cm. Mean limiting magnitude B $\sim$ 18.5. With this telescope we have obtained since 1963 more than 12,000 plates, each covering a field of 28 sq.degrees. The telescope is equipped with an UBK7 prism of 4° and with a second prism of flint of 1°.1;

combining the two prisms we can obtain the following dispersions (at H_γ):39.5, 65, 100 and 186 nm/mm.

In conclusion more than 27,000 Schmidt plates are available at Asiago with some thousand objective prism photographs. These plates are catalogued in order of R.A. and are easily available for studies and consultation.

2. SUPERNOVA SURVEY

The systematic search of supernovae at Asiago began in 1959 with the 40 cm telescope. Later, also the 67 cm telescope was occasionally employed. In the first years the search was limited to a dozen fields rich of galaxies, but gradually in the following years the number was increased to 66. The distribution of these fields in the sky relative to the galactic equator is shown in fig.1. Dark square represent fields covered by more than 200 photographs. White squares are fields recently introduced in the survey, covered by a minor number of plates.

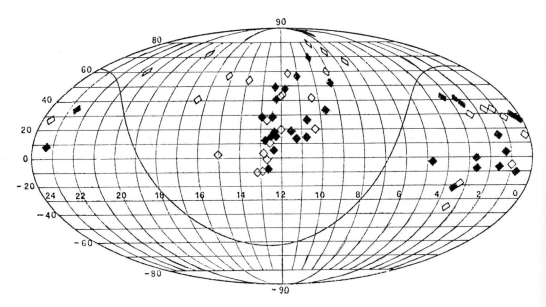

Fig.1 - Distribution of 66 fields for supernova survey.

The plates are examined as soon as possible or with the negative on positive method (alternatively, negative on negative) which is rapid and effective, or with the blink microscope. A supernova is accepted only when it is visible at least on two different plates and when all controls are made in order to exclude that the object may be a spurious image or a normal variable star or an asteroid.

All the supernovae discovered at Asiago or elsewhere which may be accessible to the instruments and latitude of Asiago are followed as long as possible in UBV with the Schmidt telescopes and the 182 cm parabolic telescope, in order to obtain an extended light curve and the spectrum at different phases.

Supernova search is not an easy task. Supernovae are rare in galaxies, perhaps less frequent than generally estimated. In several thousand plates or films obtained for the supernova survey in the last 24 years we were able to discover only 25 SNe, which is a rather low rate. This is partly due to the fact that the period from April to June, when the Coma and Virgo fields are passing in the meridian during the night, is the worst at Asiago for observing conditions. The detection threshold in the 40 cm Schmidt plates is about 15.5 and in the 67 is about one magnitude more. Supernovae weaker than this, even if their images are present in the survey plates, are generally overlooked. For a thoroughly discussion see Rosino et al. (1974). The increasing illumination of Asiago is another negative factor. Now that the 40 cm Schmidt has been moved from Asiago to the Mount Ekar where the sky is darker, we hope to improve the efficiency of the research. The quick reduction of the plates is another problem. To intercompare a pair of plates we employ from 30 to 60 minutes, even more if the plates are taken with the larger Schmidt. It is a fatiguing work and it is possible that some weak supernova in the plates may have been missed. A revision of the plates hitherto obtained is in programme.

The situation in the other few Observatories which still are continuing the photographic supernova search is not much better. As shown in fig.2 the number of supernovae discovered every two years has had a sudden drop after the death of Dr. Zwicky and the interruption of the survey with the 48-inch Palomar Schmidt. The drop would have been even steeper without the contribution (after 1979) of Maza et al. at the Cerro El Roble Observatory in Chile.

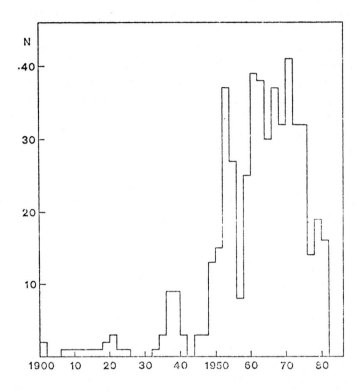

Fig.2 - Number of supernovae discovered every two years starting from 1901.

An important question at this point is whether it may be worthwhile to continue a survey which implies an enormous amount of telescope time, expensive photographic material and hard reduction work. Personally I think that every effort should be made to carry out the systematic search of supernovae at least for some more years. Several of the brightest and most interesting supernovae in the last years were found in the course of surveys with relatively small telescopes. Moreover, the occurrence of types I and II in galaxies of different classes, their frequency, the absolute magnitude at maximum, the shape of the light curve of some peculiar supernovae and the possible existence of new types are problems still partly open.

Anyway, the material obtained for the supernova survey is far from being wasted. I mention here some by-products of this material:
a) Control of SNe discovered by other observers and photometry of SNe below the detection threshold.
b) Variability of compact galaxies, quasars, BL Lac objects, etc.
c) Study of variable stars of high galactic latitude and discovery of new variables.
d) Study and discovery of asteroids and comets.

Finally I should like to mention that extensive searches of SNe with electronic and automatic devices have been recently developed. They should reduce the work and allow the discovery of many weak supernovae. However, these techniques are still in the experimental stage. The automatic search of SNe has been successful only at the Corralitos Observatory, where 12 SNe were discovered from 1968 to 1976, when this programme was interrupted. For the moment the supernova survey is entirely based on the conventional photographic techniques.

3. FLARE STARS IN YOUNG ASSOCIATIONS.

Another interesting field of researches with Schmidt telescopes is the survey of flare stars in young galactic clusters and stellar associations, particularly in the Orion Complex, in the Monoceros region (NGC 2264) and also in the Pleiades, Iades and Praesepe. The technique for the finding of flare-ups is that of multiple exposures firstly introduced by Haro. With this method hundredths of flare stars and repeated flare-ups in active stars have been detected. In a recent paper of Haro, Chavira and Gonzales (1982) a Catalogue has been presented of 519 flare stars discovered in the Pleiades through the coordinate work of the Observatories of Tonantzintla, Byurakan and Asiago.

An interesting result is that in the Pleiades region at least 37% of the flare stars are non-members of the cluster. This means that the occurrence of the flare phenomenon is high in low luminosity stars of late spectral type independently on the fact that they belong or not to a cluster.

From the programme of survey of flare stars in young associations, which has been recently joined also by the Observatories of Budapest, Abastumani, Rojen and Sonneberg, it is possible to derive a number of valuable informations on the physical characteristics of the flare stars, their possible relation to nebular variables, the occurrence in clusters of different age, the relative frequency of the flare-ups, the absolute magnitudes and spectral types.

The systematic searching of flare stars, although important for the obvious implications in the theories of stellar formation and early evolution, implies an enormous amount of telescope time and requires a strenous work of observation. So in these last years we have progressively reduced at Asiago the time dedicated to the flare survey, continuing however the photographic observations of the variables of the RW Aur or T Tau type embedded in nebulosities. The researches have been mostly concentrated in the Orion Nebula and the neighbourhood regions including the Horsehead Nebula, NGC 1999 with the Herbig-Haro objects. Hundredth of new variable stars have been identified in blue and infrared plates and partly published. The number of variables in the Trapezium Region is so high that it is very difficult to find in the rich field non-variable stars for comparison.

But this leads me to speak briefly of another field of interest for Schmidt surveys, that of variable stars in rich Milky Way areas.

4. VARIABLE STARS IN MILKY WAY AREAS

Large field telescopes are the best instruments for the discovery, classification, photographic and spectroscopic study of variable stars. In the past the surveys were made with astrographs. But the Schmidt telescopes are best suited because in general they are more powerful and allow the discovery of weak variable stars and the analysis of their light curves not only in the photographic and visual regions, but also in the near infrared. The use of objective prisms gives also large possibilities of identifying variable stars by their spectra.

At Asiago we have selected a number of Milky Way Areas which we are systematically surveying in B,V and I (Kodak IN plates + RG5) mostly with the 67 cm Schmidt. They are shown in fig.3. Some significant results have been already obtained by Maffei and by the writer.

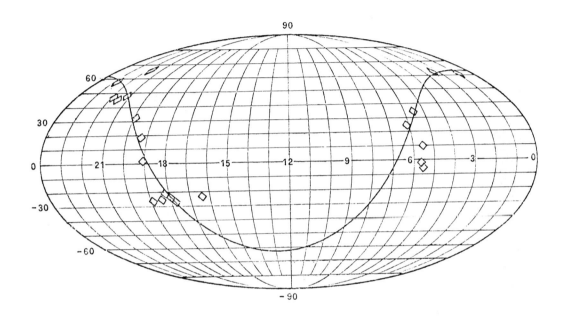

Fig.3 - Distribution of variable star fields in the Milky Way.

In an area with its centre at $18^h 15^m$ $-15°$ (1950) Maffei (1975) has found in the infrared 198 new variable stars, of which at least 108 are Mira-type. Only a few Mira variables were previously known in the same area (28 sq.degr.). Similar results have been obtained by Rosino et al. (1976,1978) examining blue and infrared plates centered in Cassiopeia at $23^h 18^m$ $+61°$ and Sagitta at $19^h 15^m$ $+18°$ which led to the discovery of more than 160 Mira stars. Another area which after a preliminary survey appears extremely rich of Mira variables is centered at

$19^h 00^m$ +2°.0 (1950). In this area 150 new Mira variables have been identified. The photographic researches in B,V and I are integrated by determinations of spectral types on objective prism plates also obtained in the near infrared. Most of the Mira variables discovered in the surveys have spectral types later than M5 and in the blue are barely perceptible and sometimes invisible even near maximum. The period distribution show a maximum frequency for stars with periods between 350 and 400 days, somewhat longer than found for the Mira variables (with period larger than 250 d) listed in the General Catalogue of Variable Stars of Kukarkin et al.(1968). It should be finally remarked that the period distribution and the number of Mira variables strongly depend on the galactic longitude and the presence of obscuring clouds in each selected region. But this is a point which is now under investigation.

The enormous increase, at least by a factor 50-100, in the number of Mira variables in some Milky Way areas, when observed in the near infrared, indicates that these stars represent an important component of the disk population and gives new insights on the possible course of star evolution. The shift of the periods of maximum frequency, on the other hand, particularly in the direction of the galactic center, suggests that low temperature Mira of advanced spectral type are likely to be much more frequent than estimated in the past.

In addition to the Mira stars these surveys of Milky Way areas operated with large field Schmidt telescopes in B and V give also the possibility of detecting other types of variable stars, from RR Lyr to U Gem, and offer the possibility of a comparison with other fields of high galactic latitude.

REFERENCES

Haro,G.,Chavira,E., Gonzales,G.,1982,Tonantzintla Bull.3,1.
Maffei,P.,1975,IBVS 985,986.
Rosino,L.,Di Tullio,G.1974,Supernovae and Supernova Remnants
　　　　　　　　　　Ed. Cosmovici,pp.19-27,Reidel.
Rosino,L.,Bianchini,A., Di Martino,D.1976,Astr.Astroph.S.24,1.
Rosino,L.,Guzzi,L.1978,Astr.Astroph.S.,31,313.

DISCUSSION

P. WILD: I should like to give cordial thanks to all our colleagues at Asiago for much encouragement and for prompt and very friendly help that I have received from them in our supernova search of Bern since more than 20 years. You were able to confirm quickly several supernovae, when our weather had deteriorated right after a discovery. I also want to report the first case (to my knowledge) of radio astronomers having outrun the optical supernova searches: comparison of a Berkeley radio picture (January 1982) and a newer Westerbork radio picture (spring 1983) showed that a point source in NGC ... had disappeared (or nearly so) in the meantime. We were asked to go through our optical files, and sure enough, there had been a supernova of mag. 17 in fall of 1981, which we had missed. It proves that the optical search could well be checked and complemented by radio observations.

HAMBURG OBSERVATORY NORTHERN MILKY WAY SPECTRAL SURVEY FOR EMISSION OBJECTS

L. Kohoutek
Hamburg Observatory
Hamburg-Bergedorf, W. Germany

1. INTRODUCTION

Objective-prism spectral surveys open the possibility to search for faint emission-line objects with the aim to complete their statistics and to pick out most interesting individual objects for further study. In the years 1964 - 1970 the Hamburg H_α Spectral Survey of the Northern Milky Way was accomplished using the Schmidt camera (80/120 cm, f = 240 cm) in Bergedorf with the following parameters: area 1 $32^\circ - 214^\circ$, $-10^\circ < b < +10^\circ$, 160 fields, Kodak 103aF + RG 1, exp. 60 min, widen. 10", 4° prism (580 Å/mm at H_γ). As a main result the list of about 140 faint objects classified as planetary nebulae or possible planetary nebulae (Kohoutek, 1965, 1969a, 1972), and the identification of about 1500 new stars having H_α in emission (Kohoutek, Wehmeyer, in preparation) can be reported. The best known examples of this survey are K 3-50, a prototype of a compact H II region, and the symbiotic variable HBV 475 = V 1329 Cyg (Kohoutek, 1969b), which is also classified as a protoplanetary nebula.

As the H_α-survey is not adequate for satisfactorily classifying the emission objects we have started a new survey in two colours (SPS), which in addition covers an area of the Milky Way larger than before. The opportunity to realize this project was given after the Schmidt camera had been moved from Bergedorf to the German-Spanish Astronomical Center, Calar Alto, Spain.

2. AIM AND MAIN PARAMETERS OF THE SPECTRAL SURVEY

The main aim of the SPS is the investigation of faint emission objects up to 17 - 18 mag: search for new objects, and verification of the classification of the objects already catalogued:

(1) Search for planetary nebulae and small (compact) H II regions.
We expect to find new candidates which either lie beyond our H_α-survey or which are very faint. The presence and the intensity of [O III] 5000, 4959 emission lines will serve as a main criterion for classification.

(2) Search for faint WR-stars.
The classification criteria are emission lines H_α, He II 4686, C III 4650 or N III 4638 and N V 4609 Å.

(3) Search for novae and variable emission stars.
Comparison with the old H_α-survey (1964-70). Attention will be paid to young T Tauri stars (type V 1057 Cyg), to protoplanetary nebulae (type V 1016 Cyg, HBV 475) and to symbiotic stars.

(4) Stars having H_α in emission.
In this wide cathegory all emission stars are meant which are not included in (2) and (3), especially Be and Me stars. The Catalogue of H_α emission stars in the northern Milky Way (in preparation) will be supplemented.

In addition supplementary lists of faint stars having particular spectral types (e.g. carbon stars) will be prepared. Moreover, if necessary the new plate material will enable a rough spectral classification of stars up to about 17 mag, and it may serve for later comparison.

SURVEY PARAMETERS:

$l\ 12° - 234°$, $-15° < b < +15°$, 330 fields ($5.5° \times 5.5°$)
$4°$-prism (580 Å/mm at H_γ)

Red region:
103aE + RG 610   (6100 - 6700 Å), exp. 15 min, widen. 10"

Blue region:
IIaO + GG 455   (4550 - 5150 Å), exp. 75 min, widen. 5"

For the red region we have chosen this plate + filter combination in order to have possibility to compare directly the present survey with the H_α-survey from 1964-70. The red plates are short exposed but later on we intend to extend this material by very deep plates (IIIaF emulsion).

In the blue region the range 4550 - 5150 Å can be considered as most interesting for emission objects; we have realized it using Kodak IIaO plates and Schott GG 455 filter. In order to reach the stellar limiting magnitude of at least 17 mag, and to avoid the overlapping of the spectra, it was not possible to extend this narrow spectral range (and to use the IIIaJ emulsion).

NORTHERN MILKY WAY SPECTRAL SURVEY FOR EMISSION OBJECTS 313

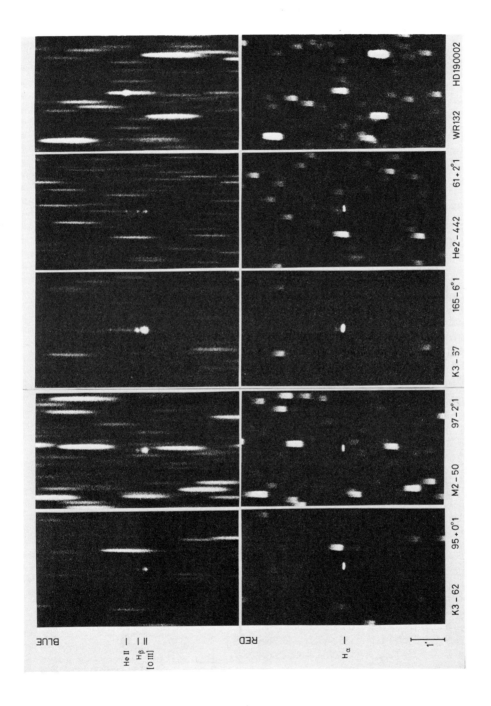

Figure 1   Examples of some individual objects seen on survey plates.

3. PRESENT STATUS OF THE SPECTRAL SURVEY AND EXAMPLES

We started the SPS in the year 1981, and till now (August 1983) plates of about 50 fields have been taken. At present the visual inspection of the material is in progress; the possibility to analyse our plates automatically would of course be supported in case this reduction technique would turn out to be effective enough.

In this contribution examples are given showing how some individual objects appear on the red and blue survey plates:

<u>K 3-62</u> ($95+0°1$): stellar planetary nebula showing bright lines of H_α, [O III] 5007, 4959 > H_β and no continuum.

<u>M 2-50</u> ($97-2°1$): small planetary (~4") having extremely bright [O III] 5007, 4959 lines.

<u>K 3-67</u> ($165-6°1$): stellar planetary with H_α and [O III] 5007, 4959 > H_β; a trace of H_γ and a faint continuum is visible.

<u>He 2-442</u> ($61+2°1$): probably stellar planetary having the lines [O III] 5007, 4959 comparable with H_β and a bright He II 4686 line. Allen (1974) described it as "compact PN with circumstellar dust emission (class D)". This object should be investigated in more detail because both the low and high excitation characteristics are present.

<u>WR 132</u> (HD 190002). The emission features (H_α, C III, IV 4650, He II 4686) of this WC 6 Wolf-Rayet star of $v = 11.55$, $b - v = +1.13$ (see van der Hucht, et.al., 1981) are overexposed, the continuum is strong.

We intend to publish the results of the SPS in numerous contributions which will contain single lists of objects of various cathegories. The present part of the Spectral Survey is financially supported by Deutsche Forschungsgemeinschaft.

REFERENCES

Allen, D.A., 1974, Monthly Notices Roy. Astron. Soc. <u>168</u>, 1.
Kohoutek, L., 1965, Bull. Astron. Inst. Czech. <u>16</u>, 221.
Kohoutek, L., 1969a, Bull. Astron. Inst. Czech. <u>20</u>, 307.
Kohoutek, L., 1969b, I.B.V.S. Budapest No.384.
Kohoutek, L., 1972, Astron.Astrophys. <u>16</u>, 291.
van der Hucht, K.A., Conti, P.S., Lundström, I., Stenholm, B., 1981, Space Sci. Rev. <u>28</u>, 227.

STAR COUNTS

Richard G. Kron
Yerkes Observatory
The University of Chicago
Williams Bay, Wisconsin 53191 USA

I. INTRODUCTION

The number of stars counted along a particular line of sight depends on the spatial distribution of stars, the luminosity function, and the absorption. Thus star count programs are designed to constrain or determine one or more of these functions. Early efforts to understand the structure of our Galaxy, including the fundamentals of stellar statistics, were largely based on work that involved star counts. Since then a growing appreciation has developed for the variety of forms the density function $D(r)$ and the luminosity function $\phi(M)$ can take, especially the recognition of different stellar populations, each with different density and luminosity functions. In the simplest formulation two distinct populations are considered: disk and halo. This suggests two distinct formation histories, but uncertainty in the picture remains (Eggen, Lynden-Bell and Sandage 1962; Ostriker and Thuan 1975; Saio and Yoshii 1979; Jones and Wyse 1983). To discriminate between various models, more information is needed. The problem is that star counts integrate over the details of the stellar statistics to a greater or lesser extent, depending on how much information is available to separate the populations. The challenge to observers is to obtain sufficient resolution. Population type indicators are kinematics, metal abundance, and age distribution, so that star counts <u>combined with</u> proper motions, or radial velocities, or colors, or line strength indices, etc. would naturally provide the greatest constraints. A basic question in this context is whether or not the two-population idealization retains its usefulness when considered in detail, and this issue could be regarded as a point of departure for further work (e.g., Strömgren 1976).

There are two ways to use star counts to develop a picture for what is going on. The first method is to assume forms for the functions $D(r)$ and $\phi(M)$ for each population, along with absolute magnitude - color relations, etc., and then model the expected observations along a line of sight in order to compare with data (e.g., Bahcall and Soneira 1980). The second method is to attempt to derive $D(r)$ and $\phi(M)$ directly from the data via the fundamental equation of stellar statistics (McCusky 1965). To a large extent it can be said that the success of the second method depends on how confidently one can assign a photometric parallax to each star -- and this depends on knowing the population types of the stars, which is often one of the questions being investigated. An example is the problem of the subgiant/turnoff star ratio (Yoshii 1982), since these stars can have quite different luminosities at about the same color.

II. STAR COUNTS WITH SCHMIDT TELESCOPES

Reckoned in terms of information gathered per unit time, Schmidt-type telescopes are unrivaled, but some types of star count programs benefit from the unique capabilities of Schmidt telescopes more than others. The number of stars seen at magnitude m_2 within a distance shell of width Δr is proportional to $\phi(M)D(r)r^2$, where m, M, and r are related in usual way, taking into account the effects of absorption (see e.g. van Rhijn 1965). This function may have a simple form -- peaked at some distance r -- or it could be more complex. Where the function $\phi(M)D(r)r^2$ is <u>small</u> can be of great astrophysical importance, and this is where Schmidt telescopes can contribute, since the large field can produce usable statistics even for types of stars of low surface density on the sky.

a) r small

The stars of lowest luminosity can be observed only at close range, within which the volume is necessarily small. Even though these stars never contribute significantly to the general star counts, it is of considerable interest to know whether or not the (bolometric) luminosity function declines at very faint absolute magnitudes. In order to find the few stars of very low luminosity, a "pointer" is required, like red colors or large proper motion.

A comment can be made here concerning the relative efficiency of larger telescopes at finding stars of low luminosity. The gain in volume sampled due to the increased depth of a 4m telescope compared with a 1.2m Schmidt does not compensate for the much larger field area of the Schmidt. However, with the larger telescope the "S/N" is higher: at high latitudes, the thinning out of the stellar distribution suppresses the number of distant stars, so the <u>relative</u> number of low-luminosity stars at a fainter apparent magnitude is larger. Another method of suppressing background stars is to observe a field in front of a dark cloud (Haro and Luyten 1961; Herbst and Sawyer 1982). However, in this case the wide field of the Schmidt is not as useful, and the fainter limiting magnitude obtainable with the larger telescope may arguably produce a better result.

b) $\phi(M)$ small

Intrinsically rare stars are also best discovered with wide field instruments. This category includes degenerate dwarf stars, OB subdwarfs, blue stragglers, and various types of variable stars. Most of these stars share the property of being blue, and Schmidt telescopes completely dominate the discovery of faint blue stars. The following is an incomplete but (I trust) representative list of some of these surveys: Humason and Zwicky (1947), Luyten and collaborators (1955), Iriarte and Chavira (1957), Feige (1958), Haro and Luyten (1962), Rubin, Moore and Bertiau (1967), Richter, Richter, and Schnell (1968), Jaidee and Lynga (1969), Barbieri and Rosino (1972), Berger and Fringant (1977), Steppe (1978), Green (1980), and Noguchi, Maehara and Kondo (1980). Interest in faint blue stars accelerated in the mid-1960's after the identification of extragalactic objects in extant lists. Still, the Galactic problems were exciting enough, as documented in the Strasbourg conference of 1964 (Luyten 1965). (The proceedings of this conference include an interesting historical discussion by Zwicky on the early role of Schmidt telecopes for survey work.) Generally omitted from the list of papers cited above are surveys explicitly designed to discriminate against Galactic stars.

An important case is the event when $\phi(M)$ is relatively small **and** r is small -- this is true for white dwarf stars, for which the Schmidt surveys of Luyten and Green have been especially productive at generating finding lists. One key question is the shape and amplitude of the white dwarf luminosity function, especially for the cooler stars (Sion and Liebert 1977). Discovery by proper motion appears to be by far the most efficient technique (Liebert et al. 1979).

c)   D(r) small

At very large distances, the density function D(r) falls to small values. Thus stars at very great distance are expected to have low surface density, and Schmidt telescopes can contribute towards mapping the outer regions of our Galaxy. An example of the discovery of a faint carbon star at high latitude is Sanduleak and Pesch (1982); indeed, Haro and Luyten (1962) had earlier remarked on the discovery of five extremely red stars, three of which they took to be N stars. (See also Weinberger and Poulakos 1977 and Poulakos 1978.) A good statistical sample of very distant stars does not yet exist (other than RR Lyraes) because of the lack of luminosity indicators easily applicable to faint stars.

III.   COLOR AND PROPER MOTION DISTRIBUTIONS

Two main generalizations have been made so far: star counts need to be accompanied by some other measurement, such as proper motion or color, in order to separate different groups of stars; and Schmidt telescopes are uniquely capable of finding stars of low surface density.

In this section I review a variety of Schmidt studies which have included either color or proper motion as an observed quantity, with emphasis on complete, magnitude-limited samples. Since other speakers at this conference will develop the scientific content of at least some of these surveys, I shall restrict myself to one or two remarks in each case about special features of the data. Also, this discussion is biased in favor of high-latitude studies because the interpretation is relatively insensitive to the adopted absorption and reddening. Examples of star count surveys at low latitude are the Case series (McCusky 1965), Becker's (1979) work towards the Galactic center, and Morales-Duràn's (1982) study of a region in the anticenter.

Luyten (1960) produced a color-magnitude array derived from five high-latitude fields. It indicated many more faint blue stars than predicted by a simple model. The model did not include evolved stars, which as realized by Luyten could well help to resolve the problem. Attention was given at the Strasbourg conference to the possibility of systematic color errors near the plate limit (see especially Sandage and Luyten 1967). In Figure 1 Luyten's color distribution is presented for the interval one magnitude brighter than the nominal plate limit.

[The issue of "the nature of the faint blue stars" is relatively convoluted. In part it depends on exactly what is meant by blue: blue in B-V may not be equivalent to blue in U-B, especially as far as the discovery techniques are concerned. Also, it is clear that there is no single identity to these stars, but rather several stellar types appear in the counts (Luyten and Anderson 1962, Kinman 1965), depending sensitively on apparent magnitude and, for lower latitudes, on galactic latitude.]

Upgren (1962, 1963) made an objective-prism survey of the North Galactic Pole, yielding the numbers of stars as a function of both apparent magnitude ($m_{pg} < 13$) and spectral type (including luminosity class). This work contributed to the picture of differing scale heights for disk stars of differing characteristic ages. There have in addition been several objective-prism surveys for specific types of stars for specific purposes. Among these could be mentioned the Case surveys (e.g., Sanduleak 1976), and Thè and Staller (1974), both aimed at the problem of the faint end of the luminosity function. Proper motions are available for many of the Thè-Staller stars, and all of the Sanduleak stars have been measured for motion (Luyten 1976; Jones and Klemola 1977). The small motions have been found to be largely due to distance, rather than due to low space velocities.

The large RGU survey by the Basel group began with SA 51 (Becker 1965), and the North Galactic Pole was observed by Fenkart (1967) to a limit G = 19.5. Their principal analysis technique is the separation of Pop I from Pop II via the positions of stars in the U-G, G-R diagram, according to apparent magnitude. The distribution of stars in the two-color diagram changes markedly with increasing apparent magnitude, with

STAR COUNTS

Figure 1. Color histograms for each of six complete high-latitude samples. All counts reduced to 1 deg^2, 1 mag interval in the indicated waveband (except for Morton and Tritton), and 0.25 mag interval in the indicated color.

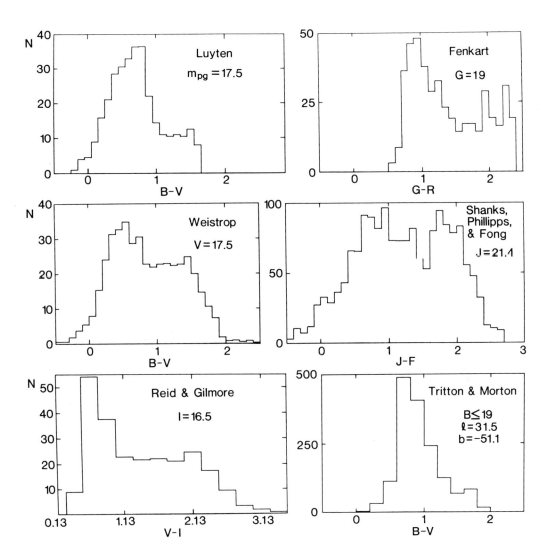

(among other effects) progressively greater ultraviolet excess. The morphology of the two-color diagram is rich in structure and thus provides excellent constraints on star count models.

Subsequently Weistrop (1972) made an accounting of the stellar population towards the North Galactic Pole using similar observational techniques (UBV instead of RGU, V < 18) but a rather different analysis procedure. Globular cluster CM diagrams were assumed to apply for halo stars, and for conventional halo parameters reasonable agreement with the data was obtained, at least for the bluer stars.

Shanks, Phillipps, and Fong (1980) used the UKST for star counts and colors at the South Galactic Pole. The data were obtained on finegrained emulsions, and automatic machine measurements (COSMOS) were used instead of an iris photometer. At their limit of B ~ J = 21.5, star/galaxy separation becomes a problem because the Schmidt scale is small and the number of galaxy images is at least comparable (Hubble 1936). Godwin and Peach (1982), in a work dealing with similar plate material, were more conservative in their assessment of the reliability of star/galaxy discrimination at faint limits, even though they used a PDS machine to digitize their plates. The Shanks, Phillipps, and Fong data are at least qualitatively in agreement with other studies, showing a bimodal color distribution at faint limits (Figure 1), in agreement with Fenkart's (1967) findings.

Reid and Gilmore (1982) have made V-I measurements at the South Galactic Pole, I < 17. The data are especially valuable for studies of red main sequence stars, since M_I is well-correlated with V-I. A subsequent analysis (Gilmore and Reid 1983) advocates a fat-disk model to describe the distribution of the bluer stars, but it is not clear that their interpretation is unique (Bahcall and Soneira 1983).

Bahcall et al. (1983) have discussed Tritton and Morton's UKST color measurements for a somewhat lower-latitude field; these are included in Figure 1 for comparison with the other B-V histograms.

As far as proper motion surveys with Schmidt telescopes are concerned, Luyten's (1963) work provides an extensive data base for the analysis of stellar statistics. A graphical summary has been given by Luyten (1977), in which he presents reduced proper motion diagrams for a total of over a hundred thousand stars. These diagrams, like the two-color diagrams, contain much structure, having in addition dynamical information. Even more structure would no doubt be apparent if the reduced proper motion diagrams were constructed for different regions of the sky separately. Two other Schmidt proper motion studies of faint stellar samples are Schilbach (1982) and Noguchi, Yutani, and Maehara (1982), but both of these were restricted to blue stars.

To fix some idea of the numbers involved, Table 1 gives motions on the basis of a very simple model: the line of sight is perpendicular to the plane, the density distribution is assumed to be exponential with a scale height z_o = 350 pc, and the number of stars pc^{-3} mag^{-1} in the plane is assumed to be 0.01 (approximately valid for main sequence stars in the range 10 < M_V < 15). These stars would have $\langle Z^2 \rangle^{1/2}$ ~ 20 km sec^{-1}, so that at high latitudes the transverse velocities would be about 47 km sec^{-1} rms. Indicative proper motion dispersions in arc sec per year using this value are given. Table 1 also gives similar calculations for a stellar population like that advocated by Gilmore and Reid (1983), namely z_o = 1500 pc, $\langle T^2 \rangle^{1/2}$ = 130 km sec^{-1}, and with a density in the plane 1/50 that of the thin-disk population.

TABLE 1. Star Counts Towards the Pole per Deg^2

	D(0) = 0.01 z_o = 350 pc		D(0) = 0.0002 z_o = 1500 pc	
m-M	A(m)	μ_T	A(m)	μ_T
0	0.0015	0.99		
2	0.022	0.39		
4	0.31	0.16	0.0073	0.43
6	3.7	0.063	0.11	0.17
8	29	0.025	1.45	0.069
10	79	0.010	15	0.027
12	18.5	0.0039	83	0.011
14	0.019	0.0016	103	0.0043
16			4.1	0.0017

When A(m) in the table is in the neighborhood of 0.03 stars deg^{-2}, we expect about one star per Schmidt field. We also require proper motion greater than about 0".015 yr^{-1} for measurement over a baseline of, say, 25 years. According to Table 1, one expects to find one M_v = +19 star at V_1 = 21 if the space density of such stars is close to 0.01 star pc^{-3} mag^{-1}: moreover, that star should have substantial proper motion. Reid and Gilmore (1981) did in fact find one such star (using red-sensitive plates) in a UKST field. These remarks can be generalized easily to white dwarfs. For instance, if at M_{pg} = +13 one star out of 25 is a white dwarf (Luyten 1960), then according to Table 1 at a distance modulus of 6 there would be expected to be about five white dwarfs with measurable motions of this absolute magnitude per Schmidt plate, if they have a scale height of 350 pc.

IV. CONCLUSIONS

Counts of common stars might as well be done with other types of telescopes. Telescopes of longer focal length enable more accurate photographic photometry -- in a sense focal Schmidt images may be too good, i.e., too small for really accurate measurement (Zwicky 1965). Also, focus variations across the plate with resulting photometric difficulties may be less of a problem at slower focal ratios. For bright stars, the great speed of Schmidt cameras is not needed. For faint stars, the statistics are good enough in small areas, and the wide field is not needed. For very faint stars, greater plate scale is required to separate with high confidence galaxies from stars.

On the other hand, the statistics of uncommon faint stars can be obtained only with Schmidt-type telescopes, as long as some kind of pointer is available to isolate the interesting stars from the mass of common stars. Examples have been given of the many very substantial contributions of Schmidt telescopes in this respect. Surveys for distant luminous stars, nearby stars of low luminosity, and nearby stars of high velocity are all of great interest. These surveys should be designed to provide more than just counts; two colors and proper motions are not difficult to generate even for very large numbers of faint stars. At brighter magnitudes, automated spectral classification is a program well worth pursuing.

This review benefited from conversations with R. Wyse and W.W. Morgan, and was supported by NSF 81-21653.

REFERENCES

Bahcall, J.N. and Soneira, R.: 1980, Astrophys. J. Suppl. **44**, pp. 73-110.
Bahcall, J.N. and Soneira, R.M.: 1983, preprint.
Bahcall, J.N., Soneira, R.M., Morton, D.C., and Tritton, K.P.: 1983, Astrophys. J. **272**, pp. 627-634.
Barbieri, C. and Rosino, L.: 1972, Astrophys. Space Sci. **16**, pp. 324-335.
Becker, W.: 1965, Zeit. für Astrophysik **62**, pp. 54-78.
Becker, W.: 1979, Astron. Astrophys. Suppl. **38**, pp. 341-353.
Berger, J. and Fringant, A.-M.: 1977, Astron. Astrophys. Suppl. **28**, pp. 123-152.
Eggen, O.J., Lynden-Bell, D. and Sandage, A.: 1962, Astrophys. J. **136**, pp. 748-766.
Feige, J.: 1958, Astrophys. J. **128**, pp. 267-272.
Fenkart, R.P.: 1967, Zeit. für Astrophysik **66**, pp. 390-403.

Gilmore, G. and Reid, N.: 1983, Mon. Not. Roy. Astron. Soc. 202, pp. 1025-1047.
Godwin, J.G. and Peach, J.V.: 1982, Mon. Not. Roy. Astron. Soc. 200, pp. 733-746.
Green, R.F.: 1980, Astrophys. J. 238, pp. 685-698.
Haro, G. and Luyten, W.J.: 1961, Bol. Obs. Tonantzintla y Tacubaya No. 21, p. 35.
Haro, G. and Luyten, W.J.: 1962, Bol. Obs. Tonantzintla y Tacubaya No. 22, pp. 37-117.
Herbst, W. and Sawyer, D.L.: 1981, Astrophys. J. 243, pp. 935-944.
Humason, M.L. and Zwicky, F.: 1947, Astrophys. J. 105, pp. 85-91.
Hubble, E.P.: 1936, "The Realm of the Nebulae", Yale University, New Haven, p. 192.
Iriarte, B. and Chavira, E.: 1957, Bol. Obs. Tonantzintla y Tacubaya No. 16, pp. 3-36.
Jaidee, S. and Lynga, G.: 1969, Arkiv för Astronomi 5, pp. 345-379.
Jones, B.F. and Klemola, A.R.: 1977, Astron. J. 82, pp. 593-597.
Jones, B.J.T. and Wyse, R.F.G.: 1983, Astron. Astrophys. 120, pp. 165-180.
Kinman, T.D.: 1965, Astrophys. J. 142, pp. 1241-1248.
Liebert, J., Dahn, C.C., Gresham, M. and Strittmatter, P.A.: 1979, Astrophys. J. 233, pp. 226-238.
Luyten, W.J.: 1955-1962, "A Search for Faint Blue Stars" 1-30, University of Minnesota, Minneapolis.
Luyten, W.J.: 1960, "A Search for ..." 22.
Luyten, W.J.: 1963-present, "Proper Motion Survey with the 48-Inch Schmidt Telescope", University of Minnesota, Minneapolis.
Luyten, W.J. (ed.): 1965, "First Conference on Faint Blue Stars", Strasbourg, August 1964, University of Minnesota, Minneapolis.
Luyten, W.J.: 1976, "Proper Motion Survey with the ..." 46.
Luyten, W.J.: 1977, "Proper Motion Survey with the ..." 51.
Luyten, W.J. and Anderson, J.H.: 1962, "A Search for Faint Blue Stars" 30.
McCusky, S.W.: 1965, in "Galactic Structure", eds. A. Blaauw and M. Schmidt, University of Chicago, pp. 1-26.
Morales-Duràn, C.: 1982, Astron. Astrophys. Suppl. 48, pp. 139-152.
Noguchi, T., Maehara, H. and Kondo, M.: 1980, Annals Tokyo Astron. Obs. 43 No. 1, pp. 55-70.
Noguchi, T., Yutani, M. and Maehara, H.: 1982, Pub. Astron. Soc. Japan 34, pp. 407-415.
Ostriker, J.P. and Thuan, T.X.: 1975, Astrophys. J. 202, pp. 353-364.
Poulakos, C.: 1978, Astron. Astrophys. Suppl. 33, pp. 395-399.
Reid, I.N. and Gilmore, G.: 1981, Mon. Not. Roy. Astron. Soc. 196, pp. 15P-18P.
Reid, N. and Gilmore, G.: 1982, Mon. Not. Roy. Astron. Soc. 201, pp. 73-94.
Richter, L., Richter, N. and Schnell, A.: 1968, Mitt. Karl-Schwarzschild-Obs. Tautenburg, No. 38, pp. 3-24.
Rubin, V.C., Moore, S. and Bertiau, E.C.: 1967, Astron. J. 72, pp. 59-64.

Saio, H. and Yoshii, Y.: 1979, Pub. Astron. Soc. Pacific 91, pp. 553–570.
Sandage, A. and Luyten, W.J.: 1967, Astrophys. J. 148, pp. 767–779.
Sanduleak, N.: 1976, Astron. J. 81, pp. 350–363.
Sanduleak, N. and Pesch, P.: 1982, Pub. Astron. Soc. Pacific 94, pp. 690–691.
Schilbach, E.: 1982, Astron. Nachrichten 303, pp. 335–340.
Shanks, T., Phillipps, S. and Fong, R.: 1980, Mon. Not. Roy. Astron. Soc. 191, pp. 47P–52P.
Sion, E.M. and Liebert, J.: 1977, Astrophys. J. 213, pp. 468–478.
Steppe, H.: 1978, Astron. Astrophys. Suppl. 31, pp. 209–241.
Strömgren, B.: 1976, The ESO Messenger, No. 7, pp. 12–13.
Thè, P.S. and Staller, R.F.A.: 1974, Astron. Astrophys. 36, pp. 155–161.
Upgren, A.R.: 1962, Astron. J. 67, pp. 37–78.
Upgren, A.R.: 1963, Astron. J. 68, pp. 194–206.
van Rhijn, P.J.: 1965, in "Galactic Structure", eds. A. Blaauw and M. Schmidt, University of Chicago, pp. 27–39.
Weinberger, R. and Poulakos, C.: 1977, Astron. Astrophys. Suppl. 27, pp. 249–253.
Weistrop, D.: 1972, Astron. J. 77, pp. 366–373.
Yoshii, Y.: 1982, Pub. Astron. Soc. Japan 34, pp. 365–379.
Zwicky, F.: 1965, remark in "First Conference on Faint Blue Stars", p. 17.

STELLAR POPULATION SYNTHESIS AND STAR COUNTS TO CONSTRAIN THE GALACTIC STRUCTURE

Annie Robin, Michel Creze
Observatoire de Besançon
F-25000 FRANCE

ABSTRACT: Our model of stellar population synthesis allows to derive synthetic star counts and distribution of colors, ages, and spectral types of stars in any given direction of observations. Here we compare results of the model with the distribution of stars in space and in absolute magnitude of stars of Gilmore and Reid. We find a small disagreement between observations of GR and predictions of our model in the space density of stars of $4 < M_v < 5$. We show that this discrepancy can well be explained by a contamination of their sample of assumed main sequence stars by red giants and subgiants.

1 - DISTRIBUTION IN ABSOLUTE MAGNITUDE OF STARS AT THE SGP

Our approach of stellar populations in the Galaxy has been described in Creze and Robin, 1983, and Robin, 1983. We compare here the predictions of our model with the distribution of stars in space and in absolute magnitude obtained by Gilmore and Reid, 1983 (GR).

We present in figure 1 the distribution in absolute magnitude of stars with apparent magnitude $15 < V < 17$ in the direction of the South Galactic Pole. We compare the distribution obtained by Gilmore and Reid (1983) (broken line) with the one predicted by the model of Bahcall and Soneira (1980) (BS) (solid line) and with the distribution obtained by our model (dotted line).

The distribution of BS is sensibly different from our's: in our model, the halo luminosity function (fig. 2) results from the various halo disk ratios adopted in different parts of the HR diagram. Our halo includes both intermediate population II and extreme halo, with scale heights $z(0.1)$ ranging between 2.1 kpc and 5.5 kpc. The density law of this component is thus much flatter than the BS one.

GR measured UK Schmidt plates in V and I bands. Then they used a M_v / V-I relation from unreddened main sequence stars to estimate absolute magnitudes assuming that all stars in their sample are on the

Main Sequence. Then they deduce the distribution of stars in absolute magnitude. We are going to show that, using both the predicted distribution of stars in absolute magnitude and the space density, our model can be useful to determine the cause of the disagreement and that the assumption of GR leads to a non-negligible misclassification of the giants and subgiants.

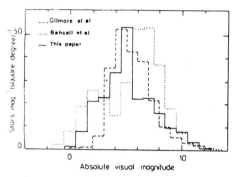

Figure 1: Distribution in M_V of stars with $15 < mv < 17$ (SGP).

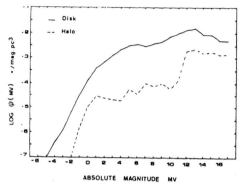

Figure 2: Luminosity function in the solar neighborhood adopted in our model.

2 - SPACE DENSITY OF STARS AT THE SOUTH GALACTIC POLE

From their observations, GR deduce the space density of stars of absolute magnitude M_V between 4 and 6 in the direction of the South Galactic Pole. The comparison between their space density and the one from our model is shown in fig. 3 a and b for stars of absolute magnitude $4 < M_V < 5$ and $5 < M_V < 6$ respectively. A very good agreement is found for this last range. But for stars with $4 < M_V < 5$, our model predicts much less stars than GR. Since both investigations use the same luminosity function for disk stars and since the agreement is good for stars of $M_V > 5$, the main sequence stars and their density law cannot be the cause of this discrepancy.

Through their M_V / V-I relation, stars with $4 < M_V < 5$ are in fact in the range V-I = 0.61 to 0.75 (see Reid and Gilmore, 1982, appendix B). In this range of colors one finds not only main sequence stars but also giants and subgiants. Contaminating stars are F5 to F8 giants with absolute visual magnitude ranging between 1.35 - 1.70 and F5 to G0 subgiants of absolute magnitude 2.5 - 3.0 (from calibrations of Deutchman et al, 1976, and our calibrations from photoelectric photometry catalogues). In the GR sample these misclassified giants slightly contaminate the counts in the range $4 < M_V < 5$ and a lack of such stars appears at $M_V = 1.5$. The misclassified subgiants of absolute magnitude $M_V = 3$ are seen by GR at $M_V = 6.5$. It thus induces an excess of stars at $4 < M_V < 5$ and at $M_V = 6.5$, and a lack at $M_V = 1.5$ and 3 in the GR sample (see figure 1).

In the GR sample, the space density of stars with $4 < M_V < 5$ induces a density law composed of two exponentials. The first one has a scale height of 300 pc and the second one 1350 pc. Our model is consistent with a two exponential density law but the first exponential would have a scale height of 250 pc since our synthetic counts are not contaminated by giants. For the sample of stars with $5 < M_V < 6$ we find a space density very close to the GR one with the same scale heights.

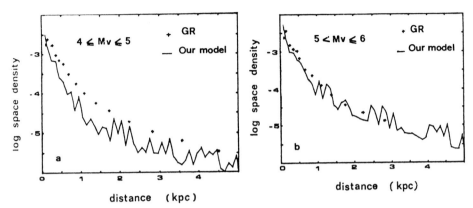

Figure 3: Density distribution of stars with distance from the Galactic Plane, in two ranges of absolute magnitude. Crosses are from Gilmore and Reid, 1983. The solid line is from our model.

3 - CONCLUSION

The analyse of Gilmore and Reid induces errors in the distribution of stars in absolute magnitude and in space. With our model we show that their assumption that all stars are on the main sequence in their sample is wrong since the number of giants and subgiants is not negligible. This misclassification of evolved stars induces an artificial increase in the scale height of disk main sequence stars from 250 pc to 300pc.

REFERENCES

Bahcall, J.N. Soneira, R.H., 1980, Astrophys. J. Suppl. Ser. 44,73
Creze, M., Robin, A., 1983, IAU coll. n° 76, Middletown, Connecticut
Deutschman, W.A., Davis, R.J., Schild, R.E., 1976, Astrophys. J. Suppl. Ser. 30,97
Gilmore, G., Reid, N., 1983, Monthly notices Roy. Astron. Soc. 202,1025
Reid, N., Gilmore, G., 1982, Monthly Notices Roy. Astron. Soc. 201,73
Robin, A., 1983, These 3eme cycle, Universite Paris 7
Twarog, B.A., 1980, Astrophys. J. 242,242

AN AUTOMATED METHOD OF GENERAL STAR COUNTS FOR DARK CLOUDS

H. Ohtani*
Department of Astronomy
The University of Manchester
England

ABSTRACT

An automated method of general star counts has been developed for the purpose of deriving the distributions of extinction of large dark clouds. The result of the application to the Southern Coalsack is given.

1. INTRODUCTION

Recently, methods of digital analysis of photographic plates of stellar fields have been developed for various particular purposes (eg. Duerr and Craine, 1982). We have developed a system of general star counts optimised to derivation of surface distribution of extinction of large dark clouds with an aim to discuss the problem of star formation in those clouds (Saito et al., 1981).

Since dark clouds are generally located in the Milky Way and the method of star counts is essentially statistical, following points have been taken into consideration in developing the system.

(1) The system is capable of treating so many stellar images as hundreds of thousands;
(2) All images may be assumed to be stellar images;
(3) Individual stellar images which compose a blended one can be counted separately;
(4) Accurate photometry of individual stars (Herzog and Illingworth, 1975; Buonanno and Corsi, 1978) is not necessary.

2. METHOD

A Schmidt plate is scanned by a two dimensional microdensitometer. The pixel size and the sampling rate are so selected that the faintest

---

* On leave from Department of Astronomy, University of Kyoto, Kyoto, Japan

star images are divided into a few pixels.

We set two criteria in detection of stellar images. One is for the photographic densities of images and the other is for their extent. Both criteria are determined empirically so that the resulting detection for a small selected area of the plate is satisfactory from comparison with the eye inspection of the area. This is a reasonable tactic since both of the sky background density and the image quality are different from plate to plate which would be processed by this system.

Each image detected is diagnosed whether it is blended or not according to the following principles. For simplicity, consider an image blended with two components. In cases where both components show individual maxima in the density profile, the locus of minima between the two maxima can be defined as the boundary of the components (Figure 1a).

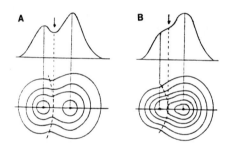

Figure 1a (left) and 1b (right). Two cases of blending of stellar images. See text for details.

On the other hand, in cases where the blending is so heavy that the fainter star does not show a corresponding maximum, it has been found that the curvature of the profile changes its sign between centres of the two components. Thus, a locus of these points is used as the boundary (Figure 1b).

These principles may be generalized to be applicable to multiplied blended images. As a practical treatment, an image is first divided into components, if any, in one direction of the pixel array. Thereafter, each resulting component is diagnosed for further blending along its ridge.

3. APPLICATION

A plate of the Southern Coalsack taken by a 50 cm Schmidt telescope of Bosscha Observatory in Indonesia was analysed. The plate size is 11cm x 11cm which covers about 22 square degrees. The limiting magnitude is $V > 17$ mag.

This plate was scanned at Kiso Observatory by using a fast two dimensional microdensitometer with a linear CCD array as its sensor. The smallest pixel size, 13μm x 24μm, was selected in scanning to give 38 megapixels over the whole plate.

The pixel data, the coordinates and the transmissions, stored on magnetic tapes were processed by using a FACOM M-200 in Kyoto and a VAX-11/780 in Manchester.

A part of the result of the image detection is shown in Figure 2; the machine result is overlaid on the reproduction of the plate. The total number of stars detected on this plate is 480 thousand.

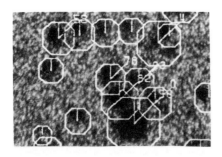

Figure 2. Detected star images are overlaid on a reproduction of a part of the original plate.

The distribution of extinction has been derived from the distribution of the star number density for the whole plate area (Figure 3a). A minimum value of the total dust mass in the Coalsack has been found to be 20 solar masses. A detailed extinction map of the region of Tapia's globules 1 and 2 (Tapia, 1973) is also shown in Figure 3b.

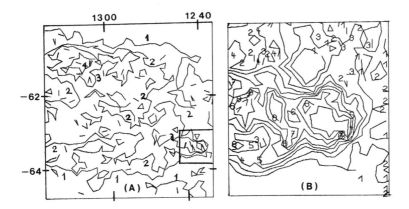

Figure 3. Distribution of extinction. (A) The whole region of the Coalsack. Contours 1 to 5 correspond to Av = 0.5, 1.5, 2.5, 3.5 and 4.0 respectively. (B) Details of a part indicated in (A). Contours 1 to 8 are for Av = 0.5 to 4.0 with step 0.5.

REFERENCES

Buonanno, R. and Corsi, C. E., 1978, in Proceedings of the 5th Colloquium on Astrophysics, Trieste, International Workshop in Image Processing in Astronomy., ed. G. Sedmak, M. Capaccioli and R. J. Allen (Observatorio Astronomico di Trieste, Trieste), pp.354-359.
Duerr, R. and Craine, E. R., 1982, Astron. J., 87, pp.408-418.
Herzog, A. D. and Illingworth, G., 1977, Astrophys. J. Suppl. Ser., 33, pp.55-67.
Saito, T., Ohtani, H. and Tomita, Y., Publ. Astron. Soc. Japan, 33, 327, 1981, pp.327-340.
Tapia, S., 1973, in IAU Symposium 52, Interstellar Dust and Related Topics, ed. J. M. Greenberg and H. C. Van de Hulst (Reidel, Dordrecht), p.43.

DISCUSSION

M. SANTANGELO: Have you tried to compare the results of your statistical analysis of star counts with the results given by multicolour photometry of stars in dark nebulae?

H. OHTANI: Yes. In the regions where Rodgers' (M.N.R.A.S. 120, p. 163, 1960) results are available, our result is consistent with his. The total dust mass obtained by us is greater than Rodgers' estimate by 40%. This difference may be attributed to the fact that Rodgers observed only few stars through the most dense parts of the clouds.

THE MAGELLANIC CLOUDS

B.E. Westerlund
Uppsala Universitets Astronomiska Observatorium
Uppsala, Sweden

ABSTRACT

Contributions to our present knowledge of the Magellanic Clouds based on observations with Schmidt telescopes are discussed.

1. INTRODUCTION

During a long period almost all the research on the Magellanic Clouds was carried out with the aid of material gathered at the Boyden Station of the Harvard Observatory, first at Arequipa, Peru (1889-1927) and then at Bloemfontein, South Africa. From 1914 on Lick Observatory contributed by measurements of radial velocities at its station in Chile. In the early 1950:s several new telescopes, including radio telescopes, were in operation in Australia, South Africa, and South America. A new epoch for astronomical observations in the southern hemisphere was beginning.

In 1932 the optical system of the "Schmidt camera" had been described (Schmidt 1932), and during the next three decades a number of Schmidt telescopes were built. In the early 1960:s 19 were ready or expected to be in use soon (apertures over 50 cm; see *Telescopes, Stars and Stellar Systems* vol. 1, p. 239, 1962). Of these, only two were located sufficiently far south to serve for Magellanic Clouds research; the ADH telescope at Boyden Station (lat.-29°) and the Schmidt telescope at Uppsala Southern Station, Mt Stromlo Observatory (-35°). At Mt Stromlo Observatory was also a 20 cm Meinel-Pearson Schmidt camera in use.

During the following two decades Schmidt telescopes were constructed for use in the south, or moved there for continuous or temporary research. Table 1 lists the southern Schmidt telescopes. It includes a reference to the Schmidt camera briefly used on the lunar surface for observations of the Clouds - lack of space makes it impossible to discuss its important contributions here (see Page, Carruthers, 1978, 1981). In a review that mainly deals with surveys we must, however, also consider other widefield telescopes with which important work on the Clouds have been carried out during the last three decades (Table 2).

Table 1. Schmidt telescopes used for research on the Clouds.

Telescope	Observatory	Dimensions/Scale	Prisms	In use
ADH	Boyden Station South Africa	81/90/303 cm 68"/mm	240 Å/mm at Hγ	1950-74
Uppsala Southern Schmidt	Uppsala Southern Station, Mt Stromlo, Australia	50/65/175 120	480	1956-
Meinel-Pearson	Mt Stromlo Observatory, Australia	20/26/20 1000		1958-
Hamburg-Schmidt	Boyden Station, South Africa	36/42/63 330	710	1967-
Curtis-Schmidt	Cerro Tololo Inter-American Observatory, Chile	61/91/214 97	580 1360	1967-
ESO Schmidt	European Southern Observatory, La Silla, Chile	100/162/306 67.5	450	1972-
SRC Schmidt	UKST, Siding Spring Observatory, Australia	122/183/307 67.5	2400 800 108	1973-
Far Ultraviolet	Naval Research Laboratory, Lunar surface (Apollo 16)	See Carruthers, G.R.: 1973, Appl.Optics 12, 2501		1972,21-23/4

Note: Uppsala Southern Schmidt was moved to Siding Spring Observatory in October, 1981.

Table 2. Other wide-field telescopes used for surveys of the Clouds.

Telescope	Observatory	Dimensions/Scale	Prisms	In use
MtWilson 10" camera	Lamont-Hussey Observatory, Bloemfontein, South Africa	25 cm 159"/mm	300 Å/mm at 5890 Å	1950
GPO Astrograph	Zeekoegat, South Africa La Silla, Chile	40 51.5	110 Å/mm at 4210 Å	1961-66 1968-

Note: A twin 7-inch f/2.5 Ektar camera was used by de Vaucouleurs (1954) at Mt Stromlo Observatory.

Recently the new large southern reflectors have been used for a variety of surveys, of the sampling kind, of the Magellanic Clouds, in particular with the powerful GRISM technique. These investigations will be referred to here only when they serve to illustrate the achievements or the limitations of the Schmidt telescopes. Our aim will thus be to establish the contributions to our present knowledge of the Magellanic Clouds that are based on Schmidt- (wide-angle-) telescope research, only. We will mainly deal with those areas where full use has been made of the special Schmidt-telescope characteristics.

2. DIRECT PHOTOGRAPHY OF THE MAGELLANIC CLOUDS

(1) Charts

(i) An *Atlas* in visual light, permitting the determination of accurate positions for objects with V < 18.5, has been prepared from Uppsala Southern Schmidt plates (Gascoigne, Westerlund 1961).

(ii) *Atlas* in B and V of the LMC and the SMC from ADH and Curtis-Schmidt plates, respectively, have been published (Hodge, Wright 1967, 1977); they contain lists and identifications of objects.

(iii) *Composite photographs* have been made by Johnson (1959a), by Walker et al. (1969), and, in colour, by Dufour and Gooding (1976).

Johnson used low-scale plates taken with the Meinel-Pearson camera. He concluded that the LMC consisted of two components: an asymmetrical Sc galaxy with its nucleus at 30 Dor, in front of an elliptical system. Walker et al. also remarked on the main components of the two Clouds. Since then, the "elliptical component" in the LMC and, in particular, the spiral structure with its nucleus at 30 Dor have appeared a number of times in the literature.

(2) HII regions

(i) *In the Clouds*. A distorted one-arm spiral was noted in the SMC by Rodgers (1959), who used the Meinel-Pearson camera to search the SMC for HII regions. His results agreed well with those by Nail et al. (1953) who used ADH red and blue plates. Johnson (1961), from Meinel-Pearson-camera plates, described the SMC as consisting of two components, one a dwarf elliptical, the other a contorted gaseous-magnetic arm with no nucleus.

Davies et al. (1976) catalogued the HII regions in the Clouds using long-exposure SRC-Schmidt plates taken with an $H\alpha$ interference filter. This was the first extensive cataloguing of emission nebulae in the Clouds since Henize (1956).

Lasker (1979, 1980) used the Curtis-Schmidt to survey the Clouds in the light of [SII]$\lambda\lambda$6713, 6731, $H\alpha$ + [NII], and [OIII]λ5007, aiming at the numerous shell nebulae but also obtaining an overview of the excitation states of all nebulae.

(ii) *Between the Clouds*. The region between the Clouds was searched for emission regions by Johnson (1959b) with the Meinel-Pearson camera. He found a faint, sinuos ribbon curving across his field and suggested that it was probably an outlying portion of the Gum nebula. P.G. Johnson et al. (1982) obtained deep Hα plates along the Magellanic Stream with a wide-field camera (30°). They combined them with high-contrast prints from IIIa-J plates and with a deep Hα plate taken with the SRC Schmidt. A diffuse Hα region was found to be correlated with the Stream; the other detected filaments were faint and blue, possibly reflection nebulosity in the galactic plane (cf. de Vaucouleurs 1954, 1960). The filaments detected by Johnson (1959b) appear to coincide with a part of these filaments.

(3) Giant and supergiant shells in the Clouds

Westerlund and Mathewson (1966) found that the two supernova remnants, N49 and N63A, belonged to a shell-like structure forming the boundary of Constellation III in the LMC and centered on the emission nebula N55. They suggested that this might be a super-supernova remnant, and that many such remnants existed in the LMC. The third supernova remnant known at that time, N132D, was in a supergiant shell centered on Constellation II. The available Schmidt plates showed that these shells were the location for the formation of associations and clusters as well as for supernova explosions.

Shells of this kind and smaller ones have been discovered on the excellent interference-filter photographs taken with the SRC Schmidt (Goudis, Meaburn 1978). A total of 85 giant shells (20-260pc) are now known in the LMC, and 9 supergiant shells are proposed in the LMC and 1 in the SMC (Meaburn 1980).

(4) Dust in the Clouds

Surveys for dark nebulae in the Clouds have been carried out by Hodge (1972, 1974c) and by van den Bergh (1974). Hodge used Curtis-Schmidt and ADH plates in B and V for studying the LMC, and Curtis-Schmidt plates for the SMC. Van den Bergh used Curtis-Schmidt plates in R, V, B, and U of both Clouds. Not surprisingly, the two surveys agree well. The visual inspection is, however, a rather subjective method; for discovery the dust cloud has to be well silhouetted against the stellar population.

The absorption in the SMC has also been investigated with the aid of background galaxies. Hodge (1974a) identified 2 200 galaxies in an area of 85 square degrees, using Curtis-Schmidt plates. From their distribution he concluded that the dust in the SMC is distributed similarly to the neutral hydrogen (Hodge 1974b). The maximum absorption is $A_V = 1.3$ mag in the core of the SMC. His results confirm those previously obtained by Wesselink (see Hodge 1974a). They have been further confirmed by MacGillivray (1975), who used deep SRC Schmidt IIIa-J plates to reach B = 23 mag.

(5) Clusters in the Clouds

Several surveys for clusters in the Clouds have been carried out with Schmidt telescopes (Table 3). Discrepancies between the catalogues exist; they are generally due to the definition of an open cluster near the

Table 3. Surveys for clusters in the Magellanic Clouds.

Observer	Telescope	Clusters	Remarks
	In the SMC		
Kron (1956)	ADH	69	
Lindsay (1956b)	ADH	94	
Lindsay (1958)	ADH	116	
Westerlund, Glaspey (1971)	Uppsala S. Schmidt	18	In the Wing
Hodge, Wright (1974)	Curtis-Schmidt	220	
Brück (1975, 1976)	SRC Schmidt	330	
	In the LMC		
Shapley, Lindsay (SL) (1963)	ADH	897	
Lyngå, Westerlund (LW) (1963)	Uppsala S. Schmidt	483	In outer regions
Hodge, Sexton (HS) (1966)	ADH	1603	

plate limit. Lindsay's comments (1964) on the 184 LW clusters, which were not in the SL list in spite of being in the surveyed area, often agree with the LW definition of an old open cluster. It is difficult to identify a star-poor cluster also on large-scale plates. Thus, Brück's cluster E165 (1975) is in the field of NGC602, observed by Westerlund (1964a) with the Mt Stromlo 74-inch telescope in B and V to V = 20 mag. It was not noted by him as a cluster, though its 5 brightest stars were among those measured. 4 of them form a vertical main sequence down to V = 19.3 mag; the fifth and brightest has V = 17.15, B-V = +1.84. It may well be an open cluster of the age of M41 and M11 and typical for many of the faint open clusters in the Clouds.

It is important for judging the efficiency of Schmidt telescopes to establish the completeness of the surveys in Table 3. Hodge (1975), using the CTIO 1.5-m reflector, carried out deep surveys for clusters in fields in the LMC. He concluded that "the catalogues ... using Schmidt plates, are surprisingly complete". 65 percent of the clusters found in his deep search were already catalogued. By extending the surveys to B = 22.5 mag in 5 fields in the LMC with the aid of the CTIO 4-m reflector he found an average increase over the Schmidt surveys with a factor of 3 (Hodge 1980). The total number of clusters in the LMC was estimated to about 6 500.

The distribution of the clusters in the Clouds have been analyzed by a number of authors. For the SMC we refer to Brück (1975), who concluded that the disk defined by the clusters has the same effective size as that defined by the stars. There is, however, a lack of clusters in the centre of the SMC, noted also by Hodge (1974d).

The SMC is suspected of having a pronounced extention in depth; this has been seen in the HI distribution as well as for many classes of stellar objects. It has been given little consideration in the analysis of the cluster distribution. Hodge (1974d) examined the possibility of the cluster system being spheroidal, but came to the conclusion that the clusters are arranged in a plane together with the young stars. Azzopardi and Vigneau (1977) confirmed this: "the star cluster and ionized hydrogen distributions are well correlated with the O-B2 regions".

The cluster distribution in the LMC was discussed by Lyngå and Westerlund (1963) who found an elliptical distribution, leading to a tilt of 45° for the LMC. This value would make the plane of symmetry of the LMC pass through the main body of the SMC, an attractive idea. The high value of 45° for the inclination is strongly objected to by de Vaucouleurs (de Vaucouleurs, Freeman 1973).

Using ADH and Curtis-Schmidt plates Hodge et al. (1970) estimated the magnitudes of the brightest stars in many LMC clusters. Hodge (1973) used 509 clusters with the brightest star brighter than V = 15.4 to discuss the evolutionary history of the cluster system. 60 percent of the clusters occur in groups isolated in space and time. The mean dimensions of the groups are about 1.5 kpc; the mean interval of formation is of the order of 10^6 yr. Undoubtedly, these groups of clusters are intimately connected with the super-associations, supergiant shells and bursts of star formation (Westerlund 1964b).

(6) Stellar associations in the Clouds

The SMC has few associations, the most pronounced ones are found in the Wing, from NGC 456-465 and out. The most complete listing of the associations in the LMC has been compiled by Lucke and Hodge (1970). They used ADH plates to identify 122 stellar associations. The dimensions vary from 15 to 350 pc. The definition of the boundaries of an association is always subjective, however, also in the LMC. Lucke (1974) has carried out B, V photometry on Curtis-Schmidt plates of the stars in the areas of the associations and produced very useful colour-magnitude diagrams.

(7) Star counts in the Clouds

(i) *The Wing of the Small Cloud* was searched for blue stars by Westerlund and Glaspey (1971) on Uppsala Southern Schmidt plates in U, B, V. Over 2 000 objects were identified and the structure of the Wing was outlined. Scattered blue stars, likely members of the SMC, were found outside the Wing.

(ii) *The disk and halo populations of the SMC* have been studied by Brück and Marşoğlu (1978) on SRC Schmidt plates. They carried out B, V photometry to 21.2 mag in two fields in the eastern part of the SMC. Their field I, in the northern-most part of the Bar, has a young population, similar to the one in the Wing (cf. Westerlund 1964a). Their field II contains a halo population similar to the one near 47 Tuc (Tifft 1963).

Brück (1980) studied the radial distribution of 163 000 stars brighter than B = 21 mag in the south-west and north-east outer parts of the SMC. Disk stars and halo stars were separated. The disk stars are closely related with neutral hydrogen; the halo stars are strongly biazed towards the western parts of the SMC. This is somewhat surprising as previous studies have indicated that the old population may have its centroid to the east of the Bar (Johnson 1961, Westerlund 1970).

(iii) *The major spiral arm of the LMC.* The existence of a spiral structure in the LMC has been much debated. Dixon and Ford (1972) used Uppsala Southern Schmidt plates in B and V to divide 8 000 stars, located in the "major spiral arm" and more luminous than $M_V = -3.5$ mag, into four age groups. If the composition of the arm corresponded to the one expected from the density-wave theory, or if the arm had been formed by ejection (Schmidt-Kaler 1977), the youngest stars should be at the inner edge of the arm and the earlier generations successively displaced outwards. All four groups showed the same distribution over the spiral arm, the only possible variation being *along* the "arm" with the youngest stars further from the centre of the LMC. The distribution is better understood as that of a super-association, in particular as the youngest stars belong to a pronounced nebular complex.

(8) The variables in the Magellanic Clouds

The available space does not permit the extensive discussion that this important field of research deserves. Therefore, I will only mention some of the many lists.

Hodge and Wright (1967, 1977) have identified the variables found in the extensive searches carried out of the Harvard collection of plates. Other studies of variables based on ADH plates exist (Lindsay 1974, Butler 1978, also for references).

The distributions and the evolution of the variables in the Clouds have been discussed by Payne-Gaposchkin and Gaposchkin (1966) and by Gaposchkin (1972). The total number of variables then was 1592 in the SMC and 1830 in the LMC. Hodge and Wright expected this to be less than a half the number of existing variables.

For the recent search for novae in the Clouds with the Curtis-Schmidt I refer to Graham (1979).

3. OBJECTIVE-PRISM RESEARCH OF THE MAGELLANIC CLOUDS

The Magellanic Clouds have been observed with Schmidt telescopes equipped with objective-prisms with great success. As the sites of the southern Schmidt telescopes often have good or excellent seeing high-quality spectra have been obtained. Overlapping of spectra in the crowded regions of the Clouds has been a severe problem. The effect has been diminished by the use of very short spectra, either by going to very low dispersions or by limiting the spectral ranges with the aid of suitable filters.

Objective-prism spectra are usually not suited for radial-velocity determinations. (See for instance Butler, Norris 1969 and Fehrenbach 1970, Wood 1970.) The only radial-velocity determinations of survey-type of the Clouds are those by Fehrenbach and his collaborators with the aid of the GPO astrograph.

Most objective-prism plates have not been fully utilized. They have been inspected visually, the objects of interest identified and slit spectroscopy and/or photoelectric photometry carried out of some representative objects. Modern measuring techniques make, however, the plates useful for accurate spectrophotometry to an extent not previously possible.

(1) Surveys for bright member stars of the Clouds

The identification of stellar members of the Clouds has been done by the determination of spectral types and luminosity classes and of radial velocities with the Fehrenbach technique (Table 4).

Table 4. Surveys for bright stellar members of the Clouds.

Observer	Telescope	No. of stars	Method	Remarks
	In the SMC			
Sanduleak (1968)	Curtis-Schmidt	169	Sp.class.	
Sanduleak (1969)	Curtis-Schmidt	47	Sp.class.	In Wing
Florsch (1972)	GPO astrograph	100	Rad.vel.	
Azzopardi, Vigneau (1982)	GPO astrograph	524	Sp.class.	
	In the LMC			
Sanduleak (1970)	Curtis-Schmidt	1272	Sp.class.	
Rousseau et al. (1978)	GPO astrograph	1822	Sp.class.	
Fehrenbach, Duflot (1982)	GPO astrograph	711	Rad.vel.	
Philip, Sanduleak (1979)	Curtis-Schmidt	312	Sp.class.	2 fields

A particular advantage of the radial-velocity technique over the spectral classification is that it permits the identification of A and F stars among the members of the Clouds; the classification gives generally only OB stars. An attempt to identify B7-G5 stars in the LMC by spectral classification on objective-prism Curtis-Schmidt plates was made by Stock et al. (1976). See, however, Fehrenbach, Duflot (1978).

(2) Surveys for red member stars of the Clouds

Spectral surveys in the near-infrared spectral region were introduced by Nassau and his collaborators (see Mavridis 1967 for references). The technique was applied by Westerlund (1961) in a study of the central regions of the LMC. The survey was extended to cover most of the LMC; catalogues of the identified stars have been puslished (Westerlund et al. 1978, 1981).

With the appearance of the fine-grain IIIa-J emulsion the range Hβ-5300 Å became available for objective-prism classification. It has been used on the Clouds by Sanduleak and Philip (1977), Rebeirot et al. (1983) and Prévot et al. (1983). Table 5 summarizes the data of the main lists.

The surveys of the SMC are few and mainly unpublished. There are no supergiants in the SMC later than M1 (Humphreys 1979). This has been confirmed by Prévot et al. (1983); their stars are all in the range K5-M1.

A comparison of the two catalogues of carbon stars in the LMC shows only 75 objects in common. This is mainly due to the fact that identification by the near-infrared CN bands gives cooler stars than the one by the green C_2 bands. (Cf. Richer et al. 1979).

Table 5. Lists of red stars in the Clouds.

Observer	Telescope	M I stars	C stars	Remarks
In the SMC				
Sanduleak (unpubl.)	Curtis-Schmidt	101		
Sanduleak, Philip, Albers (unpubl.)	Curtis-Schmidt		yes	
Prévot et al. (1983)	ESO Schmidt, GPO	199		+ 23 gal.st.
In the LMC				
Sanduleak, Philip (1977)	Curtis-Schmidt	609	474	
Westerlund et al. (1978, 81)	Uppsala S. Schmidt	480	302	+ 52 gal.M
Rebeirot et al. (1983)	ESO Schmidt, GPO	839		

The three lists of supergiants in the LMC show also certain discrepancies. The reason appears to be in the classification criteria: In the near-infrared the definition of class M0 is difficult in low dispersion; gradients are difficult to estimate visually; late K stars may be included when the green TiO bands are used.

The total number of M supergiants in the LMC from the three lists amounts to 894 objects (Rebeirot et al. 1983). The 626 carbon stars listed represent only the bright part of this class. GRISM surveys have led to the detection of a large number of fainter carbon stars, and to the discovery of more objects in the crowded regions of the LMC. Similar results exist for the SMC.

A number of M giants are listed by Westerlund et al. (1981) as possible members of the LMC. Blanco and McCarthy (1975) used the Curtis-Schmidt for a near-infrared survey of a field in the LMC with a dispersion of only 6 700 Å/mm. Over 800 late M stars were seen.

(3) Surveus for emission-line objects in the Clouds

The surveys for emission-line objects in the Clouds are still rather incomplete. The most extensive ones are listed in Table 6.

Table 6. Emission-line objects in the Clouds.

Observer	Telescope	Objects	in	Remarks
Heinze (1956)	MtWilson 10"	172	LMC	+ 415 neb.
		65	SMC	+ 117 neb.
Lindsay (1961)	ADH	593	SMC	49 P or P?
Lindsay (1963)	ADH	358	LMC	
Andrews, Lindsay (1964)	ADH	446	LMC	
Bohannan, Epps (1974)	Curtis-Schmidt	446	LMC	all new iden.

(4) Planetary nebulae in the Clouds

Table 7 lists surveys for planetary nebulae in the Clouds which have been carried out with Schmidt telescopes.

Table 7. Surveys for planetary nebulae in the Clouds.

Observer	Telescope	Number of P:s
In the SMC		
Lindsay (1955)	ADH	17
Lindsay (1956a)	ADH	20 + 9P?
Lindsay (1961)	ADH	30 + 19P?
Sanduleak et al. (1978)	Curtis-Schmidt	28
In the LMC		
Westerlund, Rodgers (1959)	Uppsala S. Schmidt	34
Lindsay, Mullan (1963)	ADH	65 + 44 no continuum
Westerlund, Smith (1964a)	Uppsala S. Schmidt	42
Sanduleak et al. (1978)	Curtis-Schmidt	102

Jacoby (1983) has discussed these surveys and related them to his deep samples with the CTIO 4-m telescope (Jacoby 1980). He concludes that the total numbers of planetaries are 285 ± 78 in the SMC, and 996 ± 253 in the LMC. The estimate for the SMC is in fair agreement with the value of 300 derived by Henize and Westerlund (1964), whereas the one for the LMC is about twice as large as the value of 450 derived by Westerlund and Smith (1964a). The estimates are sensitive to the definition of the magnitude interval used for the counts.

Sanduleak et al. (1972), using unwidened Curtis-Schmidt objective-prism spectra with a dispersion of 420 Å/mm at Hα, resolved the [NII] line $\lambda 6584$Å and Hα, and found the SMC planetaries to be nitrogen deficient.

The surveys of the Clouds for planetaries have also led to discoveries of "compact HII regions" (see Henize, Westerlund 1964, Davies et al. 1976, Sanduleak, Philip 1977, and Sanduleak et al. 1978).

(5) Surveys for Wolf-Rayet stars in the Magellanic Clouds

A summary of the discoveries of Wolf-Rayet stars in the Clouds is given in Table 8. References to additional papers may be found in those given in the Table.

Westerlund (1961) pointed out that the Wolf-Rayet stars in the colour-magnitude diagrams of the clusters and associations always were at the turn-off points. This fact should be considered in attempts to understand the range of 3 mag spanned by the early WN stars as well as by the late ones. Most likely all Wolf-Rayet stars have formed in associations; the older stars in an association may be found in its outer parts, and thus also many evolved Wolf-Rayet stars. There should, however, be a cut-off for the Wolf-Rayet stars belonging to the youngest Population I in

Table 8. Surveys for Wolf-Rayet stars in the Clouds.

Observer	Telescope	Number of W-R:s	Remarks
In the SMC			
Azzopardi, Brysacher (1979)	GPO astrograph	8	
In the LMC			
Westerlund, Rodgers (1959)	Uppsala S. Schmidt	50	
Westerlund, Smith (1964b)	Uppsala S. Schmidt	58	
Fehrenbach et al. (1976)	GPO astrograph	78	all new WN3,4
Breysacher, Azzopardi (1982)	GPO astrograph	101	

the LMC. The value of V = 15 mag, proposed by Westerlund and Smith (1964b) should be increased somewhat following the discovery of the faint WN3 and WN4 stars. This does, however, not change the evolutionary pattern proposed by them. It is now also attractive with the proposed evolutionary pattern, WN7 → WN3 (see IAU Symposium No. 99). It is more difficult to see, from the stars in the LMC, how this sequence may continue into the WC:s.

4. CONCLUSIONS

The surveys of the Magellanic Clouds with wide-field telescopes, in particular Schmidt telescopes, have during the past three decades made fundamental contributions to our knowledge of the Clouds, their structure and their evolution. They have also identified objects that have served for determining the chemical composition of the various populations of the Clouds as well as for studies of the kinematics and dynamics of the Clouds and of the Magellanic System. It is to be expected that "Schmidt"-surveys in the future will, with the aid of suitable filter-, detector- and reduction-techniques, reach fainter objects as well as overcome the problems of overlapping in the crowded regions.

REFERENCES

Andrews, A.D., Lindsay, E.M: 1964, Irish Astron. J. 6, 241.
Azzopardi, M., Breysacher, J.: 1979, Astron. Astrophys. 75, 120.
Azzopardi, M., Vigneau, J.: 1977, Astron. Astrophys. 56, 151.
Azzopardi, M., Vigneau J.: 1982, Astron. Astrophys. Suppl. 50, 291.
Blanco, V.M., McCarthy, M.F.: 1975, Nature 258, 407.
Bohannan, B., Epps, H.W.: 1947, Astron. Astrophys. Suppl. 18, 47.
Breysacher, J., Azzopardi, M.: 1982, IAU Symp. 99, 523 (eds. C.W.H. de
    Loore and A.J. Wills; D. Reidel, Dordrecht, Holland).
Brück, M.T.: 1975, Mon. Not. R. Astr. Soc. 173, 327.
Brück, M.T.: 1976, Occ. Rep. Roy. Obs. Edinburgh No. 1.
Brück, M.T.: 1980, Astron. Astrophys. 87, 92.
Brück, M.T., Marsoglu, A.: 1978, Astron. Astrophys. 68, 193.

Butler, C.J.: 1978, Astron. Astrophys. Suppl. 32, 83.
Butler, C.J., Norris, M.V.: 1969, Mon.Not.Astr.Soc.S. Africa 28, 107.
Davies, R.D., Elliott, K.H., Meaburn, J.: 1976, Mem.R.Astr.Soc. 81, 89.
De Vaucouleurs, G.: 1954, Obs. 74, 158.
De Vaucouleurs, G.: 1969, Obs. 80, 106.
De Vaucouleurs, G., Freeman, K.C.: 1973, Vistas in Astron. 14, 163.
Dixon, M.E., Ford, V.L.: 1972, Atrophys. J. 173, 35.
Dufour, R.J., Gooding, R.A.: 1976, Southwest Reg.Conf.Astron.Astrophys. 1, 71.
Fehrenbach, Ch., Duflot, M., Acker, A.: 1976, Astron.Astrophys.Suppl. 24, 379.
Fehrenbach, Ch., Duflot, M.: 1978, Astron. Astrophys. Suppl. 32, 159.
Fehrenbach, Ch., Duflot, M.: 1982, Astron. Astrophys, Suppl. 48, 409.
Florsch, A.: 1972, Publ. Obs. Strasbourg 2, no. 1.
Gaposchkin, S.: 1972, SAO Spec.Rep. No. 310.
Gascoigne, S.C.B., Westerlund, B.E.: 1961, Uppsala-Mt Stromlo Atlas of the Magellanic Clouds. Austral. Nat. Univ., Canberra.
Goudis, C., Meaburn, J.: 1978, Astron. Astrophys. 68, 189.
Graham, J.A.: 1979, IAU Coll. no. 46, 96 (eds. F.M. Bateson, J. Smak, I.H. Urch; Hamilton, New Zealand).
Henize, K.G.: 1956, Astrophys. J. Suppl. 2, 315.
Henize, K.G., Westerlund, B.E.: 1963, Astrophys. J. 137, 747.
Hodge, P.W.: 1972, Publ. Astron. Soc. Pacific 84, 365.
Hodge, P.W.: 1973, Astron. J. 78, 807.
Hodge, P.W.: 1974a, Astrophys. J. 192, 21.
Hodge, P.W.: 1974b, Suppl. No. 1 to the Small Magellanic Cloud Atlas.
Hodge, P.W.: 1974c, Publ. Astron. Soc. Pacific 86, 623.
Hodge, P.W.: 1974d, Astron. J. 79, 860.
Hodge, P.W.: 1975, Irish Astron. J. 12, 77.
Hodge, P.W.: 1980, Astron. J. 85, 243.
Hodge, P.W., Lucke, P.B.: 1970, Astron. J. 75, 933.
Hodge, P.W., Sexton, J.: 1966, Astron. J. 71, 363.
Hodge, P.W., Welch, G.A., Wills, R., Wright, F.W.: 1970, SAO Spec.Rep. No. 320.
Hodge, P.W., Wright, F.W.: 1967, The Large Magellanic Cloud. Smithsonian Press, Washington, D.C.
Hodge, P.W., Wright, F.W.: 1974, Astron. J. 79, 858.
Hodge, P.W., Wright, F.W.: 1977, The Small Magellanic Cloud, Univ. of Washington Press, Seattle.
Humphreys, R.M.: 1979, Astrophys. J. 231, 384.
Jacoby, G.H.: 1980, Astrophys. J. Suppl. 42, 1.
Jacoby, G.H.: 1983, IAU Symp. 103, 427 (ed. D.R. Flower).
Johnson, H.M.: 1959a, Publ. Astron. Soc. Pacific 71, 301.
Johnson, H.M.: 1959b, Publ. Astron. Soc. Pacific 71, 342.
Johnson, H.M.: 1961, Publ. Astron. Soc. Pacific 73, 20.
Johnson, P.G., Meaburn, J., Osman, A.M.I.: 1982, Mon.Not.R.Astr.Soc. 198, 985.
Kron, G.E.: 1956, Publ. Astron. Soc. Pacific 68, 125; 230; 326.
Lasker, B.M.: 1979, Publ. Astron. Soc. Pacific 91, 153.
Lasker, B.M.: 1980, CTIO Contr. No. 127.
Lindsay, E.M.: 1955, Mon. Not. R. Astr. Soc. 115, 248.

Lindsay, E.M.: 1956a, Mon. Not. R. Astron. Soc. 116, 649.
Lindsay, E.M.: 1956b, Irish Astron. J. 4, 65.
Lindsay, E.M.: 1958, Mon. Not. R. Astron. Soc. 118, 172.
Lindsay, E.M.: 1961, Astron. J. 66, 169.
Lindsay, E.M.: 1963, Irish Astron. J. 6, 127.
Lindsay, E.M.: 1964, Irish Astron. J. 6, 233.
Lindsay, E.M.: 1974, Mon. Not. S. Astr. Soc. 169, 343.
Lindsay, E.M., Mullan, D.J.: 1963, Irish Astron. J. 6, 51.
Lucke, P.B.: 1974, Astrophys. J. Suppl. 28, 73.
Lucke, P.B., Hodge, P.W.: 1970, Astron. J. 75, 171.
Lyngå, G., Westerlund, B.E.: 1963, Mon.Not.R.Astr.Soc. 127, 31.
MacGillivray, H.T.: 1975, Mon. Not. R.Astr.Soc, 170, 241.
Mavridis, L.N.: 1967, Coll. Late-Type St ars, p. 420 (ed. M. Hack).
Meaburn, J.: 1980, Mon. Not. R. Astr. Soc. 192, 365.
Nail, V.McK., Whitney, C.A., Wade, C.M.: 1953, Proc.Nat.Acad.Sci, 39,1161.
Page, Th., Carruthers, G.R.: 1978, NRL Report 8206.
Page, Th., Carruthers, G.R.: 1981, NRL Mem.Report 5660.
Payne-Gaposchkin, C., Gaposchkin, S.: 1966,Smithsonian Contr.Astroph. 9, 1.
Philip, A.G.D., Sanduleak, N.: 1979, Astron.Astrophys. Suppl. 35, 347.
Prévot, L., Martin. N., Maurice, C., Rebeirot, E., Rousseau, J.: 1983,
    Obs. Lyon Preprint No. 12.
Rebeirot, E., Martin, N., Mianes, P., Prévot, L., Robin, A., Rousseau,
    J., Pegrin, Y.: 1983, Astron. Astrophys. Suppl. 51, 277,
Richer, H.B., Olander, N., Westerlund, B.E.: 1979, Astrophys. J, 230,724.
Rodgers, A.W.: 1959, Obs. 79, 49.
Rousseau, J., Martin, N., Prévot, L., Rebeirot, E., Robin, A., Brunet.
    J.P.: 1978, Astron. Astrphys. Suppl. 31, 243.
Sanduleak, N.: 1968, Astron. J. 73, 246.
Sanduleak, N.: 1969, Astron.J. 74, 877, 973.
Sanduleak, N.: 1970p CTIO Contr. No. 89.
Sanduleak, N., MacConnell, D.J., Hoover, P.S.: 1972, Nature 237, 28.
Sanduleak, N., MacConnell, D.J., Philip, A.G.D.: 1978, Publ. Astron.
    Soc. Pacific 90, 621.
Sanduleak, N., Philip, A.G.D.: 1977, Publ.Warner & Swasey Obs. 2, No. 5.
Schmidt, B.: 1932, Mitt. Hamburger Sternw. 7, 15.
Schmidt-Kaler, Th.: 1977, Astron. Astrophys. 54, 771.
Shapley, H., Lindsay, E.M.: 1963, Irish Astron. J. 6, 74.
Stock, J., Osborn, W., Ibanez, M.: 1976, Astron.Astrophys. Suppl. 24,35.
Tifft, W.G.: 1963, Mon. Not. R. Astr. Soc. 125, 199.
Van den Bergh, S.: 1974, Astrophys. J. 193, 63.
Walker, M.F., Blanco, V.M., Kunkel, W.E.: 1969, Astron. J. 74, 44; 966.
Westerlund, B.E.: 1960, Uppsala Astron. Obs. Ann. 4, no. 7.
Westerlund, B.E.: 1961, Uppsala Astron. Obs. Ann. 5, no. 1.
Westerlund, B.E.: 1964a, Mon. Not. R. Astr. Soc. 127, 429.
Westerlund, B.E.: 1964b, Obs. 84, 253.
Westerlund, B.E.: 1980, Vistas in Astronomy, 12, 335 (ed. A. Beer, Pergamon Press).
Westerlund, B.E., Glaspey, J.: 1971, Astron. Astrophys. 10, 1.
Westerlund, B.E., Mathewson, D.S.: 1966, Mon. Not. R. Astr. Soc. 131,371.
Westerlund, B.E., Olander, N., Richer, H.B., Crabtree, D.R.: 1978, Astron.
    Astrophys, Suppl. 31, 61.

Westerlund, B.E., Olander, N., Hedin, B.: 1981, Astron.Astrophys.Suppl. 43, 267.
Westerlund, B.E., Rodgers, A.W.: 1959, Obs. 79, 132.
Westerlund, B.E., Smith, L.F.: 1964a, Mon.Not.R.Astr. Soc, 127, 449.
Westerlund, B.E., Smith, L.F.: 1964b, Mon.Not.R.Astr. Soc. 128, 311.
Wood, R.: 1970, Mon. Not. Astro. Soc. South Africa 29, 37.

STAR COUNTS AND DYNAMICAL PARAMETERS IN THE SMC

M. Kontizas and E. Kontizas
University of Athens, Observatory of Athens

ABSTRACT

Tidal radii, masses and relaxation times of star clusters in the SMC have been derived by means of star counts using Schmidt plates. All 43 clusters have been found to behave like true globulars although their evolutionary ages vary from $10^6 - 9 \times 10^9$ years.

INTRODUCTION

Star counts have always been an important method of studying the dynamical structure of star clusters. King (1962) using theoretical models has shown that the core radius (r_c), the tidal radius (r_t) and the richness factor (K) are needed to describe the structure of the clusters.

Freeman (1974) and Chun (1978) have used the above theory to derive masses of populous clusters in the LMC.

In this study 43 old and young globular clusters of the SMC have been measured in order to study their dynamical behaviour and find their masses.

OBSERVATIONS

Plates taken with the 1.2 m U.K. Schmidt Telescope were measured on IIIaJ, IIaO and IIaD emulsion. Two plates taken with the AAT 3.8 m telescope were also measured and for 5 clusters in common the agreement was found to be very satisfactory (Kontizas, Danezis, Kontizas, 1983).

The star counts were obtained either by placing a circular reseau on the screen of an irisphotometer or by placing a square reseau under a binocular microscope. Each reseau was centered on the cluster by eye.

STRUCTURAL PARAMETERS

The observational material allowed star counting far beyond the radii of the clusters and the density of the field was safely reached in all cases. The density profiles of the clusters were used for the derivation of r_t, assuming all hypotheses of the King's models for the galactic globulars. Fig 1 (a,b) shows the density profiles of the clusters L15 and L6.

The different sources of error give an accuracy of 15% to the derived r_t.

Comparing a grid of King's models with our density profiles the concentration parameters (c) and core radii were found. The solid line in Fig 1b is the adopted surface density curve from the model (King, 1962) for c = 2.0.

From the derived r_t and c for each cluster the masses and the relaxation times were calculated.

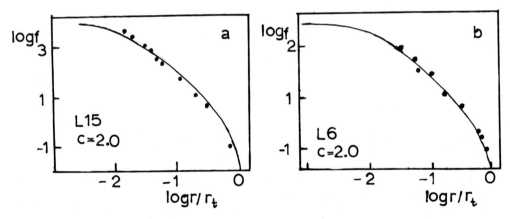

Fig. 1 Surface density profiles derived from star counts for the SMC clusters (a) L15 and (b) L6. The solid line show the adopted theoretical profile from the King model.

DISCUSSION

Previous classifications of the SMC clusters, their integrated colours, the existing c-m diagrams and the derived structural parameters found in this investigation were used as criteria for dividing the 43 studied clusters into two main groups : (i) the disk clusters and intermediate in colour and (ii) the halo "red", old clusters.

The disk clusters were found to have large tidal radii compared to the galactic open clusters with values accumulated at 30pc and concentration parameters higher than their galactic counterparts. Their masses are also found at least 10 times higher than those of our galaxy. Their relaxation times, much higher than their evolutionary ages mean that the clusters are not relaxed, although their density profiles favour well relaxed systems. The LMC, "blue", clusters were found to behave in the same way (Freeman, 1974; Geyer and Hopp, 1982).

The "halo" clusters are found to have tidal radii and concentration parameters similar to those of the galactic globulars. The masses are at least 10 times less massive than their galactic counterparts and the M/L ratios rather small for old clusters. Their relaxation times are smaller than their evolutionary ages (where exist) showing that the old SMC clusters are well relaxed system and being in agreement with their density profiles.

ACKNOWLEDGEMENTS

The authors would like to express their sincere thanks to the 1.2m, U.K. Schmidt Telescope Unit.

REFERENCES

Chun, M. S. : 1978, Astron. J., $\underline{83}$, 1062.
Freeman, K. C. : (1974). ESO/SRC/CERN Conference on Research
    Programs for the New Large Telescope, Geneva, May, 1974.
Geyer, E. H., Hopp. U. : (1982). IAU Colloquium No. 68, 235.
King, I. R. : (1962), Astron. J., $\underline{67}$, 471.
Kontizas, M., Danezis, E., Kontizas, E. (1982). Astron. and
    Astrophys. Suppl. Ser. $\underline{49}$, 1.
Kontizas, E. and Kontizas M. (1983). Astron and Astroph.
    Suppl. Ser. $\underline{53}$, 143.

NARROW SPECTRAL RANGE OBJECTIVE-PRISM TECHNIQUE APPLIED TO A SEARCH FOR SMALL MAGELLANIC CLOUD MEMBERS

Marc Azzopardi *
Dept. of Astronomy, University of Texas, Austin, TX 78712, USA

* Currently at the European Southern Observatory,
  8046 Garching bei München, FRG

I. INTRODUCTION

The Magellanic Clouds are exceptional galaxies for the study of massive stars and other luminous objects, because they are close and seen in directions where the galactic interstellar extinction is relatively low. In addition the Small Magellanic Cloud (SMC) is of special interest because of the possible peculiar effects that its low metal abundance, compared to the Solar neighbourhood, may have on the evolution of its stellar population. Consequently, a search for the most luminous SMC members has been the subject of many studies. Among the different means used to detect SMC members, the wide field cameras equipped with objective prisms, Grisms or Grenses have played a prominent part. The method consists of selecting stellar objects showing high luminosity or typical spectral features. Such surveys have been especially successful in detecting SMC Hα emission-line objects (Henize, 1956; Lindsay, 1961), OB and blue supergiant stars (Sanduleak, 1968, 1969), planetary nebulae (Sanduleak et al., 1978; Sanduleak and Pesch, 1981) or carbon and late M-type stars (Blanco et al., 1980).

II. A SEARCH FOR SMC MEMBERS

One of the major difficulties encountered in surveying SMC members comes from the projected high stellar density of this galaxy which is due to its orientation along the line of sight, and the presence of numerous nebulosities. To minimize this difficulty the classical method usually consists of reducing the length of the spectra using the lowest dispersion possible. This is done in order to reduce the number of overlapping images and to reach fainter stellar objects in a given exposure time. Although adequate, this method has the serious disadvantage of lower spectral resolution, reducing the quality and the accuracy of the spectral classifications.

Another technique for reducing the length of the spectra, and therefore the number of overlapping images, consists of using an interference filter to select the spectral range which includes the more characteristic spectral features of a given type of object. In addition, the use of interference filters, by reducing the background fog due to the sky, allows longer exposure times and hence the possibility of reaching fainter stars. This method, which Martin and Rebeirot (1972) initially used to detect OB stars in some crowded Large Magellanic Cloud regions, has been used intensively with success by the author and associates to survey different kinds of luminous stars in the SMC.

i) OB stars and blue supergiants.

The survey of OB and blue supergiant stars was carried out at La Silla with the 40 cm objective prism astrograph (Fehrenbach, 1966) of which the dispersion is roughly 121 Å mm^{-1} at Hγ. Two interference filters of about 260 Å band width centered at $\lambda 3940$ and $\lambda 4390$ respectively were used. The former filter was intended to detect OB stars and the latter blue supergiants. Exposures of 8 and 6 h respectively on IIa-O nitrogen baked plates permitted us to reach stars up to a limiting photographic magnitude of about 15. Within these limited spectral ranges the available MK spectral classification criteria adapted to objective prism spectra by Fehrenbach (1958) were used. Four fields were explored, 3 square degrees each, located in the bar and in the wing (region of NGC 456, 460 and 465), and all partially overlapping. The survey resulted in the detection of 520 stars showing high luminosity spectral features, 327 of them being new SMC members. A complete description of this survey as well as the list and identification of the objects has already been published by Azzopardi and Vigneau (1975, 1979). A third interference filter of 250 Å band width centered at $\lambda 4350$ having a homogeneous transmission of 85 ± 1% throughout permitted the measurement of the Hγ line intensity. For this purpose, interactive programs for computing equivalent widths on objective prism spectra were written by Azzopardi et al. (1978). Using standard galactic stars, this method allowed us to determine the MK luminosity class or subclass of 172 O9 to A7 stars when their spectral types were known (Azzopardi, 1981; Azzopardi and Vigneau, 1982).

ii) Wolf-Rayet stars.

The detection of Wolf-Rayet stars was carried out at La Silla with the same instrument (dispersion 150 Å mm^{-1} at $\lambda 4650$) using an interference filter centered at $\lambda 4650$ and having a pass band width of 120 Å. Wolf-Rayet stars show up strongly in this spectral region due to the emission mainly from either $\lambda 4650$ CIII (WC) or $\lambda 4686$ HeII (WN). Five circular fields of 85 arc min diameter each, all partially overlapping, were required to cover the bar and the wing of the SMC. IIa-O nitrogen baked plates were used and 6 h exposures for each field permitted us to reach the continuum of 16.5 mpg stars in the most crowded SMC regions. This survey resulted in the identification of the 3 already known WR stars and the detection of 4 new faint WR stars, afterwards confirmed by slit spectrography (Azzopardi and Breysacher, 1979). Recent observations at the prime focus of the 3.6m ESO telescope equipped with a Grism did not discover any new WR stars down to magnitude 20 confirming, the completeness of this survey at least in the regions explored.

iii) Hα emission-line objects.

More recent observations with the Curtis Schmidt Telescope at C.T.I.O. allowed us to secure good to very good SMC plates using an interference filter of 110 Å band width centered at λ6565. Exposures of 4 and 2 h were taken permitting us to reach the continuum of stars up to about 18 photographic magnitude in very crowded fields. A preliminary examination of the plates has already revealed, in the central regions of the SMC, a large number of new faint Hα emission line objects not identified in the previous surveys by Henize (1956) and Lindsay (1961). In order to discriminate the Hα emission line stars from planetary nebulae or unresolved H II regions an additional set of IIIa-J forming gas baked plates were taken through an interference filter of 80 Å band width centered at λ5000. Although a substantial number of faint objects show λλ4959, 5003[OIII] emission lines, we have so far not detected any other candidate than the planetary nebulae discovered by Sanduleak and associates.

III. WORK IN PROGRESS

From the foregoing, it is clear that objective prisms (astrographs or Schmidt Telescopes) equipped with various interference filters properly chosen for the detection of different types of objects make possible very deep surveys in fields of high stellar density. However, this technique was until now only applied to the study of the stellar populations of the Galaxy and the Magellanic Clouds. To survey other external galaxies a variant of this technique can be adapted to large telescopes using Grisms or Grenses (dispersion ≃ 2000 Å mm^{-1}) and suitable colored filters. At present, we are using this technique, described in detail by Breysacher and Lequeux (1983), to survey Wolf-Rayet stars, planetary nebulae and carbon stars in nearby galaxies.

REFERENCES

Azzopardi, M.: 1981, Ph.D. thesis, Paul Sabatier University of Toulouse, No. 979, Vol. 1.
Azzopardi, M., Bijaoui, A., Marchal, J., Ounnas, Ch.: 1978, Astron. Astrophys. 65, 251.
Azzopardi, M., Breysacher, J.: 1979, Astron. Astrophys. 75, 120.
Azzopardi, M., Vigneau, J.: 1975, Astron. Astrophys. Suppl. 22, 285.
Azzopardi, M., Vigneau, J.: 1979, Astron. Astrophys. Suppl. 35, 353.
Azzopardi, M., Vigneau. J.: 1982, Astron. Astrophys. Suppl. 50, 291.
Blanco, V.M., McCarthy, M.F., Blanco, B.M.: 1980, Astrophys. J. 242, 938.
Breysacher, J., Lequeux, J.: 1983, The Messenger 33, 21.
Fehrenbach, Ch.: 1958, Handbuch der Physik, Vol. 50, p. 1.

Fehrenbach, Ch.: 1966, Eur. South. Obs. Bull. $\underline{1}$, 22.
Henize, K.G.: 1956, Astrophys. J. Suppl. $\underline{2}$, 315.
Lindsay, E.M.: 1961, Astron. J. $\underline{66}$, 169.
Martin, N., Rebeirot, E.: 1972, Astron. Astrophys. $\underline{21}$, 329.
Sanduleak, N.: 1968, Astron. J. $\underline{73}$, 246.
Sanduleak, N.: 1969, Astron. J. $\underline{74}$, 877.
Sanduleak, N., MacConnel, D.J., Davis Philip, A.G.: 1978, Publ. Astron. Soc. Pacific $\underline{90}$, 621.
Sanduleak, N., Pesch, P.: 1981, Publ. Astron. Soc. Pacific $\underline{93}$, 431.

PHOTOGRAPHIC PHOTOMETRY WITH SCHMIDT PLATES OF STAR CLUSTERS IN THE SMC

M. Kontizas
Laboratory of Astrophysics, University of Athens

INTRODUCTION

The star clusters have always been the most important tools of testing the stellar evolution theories and when these stellar systems belong to other galaxies, then our knowledge on stellar evolution can be extended for different initial conditions than those of our Galaxy. The Magellanic Clouds being our nearest neighbour galaxies offer ideal conditions for such studies.

Since the powerful southern telescope came into operation deep plates of these two galaxies revealed a large number of new star clusters. The Schmidt plates are most useful since they have the advantage of large field where many objects can be investigated homogeneously on the same plate.

A survey of colour-magnitude (c-m) diagrams of star clusters in the SMC has been studied using plates taken with the 1.2 m. U.K. Schmidt Telescope. Twenty clusters at the west-north-east periphery have been studied. New SMC plates have been planned for the study of the south-west clusters.

OBSERVATIONS

Three V and three B plates were measured with an irish-photometer at the Royal Observatory of Edinburgh. The photoelectric sequences available do not always reach the very faint magnitudes that Schmidt plates can detect and in this case the electronographic sequences reaching the faint limit of the plate have been used very successfully (Walker, 1970; Hawkins and Brück, 1981). A detailed discussion of the photometric accuracy has been reported by Kontizas (1980).

DISCUSSION

The selected clusters are located in the halo or the SMC and their c-m diagrams show features that make them different from the clusters of

M. Capaccioli (ed.), Astronomy with Schmidt-Type Telescopes, 355–357.

our Galaxy. The most important points are outlined below.

1. Halo, "red", clusters.

There is a variety of old globular clusters all over the whole studied area; in the northern part, the cluster members suffer severe contamination with the bar or arm's stellar content and therefore the western clusters seem to be a suitable sample of the old SMC clusters.

One typical feature of their c-m diagrams is the horizontal branch (HB) which is from very weak to very conspicuous but with preference to the red side of the RR variable star strip. The fact that the red part of the HB is quite populous whereas the blue part is almost non existent does not necessarily mean high metallicity as it is explained in our galaxy.

Another characteristic is the existence of very red stars which in most cases have been proved to be carbon stars (Feast and Evans, 1973; Aaronson and Mould, 1982) and most likely cluster members as it was found from luminosity functions of some clusters (Kontizas and Kontizas, 1982).

Many faint blue stars seem to be an unusual feature of these clusters and give indication of being "blue stragglers" (Tifft, 1963; Gascoigne, 1980) have found the same c-m structure for the old SMC clusters NGC 121, L1 and K3.

These characteristic features found in the c-m diagrams support the argument that the old SMC clusters are younger than their galactic counterparts.

The resemblance of these c-m diagrams with some diagrams of remote globular clusters of our own galaxy and some dwarf galaxies might mean that galaxies like SMC with very little activity in their nucleus have different evolutionary history and lower metal content because of an evolutionary slower metal production in their nucleus. A composite c-m diagram of the central areas of six western old clusters is illustrated in Fig. 1a.

2. Disc, "blue" clusters.

The disc clusters which are the "blue" and intermediate in colour clusters (Brück, 1975) are mainly globular (Kontizas et al., 1982) and have various ages according to the subsystem in which they belong (spiral arm, bar).

The two clusters located on the north east spiral arm produce c-m diagrams with a vertical main sequence giving evidence of young age (5×10^7) and resemble the c-m diagram of the field studied by Brück and Marsoglu (1978).

A group of northern young clusters shows that the central region which is an extension of the bar contains objects younger than the ones of the arm. The main sequences of all young clusters compared to those of our galaxy are found to be much bluer in the SMC, that may again be an indication of their low metallicity.

A composite c-m diagram of the central areas of the young clusters is illustrated in Fig. 1b. The main sequence and some of the very bright red stars have been found to be the stellar content of the clusters whereas the lower part of the red giant branch is the stellar content of the halo superimposed in this area. The existence of some evolved stars in the young SMC clusters shows that the populous "blue" clusters are not very young compared to the very young clusters of our Galaxy.

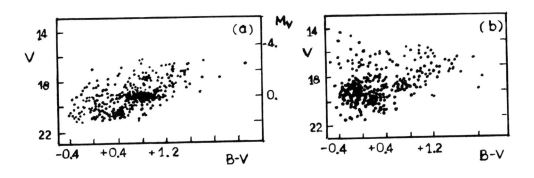

Fig. 1. Composite c-m diagrams for the central regions: a) of six SMC halo clusters b) of eight SMC disk clusters.

ACKNOWLEDGEMENTS

The author would like to address her sincere thanks to the 1.2 m U.K. Schmidt telescope Unit for providing the necessary plates and facilities.

REFERENCES

Aaronson, M. and Mould, J. 1982, Astrophys. J. Suppl. 48, 161
Brück, M.T. 1975, M.N.R.A.S. 173, 327
Brück, M.T. and Marsoglu, A. 1978, Astron. and Astrophys. 68, 193
Feast, M.W. and Evans, L. 1973, M.N.R.A.S. 167, 15p
Gascoigne, S.C.B. 1980, IAU Symp. N° 85 (ed. J. Hesser), p. 305
Hawkins, M.R.A. and Brück, M.T. 1981, M.N.R.A.S. 198, 935
Kontizas, M. 1980, Astron. and Astrophys. Suppl. 40, 151
Kontizas, M. and Kontizas, E. 1982, Astron. and Astrophys. 108, 344
Kontizas, M., Danezis, E., Kontizas, E. 1982, Astron. and Astrophys. Suppl. 49, 1
Tifft, W.S. 1963, M.N.R.A.S. 125, 189
Walker, M.F. 1970, Astrophys. J. 161, 835

LUMINOSITY FUNCTION OF OLD GLOBULAR CLUSTERS IN THE SMC

M. Kontizas and E. Kontizas
University of Athens, Observatory of Athens

INTRODUCTION

Luminosity Functions (LFs) provide useful information on the stellar content of a star cluster. One of the problems of studying the colour-magnitude (c-m) diagrams in remote clusters is the contamination of cluster members with field stars. If all the stars of a cluster are measured in concentric rings one can reach the field and compare the c-m diagram of the field with that of the central areas. For the SMC where the field and cluster areas seem to have similar c-m diagrams, this method is not always useful and then the LF becomes an important tool of overcoming this obstacle.

LFs of 10 SMC star clusters have been derived using photographic plates taken with the 1.2 m U.K. Schmidt telescope. The large field Schmidt offers the advandage of having homogeneous photographic material for both the clusters and their adjoining fields.

The photometric values were found by means of an iris-photometer and the accuracy of the measurements was discussed by Kontizas (1980). Star counts in the central areas of the clusters and their adjoining fields have provided the necessary normalising factors which were used to substract the LF of the field from the corresponding LF of the central area.

The LF of NGC152 derived from the Schmidt plates was found to be in good agreement with that found from AAT plates.

DISCUSSION

The study of the LFs of 10 old globular clusters, located in the western and the northern areas has shown the following :

(i) the bright red stars belong to the clusters and if there are red field stars they are certainly much fainter.

(ii) in most cases comparison, with the available theoretical models of globular clusters, has failed to match mainly because the subgiant part is very steep (may be due to helium content; Kontizas and Kontizas, 1982) and the horizontal branch (HB) is very populous and has only the red part of the RR Lyrae strip. This HB structure would mean high metallicity but we know that this is not the case for the SMC. So it can be assumed that the populous HB could be produced by a clump of intermediate age stars (more massive than those found in old globulars) as is the case of old open clusters of our Galaxy. In our sample the clusters L11 and NGC152 have this characteristic particularly prominent. From the c-m diagram of these ten clusters, it was found that L11 and NGC152 are amongst the youngest (Kontizas, 1976; Hodge, 1982). So the structure of the LF where the HB is expected could be an age indication and therefore the old clusters of the SMC seem to be younger than the corresponding old globulars of our Galaxy and contain more massive star members.

(iii) The field LF seems to resemble the LFs of the oldest SMC clusters of our sample. This field is purely SMC halo stars since the stars of our Galaxy give almost negligeable contribution at this part of the sky (Brück and Marsoglu, 1978).

ACKNOWLEDGMENTS

The authors are very much obliged to the 1.2 m. U.K. Schmidt telescope Unit for providing the necessary phtographic material.

REFERENCES

Brück, M. T. and Marsoglu, A., : 1978, Astron. and Astrophys. 68, 193
Hodge, P. : 1982, IAU Colloquium No 68, pp. 205
Kontizas, M. : 1976, PhD Thesis, Edinburgh University.
Kontizas, M. : 1980, Astron. and Astrophys. Suppl. Ser., 40, 151.
Kontizas, M., and Kontizas, E. : 1982, Astron. and Astrophys 108, 344.

DISCUSSION

V.M. BLANCO: In one of the viewgraphs you showed, a magnitude limit was indicated for the available information about the clusters luminosity functions. I would like to remark that in the forthcoming Magellanic Clouds observing season many SMC and LMC globular clusters will be observed at CTIO with a CCD camera on the 4-meter telescope. From work done during the past season on a few clusters, for example Kron 3, we expect the observations to extend well past the mean sequence turnoffs of the clusters. Thus, within a year we should have greatly improved information about the luminosity functions of the clusters. Also, you remarked that some clusters showed carbon stars. Generally those are the Searle et al. class IV and V and perhaps VI clusters. It is interesting that Perssons and Frogel find that in such clusters the carbon stars contribute about 50% of the total bolometric luminosity. By comparison, in the field (actually a sample field in the LMC) the carbon stars contribute only about 3% of the total luminosity. Finally, I would like to congratulate you for your beautiful work.

M. KONTIZAS: Thank you very much for your comments. I'll be glad to see the new material.

ELLIPTICITIES OF "BLUE" AND "RED" GLOBULAR CLUSTERS IN THE SMC

E. Kontizas, D. Dialetis, Th. Prokakis and M. Kontizas
Observatory of Athens, University of Athens

ABSTRACT

The projected ellipticities of twenty four "blue" and "red" globular clusters of various ages and positions in the SMC have been found using isodensity contours. The derived ellipticities have shown that the globular clusters of the SMC are more elliptical than their counterparts of the Large Magellanic Cloud (LMC) and our Galaxy. The observed ellipticities for both cluster types, do not support any age dependence.

OBSERVATIONS

Twenty five "blue" and "red" globular clusters were studied on film copies of B and V plates taken with the 1.2 m U.K. Schmidt Telescope in Australia. The quality of both plates was discussed by Kontizas (1980). Each cluster was scanned with an isodensitometer at the National Observatory of Athens. Two to three isophotes were produced for each cluster for $r > r_c$ (where r_c is the core radius of the cluster) and the mean values of the corresponding ellipticities ($\varepsilon - \frac{b}{a}$) of the best fit ellipses were found for B and V plates respectively.

For comparison the ellipticity of the galactic globular cluster 47 TUC was calculated as well. The derived value $\varepsilon = 0.046 \pm 0.05$ agrees very well with that given by Da Costa (1982).

ELLIPTICITIES

The relative frequency distributions of the ellipticities of the SMC clusters studied here is given in Fig. 1a by solid line whereas the relative frequency distribution of the ellipticities of the LMC clusters studied by Frenk et al (1982) is given in the same diagram by dashed line. From this diagram it can be seen that although the range of

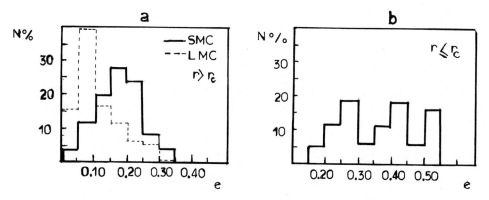

Fig. 1. Relative distributions of the ellipticities for the SMC globular clusters (solid line) and the LMC (dashed line). a) For isophotes with $r > r_c$ and b) For isophotes with $r < r_c$.

ellipticities for LMC and SMC is the same, the SMC clusters are more elliptical than the LMC. This result is verified by a statistical test. The analysis of the frequency distribution of the "blue" and "red" clusters do not show any statistical significant difference and their mean ellipticities are ϵ_{mean} (Halo) = 0.18 ± 0.07 and ϵ_{mean} (disk) = 0.17 ± 0.07. Geyer and Richtler (1981) has reached the same conclusion for the "red" and "blue" LMC clusters. For sixteen of the clusters studied here, ellipticities were found for $r \simeq r_c$, and their frequency distribution is plotted in Fig. 1b. Evidently there is no significant departure from uniformity and the ellipticities of the central part of the clusters are higher than the ellipticities of the outer parts.

ACKNOWLEDGEMENTS

The authors would like to express their sincere thanks to the 1.2 m U.K. Schmidt Telescope Unit for providing the necessary photographic material.

REFERENCES

Da Costa, G.S. : 1982, A.J., 89, 990.
Frenk, C.S. and Fall, M.S., : 1982, Mon. Not. R. Astr. Soc., 199, 565.
Geyer, E.H. and Richtler, T. : 1981, IAU Colloquium No. 68 p.p. 239.
Kontizas, M. : 1980, Astron. and Astroph. Suppl. Ser. 40, 151.

DISCUSSION

M. SANTANGELO: Coming inward to the core from the outer regions, have you found any twisting of the isophotes?

E. KONTIZAS: Yes, there is a random twisting of the isophotes going inward of the cluster. The orientation of all the major axes for all isophotes and for all the studied clusters (twenty five so far) is within a cone of 30°, that might have some cosmological meaning.

H.T. MACGILLIVRAY: What is the resolution at which you have scanned the plates and could this have affected your isodensity contours?

E. KONTIZAS: The isodensity contours are a smooth picture of the clusters and the used beam was 1-10 times bigger than the size of a bright stellar image of the cluster.

PHOTOMETRY OF EXTENDED SOURCES

G. de Vaucouleurs
The University of Texas
Austin, Texas   78712

As a contribution to the continuing struggle for precision and reliability in photographic surface photometry with Schmidt telescopes the following sections review some of the main sources of errors and ways to avoid them, where possible.

1.   ACCIDENTAL AND SYSTEMATIC ERRORS

The following sources of error will be particularly discussed:
(a) photographic calibration and local errors, microphotometer errors, sky level errors; zero point and integration errors;
(b) instrumental and atmospheric convolution effects, including the structure of the point spread function (PSF), effects of internal reflections in the corrector and the influence of scattered light (aureole);
(c) galactic and extragalactic effects, including interstellar extinction and diffuse nebulosities at high latitudes, bright field stars and galaxies; statistical fluctuations in the distribution of sub-threshold stars and galaxies.

2.   PHOTOGRAPHIC AND CALIBRATION ERRORS

The luminosity distribution $I(x,y)$ derived from the plate density matrix $D(x,y)$ is affected by errors in the photographic calibration curve $I = f(D)$ and, at low I levels, by errors in the adopted "blank sky" level. The latter is usually determined by polynomial interpolation using "numerical mapping" techniques (Jones et al. 1967, Barbon et al. 1976, Okamura 1977).

(a) External calibration. Tube sensitometers, penumbra sensitometers, step wedges and other calibration devices are often poorly designed or implemented, improperly standardized or incorrectly used. In particular, the calibration spots are often much too close to the plate edge; most Schmidt cameras suffer from this defect. Processing irregularities are largest at the edge of the plate; the outermost 2-3 cm should never be used for calibration. The spots should be spaced widely enough (at least several millimeters) that scattered light from the high intensity steps will not contaminate the low intensity steps or the plate background in their vicinity.

(b) Internal calibration. It is often the case, particularly for some of the older Schmidt plate collections, such as the POSS, that no external calibration scale is available. In such cases, it is possible to construct a characteristic curve by combining the density profiles D(r) of several stars whose magnitudes (or differences δm) are known (Kormendy 1973; Kormendy and Bahcall 1974). This internal calibration is possible because all stars have the same relative intensity profile (point spread function), $I(r)/I(0)$, so that $D(I)$ can be derived by a point-by-point comparison of two or more D(r) profiles of known δm (Figure 1). To work well this method requires that adjacency effects (Eberhard, Kostinsky effects) be negligible, which can be achieved by pushing the development to (or close to) γ_∞. Otherwise, the characteristic curve derived from stars would not be applicable to extended sources, because adjacency effects vary with the steepness of the density gradient in the image. Consequently, this internal calibration method works better with the soft star images given by long-focus reflectors than with the sharp images given by short-focus Schmidt cameras. An alternative and preferable method in the latter case is to use standard galaxies having precisely known luminosity profiles; one such standard galaxy is NGC 3379 (de Vaucouleurs and Capaccioli 1979) which can be used also to check calibration procedures. Another galaxy usable as a luminosity distribution standard is NGC 4486 (M87) (de Vaucouleurs and Nieto 1978, 1979). There is a need for more standard galaxies well-distributed around the sky.

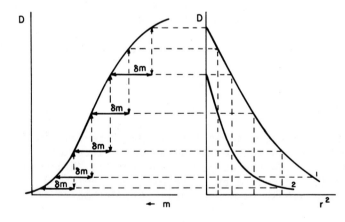

Figure 1
Internal calibration: derivation of characteristic curve D(m) from density profiles $D(r^2)$ of two stars of known magnitude difference δm.

(c) Photographic emulsions and processing techniques introduce photometric irregularities or "local errors" (de Vaucouleurs 1943, 1945, 1946, 1948c, 1976) whose origin is poorly understood. Processing should be as uniform as possible; if brush development is not practical for large plates, at least the performance of the developing machine on uniformly exposed plates should be checked. Ditto for the drying. Local errors

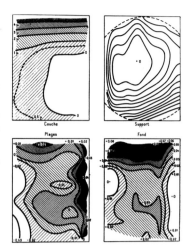

Figure 2
Local errors: comparison of variations of thickness of emulsion ("couche")(a), departures from flatness of plate ("support") (b), density (above fog) of spots exposed to uniform illumination ("plages")(c), and density variations of plate fog ("fond")(d). Note correlation between (a) and (d), and absence of correlation between (c) and all others (de Vaucouleurs 1946).

are not random errors (noise) on a given plate and so can be mapped and, to a large extent, corrected by interpolation (Figure 2), but they vary from plate to plate in an uncorrelated fashion, so that on the average the differential error increases approximately as the square root of the separation in the plane of the emulsion (random walk). The photometric errors resulting from the local errors in density vary with density (Figure 3). The optimum range is the lower part of the H & D curve (not its inflection point). Procedures for the correction of local errors by interpolation between a network of control spots (de Vaucouleurs 1943, 1946) or by the related technique of "grid photography" (Davis et al. 1980) have been developed and are essential to achieve the potential precision of photographic photometry (0.1% under laboratory conditions, 0.5% under astronomical conditions; see de Vaucouleurs 1948d). Under normal conditions interpolation of the sky background by numerical mapping techniques (Jones et al. 1967) will serve the same purpose more simply, but at some loss of precision.

(d) Vignetting. Clearly the vignetting function must be taken into account. This is done automatically when the local sky background determined by interpolation is used as a unit of specific intensity since all fluxes are reduced in the same proportion. However, allowance must be made for apparent changes of the vignetting function due to variable desensitization by water vapor trapped in the air space between filter and plate holder (Dawe and Metcalfe 1982). This effect can be eliminated by nitrogen-flushing (Figure 4).

3. MICROPHOTOMETER AND SKY LEVEL ERRORS

(a) Significant random and systematic errors are often introduced by the microphotometer or densitometer used to measure the plates. This is especially frequent for the common type of uncompensated single beam

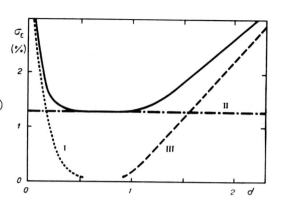

Figure 3
Local errors: average photometric errors of spots exposed to uniform illumination as a function of density. Note three components: (I) at low densities microphotometer and plate errors are dominant because of low gradient of characteristic curve, (II) at intermediate densities processing errors are dominant, (III) at high densities non-uniform thickness of emulsion makes increasing contribution. The optimum region is at the base of the linear part of the characteristic curve, not near its inflection point (de Vaucouleurs 1946).

instruments whose readings depend directly on the stability of the light source and of the electronic amplifiers, both notorious for their drift properties. The high speed digitized instruments, such as the PDS machine, add to these defects high noise at high densities, a non-linear (or non-logarithmic) response, sensitivity to scattered light, and other sources of error (Bozyan and Opal 1983). It is not uncommon that densities measured on two different days on the same plate even with identical instrumental settings will not repeat precisely, unless special precautions are taken to monitor the response of the instrument. If available, the dual beam, compensated systems are much to be preferred.

(b) All nebular photometry is, per force, done in the presence of a superimposed strong noise signal, the luminosity of the night sky, I_s, so that, for example, the galaxy signal (specific intensity), I_g, is obtained as the difference, $I_g = I_{g+s} - I_s$, between the measured intensity, I_{g+s}, of galaxy + sky and the intensity of the local sky light, I_s, the sum of the airglow, the zodiacal light, and atmospheric and instrumental scattering. This local "sky level" can only be estimated by interpolation across the object field between adjacent "blank sky" areas surrounding it. This interpolation is generally performed by fitting a low-order mapping polynomial in x,y which necessarily assumes a smooth underlying distribution of I_s. High spatial frequency random fluctuations can not be detected or avoided.

At high intensities, when $I_g \gg I_s$, small departures from the interpolating surface have only a minor effect on I_g. For example, if the error in the interpolated sky level is $\delta I_s/I_s \simeq 0.01$, the relative error is a negligible 0.1% = 0.001 mag where $I_g = 10\ I_s$, or $\mu_g(B) \simeq 19.8$ mss ($\equiv$ mag per square arc second) if $\mu_s(B) \simeq 22.3$ mss. However, in the faint outer parts of a galaxy where $I_g \ll I_s$ and where $D_{g+s} - D_s$ tends asymptotically to zero the error becomes dominant; thus, if $I_g = 0.1\ I_s$, or

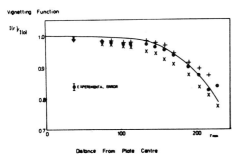

Figure 4
Vignetting: apparent vignetting functions for UK 1.2 m Schmidt telescope. Line: comparison plates (geometric effects and scattering); +: hypersensitized plates without filter, no N_2-flushing of plateholder; x: hypersensitized plates with filter, no N_2- flushing of plateholder; o: hypersensitized plates with filter and N_2-flushing of plateholder (Dawe and Metcalfe 1982).

$\mu_s(B) \simeq 24.8$ mss, the same 1% error in I_g introduces an error of 10% or 0.1 mss in I_g, and if $I_g = 0.01\ I_s$, or $\mu_g(B) \simeq 27.3$, $\delta I_g/I_g \simeq 1$ and the magnitude error diverges. This effect can drastically change the apparent photometric profile of the faint outer coronae of galaxies (Figure 5). In particular, it can introduce fictitious "tidal radii" truncations or, conversely, "tidal extensions" to elliptical galaxies, depending on whether the sky level is placed slightly too high or too low. In the best case, current photoelectric techniques define the sky level with mean errors of $\sim 0.1\%$ (equivalent to $\mu_B \simeq 30$ mss), but photographically it is difficult to do better than 0.5% ($\mu_B \simeq 28$ mss) on single plates.

4. ZERO POINT AND INTEGRATION ERRORS

The zero point is usually fixed by photoelectric photometry, either by direct photoelectric scans calibrated versus standard stars, or by equating the photographic luminosity integral to the flux measured photoelectrically within the corresponding integration area, generally a circular aperture centered at the nucleus in the case of galaxies. In either case the following errors may be present:

(a) Zero point errors. The zero point of the magnitude scale is affected by errors in the magnitudes of the standard stars, in the calibration of the amplifier in the d.c. mode (or improper resolution corrections in the pulse counting mode) and in the transformation equations to the standard system. The possibility of error in measuring the diameters (or areas) of the field apertures and in the plate scale used to transform them to angular measure should not be ignored. Both should always be checked by timing transits of polar stars. In most cases, it is difficult to reduce the total mean error of the zero point from all sources to less than 0.02-0.03 mag.

(b) Integration errors. These are introduced by errors in the adopted sky level, by the finite width of the integration intervals, by inclusion of field stars or background galaxies. In the case of NGC 3379 (de Vaucouleurs and Capaccioli 1979), the total mean error (zero point and

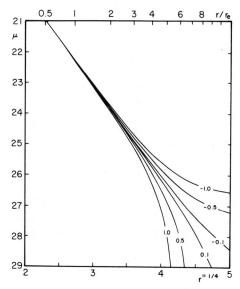

Figure 5
Effect of errors in adopted sky level on apparent luminosity profile of an E0 galaxy obeying the $r^{1/4}$ law. Note that depending on its sign the systematic error δD_s (in units of 0.01 when $D_s = 0.8$) introduces an apparent "tidal cutoff" or a "tidal extension". Surface brightness scale μ and radial scale r are for NGC 3379 (Capaccioli and de Vaucouleurs 1983).

integration) of photoelectric aperture photometry by several observers was 0.065 mag, much larger than is generally quoted by individual observers from internal evidence, but in line with the rule of thumb that on the average true external mean errors are 2-3 times larger than internal errors.

(c) Extrapolation errors. In the absence of systematic errors in the photometry the integrated magnitude of a diffuse object, such as a galaxy, that fades insensibly into the night airglow, the calculated integrated magnitude within a specified isophote, refers unavoidable to some unknown fraction of the total luminosity of the galaxy. Procedures for the derivation of the total (or asymptotic) magnitude have been developed (de Vaucouleurs 1948b, 1977); the incompleteness correction is often in the range of 0.3 to 0.03 mag, depending on the detection threshold.

5. INSTRUMENTAL AND ATMOSPHERIC CONVOLUTION EFFECTS

Because of their small plate scale, Schmidt telescopes in general are not suitable for high-resolution studies of small structures, such as the nuclei of galaxies, particularly if coarse-grained photographic emulsions are used. However, if wide field electronographic cameras can be successfully adapted for use with Schmidt cameras, as proposed by Griboval (this conference), and especially if such systems can be operated in space, the enormous gain in resolving power will make them surpass the resolution presently available at the Cassegrain foci of large ground-based reflectors. Then the same problems of convolution by the instrumental point spread function (PSF) will need to be considered.

PHOTOMETRY OF EXTENDED SOURCES   373

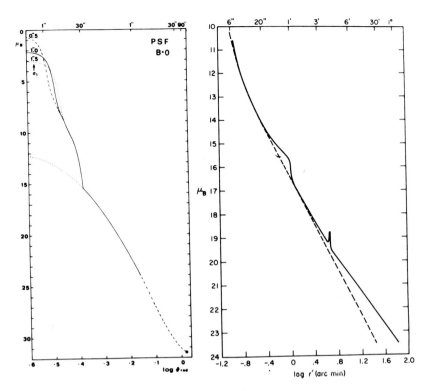

Figure 6
Typical point spread functions for (a) McDonald reflectors after Capaccioli and de Vaucouleurs (1983) and (b) Palomar 1.2 m Schmidt telescope after Kormendy (1973). Note effects of variable seeing in (a), of internal reflections and halation in (b). Surface brightness scales are normalized to star of total magnitude B = 0.

(a) Point Spread Functions. A combination of high-resolution microphotometric analysis of stellar images of different magnitudes, and of photoelectric scans of very bright stars gives the typical PSF of medium sized reflectors at $\sim$ 2 km elevation shown in Figure 6a. This may be compared with the mean PSF for the Palomar 1.2 m Schmidt telescope (Figure 6b); except for the effect of multiple internal reflections and the lack of high resolution data in the center of Figure 6b, the curves are generally similar. Two distinct regimes are in evidence:
(I) a steep decline, extending typically to r $\simeq$ 30", is dominated by telescopic diffraction and atmospheric turbulence; this is the "image" proper;
(II) a shallower decline extending out to r $\simeq$ 1°.5, dominated by distant telescopic and atmospheric (particle) scattering, may be described as the "aureole" (Dermendjian 1957, 1959). The latter eventually merges smoothly into the Rayleigh scattering component (dashed line in Figure 6a). Under average seeing conditions the total range of surface brightness is $\sim$ 30 magnitudes.

The innermost part of the image which is dominated by atmospheric turbulence does not have a fixed profile but varies with seeing conditions. In general, the core of the image is well-approximated by a Gauss function (de Vaucouleurs 1948b; King 1971; Brown 1978) $G_1(r;\sigma_1)$ which includes something like 80 to 90% of the energy in the star image (component I). Beyond a few seconds from the center of the image the wings are brighter than expected from G_1 alone. This excess, which can be represented by the sum of gaussians (Brown 1978) or exponentials (Newell 1980), contains typically $\sim$ 10 to 20% of the energy in the image. This fraction varies with the condition of the optics and the turbidity of the atmosphere (Piccirillo 1973).

(b) Effect of scattered light (aureole). Beyond $r \simeq 30"$ the PSF is dominated by scattered light forming an "aureole" around the central image (Figure 6a). Within the "image" the contribution of the aureole to the observed surface brightness is essentially negligible; for example, at the center the aureole is roughly 10 magnitudes fainter than the image. But the contribution of the aureole to the observed surface brightness of the outer parts of a galaxy may become increasingly significant at distances $r > 30"$, depending on the luminosity law in the object and the scaling factor. An example is given in Figure 7.

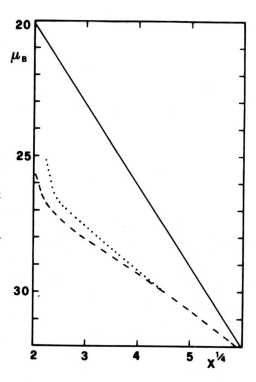

Figure 7
Scattered light (dashed curve) of model of NGC 3379 (solid line) and of central point source having the same total magnitude (dotted curve) for the PSF of Figure 6a. Note that luminosity excess due to scattered light is less than 0.14 mss at all x > 459" where μ_B(obs.) > 28.0 (Capaccioli and de Vaucouleurs 1983).

6. GALACTIC AND EXTRAGALACTIC EFFECTS

External galaxies are observed through the irregularly absorbing and scattering interstellar dust layer, among the field stars and (at least in clusters) among other galaxies, which cause more or less random fluctuations in the luminosity distribution in the galaxy image and in the sky "background." Galactic nebulosities are similarly observed against a foreground and background of other nebulosities, stars and the pervading interstellar medium. Further, undetected (sub-threshold) stars and galaxies cause additional fluctuations in the "blank sky" level.

(a) Galactic extinction effects at high latitudes. While it may be assumed that galactic extinction is uniform over the small areas covered by most galaxies on large scale reflector plates, this assumption is not warranted in the case of the larger fields covered by Schmidt plates nor in the case of large galaxies at low galactic latitudes, such as M31 and the Magellanic Clouds. While the assumption of zero extinction within 30 to 40 degrees from the galactic poles is demonstrably untenable (de Vaucouleurs and Buta 1983), and the cosec b law is definitely valid (statistically) up to both galactic poles, it is very difficult to make realistic estimates of the small scale fluctuations of galactic extinction at high latitudes. The most direct evidence comes from the detection of reflection nebulosities.

(b) High-latitude reflection nebulosities. It has been known for many years that faint emission and/or reflection nebulosities can be detected by deep photography at intermediate galactic latitudes (de Vaucouleurs 1955, 1960; King, Taylor and Tritton 1979). Spectacular examples have been detected in recent years thanks to the development of the high-contrast IIIa-J emulsion (Arp and Kormendy 1972, Arp and Lorre 1976, Sandage 1976, Cannon 1979). All fields show much fine filamentary, cirrus-like structure down to the resolution limit ($\sim 2"$) of the 1.2 m Schmidt telescopes. At latitudes $|b| < 45°$ available data indicate a range of surface brightness $27 > \mu_B > 25$ mss in regions where the estimated mean extinction is $A_B \simeq 0.3$ mag. At latitude $|b| > 45°$ corresponding information is not yet available, except that the maximum surface brightness is probably much lower, perhaps $\mu_B > 27$ mss and its angular power spectrum is unknown.

Note that only the fluctuations about the mean surface brightness of the foreground sky in a given field introduce errors in galaxy photometry (a uniform veil would not be detected on direct photographs nor would cause errors, but would simply raise the detection threshold). In the case of photographic surface photometry and to the extent that the diffuse galactic component is a continuous function, the lower frequencies of the brightness fluctuations are effectively removed when numerical mapping techniques are used to define the sky "background" density level across the object field.

(c) Bright field stars and galaxies. Each star or galaxy in the field of the telescope is a source of scattered light superimposed on the airglow and galactic foreground. In principle, this component can be calculated via the PSF as the sum of the convolved images (particularly the aureoles) of all sources in the field. The problem is particularly severe in galaxy clusters where the local sky level is determined to a large extent by the overlapping coronae of individual galaxies and their optical aureoles as well as by the intergalactic stellar "sea" permeating the whole cluster (de Vaucouleurs and de Vaucouleurs 1970, Thuan and Kormendy 1977). Spectacular examples of this problem have been brought out by photographic contrast enhancement techniques (Malin 1981, Malin and Carter 1980).

In the case of detailed photographic surface photometry the major (low frequency) part of the stellar aureole component is automatically removed by numerical mapping of the "sky" background (Capaccioli and D'Odorico 1980). The procedure fails when a bright star is directly superimposed on the image of the galaxy or nebula, in which case there is no satisfactory way of restoring the lost information.

(d) Sub-threshold stars and galaxies. As was first noted by Miller (1963) statistical fluctuations in the number density of stars and galaxies which are individually below the detection threshold introduce another type of noise which may be significant and impose still another limit to the precision of the photometry at faint light levels. The first and second moments of the contributions of subthreshold stars and galaxies to the luminosity of the "blank sky" near the galactic poles have been calculated by numerical integration of their respective luminosity functions under the assumption of Poisson statistics (i.e. ignoring clustering). An application to photoelectric scans of NGC 3379 (Capaccioli and de Vaucouleurs 1983) shows that (i) the contributions to the total noise of the stars and galaxies fainter than $B \simeq 20.5$ are very nearly equal, (ii) the major contributors to the noise in each case are the stars and galaxies within one magnitude of the detection threshold, and (iii) the total noise due to subthreshold stars and galaxies is equivalent to $\mu_B \simeq 29.8$ mss.

7. CONCLUSIONS

The various sources of accidental and systematic errors in photographic (and photoelectric) surface photometry of faint extended sources, particularly galaxies, are such that it is difficult, if not impossible, to obtain significant quantitative information at brightness levels fainter than $\mu_B \simeq 28$ mss ($\mathcal{L} < 0.3 \mathcal{L}_\odot pc^{-2}$). Even at brighter levels errors less than 0.1 mag are difficult to achieve and in the nuclear regions ($\mu_B < 18$ mss) larger errors are introduced by convolution effects. Most of these errors are inherent to the nature of things, and except for improved resolution and lower sky background with space telescopes, will not be easily reduced by future technical advances.

This review paper, based mainly on work done in collaboration with Prof. M. Capaccioli, was supported in part by NSF Grant INT 8022847 and CNR Grant 100601C under the US-Italy Cooperative Science Program.

REFERENCES

Arp, H. and Kormendy, J. 1972, Astrophys. J. Letters 178, L101.
Arp, H. and Lorre, J. 1976, Astrophys. J. 210, 58.
Barbon, R., Benacchio, L. and Capaccioli, M. 1976, Mem. Soc. Astr. Italiana 47, 263.
Bozyan, E.P. and Opal, C.B. 1983, preprint.
Brown, G.S. 1978, Univ. Texas Publ. Astr. No. 11.
Cannon, R.D. 1979, in D.S. Evans (ed.), Photometry, Kinematics and Dynamics of Galaxies, Dept. Astr., Univ. Texas, Austin, p.27.
Capaccioli, M. and de Vaucouleurs, G. 1983, Astrophys. J. Suppl. 52,
Capaccioli, M. and D'Odorico, S. 1980, in P.L. Bernacca and R. Ruffini (eds.), Astrophysics from Spacelab, D. Reidel Publishing Company, Dordrecht, p.317.
Davis, M., Feigelson, E. and Latham, D.W. 1980, Astr. J. 85, 131.
Dawe, J.A. and Metcalfe, N. 1982, Proc. Australian Soc. Astr. 4, 466.
Dermendjian, D. 1957, Ann. Geophys. 13, 286; 1959, Ann. Geophys. 15, 218.
de Vaucouleurs, G. 1943, Sci. Industries Photog. (2), 14, 149; 1945, J. Physique (8), 6, 205; 1946, Sci. Industries Photog. (2), 17, 257; 1948a, J. des Observateurs 31,113; 1948b, Ann. d'Astrophys. 11, 247; 1948c, Revue d'Optique 27, 541; 1948d, Com. Nat. Fr. Astr., Comm. et Mem., 160; 1955, Observatory 75, 170; 1958, Astrophys. J. 128, 465; 1960, Observatory 80, 106; 1976, in J.L. Heudier (ed.), I.A.U. Working Group on Photographic Problems, Proc. Grenoble meeting, p. 93; 1977, Astrophys. J. Suppl. 33, 211.
de Vaucouleurs, G. and Buta, R. 1983, Astr. J. 88, 939.
de Vaucouleurs, G. and Capaccioli, M. 1979, Astrophys. J. Suppl. 40, 699.
de Vaucouleurs, G. and de Vaucouleurs, A. 1970, Astrophys. Letters 5, 219.
de Vaucouleurs, G. and Nieto, J.L. 1978, Astrophys. J. 220, 449; 1979, Astrophys. J. 230, 697.
Griboval, P. 1984, these proceedings.
Jones, W.B., Obitts, D.L., Gallet, R.M. and de Vaucouleurs, G. 1967, Publ. Dept. Astr., Univ. Texas, Ser. II, Vol. I., No. 8.
King, D.F., Taylor, K.N.R. and Tritton, K.P. 1979, Monthly Notices Roy. Astron. Soc. 188, 719.
King, I. 1971, Publ. Astr. Soc. Pacific 83, 199.
Kormendy, J. 1973, Astr. J. 78, 255.
Kormendy, J. and Bahcall, J.N. 1974, Astron. J. 79, 671.
Malin, D.F. 1978, Nature 276, 591; 1981, Amer. Astron. Soc. Photo-Bulletin No. 27, 4.
Malin, D.F. and Carter, D. 1980, Nature 285, 643.
Miller, R.H. 1963, Astrophys. J. 137, 733.
Newell, E.B. 1980, in G. Sedmak, M. Capaccioli and R. Allen (eds.), Image Processing in Astronomy, Osservatorio Astr., Trieste, p. 100.
Okamura, S. 1977, Ann. Tokyo Astr. Observ., Vol. XVI, No. 3, p.11.
Piccirillo, J. 1973, Publ. Astr. Soc. Pacific 85, 278.
Sandage, A.R. 1976, Astr. J. 81, 954.
Thuan, T.X. and Kormendy, J. 1977, Publ. Astr. Soc. Pacific 89, 466.

HIGH-RESOLUTION NARROW-FIELD VERSUS LOW-RESOLUTION WIDE-FIELD
OBSERVATIONS OF GALAXIES

M. Capaccioli[1], E. Davoust[2], G. Lelièvre[3], J.-L. Nieto[4]
1: Istituto di Astronomia, Università di Padova
2: Observatoire de Besançon
3: CFHT Corporation and Observatoire de Paris
4: Observatoire du Pic du Midi et de Toulouse

Introduction

There is an increasing evidence that small-scale phenomena occuring in the inner regions of galaxies are related to large-scale phenomena such as, e.g., merging or violent interactions between galaxies. Plausible scenarios (e.g. Rees, 1978) involve, for instance, accumulation of material from the outside along the accretion disk of a black hole and subsequent ejection in two opposite directions into the intergalactic medium. Active galaxies and QSO's may be the extreme examples of the link between large-scale phenomena contributing to the evolution of galaxies. Moderately active or even normal galaxies are also submitted to entangled events on both scales. The aim of this communication is to illustrate the complementarity between high-resolution, small-field telescopes and Schmidt-type telescopes for the study of this phenomenology, and to stimulate further research by a few challenging examples.

1. NGC 3379 and normal elliptical galaxies.

a) de Vaucouleurs and Capaccioli (1979) reported some evidence for the existence of a spike of light at the center of NGC 3379. This was subsequently confirmed by observations at higher resolution (Nieto and Vidal, 1983) which ruled out Kormendy's (1982) interpretation of the central region of NGC 3379 in terms of an isothermal core. But the shape of the spike of light, its total luminosity, and therefore its physical interpretation, strongly depend on the model assumed to fit the rest of the galaxy.

b) Another possible correlation between both scales is found in the wavy residuals with respect to any smooth luminosity profile of the main body of the galaxy (de Vaucouleurs and Capaccioli, 1979; Nieto and Vidal, 1983), detectable on medium-scale material, and in the shells revealed with wide-field material (see Malin, this volume).

c) A further aspect is the dependence of geometry on radius possibly due to triaxiality: is a phenomenon intrinsic to the galaxy or is it due to

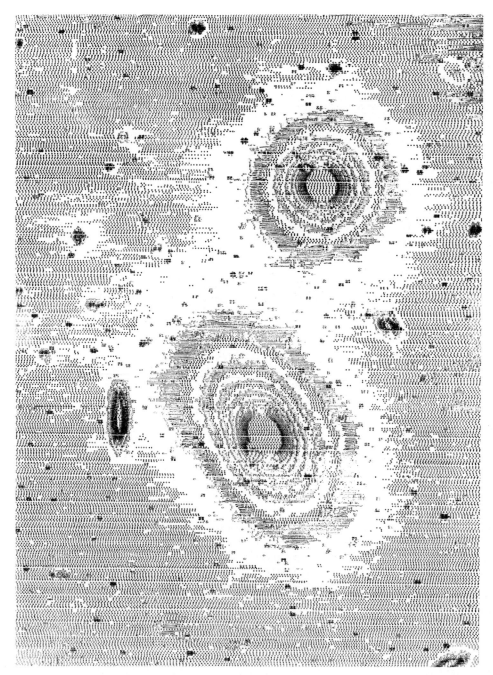

Figure 1: Joyce-Loebl isodensity tracings of the halo surrounding the two galaxies NGC 4374 and 4406, from a one hour exposure on IIIaJ emulsion taken by J. Sulentic with the Palomar 48-inch Schmidt telescope. (North at left, East at top).

perturbations produced by massive neighbours or both ? Evidence for gravitational interactions must be searched for with small-scale telescopes whereas the study of the changing geometry (at least in the inner parts) requires the use of large-scale observations at least for technical reasons (Capaccioli, 1983).

2. Substructures and the case of NGC 3384.

Polar rings in lenticular galaxies (e.g. A0136-0801) could be either intrinsic or due to interactions with the environment. Schweizer et al. (1983) favor the second possibility; but NGC 2685, the prototype of the subclass, is an isolated object. In a high-resolution study of NGC 3384, Davoust, Lelièvre and Nieto (1983) detected an elongated feature perpendicular to the main body of the galaxy for which several arguments prompted to assume that it is a polar ring and not a weak bar. In addition, Malin (this volume) showed a well defined spiral arm at the N.E. end of this galaxy. Although there is no definite evidence that this outer arm is due to interaction with the companion galaxy NGC 3379, the presence of both peculiar features detected via small-and large-scale observations deserves further attention. Note also that several S0's (e.g. NGC 4036) present unexplained structures in the outerparts (see, e.g., Barbon, Benacchio and Capaccioli, 1978).

3. The encounter of NGC 4374 and NGC 4406.

Benacchio et al. (1983) reported a very poor agreement between photographic photometry (for which the large field allowed an accurate determination of the sky background) and photoelectric aperture photometry from the catalogue of Longo and de Vaucouleurs (1983). The sky determination was suspected to be the cause of the discrepancy. A study of an extensive collection of small-scale plate material taken at different telescopes (Palomar-Schmidt, Tautenburg, UKST) reveals that a huge halo surrounds the two galaxies (Fig. 1). We suggest that i) intergalactic material was accreted by NGC 4406 (V_o= -419 kms^{-1}) during its high velocity passage through the center of the Virgo cluster ($V_o \cong$ +1000 kms^{-1}) and ii) forced star formation induced by stripping of gas due to ram pressure as suggested by the similar morphologies of the X-ray and stellar haloes (Jones and Forman, 1981) and by the existence of a type I supernova (Smith, 1981). Evidence for interaction between the two galaxies suggests, and time scale arguments are consistent with, the hypothesis that the close passage of NGC 4406 in the center of the Virgo cluster has produced a sudden dumping of gas in the nucleus of NGC 4374 of which the inner dust lane is a tracer (Fig. 2), able to supply material to the jet point source.

Acknowledgements:

We wish to thank Dr. J. Sulentic for allowing us the use of a very deep 48" Palomar-Schmidt plate, and Prof. G. de Vaucouleurs for his critical reading of the manuscript.

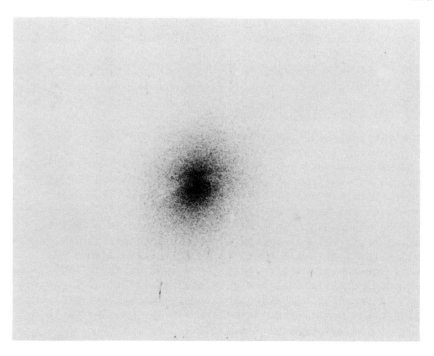

Figure 2: The dust complex in the nuclear region of NGC 4374, from a high-resolution (FWHM ∼ 0."75) photograph taken at the prime focus of the CFH Telescope (exp. time 5 minutes).

REFERENCES

Barbon, R., Benacchio, L., Capaccioli, M., 1978, Astron. Astrophys. 65 165.
Benacchio, L., Capaccioli, M., de Biase, G., Santin, P., Sedmak, G., 1983, Preprint.
Capaccioli, M., 1983, Mem. SAIt., in press.
Davoust, E., Lelièvre, G., Nieto, J.-L., 1983, in preparation.
de Vaucouleurs, G., Capaccioli, M., 1979, Astrophys. J. Suppl. 40, 699.
Jones, C., Forman, W., 1982, IAU Symp. 97, 97.
Kormendy, J., 1982, in "Morphology and dynamics of galaxies", 12th advanced course of the Swiss Society of Astronomy and Astrophysics.
Longo, G., de Vaucouleurs, A., 1983, The University of Texas Monograph in Astronomy n. 3.
Nieto, J.-L., Vidal, J.-L., 1983, preprint.
Rees, M., 1978, Nature, 275, 516.
Schweizer, F., Whitmore, B.C., Rubin, V.C., 1983, Astron. J. 88, 909.
Smith, H.A., 1981, Astron. J. 86, 998.

GLOBAL STRUCTURE OF GALAXIES AND QUANTITATIVE CLASSIFICATION

S. Okamura, K. Kodaira, and M. Watanabe
Tokyo Astronomical Observatory, University of Tokyo
Mitaka, Tokyo 181, Japan

We have been carrying out digital surface photometry of galaxies using the plates taken with the 105-cm Schmidt telescope at the Kiso Observatory, a branch of the Tokyo Astronomical Observatory. Two-dimensional V-band luminosity distributions of 261 galaxies in the Virgo and the Ursa Major regions have been compiled so far (Watanabe et al. 1982, Watanabe 1983). They form one of the largest samples of homogeneous surface photometric data. This discourse is to draw attention to some new results, which are summarized below, on the structure of galaxies derived from the analysis of the homogeneous sample. Detailed discussion will be given elsewhere (Kodaira et al. 1983, Okamura et al. 1983, Watanabe et al. 1983). Most of the sample galaxies are members of either the Virgo or the Ursa Major Clusters, both lying at nearly equal distance (Aaronson and Mould 1983). Their morphological types range from elliptical (T=-5) to Magellanic irregulars (T=10).

(1) Several photometric parameters are derived on the basis of the generalized radial profile, that is, a radial profile which would be obtained if a galaxy were to be seen face-on (Watanabe et al. 1982), within a limiting surface brightness of 26 mag arcsec^{-2}. They include the integrated magnitude, V_{26}, diameter, D_{26}, mean surface brightness, SB = V_{26} + 5 log D_{26} + const, and the mean concentration index, $X1(P)$. The last quantity, $X1(P)$, is a weighted mean of five concentration indices, P_n (n=1,2,...,5), which represent the fractional luminosity included within $D_{26}/2^n$ relative to the luminosity within D_{26}. In their definition, the indices P_n are similar to the concentration indices C_{21} and C_{32} introduced by de Vaucouleurs (1977a, b).

(2) A good correlation is found between the mean concentration index and the morphological type although the scatter is relatively large. One cause of the scatter is identified with the variation of the intrinsic surface brightness among galaxies. It is suggested that the mean concentration index, when combined with the mean surface brightness, is sensitive to the bulge-to-disk ratio of a galaxy. The $X1(P)$ versus SB diagram may provide a useful tool to investigate the

luminosity structure of a galaxy and to characterize galaxy content of clusters of galaxies out to ∼100 Mpc.

(3) The principal component analysis of the above photometric parameters, V_{26}, $\log D_{26}$, SB and $Xl(P)$ shows that there are two independent variables, in agreement with the results by Brosche(1973)'s pioneering work for 31 nearby disk galaxies. The two independent variables derived for the present sample closely coincide with those by Brosche who included in his analysis dynamical parameters as well as photometric parameters. This suggests that the global structure of galaxies may be characterized by photometric parameters alone without dynamical parameters.

(4) It is shown from the principal component analysis that the diameter and the surface brightness can be used as fundamental parameters to characterize the structure of galaxies. The diameter versus surface brightness diagram is constructed for 149 galaxies in the sample. In the diagram, galaxies form two distinct branches consisting mainly of elliptical and S0 galaxies and late-type galaxies respectively, which run almost perpendicular to each other. The elliptical-S0 branch corresponds to the μ_e - r_e relation of early-type galaxies found by Kormendy (1977,1980). This diagram may be used as a possible distance tool and the basis of a quantitative classification of galaxies and of a characterization of clusters of galaxies.

REFERENCES

Aaronson, M. and Mould, J. R. 1983, Ap. J., <u>265</u>, 1.
Brosche, P. 1973, Astr. Ap., <u>23</u>, 259.
de Vaucouleurs, G. 1977a, Ap. J. Suppl., <u>33</u>, 211.
────────────── 1977b, The Evolution of Galaxies and Stellar Populations, ed. B.M.Tinsley and R.B.Larson (Yale Univ. Obs., New Haven), p. 43.
Kodaira, K., Okamura, S., and Watanabe, M. 1983, Ap. J. Letters (in press)
Kormendy, J. 1977, Ap. J., <u>218</u>, 333.
────────────── 1980, Proc. ESO Workshop on Two-Dimensional Photometry, ed. P.Crane and K.Kjär, p. 191.
Okamura, S., Kodaira, K., and Watanabe, M. 1983, Ap. J. (in press).
Watanabe, M. 1983, Annals Tokyo Astr. Obs., 2nd Ser., <u>19</u>, 121.
Watanabe, M., Kodaira, K., and Okamura, S. 1982, Ap. J. Suppl., <u>50</u>, 1.
Watanabe, M., Kodaira, K., and Okamura, S. 1983, in preparation.

SURFACE PHOTOMETRY OF PURE DISK GALAXIES

C. Carignan
Mount Stromlo and Siding Spring Observatories
and
Kapteyn Astronomical Institute

ABSTRACT:

Surface photometry from U.K.S.T.U. plates in U, B_J and R_F is presented for the three Sculptor Group galaxies NGC 7793, NGC 247 and NGC 300. Their photometric parameters are derived and compared to those of M33. Our color maps are used to sort out the data into blue and red pixels where the blue pixels profiles are expected to be rather irregular, representative of the clumpy locations of areas of recent star formation while the red pixels profiles should approximate the smoother distribution of the old disk population.

What do we mean by "pure" disk galaxies? We call pure disk galaxies (PDG), late-type spirals (Scd-Sd-Sdm) showing no apparent bulge or in which the bulge contributes very little (<2-3%) to the total luminosity. Our aim is to combine photometry with kinematical data to study the mass distribution of PDG and especially the ratio of unseen halo mass to luminous disk mass (Carignan, 1983; Carignan and Freeman, 1983) using the light distribution as a tracer of the luminous mass component. Because they have no bulge, PDG are ideal candidates for that kind of work since we only have one luminous component to model.

In this paper, we want to concentrate on the results of the surface photometry for three such systems in the Sculptor Group. The photometric parameters for NGC 7793, NGC 247 and NGC 300, derived from their total B_J light profiles are given in Table 1. As we can see, the three galaxies exhibit an exceptionnally wide range in central disk surface brightness from the bright disk of NGC 7793 at $B(0)_c = 20.33$ to the low surface brightness disk of NGC 247 at 23.44. This is even fainter than Romanishin and Strom (1979) mean value of 22.88 for their sample of 12 LSB galaxies. NGC 300 has a more "normal" $B(0)_c = 22.23$. However, this is still within the observed range when one considers that in Freeman's (1970) sample, seven galaxies had a $B(0)_c$ 1.5 magnitude brighter than its mean value of 21.65 and one system IC 1613 was even fainter than NGC 247 at $B(0)_c = 23.7$.

The three Sd galaxies being of equivalent absolute magnitude ~-18, the fitted scale lengths of their exponential fall-off vary accordingly to their surface brightness from 1 to 3 kpc. Again, this is very similar to Freeman's (1970) mean for all morphological types of 2.1 ± 1.3 kpc, Schweizer's (1976) value of 2.7 ± 0.5 kpc mainly for Sc's and Romanishin and Strom (1979) 2.1 ± 0.7 kpc where all the distances are adjusted for H_0 = 100 km/sec/Mpc.

When compared to the northern Scd galaxy M33, the Sculptor Group galaxies appear to be slightly smaller with a mean face-on diameter of $\langle D(0) \rangle$ = 9.9 ± 1.0 kpc compared to 12 kpc for M33 and a mean absolute magnitude of -18.11 ± 0.17 compared to -18.51. In that respect, they are probably closer to the LMC which has D(0) = 8.25 kpc and M_T^o=-18.17.

As for the radial dependance of colors, no significant gradient greater than 0.2 magnitude has been seen. While this does not tell us anything on the run of $(M/L)_{disk}$ with radius, it is at least consistent with it being constant.

Since we want to use our luminosity profiles as tracers of the luminous mass component, we had to study the effects of recent star formation and "patchy" internal absorption. To do this we followed Talbot et al. (1979) and sorted out the data into red and blue pixels profiles adopting a separating color of $(B_J - R_F)$ = 1.0 (~(B-V) = 0.65). However, it should be emphasized that this cannot be considered as a complete decomposition which would require a more thorough analysis (cf. Jensen et al., 1981 for M83). What we are doing here is just plotting the mean surface brightness of blue and red pixels as a function of radius.

The most important result in the case of NGC 7793 is the persistence of the type II (Freeman, 1970) profile for the red pixels which confirms the dynamical significance of such a feature. This goes against Talbot et al. (1979) suggestion that most type II profiles could be due to an excess of young blue light producing a "hump" that mimics a type II profile as in the case of M83.

As for NGC 247, the red pixels profile is almost a pure exponential on the whole radius range. On the other hand, the blue pixels profile shows three interesting features. First, it confirms the region of higher internal absorption between 5' and 7' which is seen on short exposure plates and in the isophotes map. In that region, the blue pixels are 0.3 - 0.4 magnitude fainter while the red pixels are almost unaffected. This is followed by a region where the blue pixels surface brightness stays more or less constant while crossing the broad northern spiral arm before reaching the exponential fall-off region.

We also obtained very interesting results for the more complicated profile of NGC 300. First, the hump seen around 3!5 again seems to be a feature of the old disk being clearly present in the red pixels profile. Because at high surface brigthness, this feature is quite important dynamically. On the other hand, the hump ~ 10' seems mainly due to young blue light so that the true scale length of the disk, when taking out the effect of this hump, is substantially smaller. As for the blue pixels profiles, it illustrates very well Schweizer's (1976) results that the arms light falls-off more gradually than the disk light.

I would like to thank the staff of the U.K.S.T.U. for kindly taking the plates used in this work and Dr. K.C. Freeman for many valuable discussions.

TABLE 1:
Photometric Parameters for the Scultor Group Galaxies and Comparison with M33.

	M33*	NGC7793	NGC247	NGC300
Morphological type	SA(s)cd	SA(s)d	SAB(s)d	SA(s)d
Distance (Mpc)	0.72	3.38	2.52	1.90
Face-on diameter D(0) (kpc)	12.0	8.8	10.8	10.1
Corrected total magnitude B_T^o	5.79	9.33	9.00	8.38
Absolute magnitude M_T^o	-18.51	-18.31	-18.01	-18.01
Exponential disk parameters				
Central surface brightness $B(0)_c$	21.05	20.33	23.44	22.23
Scale length α^{-1} (kpc)	1.65	1.08	2.93	2.06
Maximum rot. vel. V_{max} (km/sec)	107	95	108	87
Hydrogen mass to luminous mass ratio M_{HI}/M_{lum}	0.05	0.12	0.09	0.25
Luminous mass to blue luminosity ratio M_{lum}/L_B	5.0[1]	2.0[2]	5.0[2]	2.5[2]

* parameters from de Vaucouleurs, 1959 and de Vaucouleurs and Caulet, 1982.
[1] Kalnajs, 1983.
[2] Carignan, 1983.

REFERENCES

Carignan, C. 1983, in The Milky Way Galaxy, I.A.U. Symposium no. 106, in press.
Carignan, C. and Freeman, K.C. 1983, in preparation.
de Vaucouleurs, G. 1959, Ap.J., 130, 728.
de Vaucouleurs, G. and Caulet, A. 1982, Ap. J. Suppl. Ser., 49, 515.
Freeman, K.C. 1970, Ap. J., 160, 811.
Jensen, E.B., Talbot, R.J. and Dufour, R.J. 1981, Ap. J., 243, 716.
Kalnajs, A. 1983, in Internal Kinematics and Dynamics of Galaxies, I.A.U. Symposium no. 100, 87.
Romanishin, W. and Strom, S.E. 1979, in Photometry, Kinematics and Dynamics of Galaxies, ed. D.S. Evans, Univ. of Texas at Austin, p. 151.
Schweizer, F. 1976, Ap. J. Suppl. Ser., 31, 313.
Talbot, R. J., Jensen, E.B. and Dufour, R.J. 1979, Ap. J., 229, 91.

ELLIPTICAL GALAXIES WITH SHELLS

D. Carter
Mount Stromlo and Siding Spring Observatories
Australian National University, Canberra

A considerable number of otherwise normal elliptical galaxies show sharp edged, low surface brightness concentric features, often at very large radii from the parent galaxy. Features like this are visible in some of the photographs presented by Arp (1966). More recently extensive lists of galaxies with such features have been prepared from searches of sky survey material by Malin and Carter (1980, 1983) and Schweizer (1983), and the implications of these features for models of ellipticals and for peoples ideas about their formation have been considered in detail.

THE PROPERTIES OF SHELL GALAXIES

Before considering what the shell forming process is it is important to define what the properties of shell galaxies are. The data base is the sample of 137 shell galaxies south of $-17°$ found on the SRC IIIaJ survey by Malin and Carter (1983). These galaxies are galaxies at elliptical morphology in which Malin and I could see one or more sharp edged concentric features, these are the defining characteristics of the sample, but the sample also has the following properties:

1) In elongated galaxies the shells tend to lie along the major axis.
2) The shells are concentric but not complete.
3) In elongated galaxies the shells will alternate with radius, if the galaxy is elongated North South and the innermost shell lies on the northern semi axis, then the second shell will be to the south, the third to the north, the fourth to the south, and so on.
4) Shell galaxies tend not to occur in clusters or rich groups of 13 shell galaxies classified as T = -5 or -4 by the second reference catalogue (de Vaucouleurs et al. 1976, hereafter RC2) only one lies in what RC2 calls a group, whereas 40% of all ellipticals lie in rich groups or clusters. Of the 137 shell galaxies in our prime sample only five (including NGC1316 and

IC4329, neither of which is classified as an elliptical by RC2) lie in clusters

5) Of the isolated ellipticals in RC2 we detect shells around 12 (16.5%).
6) Of the 137 galaxies in our prime sample only two, NGC1316 and NGC5128, are coincident with radio sources in the Parkes catalogue. This is a smaller number than would be expected of a similar sample of randomly chosen ellipticals. Shell galaxies are not radio sources.
7) Sensitive searches of a small sample of shell galaxies have revealed no strong emission lines, and, with one exception no neutral hydrogen. The exception is NGC2865 which does show a considerable quantity of H I.
8) BVRJH colours for the outer shell of NGC1344 (Carter, Allen and Malin 1982) look like those of stars of spectral types G to K, but there is evidence that the shells are bluer than the parent galaxy.

THE NATURE OF THE SHELLS

Two classes of model have been proposed to account for the observed properties of shells; models in which stars are formed in a galactic wind (Fabian, Nulsen and Stewart 1980); and models in which a small, dynamically cold galaxy falls in to the more massive parent elliptical. Although, as Fabian et al. have pointed out, the mass function of stars which would be formed in a galactic wind is essentially unknown, we feel that the colours of the outer shell of NGC1344, together with the detailed agreement of the dynamical models with the observed properties, favour these models over the stellar wind models. Accordingly I will discuss the dynamical models in more detail.

It is clearly a problem for a model of a collision to produce as many features as sharp edged as we see; in NGC3923 for example we can identify between 19 and 23 separate sharp edged shells. For the features to be sharp edged the galaxy which falls in must be very cold, any velocity dispersion will tend to smear out the features. Another problem encountered by the models is that the number of particles in an n body simulation is limited to maybe 1000 by the available computer time, and it is rather difficult to simulate 20 sharp edged features with only 1000 particles.

Two distinct processes have been suggested to account for shells. The first is due to A. Toomre, and is described by Schweizer (1983). In this process the disc galaxy falls in on a non radial orbit and wraps spatially around the centre of mass of the elliptical. The disc galaxy forms arcs when viewed from out of the orbit plane of the disc. The structures do not resemble the shells seen in galaxies such as NGC3923 and NGC1344, being confused and sometimes overlapping, but they may resemble the structure seen in NGC1316 and NGC5018 (Fig. 1).

Fig. 1   NGC 5018. A galaxy with shells and dust. This is a CCD image at 5200 Å from the Canada-France-Hawaii Telescope.

The second process is due to Quinn (1982, 1983), who discusses radial encounters between disc galaxies and a potential well representing an elliptical. In Quinn's process wrapping takes place in phase space, which can be illustrated for a one dimensional system, and the maximum spatial excusion of each phase wrap at a particular time produces a sharply defined density maximum. The density maxima are radially propogating density waves, and the outer shells are the first to form. These models, although limited by the number of particles it is possible to reproduce many of the observed features of shells, specifically the density maxima are interleaved in the manner we observe in the elongated shell galaxies, such as NGC 3923 and NGC 1344

FURTHER WORK

With Dr. B. P. Fort of Toulouse Observatory I have begun a project to obtain much better colours for some shells, we obtained images in four passbands with a CCD on the Canada-France-Hawaii Telescope, an example is shown in Figure 1. The colours will tell us about the galaxy which has fallen in, and possibly about the merger process itself.

Quinn (1983) has pointed out that shells may well tell us about the potential of an elliptical. At present we are limited by the small size of the sample for which detailed observations are possible, but only a third of the sky has been surveyed to deep enough levels to uncover these features, and we hope that the SRC equatorial and Palomar northern surveys will increase our sample by a factor of three or so.

REFERENCES

Arp, H.C.: 1966, "Atlas of peculiar galaxies", Caltech.
Carter, D., Allen, D.A., and Malin, D.F.: 1982, Nature, 295, 126.
Fabian, A.C., Nulsen, P.E.J., and Stewart, G.C.: 1980, Nature, 287, 613.
Malin, D.F., and Carter, D.: 1980, Nature, 285, 643.
Malin, D.F., and Carter, D.: 1982, Astrophys. J., in press.
Quinn, P.J.: 1982, Thesis, Australian National University.
Quinn, P.J.: 1983, preprint.
Schweizer, F.: 1983, IAU Symposium 100 (Dordrecht: Reidel) p. 319.
Vaucouleurs de, G., de Vaucouleurs, A., and Corwin, H.G.: 1976,
    "Second Reference Catalogue of Bright Galaxies", University
    of Texas.

DISCUSSION

T. SHANKS: If the dynamical model is correct, why do we see no residual gas from the disk galaxy?

D. CARTER: In some cases we see dust. The galaxy which is swallowed probably has to be of very low gas, so perhaps all of the gas is used up in a short time, whereas the shells last longer. NGC 2865 does have quite a lot of neutral hydrogen.

W.P. BIDELMAN: What is wrong with the simple-minded idea that your shells are just remnants of spiral (or circular) arms that the elliptical galaxy might have once possessed?

D. CARTER: Shells are concentric, also they have sharp outer edges which suggest a spherical structure seen in projection. Nevertheless there are galaxies with weak spiral arms, and at a distance it is difficult to tell them from shell galaxies, but we try to throw them out of our sample.

J.-L. NIETO: About the possible relations existing between outer parts and inner parts of galaxies, we have preferentially geometry changes in shell elliptical galaxies.

D. CARTER: The sample of shell galaxies for which we have data on the inner parts is very small, one shows a clear isophote twist and some have dust lanes, but the numbers are not statistically significant.

BRAND NEW CANDIDATES FOR DEEP SCHMIDT PLATES

M.Capaccioli[1], E.Davoust[2], G.Lelièvre[3], J.-L.Nieto[4]
1: Istituto di Astronomia, Università di Padova
2: Observatoire de Besançon
3: CFHT Corporation and Observatoire de Paris
4: Observatoire du Pic du Midi et de Toulouse

With reference to the complementarity between large-scale, low-resolution and small-scale, high-resolution telescopes (Capaccioli et al., 1984), we briefly report here on very recent CFH telescope photographic observations of early-type galaxies which are good candidates for further investigations with Schmidt-type telescopes. The photographs were taken at the newly installed f/8 Cassegrain focus with good seeing conditions. The large scale (~ 7 arcsec mm^{-1}) allows to overcome some of the major difficulties encountered in the reduction of the prime focus high resolution (FWHM$\sim$0".75) material (Nieto, 1983).

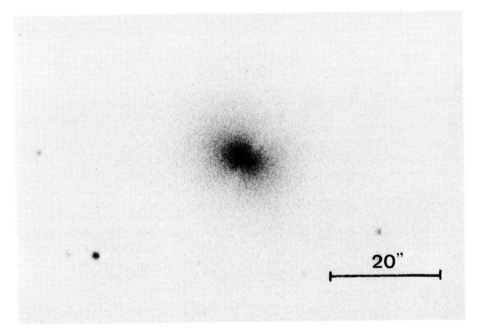

Figure 1. NGC 6702 from a IIaO + GG385 plate (exposure time 30 minutes).

Visual inspection of the plates already provides some new astrophysical information on two galaxies:

- NGC 584. This galaxy, known to be a good example of the class of ellipticals with isophotal twisting (Williams and Schwarzschild, 1979), turns out to be a lenticular with bulge, lens and disk clearly visible.

- NGC 6702. This is a new prominent example of the elliptical galaxies with a dust lane along the minor axis (Figure 1).

REFERENCES

Capaccioli,M., Davoust,E., Lelièvre,G., Nieto,J.-L.,1984,this vol.,p.393.
Nieto,J.-L., 1983, Astron.Astrophys.Suppl.Sr., 53, 383.
Williams,T.B., Schwarzschild,M., 1979, Astrophys.J.Suppl.Sr., 41, 209.

A SEARCH FOR "YOUNG GALAXIES"

R. McMahon, R. Terlevich, C. Hazard, M. Irwin
Institute of Astronomy, Cambridge
J. Melnick
University de Chile

1. INTRODUCTION

Among the most interesting objects found on low dispersion IIIaJ objective prism plates taken with the UK Schmidt telescope are a large number of compact and intrinsically very bright 'extragalactic HII regions'. Spectroscopically they are indistinguishable from the giant HII regions found in spiral and irregular galaxies except that they are much more luminous with absolute magnitudes in the range M_V -14 to -23. The stellar continuum is extremely blue indicating a predominantly young stellar population. The Hβ luminosities, 10^{38} to 10^{42} ergs/sec indicate that they must contain $10^5 - 10^7$ OB stars with a total mass in stars of $10^6 - 10^9$ $M_\odot$. The associated high rate of star formation cannot have continued for more than 10^7 years, which implies that these systems are either undergoing star formation for the first time, or that they undergo intermittent bursts of star formation. The chemical composition of these systems is also remarkable in that the oxygen abundance is invariably less than solar.

Terlevich and Melnick (1981) have shown that for isolated extragalactic HII regions there is a strong correlation between the velocity dispersion (σ) and Hβ luminosity. More importantly they show that the velocity dispersion is also a function of metallicity and if this is allowed for, it considerably reduces the dispersion in the σ/Hβ relationship permitting the use of HII regions as reliable distance indicators. The HII regions are therefore not only important for studies of star and galaxy formation but promise to provide a valuable new cosmological probe.

2. LOW REDSHIFT SURVEY

Our initial laborious visual searches of objective prism plates have led to the discovery of a few hundred extragalactic HII regions. A large spectroscopic survey of all Tololo, Cambridge, and Markaryan emission line galaxies is under progress. At present the sample is 50%

complete, with spectra of more than 250 galaxies already collected. This data is presently under analysis but one interesting result has been that none of our sample has an oxygen abundance less than 1/20 solar. One possible explanation is that many of our objects are more luminous and have stronger lines than IZw18, which has an oxygen abundance 1/40 solar.

IZw18 the prototype "Young Galaxy" was discovered independently by both Zwicky in his searches for compact galaxies and by Markarian in his search for galaxies with ultraviolet excess. It is now possible to combine both these search techniques, in an automatic survey using the APM at Cambridge. A sample of compact galaxies is selected from the direct plates and the objective prism spectrum analysed either automatically using PRS (Hewett et al. this conference) or manually. This sample should contain the weak-lined and fainter objects absent from our manual searches. A search for blue compact galaxies in one field has led to the discovery of several extragalactic HII regions missed in our visual survey. It is from this sample that we hope to find young galaxies with metal abundances comparable to, or less than IZw18.

3. HIGH REDSHIFT SURVEY

The HII regions are discovered by their strong [OIII] 5007 emission and hence the IIIaJ surveys are limited to redshifts < 0.06. We have recently initiated a search using IIaD and IIIaF emulsions. The use of these emulsions in manual searches for emission lines presents some difficulties because of variations with wavelength in the emulsion sensitivity and the low dispersion of the UKST low dispersion prism above 5000 angs. This is not a problem with the new higher dispersion prism or the ESO prism. However, we have had some success with a few candidates in the range 0.07 to 0.2 (0.08 and 0.10 confirmed spectroscopically).

Recently we have discovered two HII regions with redshifts of 0.18 on IIIaJ plates by virtue of their strong OII 3727 emission. HII regions in the range 0.2 to 0.31 have been discovered by Arp (1983) and Osmer (1982) in their searches for quasars. We now have a sample of HII regions from zero redshifts out to 0.3.

Thus we have a powerful new tool for studies out to cosmologically significant distances and an easy method of studying the Hubble flow. A great advantage of these extragalactic HII regions over the QSO's is that their physics is relatively well understood.

REFERENCES

Arp, H. 1983, Ap.J., 271, 479.
Osmer, P. 1982, Ap.J., 253, 28.
Terlevich, R., Melnick, J. 1981, MNRAS, 195, 39.

A SURVEY OF SOUTHERN COMPACT AND BRIGHT NUCLEUS GALAXIES

A. P. Fairall
University of Cape Town, Rondebosch 7700. Cape. South Africa.

During the 1960's, Zwicky made an intensive examination of the Palomar Sky Survey plates that resulted in his "Catalogue of Galaxies and of Clusters of Galaxies" (Zwicky *et al.*, 1961-1968). In the course of this study, he picked up numerous examples of what he labelled as "compact galaxies and compact parts of galaxies"; five lists were initially circulated and later presented as a catalogue (Zwicky 1971). To Zwicky the "compact" connotation suggested extremely high stellar densities, but line profiles failed to reveal the large velocity distributions expected of such concentrations of stars. Nevertheless an unexpected benefit was that many of his "compact parts" turned out to be the nuclei of actives galaxies, particularly Seyfert galaxies. Furthermore, since the nuclei have to stand out to gain "compact part" status, Seyfert nuclei from Zwicky's lists tend to be fairly extreme specimens.

In the mid 1970's, the ESO Quick Blue Survey and later the UK IIIa-J Sky Survey appeared. The author, having examined and observed many of Zwicky's galaxies, started to scan for compact and bright nucleus galaxies in the southern skies. Since spectroscopy was essential for identification of active galaxies, the policy was to scan and select just sufficient for follow up observing. Initially, grating photography (akin to the Grism) offered hope for obtaining multiple slitless specrta from a single exposure, but the accumulation of featureless spectra encouraged the shift to conventional spectroscopy. Turnover was greatly increased with the advent of IPCS and RPCS detectors.

Today, after nearly 8 years, about 200 fields have been scanned and more than a thousand galaxies have been followed up. Starting at the South Celestial Pole, the search has reached the Declination $-45°$ zone in the Southern Galactic Hemisphere and the Declination $-35°$ zone in the Northern Galactic Hemisphere. Results are published in six papers (Fairall 1977 to 1983b). Overall statistics are as follows:

Seyfert 1 galaxies	5
Seyfert 2 galaxies	9
Near Seyfert (but almost certainly Seyfert 2)	23

Narrow emission line galaxies	91
Conventional stellar content (reliable redshift obtained)	619
Conventional stellar content (but no reliable redshift obtained)	279
Superposed stars mimicking bright nuclei	38
Planetary nebula mimicking bright nucleus	1
	1065

Thus the bulk of the galaxies turn out to have normal stellar spectra, and much of the effort of the programme has been the production of redshifts. While exposure times have been compromised against the need to examine as many galaxies as possible, the redshifts are probably good to 200 km s^{-1} (as suggested by a comparison with galaxies measured by other investigators). Occasional discrepancies do occur, but these are by no means not unique to the present work.

The main incentive behind the programme has been the discovery of Seyfert galaxies, and hopefully, extreme Seyferts that might throw more light on the nature of their enimatic nuclei. Two discoveries from the present list stand out. F-9 = ESO 113-IG45, the most luminous Seyfert 1 known, is a vital link in the Seyfert 1 = QSO hypothesis (and like other extreme Seyfert 1's it shows anaemic spiral structure which may form a trend towards the even more anaemic structure surrounding quasars). F-427 = ESO 263-G13, an extreme Seyfert 2 galaxy (Fairall 1983c) appears to show rapid variability in its emission lines - currently regarded as controversial (the author has done his best to get rid of the evidence - but it refuses to go away!)

One may question the viability of this technique as a means for finding active galaxies, since objective prism searches and X-ray surveys would appear superior alternatives. The author has also participated in both of these. While the objective prism (Bohuski, Fairall and Weedman 1978) is unquestionably more productive in quantity, it failed to detect two sample Seyferts (F-9 and F-51) from the current survey. Precise X-ray positions are obviously ideal, but searches in boxes can often lead to the examination of many galaxies with negative results - a check of about 70 galaxies carried out by the author in 1981 led to one worthwhile discovery (Fairall, McHardy and Pye 1982). We must conclude that the three separate methods are perhaps more complementary than competitive and that the selection of galaxies solely from their appearance on Schmidt Sky Surveys is still a worthwhile pursuit.

References

Bohuski, T.J., Fairall, A.P., and Weedman, D.W.: 1978, Astrophys. J. 221, 776.
Fairall, A.P.: 1977, Mon. Not. R. astr. Soc. 180, 391.
Fairall, A.P.: 1979, Mon. Not. R. astr. Soc. 188, 349.
Fairall, A.P.: 1980, Mon. Not. R. astr. Soc. 192, 389.
Fairall, A.P.: 1981, Mon. Not. R. astr. Soc. 196, 417.
Fairall, A.P.: 1983a, Mon. Not. R. astr. Soc. 203, 47.

Fairall, A.P., 1983b (to be submitted to *Mon. Not. R. astr. Soc.*)
Fairall, A.P., 1983c, *Nature, 304,* 241.
Fairall, A.P., McHardy, I.M. and Pye, J.P., 1982, *Mon. Not. R. astr. Soc., 198,* 13P, 1971.
Zwicky, F., 1971, "Catalogue of Selected Compact Galaxies and of Post-eruptive Galaxies", published privately.
Zwicky, F., Herzog, E., Wild, P., Karpowicz, M. and Kowal, C.T., 1961-68 "Catalogue of Galaxies and of Clusters of Galaxies", published in 6 volumes by the California Institute of Technology.

OBJECTIVE-PRISM REDSHIFTS OF FAINT GALAXIES

J.A. Cooke, B.D. Kelly*, S.M. Beard, D. Emerson
Department of Astronomy, University of Edinburgh
*Royal Observatory, Edinburgh EH9 3HJ, Scotland

ABSTRACT: A large sample of spectra of faint galaxies has been obtained using COSMOS measurements of UKST objective-prism plates. Computer software has been developed to obtain the radial velocities of large numbers of these galaxies automatically over a magnitude range of about B = 16 to 19. Initial tests have been performed on a sample of about 1400 galaxies from an area of about 5 x 4 degrees square.

INTRODUCTION

Plates taken using the low-dispersion prism (2480 $Å mm^{-1}$ at H γ) on the UK Schmidt telescope (UKST) were measured using the COSMOS measuring machine by the Image and Data Processing Unit (IDPU) in the Royal Observatory Edinburgh (ROE). The plates were exposed to give a sky background density of about 0.9, to obtain a good signal-to-noise ratio in the machine measurements; only plates taken in good seeing ($\sim$1 arcsecond) have spectra of sufficient quality. The authors have written software to extract the spectra from raw COSMOS data.

The spectra from an area of UKST plate UJ4529P covering part of the Indus Supercluster were measured by Beard (1983). From these spectra Beard isolated a large sample of objects (about 2000 in total) which he identified as galaxies from their spectra. This sample has been used as a sample of galaxies with 'known' redshifts (determined interactively from objective prism spectra, not from slit spectra) for the development of algorithms designed to determine galaxy redshifts automatically.

AUTOMATED REDSHIFT MEASUREMENTS

Details of the ideas and techniques behind redshift measurement from objective prism spectra are given elsewhere (Cooke et al 1982). The automated technique basically uses a pattern match between the processed spectrum and a set of processed masks, which consists of a standard galaxy spectrum (Oke and Sandage 1968) reproduced at redshift

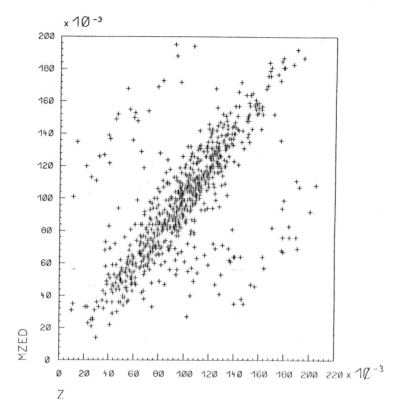

Figure 1. The comparison of automatically derived redshifts (MZED) with redshifts obtained interactively (Z). The range for pattern matching is from redshift 0 to 0.2; a selection criterion is applied which removes faint objects that have been assigned low redshifts by machine (this is justified on number-magnitude grounds). From an initial sample of 1400 objects (a subset in area of Beard's sample), about 800 appear here.

intervals of 0.001 in z. Parameters produced for the match include a signal-to-noise parameter for each spectrum. The system has been applied to Beard's Indus data, and to simulated galaxy spectra. Figure 1 shows how the system can be used to duplicate the manual measurements well.

This approach is useful, in that it replaces a lot of tedious interactive work at a computer terminal. However, the simulations show that, from signal-to-noise considerations, not all the redshifts obtained automatically can be accepted as correct; typically with the signal-to-noise parameter (SNR) defined as the mean continuum level around 4500Å divided by the rms noise per pixel, some objects are assigned incorrect redshifts for SNR = 8 and about half the objects have incorrect redshifts for SNR < 5. In 'real' data the results appear to be somewhat worse; further work is needed to establish detailed numbers.

CURRENT WORK

The problem becomes even harder when a mixed sample of stars and galaxies is the starting point. In a typical sample, the majority of the stars might be expected to have late-type spectra (Reid and Gilmore 1982) and so might be expected to correlate quite well with a late-type galaxy spectrum at zero redshift. This is not the case, and the faint, noisy stellar spectra match the galaxy standard at a range of redshifts. A better solution is needed and two approaches are being investigated: (a) initial star/galaxy separation of the sample (e.g. Parker et al, this colloquium) to enable the stars to be completely removed from the sample; and (b) a more complex pattern match, in which stars of various spectral types (as well as the galaxy) are matched. The software is structured so that it will be possible to use patterns of several spectral types of both galaxies and stars. This method has the advantage of producing a sample independent of star-galaxy separation, and also would produce an initial rough stellar spectral classification.

Astronomical results from this particular sample are presented by Beard et al (1983).

CONCLUSION

The present system can be used to produce redshifts of galaxies from objective prism spectra in a more consistent way than can be done using manual techniques. However for data with poor signal-to-noise it must be emphasised that the redshifts produced are not totally reliable; this applies also to the manual measurements. The velocities obtained can be used as distance indicators and hence used to give information on large-scale structure in the Universe.

ACKNOWLEDGMENTS

We thank UKSTU for plate material and IDPU, ROE for COSMOS measurements. Data processing was performed on the STARLINK system. SMB is supported by an SERC research studentship.

REFERENCES

Beard, S.M., Cooke, J.A., Emerson, D., Kelly, B.D.: 1983, submitted to Mon. Not. R. astr. Soc.
Beard, S.M.: 1983, PhD thesis in preparation, University of Edinburgh
Cooke, J.A., Emerson, D., Beard, S.M. and Kelly, B.D.: 1982, "Workshop on Astronomical Measuring Machines 1982", occasional reports of the Roy. Obs., Edinburgh, 10, pp. 209-218
Reid, N. and Gilmore, G.: 1982, Mon. Not. R. astr. Soc. 201, pp. 73-94
Oke, J.B. and Sandage, A.: 1968, Astrophys. J. 154, pp. 21-32.

OBJECTIVE-PRISM GALAXY REDSHIFTS IN FIELDS AROUND THE SOUTH GALACTIC POLE

Q.A. Parker[1], H.T. MacGillivray[2], R.J. Dodd[3], J.A. Cooke[4], S.M. Beard[4], B.D. Kelly[2] and D. Emerson[4].

1. Dept. of Astronomy, St Andrews University, Fife, Scotland.
2. Royal Observatory, Blackford Hill, Edinburgh, Scotland.
3. Carter Observatory, P.O. Box 2909, Wellington 1, New Zealand.
4. Dept. of Astronomy, Edinburgh University, Edinburgh, Scotland.

ABSTRACT

Measurements made with the COSMOS machine on deep objective-prism photographs taken with the UK 1.2m Schmidt Telescope are being used to obtain approximate redshifts (accurate to ~ 0.01 in z) for large numbers of galaxies in fields near the South Galactic Pole. The data are suitable for investigations of the distribution of galaxies, such as the detection of large-scale density enhancements or voids.

1. INTRODUCTION

Cooke et al (1984, these proceedings) have described a technique for obtaining approximate redshifts for galaxies from low dispersion (resolution of 2480Å/mm at Hγ) objective-prism spectra. The technique relies upon the accurate measurement of the position of the emulsion cut-off (taken as a wavelength standard at $\lambda = 5380$Å for the unfiltered IIIaJ emulsion) and the position of the so-called "4000Å feature" (at $\lambda = 3990$Å) which changes relative to its rest wavelength position with redshift. This 4000Å feature is only present in the spectra of elliptical and early-type spiral (up to type Sb) galaxies. Redshifts can only be reliably obtained for those galaxies with a reasonably well defined 4000Å feature and in a limited magnitude range (16<=B<=19). The advantage of the technique is that when properly applied, the method can yield redshifts for some thousands of galaxies from a single UK Schmidt Telescope (UKST) objective prism plate. Unfortunately, the measurement accuracy ($\Delta z = 0.007$ at $z = 0.02$, $\Delta z = 0.013$ at $z = 0.20$) is insufficient to enable detailed properties of the galaxy distribution to be investigated or cluster velocity dispersions to be determined. However, the results do enable such studies as the detection of large-scale density enhancements and voids, the 3-dimensional distribution of clusters and the verification of Bautz-Morgan (1970) classification. The latter has important consequences for the analysis of cluster redshifts, since many published redshifts rely on measurements of the brightest cluster member only, and may be incorrect in some instances (see e.g. Sarazin et al 1982).

We are currently using the technique in order to obtain redshifts for galaxies and clusters in several fields around the South Galactic Pole (SGP). The aim of this investigation is to provide a systematic survey which can be used for statistical studies of the distribution of galaxies and clusters, and searches for the presence of large-scale agglomerations in the matter distribution. To this end the survey is being carried out in conjunction with the faint galaxy survey from direct photographs (MacGillivray and Dodd 1984, these proceedings).

2. METHODS

At present, galaxy redshifts are obtained using an interactive procedure. This interactive processing is inefficient for the analysis of large areas of plate and is soon to be replaced by a completely automatic technique (Cooke et al 1984, these proceedings). This has meant that we have been restricted in the past to small areas of plates such as individual rich clusters of galaxies and small field areas.

Data are obtained from COSMOS in its mapping mode of measurement (MacGillivray 1981), with a pixel size of 16μm, on UKST objective-prism photographs. Rich clusters are identified from isoplethal maps of the distribution of galaxies on the direct plate. Areas surrounding the clusters in the data are extracted as 512 x 512 element picture arrays which are then processed using techniques which are described in detail by Parker et al (1982). The arrays are displayed on a colour monitor as digitised images (see Figure 1 of Clowes et al 1984, these proceedings), the spectra of galaxies (which are identified on the direct plate) are selected, the 4000Å feature identified with a cursor and the redshift obtained.

3. RESULTS

The southern sky survey field number 349 (with 1950 coordinates 00h 00m, 35°00') is interesting because of the large number of rich clusters it contains. Many of these can clearly be seen in the isoplethal map of the field (figure 1) for galaxies down to B = 21.8. In figure 2 we show the results obtained for the rich cluster to the East of the plate centre. This cluster has been studied by Carter (1980) who obtained a redshift of z = 0.114 for the central cD galaxy. In figure 2 we see that the peak reshift in the direction to the cluster is in the range 0.11-0.12, in good agreement with the redshift from slit spectroscopy. Several other clusters in this field likewise have peak redshifts in this range indicating that

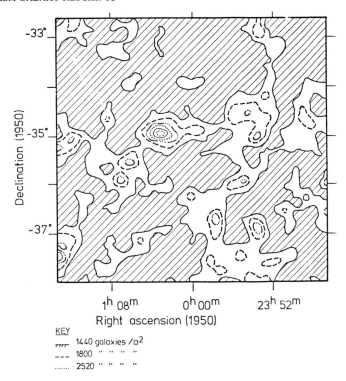

Figure 1 The distribution of galaxies down to B=21.8 in the southern sky survey field 349. Several rich clusters may be seen from the isopleths.

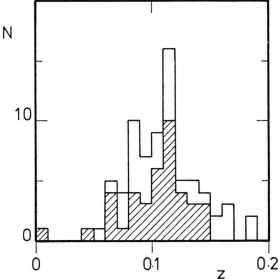

Figure 2 Histogram showing the redshifts for galaxies in the direction to the rich cluster at 00h 03.5m, -35°. The filled histogram indicates galaxies for which a high degree of confidence has been attached to the identification of the 4000Å feature.

they may form part of a large-scale structure extending over much of the field.

REFERENCES

Bautz, L.P. and Morgan, W.W., 1970. Astrophys. J., 162, L149.
Carter, D., 1980. Mon. Not. R. astr. Soc., 190, 307.
Clowes, R.G., 1984, in "Astronomy with Schmidt-type Telescopes", IAU Colloquium No. 78, ed. M. Capaccioli, Asiago, Italy, these proceedings, p. 107.
Cooke, J.A., Kelly, B.D., Beard, S.M. and Emerson, D., 1984, in "Astronomy with Schmidt-type Telescopes", IAU Colloquium No. 78, ed. M. Capaccioli, Asiago, Italy, these proceedings, p. 401.
MacGillivray, H.T., 1981, in Astronomical Photography 1981, eds. J-L Heudier and M.E. Sim, Nice, France, p. 277.
MacGillivray, H.T. and Dodd, R.J., 1984, in "Astronomy with Schmidt-type Telescopes", IAU Colloquium No. 78, ed. M. Capaccioli, Asiago, Italy, these proceedings, p. 125.
Parker, Q.A., MacGillivray, H.T., Dodd, R.J., Cooke, J.A., Beard, S.M., Emerson, D. and Kelly, B.D., 1982, in Proceedings of the Workshop on Astronomical Measuring Machines 1982, eds. R.S. Stobie and B. McInnes, Edinburgh, Scotland, p. 233.
Sarazin, C.L., Rood, H.J. and Struble, M.F., 1982. Astron. Astrophys. 108, L7.

SEARCHING FOR EMISSION-LINE GALAXIES

T. D. Kinman†
Kitt Peak National Observatory*

Survey methods for finding emission-line galaxies are reviewed. Observational selection effects are investigated by comparing different surveys and the limitations of the different techniques are discussed. The advantages of Hα surveys for finding low luminosity galaxies and those with low excitation emission spectra are emphasized.

1. INTRODUCTION

Nearly all the emission lines that we see in the spectra of galaxies are formed in hot interstellar gas. Only a few cases are known where the lines formed in stellar atmospheres are strong enough to be seen in an integrated galaxy spectrum - e.g. He II 4686 from the bright but short-lived WR stars in the galaxy Tololo 3 (Kunth and Sargent 1981). Smith (1981) has recently reviewed the methods used to find broad-lined emission objects; we here concentrate on the techniques for finding the more common and quite diverse narrow-lined types. There are a variety of ways in which the gas can be excited in these galaxies, but the discovery problem in all cases is to recognize the emission spectrum in the presence of an (often strong) absorption spectrum that comes from the stars in the galaxy.

The most important emission lines, for discovery purposes, are [OII]3727, Hβ 4861, [OIII]4959,5007, Hα 6563 and [NII]6548,6584. Early photographic spectroscopy (e.g. Humason, Mayall and Sandage 1956) used blue sensitive plates on which Hβ and the [OIII] lines would only be seen if these lines were strong and of low redshift. Early estimates of the frequency of occurrence of emission lines (Mayall 1939, Humason 1947 given in Table 1) refer to the [OII]3727 line and showed that this line

†Visiting Astronomer, Cerro Tololo Inter-American Observatory which is operated by AURA, Inc., under contract with the National Science Foundation.

*Operated by the Association of Observatories for Research in Astronomy, Inc., under contract with the National Science Foundation.

Table 1. Emission occurrence as function of galaxy type.

Galaxy type	Mayall (1939) n	%	Humason (1947) n	%	H.M.S. (1956)* n	%
E0-7	14	7	77	20	125	11
S0	-	-	48	46	91	30
Sa	10	15	37	78	71	37
Sb	20	52	46	87	93	61
Sc	27	56	25	88	69	64

*[Calculated from Humason, Mayall and Sandage (1956), Table I]

is more frequent in the later Hubble types with their larger ratio of young to old stars. Humason used only spectra that were sufficiently exposed to have shown the [OII] line and used only about half the number (n) of spectra listed in HMS Table 1; he thereby got a significantly higher frequency than is obtained from a simple analysis of the HMS data. Analysis based on a literature search is therefore likely to underestimate emission-line frequencies. Gisler's 1978 literature analysis of 1316 galaxies with velocities less than 15,000 km s^{-1} (obtained with spectral dispersions higher than 500 Å mm^{-1}) is important because it showed the difference in frequency for galaxies in clusters and in the field; for E, E-S0 types he found a frequency of 26% in the field but 0% in clusters, while for Sc and I types he found 88% in the field and 75% in dense clusters.

[OII]3727 is a good indicator of ionized gas in E galaxies because their continua are relatively weak in the ultraviolet. In later types of galaxy, the Hα + [NII] group of lines becomes a rival indicator (see the discussion in Mayall 1956) but historically most surveys for emission-line galaxies have been made in the blue where the photographic plates are fastest. Recently, Keel (1983) searched for Hα + [NII] in the nuclei of an optically complete sample of 93 spirals with $B_T \leq$ 12.0. All showed emission (five were Seyferts); his high spatial resolution (400-800 pc) allowed the [NII]6584 to be detected even when the Hα emission was swamped with the underlying absorption. Besides such compact sources of emission, there are various diffuse components in later type galaxies (Georgelin, Georgelin and Sivan 1979) that can contribute significantly to an integrated spectrum. Emission in elliptical galaxies is also complex. Gisler and Butcher (1980) searched 54 E type galaxies for Hα with the Kitt Peak video camera using interference filters and found evidence for emission in 40 of them. Only 6 of these 40 had emission in unresolved cores; the remainder had emission that was to some extent distributed – in a few cases completely so.

Clearly survey techniques must be tailored to the type of galaxy that interests us; no universal method will work on all types. For some purposes (e.g. identifying very rare galaxies or studying galaxies for evolutionary effects where a significant look-back time is needed), we are forced to work with galaxies that are so distant that little structural information can be obtained about them. More generally, our understanding of survey material is greatly enhanced if structural information is available. Consider therefore, as a working limit, a galaxy with a diameter (at μ_B = 25) of 5 arcsec; for non-Seyfert Markarian galaxies (mean $\mu_B \sim 21$) this corresponds to apparent magnitude B = 17.75. For the best ground-based observations, its image will contain about 50 pixels, while from space (with adequate signal-to-noise) one might achieve 2000. According to Morgan, Kayser and White (1975), the classification of galaxies requires about 100 pixels. If we use Huchra's empirical diameter vs. M_B relation (which assumes H_0 = 50 km s^{-1} Mpc^{-1}), we find the following limiting redshifts to go with this resolution: 44,668 km s^{-1} at M_B = -22, 17,783 km s^{-1} at M_B = -20, 7079 km s^{-1} at M_B = -18, 2818 km s^{-1} at M_B = -16 and 1122 km s^{-1} at M_B = -14. The rare but intrinsically bright active galaxies such as Seyfert type 1, will be somewhat brighter for their diameter than these non-Seyferts. To study these, we need to survey a large volume of space and therefore need a survey with a large redshift range. On the other hand, only a small redshift range will give the spatial resolution needed to survey for the more numerous and diverse types of intrinsically faint galaxies.

2. FINDING EMISSION-LINE GALAXIES BY APPEARANCE

The more active and luminous emission-line galaxies look compact and so compactness has been used as a criterion for finding them. Out of 141 of the bluer Zwicky compact galaxies, Sargent (1970) found 10 Seyferts, 53 with sharp-lined emission, 60 with absorption lines and 3 with continuous spectra. Rodgers, Peterson and Harding (1978) observed compact galaxies from Zwicky's IX list; 24 out of 30 of the bluer galaxies showed emission but only 3 out of 27 of the red ones. Samples of Arakelian's (1975) high surface brightness galaxies (considered to be 50% E - S0 types and 25% Sa types) were observed by Arakelian, Dibai and Esipov (1975) and Doroshenko and Terebizh (1975). They found that more than half showed Balmer emission. Fairall (1983 and references therein) selected high surface brightness galaxies with sharp boundaries (to avoid E types) from the ESO Quick Blue and the SRC IIIa-J surveys; 21% of these galaxies show emission.

Table 2. Surveys of Compact Galaxies

Survey	Percentage with redshifts >10,000 km s^{-1}	
	Emission lines	Absorption lines only
Sargent (1970)	25%	60%
Arakelian et al. (1975)	42%	-
Doroshenko et al. (1975)	2%	-
Fairall (1977-1983)	20%	51%
Rodgers et al. (1978)	28%	55%

Compact galaxies are a heterogeneous group but the bluer ones do contain a fair proportion of emission-line galaxies. Table 2 compares the percentages of the galaxies in these surveys that have redshifts greater than 10,000 km s^{-1} ; there are less than half as many emission-line galaxies in this redshift range as those with absorption spectra. Most of the emission-line galaxies are narrow-lined ones with only moderate luminosities and one cannot expect to select them by appearance at redshifts greater than 10,000 km s^{-1} as we saw above. The case is quite different for the intrinsically bright broad-lined galaxies; 74% of those found in these surveys have redshifts of more than 10,000 km s^{-1} which is very comparable with the percentage of Markarian galaxies in this redshift range that are Seyfert 1 galaxies and which are purely color-selected.

It is believed that interacting galaxies can induce star-formation and hence be a source of emission-lines from the gas near hot, young stars. It is therefore very interesting that in a spectroscopic survey of 43 galaxies from Verontsov-Velyaminov's 2nd Atlas of interacting galaxies (1977), Barbieri et al. (1979) found 48% definitely and 24% possibly showed emission. My own experience of emission-line galaxies is that amongst the narrow-lined less luminous types, they more often than not have rather irregular images - presumably because much of their light comes from either (a) unresolved multiple systems or (b) young and therefore not dynamically relaxed stellar components.

3. FINDING EMISSION-LINE GALAXIES BY COLOR

The correlation of uv-excess with the presence of emission lines is well known (c.f. The discovery of a planetary nebula in M15 on an ultraviolet plate by Pease 1928). Haro (1956) showed that emission-line galaxies could be found by comparing blue, visual and uv images on a single Schmidt plate and the technique has been used in a variety of ways since (see Kinman and Hintzen 1981 for references). An example from a current survey by Bushouse and Gallagher is shown in Fig. 2. Galaxies with redshifts up to the range 30,000 - 40,000 km s^{-1} have been found this way. A disadvantage is the limited accuracy that can be obtained in estimating colors by visual inspection; also the poorer uv images of single-element corrector Schmidt telescopes can be a problem with fainter images (Savage 1983). A large current survey of this type is that using U, G and R colors with 103a-E emulsion on the Kiso 1.05-m Schmidt (Takase 1980, 1983) where, so far, over 3000 uv-excess galaxies have been found in 1625 square degrees to a limiting photographic magnitude of 17-18. Spectra of 30 of these showed that 80% were emission-line galaxies.

Color can also be judged from low dispersion objective-prism images. The largest survey of this kind is Markarian's using a 1.5° prism on the 1-m Byurakan Schmidt (IIa-F emulsion, 2500 Å mm^{-1} at Hβ); some 1399 galaxies with strong uv continua had been found up to their most recent paper (Markarian, Lipovetskii and Stepanian, 1979). A similar program with the same equipment but a wider range of emulsions is reported by Kazarian (1979). Some 11% of Markarian galaxies are Seyferts according to Huchra and Sargent (1976). Huchra (1977) has studied some 200 non-Seyfert Markarian galaxies in some detail: only 4% of these have redshifts that exceed 10,000 km s^{-1}. A subset of about 100 non-Seyfert Markarian galaxies with bright nuclei (so-called "starburst" nuclei) have recently been studied by Balzano (1983) and these are also relatively nearby objects: 66% have redshifts of less than 5,000 km s^{-1} while only 9% have redshifts greater than 10,000 km s^{-1}. The majority have rather low-ionization emission spectra that are characteristic of the photoionization of gas by the radiation from hot stars. These various color surveys differ both by their color criterion and by their limiting magnitude. Thus Haro galaxies correspond roughly to that two thirds of the Markarian galaxies that has the greatest uv-excess. The Kiso survey goes deeper than the Markarian survey (inter alia) and discovers about ten times more galaxies per unit area of sky.

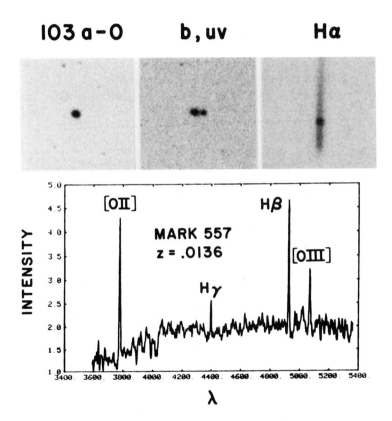

Fig. 1. Markarian 557: (top left) 10-min. unfiltered 103a-O, Palomar 1.2-m Schmidt. (top center) blue-uv, Bolton and Wall (1970). (top right) 180-min. Hα, 0.6-m CTIO Curtis Schmidt (below) 16-min blue scan, IIDS, Kitt Peak 2.1-m telescope.

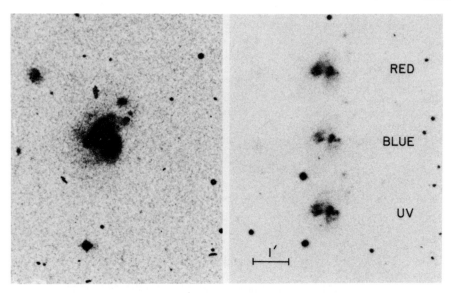

Fig. 2. Arp 299: (left) Palomar Blue Sky Survey to same scale as (right) multiple exp., red (30-min., RG610), blue (30-min., Wr47A), uv (60-min., UG2), IIIa-F, 0.6-m Burrell Schmidt.

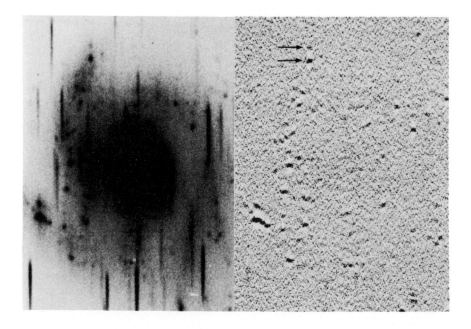

Fig. 3. M 101: (left) 180-min. Hα exp. with 0.6-m Burrell Schmidt. (right) same field after subtraction of field smoothed along dispersion so that only the emission lines remain. The arrows show Hα and [SII] for one H II region.

Table 3. Properties of Galaxies illustrated in Fig. 4.

Object redshift[a]	B B-V U-B	λ4000 break[b]	Mich. color[c]	[OIII]+Hβ flux[d] EW[e] line str.[f]		[OII] flux[d] EW[e] line str.[f]	
ZWG 384.055 +8,295	15.37 +0.77 +0.34	0.8	–	0.1	2	0.1	4
MICH III-295 +5,476	16.99 +0.41 -0.23	1.0	2:	3.2 M,m	332	0.8 M,w-	75
MICH III-296 +5,400	16.72 +0.52 -0.09	0.9	2	3.7 W,w	78	2.2 W,w-	57
HARO 0049.5+01 +13,010	17.68 +0.46 -0.18	0.8	–	0.4	23	0.4	+32
				confused with star to South on thin prism plate			
MICH III-283 +4,630	17.29 +0.53 -0.23	1.0	2	3.9 M,m	219	1.8 M,w	115
MICH III-286 +1,670	15.31 +0.61 -0.12	0.9	1-2	1.6 M,w	20	1.8 W,w-	38

(a) galactocentric redshift in km s^{-1}.
(b) Ratio: $F\lambda(3800-4000)/F\lambda(4000-4200)$ at rest wavelength.
(c) Color from Michigan list III: 1 = blue, 3 = red.
(d) Flux in units of 10^{-14} erg cm^{-2} s^{-1}.
(e) Equivalent width in Å.
(f) Line strength on thin prism plate: M = medium, W = weak from Michigan list III; m = medium, w = weak, w- = marginal on author's thin prism plate (90-min., IIIa-J), CTIO Curtis Schmidt.

All plates in Fig. 4 are 3' square, North to top and East to left. The blue-uv exposures were 8 min. with GG 13 for blue and 60 min. with UG 1 for the uv image which is 12 arc sec to the West on a Palomar 1.2-m Schmidt plate by Bolton and Wall (1970): the images would be equally strong for U-B = -0.4, The Hα exposures were 180-min on IIIa-F with an RG2 filter and the 10° prism combination of the CTIO Curtis Schmidt.

4. FINDING EMISSION-LINE GALAXIES DIRECTLY BY THEIR LINES

The most unambiguous way to discover emission-line galaxies is directly from objective-prism spectra of adequate resolution. A curious reluctance to do this until quite recently seems to have stemmed from unduly conservative estimates based on experience with absorption-line spectra (Minkowski 1972). It became apparent in the early 1970's, that about 20% of Markarian's blue galaxies showed emission lines on spectra with a dispersion as low as 2500 Å mm^{-1} at Hβ. This induced Smith (1975)

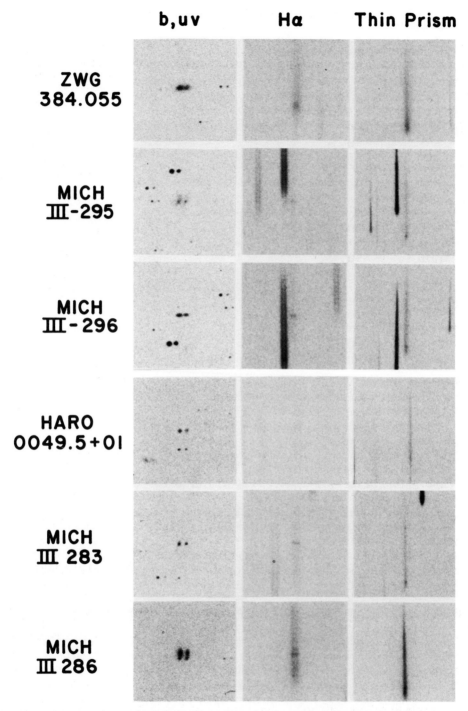

Fig. 4. Comparison of three techniques for finding emission-line galaxies. For details, see text and Table 3.

Table 4. Distribution of [OIII]/Hβ ratio for diferent surveys.

	No Hβ	_____Range in [OIII]/Hβ_____				Survey
		0 - 3	3 - 5	5 - 7	>7	
Number per 100 sq. degrees	10.7	31.2	6.5	1.6	1.6	Kinman (Hα)
	0	4.4	6.4	2.0	3.2	Wasilewski
Percentage	21	60	13	3	3	Kinman (Hα)
	24	46	18	9	2	Balzano
	16	42	33	7	2	Huchra
	18	20	20	20	22	Mich.-Tololo
	0	27	40	13	19	Wasilewski

to start a survey to detect emission-line galaxies and QSO by their lines at a somewhat higher spectral resolution (1740 Å mm^{-1} at Hβ on IIIa-J) using the uv-transmitting 1.8 "thin" prism (Blanco 1974) on the 0.6-m CTIO Curtis Schmidt. He found that both [OII] and the [OIII] - Hβ group of lines could be detected in a 90-minute exposure with a blue continuum magnitude of 17.5-18.0. Several hundred QSO and emission-line galaxies have now been detected in the South galactic cap by this method in the Michigan-Tololo survey (MacAlpine and Williams 1981 and references therein). A similar survey is being carried out in the North using the Burrell Schmidt (Pesch and Sanduleak 1983); their discovery rate of blue and/or emission-line candidates is about 0.5 per sq. degree or about 5 times that of the Markarian survey. Lewis (1983) has taken spectra of 83 of the Michigan-Tololo (MT) galaxies; 69 (83%) of them showed emission lines and of these 35% had redshifts exceeding 10,000 km s^{-1}. A small survey at comparable dispersion (2400 Å mm^{-1} at Hγ) has been made on IIIa-F emulsion with the 1.24-m UK Schmidt by Kunth, Sargent and Kowal (1981). They observed the same field containing Zwicky galaxies that was observed by Rodgers et al. (1978) and found some 23 emission-line galaxies (30% with redshifts greater than 10,000 km s^{-1}) but only two of these were in common with those found by Rodgers et al.

At higher dispersions, the emission-line visibility improves as the continuum is spread out (McCarthy and Treanor 1970, Kinman 1979) although the limiting magnitude of the continuum brightens. Wasilewski (1983) has undertaken such a survey of 825 sq. degrees of the North galactic cap. He used the 4° prism on the Case Burrell Schmidt (400 Å mm^{-1} at Hβ), and with 60 minute exposures on IIIa-J emulsion, discovered 96 emission-line galaxy candidates. His spectroscopic analysis gave redshifts for two thirds of these; only 10% have redshifts greater than 10,000 km s^{-1} but 10-15% are Seyferts. Only about 20% of the Markarian galaxies in his field were rediscovered as emission-line galaxies in this survey, however, possibly because of the low ultraviolet transmission of the 4° prism.

Finally, Kinman (1979, 1983) has undertaken a higher dispersion survey for field emission-line galaxies in Hα. The aim is to achieve a high completeness for nearby (particularly dwarf) galaxies at the expense of detecting the more distant and luminous ones. Both Curtis and Burrell 0.6-m Schmidts have been used with a 10° prism or prism-combination using IIIa-F emulsion and an RG2 filter to limit the waveband to 6400-6850 Å (400 Å mm^{-1} at $H\alpha$). Exposures are limited to 180-minutes, not by sky brightness, but by telescope stability and field rotation effects that come from guiding some 6° from the field center. Even though the sensitivity cut-off of the IIIa-F emulsion excludes galaxies with redshifts greater than 12,000 km s^{-1}, over 0.5 galaxies per sq. degree are discovered with this technique and a larger number of the redder Seyfert 2 galaxies are found per unit area of sky than in the Markarian survey. A similar $H\alpha$ survey is currently being used by Moss and Whittle (1983) to study the nearer rich Abell clusters of galaxies.

5. COMPARISONS OF DIFFERENT SURVEYS

The uneven distribution of galaxies on scales of 10 - 100 Mpc limits the accuracy with which one can compare surveys made in different parts of the sky. Nevertheless it is clear that the compact galaxy criterion, the Michigan-Tololo technique and the blue-uv Haro method pick up the largest percentages ($\gtrsim 25\%$) of distant (redshift >10,000 km s^{-1}) galaxies. The two high dispersion surveys were clearly poor in this respect and the Markarian Survey is intermediate. The overall density (candidates per sq. degree) varies from 0.1 for the Markarian Survey to ten times greater for the Kiso survey. It is clear that besides the redshift cutoff and the limiting magnitude, other factors are involved. Thus the Rodgers et al. and Kunth et al. surveys of the same field have about the same overall density and fraction of distant galaxies, but have only two galaxies in common. As another example, a 1.4 x 0.9 degree field near the Abell clusters 2197-2199 contains 19 Kiso uv-excess galaxies and 14 $H\alpha$ galaxies but only 11 of these are common to both surveys. All surveys have completeness problems in the sense that a repeat plate or a repeat inspection of the same plate does not give identical results, but little mention is made of this effect. An exception is the Palomar-Green survey for bright quasars (Green 1976); some 47% of this blue-uv survey of 10,714 sq. degrees of sky with the 0.45-m Palomar Schmidt was observed twice. The two side-by-side images on II-a-O film were automatically compared with a PDS scanner: rms errors for B and for U-B colors were estimated to be ±0.27 and ± 0.24 mag. respectively. Comparison of the results for the two independent exposures of the same field showed that on the average 61% of the uv-excess objects were detected on each exposure (Green, 1983). This may be compared with Plaut's (1966) estimate that visual comparison of two 103a-O plates taken with the 1.2-m Palomar Schmidt leads to a 40-50% chance of detecting a 1 mag. difference at 15th magnitude and only about half this chance at 18th magnitude. Clearly uncorrected results from single-pass surveys are subject to sizeable errors and the actual detection limits in any one field are likely to vary from the mean. This is illustrated in Fig. 4 and Table 3 where it is seen that there is some variation in detail in the various estimates of line strengths given in the M-T survey and from a plate taken in good seeing by the author and the actual measured fluxes and equivalent widths for these objects. It is also apparent that the line strengths are a function of the equivalent widths as well as the line fluxes and that the visibility of lines with equivalent widths less than ~30Å is rather marginal. Fig. 4 illustrates both the generalities and singularities that one finds by comparing different survey data. ZWG 384.055 was understandably only found on the Hα survey because its blue emission lines are quite weak and its U-B excess is confined to its core. Rather surprisingly Mich III-295 was missed by Haro because it has a bigger uv-excess than Mich III-296 which he did detect only a few arc minutes away. Haro 0049.5+01 was not detected in Hα because its redshift is too large and not by the M-T survey because of confusion with the star (PHL 6780) to the south and because its lines are probably too weak for the M-T technique anyway.

Table 4 gives the distribution of the excitation parameter (given by the [OIII]4959,5007/Hβ ratio) in different surveys. This shows that the two high dispersion surveys both contain about the same number of the higher excitation ([OIII]/Hβ >3) galaxies but that the blue Wasilewski survey picks out far fewer low excitation objects. The Markarian survey (exemplified by Balzano's "star-burst" subset and Huchra's non-Seyfert subset) gives a similar distribution to that of the nearby galaxies of the Hα survey, but the M-T survey has a higher percentage of high excitation galaxies. Presumably this is because Hα/Hβ≳3, and so the blue emission lines are only more conspicuous than Hα for the high excitation objects. Lewis (1983) has, however, discovered some quite low excitation [OIII]/Hβ~0.3 galaxies in the M-T survey with large Hα/Hβ ratios and with [NII]6584 comparable to Hα; he calls these galaxies weak oxygen red (WOR) types. Mark 557 (Fig. 1) may belong to this class. Although reddish [(B-V)=+0.7 and (U-B)=+0.07 in a 24 arcsec aperture], it has a bluer nucleus with [NII]6584/Hα~0.5 and Hα +[NII]/Hβ~13; its [OI]6300 is weak so it is not a LINER type (Heckman 1980). Its strongest blue emission line ([OII]3727) only has an EW of 20Å, so it would be a difficult object for a blue emission-line survey even though it is quite bright (V=14.4). It is hoped that more of these objects will be found in the Hα survey. The galaxy with the lowest known oxygen abundance (I Zw 18) has an [OIII]/Hβ ratio of ~3, galaxies with a much lower oxygen abundance than this could appear to have a lower excitation ratio and have stronger Hα than their blue emission; they may be expected to have a strong uv-excess.

Finally, it is of interest to note the ratio of galaxies with emission to the total number in the field. This data is not often given but is an aid to interpreting the quality of a survey. Takase (1980) finds that the ratio of Kiso uv-excess galaxies to all galaxies is 0.25 in the field but is smaller in clusters; this agrees with Gisler (1978). Table 5 shows the fraction of Zwicky galaxies that have emission in a small sample field of the Hα survey. Most of the nearby and brighter ones show emission because their individual H II regions are resolved and bright enough to be detected by the survey (flux in Hα $\gtrsim$ 2 x 10^{-14} erg cm^{-2} s^{-1}). For fainter galaxies, the number drops

Table 5: Emission in a sample of Zwicky Galaxies (ZWG)*

Range in m_z :	12.9	13.0-13.9	14.0-14.9	15.0-15.6	15.7
No. of ZWG :	6	8	22	44	30
No. with emission :	5	4	8	7	2

*Taken from 30 sq. deg. field near NGC 1023

rapidly and for the Zwicky galaxies as a whole, the fraction is ~24% - very similar to that found for the Kiso galaxies.

6. FURTHER COMMENTS

A variety of techniques for finding emission-line galaxies is needed because each can be used to check on the observational selection effects that are present in the others. Some surveys are more effective than others for finding certain classes of galaxy. Thus the M-T and blue compact galaxy surveys are best for finding broad-lined luminous galaxies while the Hα survey is probably the best for finding nearby low-ionization galaxies. New techniques (e.g. the detection of [OII]3727 on uv-filtered high dispersion objective-prism spectra by Moss 1983) are welcome. The [SIII]9069,9531 lines need attention. They can be stronger than Hα if the visual extinction exceeds 3 magnitudes (Hippelein and Munch 1981) and so could perhaps be used to look for Seyfert 1 galaxies (particularly when edge-on).

Most surveys are based on the visual inspection of plates, but more consistent results should be obtainable by automatic means. Particular care is needed however to choose a simple search algorithm. Likely possibilities are uv-excess from a two-color plate or the λ4000 break in an objective prism spectrum or a single strong line such as Hα. With the microphotometry and digitization of plates, one can use simple ways of enhancing the visibility of emission lines as shown in Fig. 3. The field is smoothed along the direction of its dispersion and then this is subtracted from the original field. The extended background is scarcely changed by the smoothing and so this is removed and the emission-lines (which are sharp) remain.

Surveys must be related to each other. Thus work on nearby bright galaxies (Keel 1983, Kennicutt and Kent 1983) needs to be quantitatively related to the spectroscopy of wide-field surveys. At the other extreme, work on distant galaxies (which usually refers to the entire image of the galaxy) has to be compared with work on nearby galaxies at comparable spatial resolution. There is a steep rise in counts of faint galaxies in the j passband at j ~20. Hamilton (1982, 1983) has been trying to explore the nature of these faint galaxies by working on a somewhat brighter sample of uv-excess galaxies. He finds a redshift distribution that peaks at 10,500 km s^{-1} but with a long tail to higher redshifts. Clearly many are quite nearby (and his data agree with an extrapolation of the number found in the Markarian survey), but he only found one Seyfert galaxy out of a sample of sixty. This is an important area of work that will be helped when we know more about the K-terms for these galaxies.

We certainly need more surveys, but there is perhaps an even more pressing need for the spectroscopic and photometric follow-up programs that give value to a survey. Less than half Markarian's galaxies have been observed spectroscopically and still fewer spectrophotometrically so that accurate line ratios are known. Generally this spectrophotometry is not adequate to give information about important faint lines such as [OIII]4386 or [OI]6300. Complexities such as the spectral differences between HII regions and inter HII regions (Hunter 1983) are only just becoming appreciated. This spectroscopic work is essential if the older surveys are to be properly exploited and newer ones appropriately planned. Telescope time is scarce and we should look to ways in which the published results of a wide-field survey can be made attractive to spectroscopists and photometrists. At least part of any survey should be more than one-pass so that completeness estimates can be realistic. Finally one can never emphasize too much the need for good positions (and if possible offsets) and good finding charts.

I acknowledge with thanks the use of the Curtis Schmidt telescope of the University of Michigan at CTIO and the Burrell Schmidt telescope of the Case Western Reserve University at KPNO. I am also grateful to a number of colleagues for discussions and for making available their results prior to publication: K. Davidson, J. Gallagher, R. Green, D. Hamilton, D. Hunter, W. Keel, D. Lewis, G. MacAlpine, C. Moss, B. Takase, A. Wasilewski and M. Whittle. Thanks are also due to J. G. Bolton for the use of his Palomar Schmidt plates and to C. T. Mahaffey for taking plates with the Burrell Schmidt.

REFERENCES

Arakelian, M. A.: 1975, *Byurakan Obs. Comm.* No. 47, p.3.
Arakelian, M. A., Dibai, E. A. and Esipov, V. F.: 1975, *Astrofizika* 11, p.377.
Balzano, V. A.: 1983, *Astrophys. J.* 268, p.602.
Barbieri, C., Casini, C., Heidmann, J., di Serego, S., and Zambon, M.: 1979, *Astron. Astrophys. Suppl. Ser.* 37, p.559.
Blanco, V. M.: 1974, *Publ. Astron. Soc. Pacific* 86, p.841.
Bolton, J. G. and Wall, J. V.: 1970, *Australian Jour. Physics* 23, p.789.
Bushouse, H. and Gallagher, J.: 1983, private communication.
Doroshenko, V. T. and Terebizh, V. Yu.: 1975, *Astrofizika* 11, p.631.
Fairall, A. P.: 1983, *Monthly Notices Roy. Astron. Soc.* 203, p.47.
Georgelin, Y. M., Georgelin, Y. P. and Sivan, J.-P.: 1979, in W. Burton (ed). Proc. IAU Sym. 84, *Large-Scale Characteristics of the Galaxy* Reidel, Dordrecht, p.65.
Gisler, G. R.: 1978, *Monthly Notices Roy. Astron. Soc.* 183, p.633.
Gisler, G. R. and Butcher, H. R.: 1980, *Bull. Am. Astron. Soc.* 12, p.835
Green, R.: 1976, *Pub. Astron. Soc. Pacific* 88, p.665.
Green, R.: 1983, private communication.

Hamilton, D.: 1982, *Pub. Astron. Soc. Pacific* 94, p.754.
Hamilton, D.: 1983, private communication.
Haro, G.: 1956, *Bol. Obs. Tonantzintla Y Tacubaya* 2, p.8.
Heckman, T. M.: 1980, *Astron. Astrophys.* 87, p.152.
Hippelein, H. and Munch, G.: 1981, *Astron. Astrophys.* 95, p.100.
Huchra, J. P.: 1977, *Astrophys. J. Suppl.* 35, p.171.
Humason, M. L.: 1947, *Pub. Astron. Soc. Pacific* 59, p.180.
Humason, M. L., Mayall, N. U. and Sandage, A. R.: 1956, *Astron. J.* p.97.
Hunter, D.: 1983, preprint.
Kazarian, M. A.: 1979, *Astrofizika* 15, p.5.
Keel, W. C.: 1983, *Astrophys. J. Suppl.* 52, p.229.
Kennicutt, R. C. and Kent, S. M.: 1983, *Astron. J.* 88, p.1094.
Kinman, T. D.: 1979, *Ricerche Astr.* 9, p.151.
Kinman, T. D.: 1983, *Monthly Notices Roy. Astron. Soc.* 202, p.53.
Kinman, T. D. and Hintzen, P.: 1981, *Publ. Astron. Soc. Pacific* 93, p.405.
Kunth, D. and Sargent, W. L. W.: 1981, *Astron. Astrophys* 101, p.L5.
Kunth, D., Sargent, W. L. W. and Kowal, C.: 1981, *Astron. Astrophys. Suppl.* 44, p.229.
Lewis, D. W.: 1983, Ph. D. Thesis, Univ. of Michigan.
MacAlpine, G. M. and Williams, G. A.: 1981, *Astrophys. J. Suppl.* 45, p.113.
Markarian, B. E., Lipovetskii, V. A. and Stepanian, D. A.: *Astrofizika* 15, p.549.
Mayall, N. U.: 1939, *Lick Obs. Bull* 19, p.33.
Mayall, N. U.: 1958, in N. Roman (ed.) Proc. IAU Sym 5, *Comparison of the Large Scale Structure of the Galactic System with that of other Stellar Systems* Cambridge University Press, Cambridge, p.23.
McCarthy, M. F. and Treanor, P. J.: 1970, *Observatory* 90, p.108.
Minkowski, R.: 1972, in U. Haug (ed.) *Conference on the Role of Schmidt Telescopes in Astronomy*, Hamburger Sternewarte, Bergedorf p.8.
Morgan, W. W., Kayser, S. and White, R. A.: 1975, *Astrophys. J.* 199, p.545.
Moss, C.: 1983, private communication.
Moss, C. and Whittle, M.: 1983, private communication.
Pease, F. G.: 1928, *Pub. Astron. Soc. Pacific* 40, p.342.
Pesch, P. and Sanduleak, N.: 1983, *Astrophys. J. Suppl. Ser.* 51, p.171.
Plaut, L.: 1966, *Bull. Astron. Inst. Netherlands Suppl.* 1 p.105.
Rodgers, A. W., Peterson, B. A. and Harding, P.: 1978, *Astrophys. J.* 225, p.768.
Sargent, W. L. W.: 1970, *Astrophys. J.* 160, p.405.
Savage, A.: 1983, *Astron. Astrophys.* 123, p.353.
Smith, M. G.: 1975, *Astrophys. J.* 202, p.591.
Smith, M. G.: 1981, in F. Kahn (ed.) *Investigating the Universe* Reidel, Dordrecht, p.151.
Takase, B.: 1980, *Pub. Astron. Soc. Japan* 32, p.605.
Takase, B.: 1983, private communication.
Verontsov-Velyaminov, B. A.: 1977, *Astron. Astrophys. Suppl.* 28, p.1.
Wasilweski, A. J.: 1983, *Astrophys. J.* (in press).

REDUCTION OF SLITLESS SPECTRA –
THE DETECTION OF FAINT EMISSION LINES

H.-M. Adorf, H.-J. Röser
Max-Planck-Institut für Astronomie, Heidelberg

ABSTRACT

We have developed a spectra reduction system producing relative spectrophotometry directly from prism as well as from grating spectra plates. Lines and other spectral features are detected, and their essential parameters are measured automatically. A deep quasar survey using this technique is now underway.

1. MOTIVATION AND AIMS

This work was motivated by the questions: "Why are so few quasars known with redshifts greater than 3.5? Are they overlooked by the present survey techniques?" The usual spectral survey method has been "to look at the plates" in order to detect obvious emission lines. Although the human eye definitely is an excellent "image processor", it does not work quantitatively (and fails on low contrast images). It is therefore difficult to attack cosmological problems such as the determination of the quasar space density. These problems require a well defined sample of quasars, crucially depending on the survey's completeness.

We decided to investigate the advantages of quantitative survey work, which should result in a homogeneously selected sample with a definite completeness limit and should make different quantitative surveys comparable with each other. Spectrophotometry from the survey plate is also required, because follow-up observations with big telescopes for large numbers of faint objects would cost too much time to be feasible.

An ideal reduction procedure should maximize both completeness of the sample and the reliability of the findings. It should (1) certainly be automated, (2) be "self-adaptive", i. e., work effectively very close to the noise, and (3) quickly and reliably single out the spectra with the most "unusual" features. The data together with the procedures used should define what is unusual. Specifically, the procedure should determine the limiting equivalent width of a line as a function of the wavelength, the continuum magnitude, and the line width.

2. OBSERVATIONS AND DATA MATERIAL

The search for high-redshift quasars requires deep plates in the red. In order to test our spectra reduction system (SRS) we used a blue plate instead, which contains several known quasars and quasar candidates. The plate was taken at the 80/120/240 cm Schmidt telescope on Calar Alto (Spain) in 1981, and kindly loaned to us by Dr. Haug (Hamburg), who is pursuing his own visual survey for emission line objects. The plate measures 24 cm x 24 cm and covers a projected area of 5.7 deg x 5.7 deg. The unwidened spectra, dispersed by the "thin" 1.7 deg objective prism, were recorded on a baked Kodak IIIa-J emulsion. Between 350 and 520 nm the reciprocal linear dispersion rises from 75 to 245 nm/mm, with 139 nm/mm at H_γ. Thus, the resolving power under 2" seeing declines from 200 down to 90, with 135 at H_γ. In 75 minutes a sky background of about 22.7 mag/arcsec2 produces a plate density of 1, and a background signal-to-noise ratio of 20 in an area of 1 arcsec2.

3. FIRST PHASE IN REDUCTION: FROM RAW DATA TO SPECTRAL INTENSITIES

In establishing the limits of a survey technique we are concerned with the properties of a whole sample of spectra. In order to avoid pre-selecting this set, we scanned sub-areas (or frames) of the plate instead of single spectra using our PDS-microdensitometer with an aperture of 10 μm x 10 μm and a step size of 5 μm. Thus one frame, measuring 2000 x 2000 pixels, covers less than 2 permille of the total Schmidt plate area, or roughly 1/20 deg2. Presently, each frame is then graphically displayed, and "usable" spectra, which surpass a 2 δ (background) threshold above the local background, are interactively selected. A frame contains 30 to 40 of such usable spectra. The limiting brightness of the plate exceeds 20 mag.

All usable spectra are then reduced with procedures comparable to those used by others (Clowes et al. 1980, Christian 1982, Vaucher et al. 1982), which result in a "one-dimensional, background-subtracted, noise-suppressed, wavelength-calibrated spectral intensity trace" along each of the usable two-dimensional spectra. As an example, Figure 1 shows the intensity trace of the quasar PHL 938, which one frame was centered on. Details concerning the SRS can be found elsewhere (Adorf 1984). Note that in the present work we used a constant contrast γ = 3.84 for density to intensity conversion, and did not correct for the varying spectral sensitivity of the emulsion. Also, the wavelength scale is preliminary only and rather uncertain in the green.

4. SECOND PHASE IN REDUCTION: THE DETECTION OF LINES AND OTHER FEATURES

Since we are trying to detect quasars with arbitrary redshifts by the presence of emission lines, we must be prepared to detect a line of unknown (but presumably broad) shape, of unknown intensity, at an unknown position on an unknown (but presumably flat), noisy continuum. In order to detect a line, measure its position and derive a confidence parameter we cross-correlate a given spectrum with a set of (Gaussian) template profiles differing only in their full width at half maximum (FWHM). The correlation peak determines the detected line's FWHM and also its similarity to the template. (The detection of features other than lines, such as continuum breaks, proceeds similarly, and only requires a different set of template profiles.) Next the spectrum is convolved with an appropriate linear filter to measure the line flux or its equivalent width.

Distinguishing noise bumps from true lines remains a problem: Ideally, the frequency distribution of a measured quantity, such as the "similarity" or the equivalent width, in the whole sample of spectra, should tell us which lines are probably caused by statistical fluctuations, and how to choose the right detection threshold. As an example, Figure 2 shows a scatter diagram of the measured line equivalent widths versus the J-magnitude of the object to which it belongs, for a preselected line FWHM of 10 nm. From the observed frequency distribution of equivalent widths we read a current lower limit of around 12 nm for the detection of such a line in a spectrum of an object brighter than 18th mag.

The ability of our SRS to select a complete sample of spectra has been tested against three colleagues who, using a microscope, visually examined 200 spectra within a test area on a different, red plate. Their task was to search for emission or absorption lines, and to assign a "degree of certainty" to their findings. The visual surveyors seemed to differentiate best on moderately dense spectra, but worked inefficiently at the faint and the bright ends, as has also been reported by others (Haug 1983). At the important faint end, all of the visually detected emission-line-candidate spectra were also found by the SRS (which found 50 % more spectra in all). The ability of the reduction system to reject artifacts with a definite signature (e. g. dust and spectrum overlaps), however, remains to be improved.

5. CONCLUSION

We have developed an automated spectra reduction system (SRS), which produces spectrophotometry directly from slitless spectra plates. The SRS detects and measures spectral features such as lines, and has proven not only to be complete in its findings but also to go deeper than visual surveyors. The ability of the SRS to discriminate against dust and other faults, however, remains to be improved. The SRS is currently used for a quantitative deep spectral quasar-survey on plates with photographically recorded prism spectra, and on films with electronographically recorded grating spectra. Our SRS is general enough so that other applications could be considered as well.

REFERENCES

Adorf, H.-M. 1984: Ph. D. thesis (in preparation)
Christian, C. A. 1982: Astrophys. J. Suppl. Ser. 49, pp. 555 - 592
Clowes, R. G., Emerson, D., Smith, M. G., Wallace, P. T., Cannon, R. D., Savage, A., Boksenberg, A. 1980: Mon. Not. R. Astr. Soc. 193, pp. 415 - 426
Haug, U. 1983: private communication
Vaucher, B., Kreidl, T. J., Thomas, N. G., Hoag, A. A. 1982: Astrophys. J. 261, pp. 18 - 24

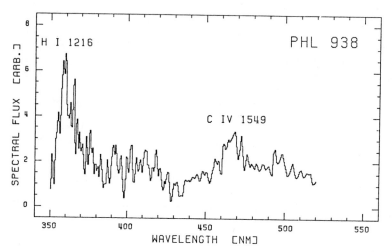

Figure 1: Relative spectral flux of the quasar PHL 938 (B = 17.49, z = 1.955) as derived by the spectral reduction system.

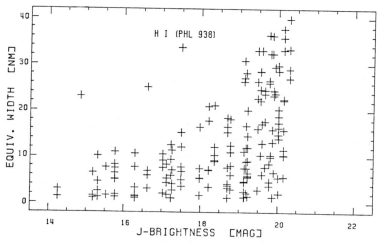

Figure 2: Scatter diagram of measured equivalent widths versus the J-magnitude of the object to which the line (candidate) belongs.

UV GALAXIES AND SUPERASSOCIATIONS

E. Ye. Khachikian
Byurakan Astrophysical Observatory
Yerevan State University
Armenia, USSR

1. INTRODUCTION

I will talk about the UV galaxies discovered on plates of the 1-m Schmidt telescope of the Byurakan Observatory by Markarian and his collaborators and by Kazarian. Up to now the number of UV galaxies is more than 2000.

It is well known that more than 85% of UV galaxies show emission spectra. Among them there have been discovered QSO's, Seyfert galaxies and galaxies with active nuclei with narrow emission lines. It turns out also that some UV galaxies show spectra typical of superassociations (SA). Some of them really turn out to be SA, connected with nearby galaxies, for example, Markarian 94 (Arp & Khachikian 1974), Markarian 5, 59, 71, 256 (Khachikian & Sahakian 1975). Altogether, in the first six lists of Markarian it has been found about 40 of these types of objects. Contrary to QSO's, Seyferts and narrow emission line galaxies, which have starlike nuclei, the central regions of some other UV galaxies entirely consist of several SA: Markarian 7 (Casini, Heidmann and Tarenghi 1979), Markarian 8 (Khachikian 1972), Markarian 325 (Coupinot, Hecquet and Heidmann 1982), Kazarian 5 (Kazarian & Khachikian 1977). In many cases UV galaxies themselves contain one or more SA: Markarian 12, 38, 307, 848, 984 (Khachikian, Petrossian & Sahakian 1983a). There are case when one or both components of double nucleus UV galaxies have the characteristics of SA: Markarian 104, 306, 111, 710, 739 (Khachikian, Petrossian & Sahakian 1979, 1980a,b; Khachikian, Korovyakovskiy, Petrossian & Sahakian 1981). It is necessary to note just one more type of SA often called 'isolated giant HII regions': Markarian 116 = I Zw 18 (Searle & Sargent 1972). Apparently, among UV galaxies there are quite a number of isolated SA. Actually, they are a particular type of compact galaxies which mainly consist of hot stars and diffuse matter.

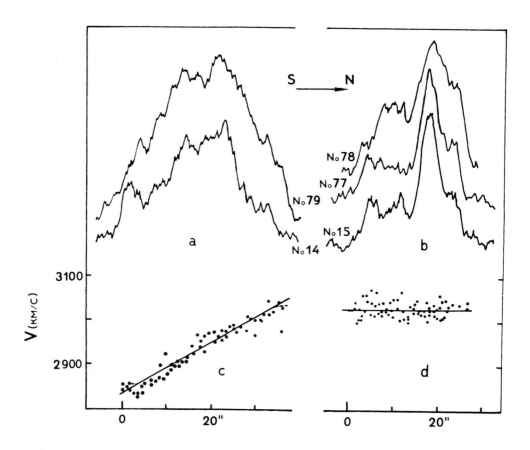

Figure 1. <u>Upper</u>: tracings of Hα along the slit (a, western side, b, eastside). <u>Lower</u>: rotation curves for the (c) western and (b) eastern sides.

Summarizing, we can conclude that among UV galaxies there are objects which are anyhow connected with SA, namely:

1. SA as a part of the nearby galaxy;
2. Galaxies whose central parts consist of several SA;
3. Galaxies containing one or more SA;
4. Double nucleus galaxies in which one or both nuclei resemble SA;
5. Isolated SA.

From the point of view of spectroscopy and photometry, there are no principal differences among these different types of SA. They differ nevertheless by sizes, luminosities and sometime by chemical composition (Petrossian 1983).

I will present some results of: 2) spectroscopy of Markarian 7 which falls under group 2, and 3) a statistical investigation of SA belonging to the group 3.

2. MARKARIAN 7

Markarian 7 consists of two almost rectilinear segments in a figure resembling the upside-down letter 'V'. Each segment consists of SA. The size of the western segment is equal to 35" or 7.1 Kpc, the eastern 25" or 5.1 Kpc. Twelve spectra were obtained with the 6-m telescope in the spectral interval from 3700 to 7000 Å. The slit of the spectrograph was directed along the segments.

Both segments show approximately the same spectra: continuum and emission lines [SII] 6731/17, [NII] 6584/48, Hα, N_1, N_2, Hβ, Hγ, and [OII] 3727. The intensities of the continuum and the lines are changing along the slit. In Fig.s 1a and 1b the tracings of Hα (along the slit) for the western (two tracings) and the eastern (three tracings) segments are presented. The rotation curves of the two segments are also plotted in Fig.s 1c and 1d. They are superposed with the structural tracings of Hα.

The western segment shows a solid body rotation with a radial velocity difference of 180 km/s between the two ends. The eastern one does not show any appreciable rotation effect relative to us. However it is very important that at the point where the two segments intersect the radial velocities are the same, about 3000 km/s. This fact speaks in favour of a common physical nature of both segments. In other words, Markarian 7 is rather a single galaxy with a complex kinematical structure than a double system.

3. SA IN UV GALAXIES (Fig. 2)

Recently about 150 SA in 57 spiral UV galaxies were found by the author and coworkers (Khachikian, Petrossian & Sahakian 1983a). Some physical parameters of these SA (size, color, absolute magnitude, distance from the nucleus) were estimated.

The statistical investigation of these data leads to the following conclusions (Khachikian, Petrossian & Sahakian 1983b):

1. The average linear dimension of SA is 1 Kpc, the mean absolute magnitude is $M = -15^m$.

2. SA are found in dwarf, giant as well as supergiant UV galaxies (mainly in barred galaxies) on the arms and at the end of the bars.

3. The number of SA is higher in barred spirals and they are distributed farther from nucleus than in normal spirals. SA are found frequently in the late type spirals; they have smaller sizes and are situated closer to the nucleus.

4. In the distribution of SA as function of the distance from the nucleus two maxima are noticeable: the first strong one appears at the distance of 0.4 R_{gal}, the second at 0.8 R_{gal}.

5. About 10% of UV galaxies containing SA are Seyfert galaxies, about 12% multinucleus galaxies. More than 30% of the galaxies of our sample are radio sources with the threshold intensity of 10 mJy.

The above mentioned data allow us to conclude that SA are closely connected with manifestation of activity in UV galaxies.

REFERENCES

Arp H.C., Khachikian E.Ye.: 1974, Astrofizika, 10, 173
Burenkov A.N., Khachikian E.Ye.: 1983, Astrofizika, in press.
Casini C., Heidmann J., Tarenghi M.: 1979, Astron.Astrophys., 73, 216.
Coupinot G., Hecquet J., Heidmann J.: 1982, M.N., 199, 451.
Kazarian M.A., Khachikian E.Ye: 1977, Astrofizika, 13, 415.
Khachikian E.Ye: 1972, Astrofizika, 8, 529.
Khachikian E.Ye., Korovyakovskiy Yu.P., Petrossian A.R., Sahakian K.A.: 1981, Astrofizika, 17, 231.
Khachikian E.Ye., Petrossian A.R., Sahakian K.A.: 1979, Astrofizika, 15, 209.
Khachikian E.Ye., Petrossian A.R., Sahakian K.A.: 1980a, Astron.J.Letters (USSR), 6, 262.
Khachikian E.Ye., Petrossian A.R., Sahakian K.A.: 1980b, Astron.J.Letters (USSR), 6, 552.

Khachikian E.Ye., Petrossian A.R., Sahakian K.A.: 1983a, Astrofizika, in press.
Khachikian E.Ye., Petrossian A.R., Sahakian K.A.: 1983b, Astrofizika, in press.
Khachikian E.Ye., Sahakian K.A.: 1975, Astrofizika, 10, 173.
Petrossian A.R.: 1983, Astron.Zirkul. (USSR), in press.
Searle L., Sargent W.L.W.: 1972, Astrophys.J., 173, 25.

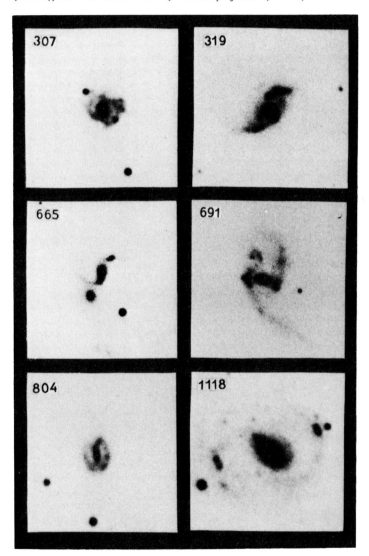

Figure 2. Photographs of UV galaxies containing SA, all obtained with the 6-m telescope except Markarian 1118 (2.6-m Byurakan telescope). Each field is about 1!5 x 1!5.

THE CERRO EL ROBLE SAMPLE OF FAINT ULTRAVIOLET EXCESS OBJECTS IN THE SOUTH GALACTIC POLE

Luis E. Campusano[1] and Carlos Torres
Observatorio Astronómico de Cerro Calán
Departamento de Astronomía
Universidad de Chile
Casilla 36-D, Santiago de Chile

ABSTRACT

The idea to develop selected regions of the sky for extragalactic research in the Galactic Polar areas is emphasized. One such region, centered at $\alpha=$00h 53m (1950) $\delta=-28°03'$, has been examined by several authors for surveys of QSO candidates. We have also searched 44-deg^2 of this region, containing the South Galactic Pole (SGP), for relative ultraviolet excess objects (UVXs) in 5 partially overlapping fields of the 70/100 cm Maksutov telescope of Cerro El Roble. The search was found to be extremely incomplete at B>19 mag, while the completeness for B $\lesssim$ 19 was estimated to be approximately 30% in non overlapping regions and 50% in overlapping regions. Only the central overlapping region was used for the comparison of the surface densities of different UVX samples. The surface density in this region is 7.2 UVX/deg^2 at B$\sim$19 mag and approximately a factor 2 larger than the one found by Savage and Bolton (1979) in two 25 deg^2 fields near the SGP. In addition, our surface density value is a factor 3/2 larger than the density found by Braccesi, Formiggini and Gandolfi (1970) near the North Galactic Pole. Twenty-seven new QSOs have been already identified in our UVX sample from spectroscopic data collected on the Las Campanas 2.5 meter telescope. A statistical test applied to the surface distribution of our UVXs in the central 25 deg^2 region, gave a slight suggestion of non-uniformity.

I. INTRODUCTION

It is well known that the regions of the sky which are at high galactic latitudes ($|b|\sim 90°$) are privileged, over other directions, for the observation of extragalactic objects due to the advantage of low obscuration and low star density; among them, the regions containing the Galactic Poles are specially noteworthy. On the other hand, it is also

[1] Visiting Astronomer, Cerro Tololo Inter-American Observatory

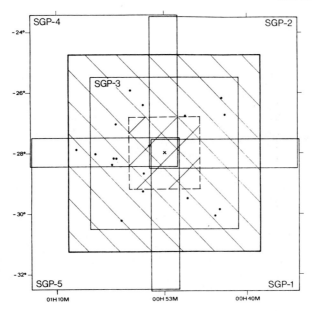

Figure 1. The approximate positions of the five 5°×5° Maksutov fields employed for the search of UVX objects are shown in this schematic (α, δ) diagram where hatching denotes a 44 - deg^2 area. The closed dots represent the UVX objects selected in more than one field. Grism plates for the central 1.4 deg^2 region, shown with crossed hatching, have been obtained with the 4 meter CTIO telescope.

well known the convenience of defining some selected areas on the sky, where a large variety of independent investigations can be concentrated. Survey works must be carried out first in order to set samples of objects in given regions, and it is desirable to include all available techniques and frequency ranges. Following stages, for example for extragalactic objects, should (hopefully) end up with counts, redshifts and magnitudes for a large number of objects, enabling us to gain knowledge of their spatial distribution and of the physical relations between the constituents. Therefore, the formation of selected areas for extragalactic research, that are located in the Galactic Polar areas, are clearly justified. An important aspect is the choice of the size of the regions (or sub-regions), and it is advantageous to have, at least, two equivalent (in principle) ones for considerations on the isotropy of the distributions.

We shall refer here to one of such regions containing the South Galactic Pole (SGP), in connection to the surveys of QSOs that have been conducted there and specifically to the search of faint ultraviolet excess (UVX) objects with the 70/100 cm Maksutov telescope located at the Cerro El Roble Astronomical Station of the Universidad de Chile. The sizes selected for the surveys have been different, but all of them are approximately centered at α=00h 53m (1950) δ=-28°03', which

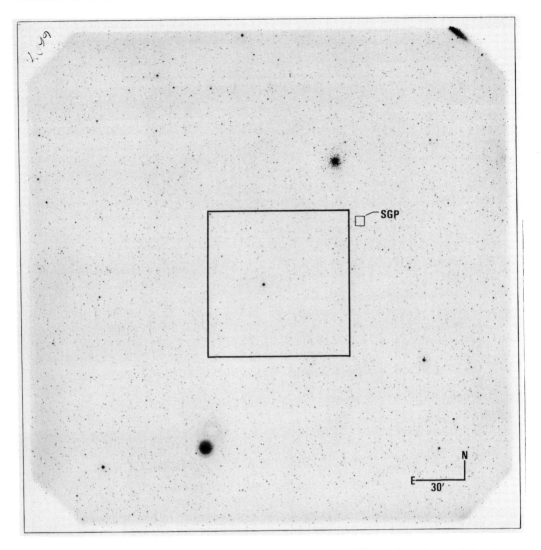

Figure 2. Reproduction of a plate (IIaO+GG 385, 25 min) taken with the 70/100 cm telescope located at the Cerro El Roble Astronomical Station of the Universidad de Chile (f/3, scale 99 arcsec/mm). Northeast is at the top left-hand corner. Three independent optical-searchs of QSOs have been conducted in this 25 deg^2 area centered at α=00h 53m (1950) δ=-28° 03', containing the SGP. The region inside the central square drawn in this illustration has been selected by us for a deeper survey of QSOs. A potentially powerful path for extragalactic research is to concentrate observational efforts in selected areas, particularly in Galactic Polar areas such as this one.

corresponds to $b \simeq -89°$. Clowes and Savage (1983) have found 162 QSOs candidates from eye-inspection of 25 deg^2 of a UK-Schmidt objective-prism plate. Shanks et al. (1983) have formed a sample of 660 UVX

Table I. Counts for the 373 UVX objects in the 44 deg² in the SGP

B	N	N(<B)	logN(<B)
9.0– 9.9	1	1	0.00
13.0–13.9	4	5	0.70
14.0–14.9	17	22	1.34
15.0–15.9	27	49	1.69
16.0–16.9	58	107	2.03
17.0–17.9	80	187	2.27
18.0–18.9	110	297	2.47
19.0–19.9	66	363	2.56
20.0–20.9	9	372	2.57
21.0–21.9	1	373	2.57

objects (U-B<-0.35) in a 11.5 deg² area using machine measurements of UK-Schmidt plates. Campusano and collaborators have conducted a search of optical counterparts in a 72 deg² area (Campusano and Torres, 1981), have recently formed a UVX sample of 373 objects in a 44 deg² region, and are presently carrying a grism search of QSO candidates in a 1.4 deg² zone (Campusano and Moreno). In this report we outline the characteristics of the Cerro El Roble sample of UVXs; a full account of this investigation has been given elsewhere (Campusano and Torres, 1983). A spectroscopic survey of QSO candidates in the SGP is being carried out by Campusano and Zamorano at the Las Campanas Observatory with the 2.5 meter telescope.

II. THE OBSERVATIONS AND THE SELECTION METHOD

The 44-deg² surveyed area was covered with five Maksutov fields. They were chosen to be partially overlapping (see Fig. 1), in order to estimate the completeness of our sample. The intersection of the fields, or overlapping region, is the central 25-deg² area; a photograph of this zone is shown in Fig. 2.

For the selection of UVXs, separate U and B plates were obtained and then inspected with an unmodified Zeiss Blink comparator. The exposure times were set to give a limiting magnitude of B$\sim$20 mag and a balance of the images for U-B $\simeq$ -0.6. The selection was performed through several independent inspections which involved the two authors and the research assistant L.E. González. Three degrees of ultraviolet excess were assigned: strong (S), medium (M) and weak (W). Blue magnitudes were estimated for the selected objects, using the Cuffey variable iris astrophotometer of CTIO, La Serena.

III. PROPERTIES OF THE CERRO EL ROBLE SAMPLE

From an analysis of our survey we have arrived to the following properties and conclusions:

1. We have established a catalogue of 363 faint UVX objects with U-B $\lesssim$ -0.6 and brighter than B$\sim$20 mag. in a region of 44 deg^2 centered in the South Galactic Pole. Counts are given in Table I.

2. The catalogue was found to be extremely incomplete at B>19 mag.

3. The completeness of the catalogue at B $\lesssim$ 19 could be estimated thanks to the choice of the field plates with overlapping regions. Our sample of UVXs is approximately 30% complete in non overlapping regions and 50% complete in overlapping regions (with two plates). Therefore, the searched area can be divided in a central overlapping region of 25 deg^2, $\sim$ 50% complete, and the remaining one with a degree of completeness of $\sim$ 30%.

4. Our central overlapping region has a surface density of 7.2 UVX/deg^2 at B$\sim$19 mag. This surface density is about twice that found by Savage and Bolton (Savage 1978; Bolton and Savage 1978; Savage and Bolton 1979) in their original selection of UVXs near the South Galactic Pole. This result would suggest a considerable degree of incompleteness of their UVX search, if errors in the magnitude scales are absent and the distribution of UVXs in the sky is isotropic.

5. The surface density of the sample of Braccesi, Formiggini and Gandolfi (1970) is approximately 1/3 of the corrected value, for incompleteness, of our central region. Therefore, a certain degree of incompleteness is possibly present in the Braccesi sample.

6. We rediscovered only one of the two known QSOs with B $\lesssim$ 19 that we should have detected as UVXs.

7. One of us (LEC) has identified already 27 new QSOs from our UVX catalogue in a fast preliminary inspection of spectroscopic data collected at the Las Campanas Observatory (Chile).

8. From statistical analysis of the surface distribution of the UVX objects in the central overlapping area, we found a slight suggestion of non-uniformity when we used cells of 0.3 deg^2. Further evidence would be desirable to reject or confirm the suspected inhomogeneity.

IV. ACKNOWLEGMENTS

We are grateful to L.E. González and M. Wischnjewsky for their assistance during the course of this work. M. Fajardo carefully typed the manuscript and C. Monsalve made the fine illustrations. LEC wishes to express his appreciation to the Directors of CTIO and of Las Campanas Observatory for their hospitality and for the allocation of telescope time. This work has been partially supported by the Departamento de Desarrollo de la Investigación de la Universidad de Chile (clave

E-1340). The participation in this conference has been possible thanks to the financial support of the European Southern Observatory

REFERENCES

Bolton J.G., and Savage A.: 1978, IAU Symp. 79, edited M.S.Longair and and J.Einasto, p. 295.
Braccesi, A., Formiggini, L., Gandolfi, E.: 1970, Astron.Astrophys. 5, p. 264.
Campusano, L.E., Torres, C.: 1981, Proceedings of the Second Latin American Regional Meeting I.A.U., Venezuela, Rev. Mexicana Astron. Astrof. 6, p. 29.
Campusano, L.E., Torres, C.: Astron. J. (in press).
Clowes, R.G., Savage, A.: 1983, Mon. Not. R. Astron. Soc. (in press).
Haro, G., Luyten, W.J.: 1962, Bol. Obs. Tonantzintla Tacubaya 3, p. 37.
Savage, A.: 1978, Ph.D. thesis, University of Sussex.
Savage, A., Bolton, J.G.: 1979, Mon. Not. R. Astron. Soc. 188, p. 599.
Shanks, T., Fong, R., Green, M.R., Clowes, R.G.: 1983, Mon. Not. R. Astron. Soc. 203, p. 181.
Usher, P.D.: 1981, Astrophys. J. Suppl. 46, p. 117.

A STUDY OF ULTRAVIOLET-EXCESS GALAXIES BASED ON THE KISO SURVEY

Bunshiro Takase, Takeshi Noguchi, and Hideo Maehara
Tokyo Astronomical Observatory

Abstract A number of ultraviolet-excess galaxies have been detected during the course of our surveys using the Kiso Schmidt telescope. In this report, a classification scheme is proposed for 142 selected objects on the basis of their morphological features, and the relation between the morphological type and the degree of ultraviolet-excess is presented. In general irregular galaxies with conspicuous H II regions and pair galaxies tend to show higher degree of ultraviolet-excess, while the degree in spiral galaxies appears to range widely.

1. INTRODUCTION

A survey for ultraviolet-excess (hereafter abbreviated as UVX) objects has been carried out as one of the programs of the Kiso 105 cm Schmidt telescope of the Tokyo Astronomical Observatory (Takase et al. 1977). They are detected by means of the multicolor image method. Either U, G, and R triple images or U and R double images are exposed on a single plate. The observational technique is explained by Noguchi et al.(1980).

Takase (1980) found some 1,100 UVX galaxies which have been called Kiso UVX galaxies (abbreviated as KUG) in 20 sky areas covering about 650 square degrees. The number of KUGs is roughly ten times that of the Markarian galaxies over these areas. This implies that our direct image photographs have much better detectability and considerably fainter limiting magnitude than the Markarian's objective prism spectrograms.

The survey for KUGs has since been extended to much wider sky areas. Further follow-up observations have also been made for randomly selected samples of these objects with the 188 cm reflector of the Okayama Astrophysical Observatory. Up to now direct photographs of 44 KUGs have been taken at the Newtonian focus where the plate scale is about three times larger than that of the Schmidt photographs. Furthermore spectrograms of 35 objects have been taken with the Cassegrain Image Intensifier spectrograph attached to the reflector.

During the course of these surveys we have become interested in the wide variety of morphological features of KUGs and attempted to classify them into several types. This report is a concise version of the full paper (Takase et al. 1983), where a list of 142 objects with their morphological type together with their degree of UVX and several other informations is included.

2. MORPHOLOGICAL CLASSIFICATION

From direct photographs taken at Okayama together with the Palomar Sky Survey prints, and the Kiso multicolor plates as well, we tried to classify our sample of KUGs. Our classification scheme is as follows:

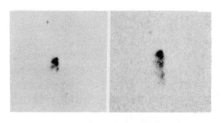

Type Ic: KUG 1626+413

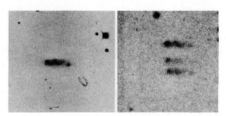

Type Ig: KUG 0225-103

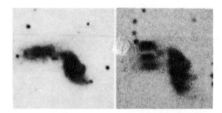

Type Pi: KUG 1047+332

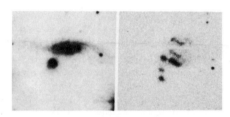

Type Id: KUG 2259+157

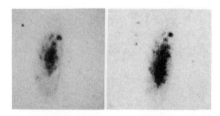

Type Sk: KUG 2257+157

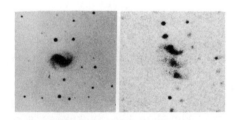

Type Sp: KUG 0239+345

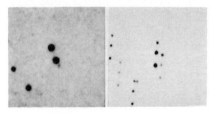

Type C : KUG 0935+407

Figure 1. Representative objects of each classification type. In each pair the left one is a direct photograph while the right one is that reproduced from the Kiso three color plate, where U, G, and R images are lined up from top to bottom.

1.1 Irregular with clumpy H II regions (denoted as Ic)
1.2 Irregular with a conspicuously giant H II region (Ig)
2.1 Pair of interacting components (Pi)
2.2 Pair of detached components (Pd)
3.1 Spiral with knotty arms (Sk)
3.2 Spiral with peculiar bar or nucleus (Sp)
4   Compact (C)

In Figure 1 representative samples of each type are given with pairs of direct and multicolor photographs. The Ic type is composed of several conspicuous conglomerations of H II regions. The Ig type has a single supergiant H II region complex usually at one edge of the elongated body of the galaxy. The Pi type is composed of mutually interacting and tidally deformed components of galaxies, while the Pd type is a pair consisting of a slug shaped elongated galaxy and a small globular galaxy, the latter of which usually has a higher degree of UVX than the former. The Sk type is morphologically not so peculiar as other types, except that there are several medium size H II region knots along spiral arms. The Sp galaxies have either a peculiar nucleus or an abnormal bar structure. For example they have a brilliant starlike nucleus, a hot spot complex, or a split or otherwise deformed bar. Seyfert galaxies may be included in this Sp type. Finally the C type are galaxies which are compact.

3. THE DEGREE OF ULTRAVIOLET-EXCESS AND RELATED STATISTICS

The degree of UVX or the color of each galaxy can be estimated from the brightness of the U image relative to the G and/or R images on our multicolor plates. Symbols H, M, and L are used for high, medium, and low degree of UVX, respectively. These correspond to the color index CI defined by Noguchi et al.(1980) approximately as

$CI \sim -0.5$ for H,
$CI \sim 0$ for M, and
$CI \sim +0.5$ for L.

Table 1 shows frequency statistics of 142 KUGs in respects of both the morphological type and the UVX degree. For Pi and Pd galaxies which have two or more components, only the one with the highest degree of UVX is included in the statistics.

It seems that the Sk and Sp types with a low UVX connect smoothly with normal S galaxies which have no UVX. Components of the Pi and Pd types with a lower UVX which are not counted in the Table have either a low or no UVX. So these may also continue to their respective normal gal-

Table 1

FREQUENCY STATISTICS IN RESPECTS OF THE MORPHOLOGICAL TYPE AND UV-EXCESS DEGREE

TYPE / UVX DEGREE	H	M	L	TOTAL
Ic	16	9	0	25
Ig	9	0	0	9
Pi	13	7	0	20
Pd	10	3	0	13
Sk	2	12	6	20
Sp	12	20	4	36
C	5	5	0	10
?	3	5	1	9
TOTAL	70	61	11	142

axies with no UVX. For the Ic, Ig and C types, however, the fact that there are no samples with a low UVX, may indicate that all of these types of galaxies are UVX objects.

4. REMARKS

Some selection effects are unavoidable in sampling and classifying the objects. For example distant galaxies are sometimes overlooked or unclassifiable due to their small sizes, and those with unfavorable orientation are apt to be misclassified due to the projection effect, Thus the statistics given above are not so complete that they represent the real distribution of the morphological type and the degree of UVX.

The portions of the objects with UVX have been in general interpreted as more or less giant H II regions where an active star formation is taking place (e.g., Benvenuti et al. 1982). The objects in our present sample range from weakly or moderately active Sk types to hyperactive Ic and Ig types. Some fundamental differences seem to exist among these various type objects.

In order to study these characteristics in more detail, spectroscopic investigations are indispensable. A preliminary study has been made on the basis of our spectrograms obtained with the Okayama telescope, and it has been found that there are appreciable differences between the excitation states of nuclear and outer H II regions as suggested by several authors, e.g., by Kazaryan et al. (1981). In addition, a number of our objects have been observed with the Nobeyama 45 m radio telescope. The results of this radio observation will be published elsewhere (Maehara et al. 1983).

REFERENCES

Benvenuti,P., Casini,C., and Heidmann,J.1982, Monthly Notices Royal
    Astron. Soc., 198, pp. 825 - 831.
Kazaryan,M.A., Petrosyan,A.R., and Tamazyan,V.S. 1981, Soviet Astron.
    Lett., 7, pp. 359 - 360.
Maehara,H., Inoue,M.,Takase,B., and Noguchi,T. 1983, in preparation.
Noguchi,T., Maehara,H., and Kondo,M. 1980, Annals Tokyo Astron. Obs.,
    ser. 2, 18, pp. 55 - 70.
Takase,B. 1980, Publ. Astron. Soc. Japan, 32, pp. 605 - 612.
Takase,B.,Ishida,K.,Shimizu,M.,Maehara,H.,Hamajima,K.,Noguchi,T., and
    Ohashi,M. 1977, Annals Tokyo Astron. Obs., ser. 2, 16, pp. 74-109.
Takase,B., Noguchi,T., and Maehara,H. 1983, Annals Tokyo Astron. Obs.,
    ser. 2, 19, No. 2, in press.

QUASI STELLAR OBJECTS AND ACTIVE GALACTIC NUCLEI

C. Barbieri
Institute of Astronomy, University of Padova, Italy

1. INTRODUCTION

The role of the Schmidt telescopes in the discovery of the Quasi Stellar Objects and of the Active Galactic Nuclei, and in the understanding of their properties was and continues to be of the greatest importance. Thousands of Radio-Sources have been quickly associated to their optical counterparts thanks to the worldwide availability of the Palomar Observatory Sky Survey plates and charts and more recently of the films of the ESO B Survey. Other thousands of QSOs and AGNs devoid of radio emission are found by the large Schmidts nowaday in operation. This wealth of data give fundamental cosmological knowledge and insight in the physical processes occuring in these objects. I'll concentrate in this Review on two specific topics, namely on the discovery techniques and on the study of the optical variability. To both subjects, the 67/92 cm Schmidt telescope here at Asiago has made significant contributions. The first topic is treated in several excellent papers, such as the one by M. Smith (1978) and the one by P. Veron (1983); the material presented in the second part is largely new. In the following, I'll use rather loosely the terms QSOs and AGNs to designate a variety of objects including Quasars (those QSOs in catalogs of Radio-Sources), high-redshift compact galaxies with emission lines, BL LACs et similia.

2. DISCOVERY OF QSOs WITH SCHMIDT TELESCOPES

Almost immediately after the identification of the first quasars, Schmidt telescopes, and in particular the 48-inch at Mount Palomar, were put to work to ascertain the existence of objects having the same optical features but devoid of strong radio emission. The technique then used was the one devised in the pioneering surveys of blue stars in high galactic latitudes carried out at Tonantzintla Observatory: two or three exposures are successively taken on the same plate, after a filter change and a small displacement of the telescope. An example is given in Fig. 1, taken from an Asiago plate; a UV quasar, 4C 49.22, is easily seen. This plate employes a most used combination constituted by the E-K 103a-0 emulsion plus the U and B filters. The exposure times are so calibrated to produce approximately equal images for an A0 star. As is well known, QSOs occupy a large area of the (U-B, B-V) plane were no main sequence

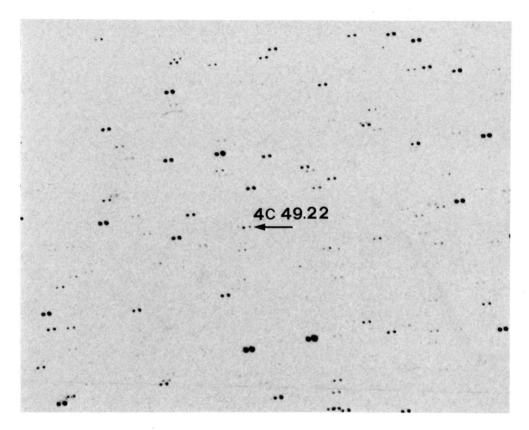

Figure 1 - A two-colour image of the Quasar 4C 49.22, from an Asiago plate of the still unpublished field around χ U Ma.

stars are found; see Fig. 2, taken from the Asiago Catalogue of QSOs, ed. 1982. With few exceptions, all objects with $z < 2.5$ have $U-B < -0.3$; therefore, if the search is limited to objects bluer than this limit, then one does not introduce systematic biases against particular z values. The situation is less clear for higher z, because reliable UBV data are lacking. At any rate, from Fig. 2 and also from theoretical considerations, redder colors are expected, so that the discovery of QSOs with $z > 2.5$ using this multicolour technique seems virtually impossible. Going back to QSOs with $z < 2.5$, we see a first limitation of the method, namely that from the colors one can obtain at most a rough estimate of their redshifts. The reason for the trend of the colors with z is qualitatively understood: the non-thermal shape of the continuum and the motion with z of the bright emission lines across the fixed color bands produce a typical pattern shown in Fig. 3 (Sandage, 1967). The estimate of z from the colors however is so uncertain to be practically useless even for statistical discussion of limited scope. A second drawback of the multicolour method is the fact that too many spurious candidates are also found, especially at $B < 18.5$ (see Veron, 1983). It becomes then imperative to supplement the $U-B < -0.3$ criterium with additional constraints. For in-

QUASI STELLAR OBJECTS AND ACTIVE GALACTIC NUCLEI

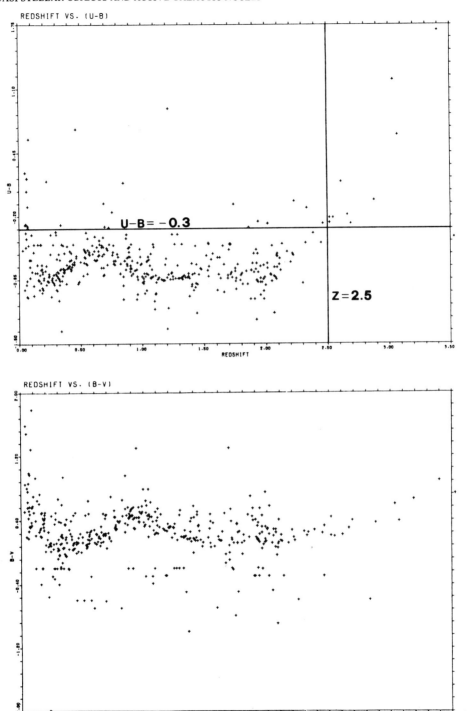

Figure 2 - The U-B, B-V colours of the QSOs in the Asiago catalogue. Very few objects have U-B>-0.4, unless z>2.5.

stance, Braccesi (1967) remarked that the non thermal continuum renders the QSOs brighter in the near-IR than normal stars with the same U-B. Results for the field 13h, +36d have been published by Braccesi et al. (1970). Unfortunately the low efficiency of the near-IR emulsions prevented a large scale application of the method; a recent survey using the IR criterium has been carried out by Kron and Chiu (1981).

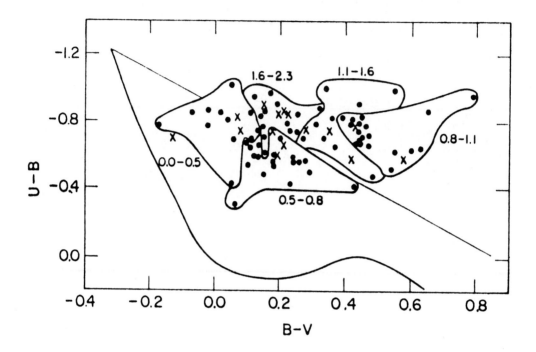

Figure 3 - The (U-B, B-V) locus of QSOs as function of z. The correlation of colors with redshift is not precise enough to permit statistical discussions or unequivocal identifications.

A second useful restriction has been the requirement of zero propermotion. This precaution was introduced by Sandage and Luyten (1967) and has been used more recently in the already mentioned survey by Kron and Chiu (1981). Once again, the very existence of the first epoch POSS pla tes has been of the greatest value. Finally, the optical variability is of great help to discriminate bone-fide QSOs from blue stars (Usher et al., 1983). Despite all ingenuity however, the multicolour metod requires subsequent spectroscopy with a large telescope, in order to determi ne the nature of the candidate and measure its redshift.

I have collected in Table 1 some surveys done with this multicolour technique. Although completeness is not claimed, the Table gives a fair idea of the great amount of research thus done. The Asiago Catalogue of QSOs, essentially complete to the end of 1981, contains some 150 objects thus discovered. Let me add a few words about the Asiago Survey. The pla tes were obtained between 1967 and 1971; three fields were published but

TABLE 1 - MULTICOLOR SURVEYS OF QSOs AND AGNs WITH SCHMIDT TELESCOPES

Designation and Main References	Schmidt Telescope and Combination
PHL: Haro and Luyten (1962) Sandage and Luyten (1967,1969)	Palomar 48 inch 103a-D + UBV, zero proper motion
Sandage and Veron (1965)	Palomar 48 inch 103a-O + UB
Rubin et al. (1967)	Palomar 48 inch 103a-D + UBV
BFG; Braccesi, Formiggini and Gandolfi (1970)	Palomar 48 inch 103a-O + UB; 103a-D + V; I-N (sens.) + I
W; Weistrop (1972)	Palomar 48 inch 103a-O + UB; 103a-D + V
PG; Green (1976)	Palomar 18 inch IIa-O (sens.) + UB
Berger and Fringant (1976)	Palomar 48 inch 103a-D + UBV
Steppe (1978)	Palomar 49 inch 103a-E + R; 103a-O + UG
Usher (1981)	Palomar 48 inch 103a-D + UBV
Arp and Surdey (1982)	Palomar 48 inch 103a-O + UB
Iriarte and Chavira (1957)	Tonantzintla 26/30 inch 103a-D + UBV
Richter and Sahakjan (1965)	Tautenburg 400/200/134 cm 103a-O + UB; 103a-G + V
A; Barbieri et al. (1968)	Asiago 92/67 cm 103a-O + UB
Naguchi et al. (1980)	Kiso 105 cm 103a-E (sens.) + UGR

the lack of adequate spectroscopic facilities prevented for long time
the proper identification of their content. But now, thanks to the generous
efforts of Laura Erculiani-Abati, data are finally coming; supplementing
the Asiago plates with material taken at the Tautenburg and
UK Schmidts and with slit spectra at the 6 m telescope in Zelenchuskaya
she has already confirmed more than a dozen QSOs. These first encouraging
results are found in her paper at this Conference and will prove
useful to determine the areal density of the brighter objects.

I move now to a second, extremely successful technique to utilize
the Schmidt telescope to discover QSOs and AGNs, namely with low dispersion
objective prisms. With them, low redshift excited galaxian nuclei
and emission line regions are easily found, as was proved in the classic
works of Markarian (1967). At much higher redshifts, the bright line
Ly-alpha enters the blue and visible regions, so that QSOs with
$1.7 < z < 2.4$ are mostly found, with a non negligeable number having
redshifts as low as 1.6 or as high as 3.4. An example is given in Fig. 4,

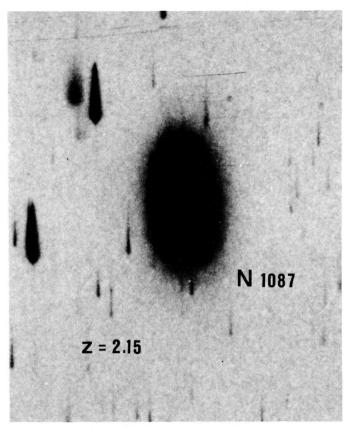

Figure 4 - A radio-quiet QSO found a UK Schmidt plate, in the still unpublished survey of the field at 2h +0d.

from a survey we have in progress with the UK Schmidt in a field at
2h 0deg. The advantages of the objective prism technique are easily understood;
the efficiency is so terribly high that hundreds of new QSOs
are added to the Catalogues every few months. Indeed the yield of a single
UK Schmidt plate is between 100 to 200 candidates (Smith, 1978), for

most of which a fair estimate of the redshift is immediately available. Again without any claim of completeness, I have listed in Table 2 some surveys done with objective prisms.

TABLE 2 - OBJECTIVE PRISM SURVEYS OF QSOs AND AGNs WITH SCHMIDT TELESCOPES.

Designation and Main References	Schmidt Telescope, Plate and Dispersion
Markarian (1967)	Byurakan 102/132 cm IIa-F 2500 A/mm at H beta
CTIO; Osmer and Smith (1980) UM; MacAlpine et al. (1977) Smith (1975)	CTIO 61/91 cm IIIa-J (sens.) 1740 A/mm at H beta
Savage and Bolton (1979)	UK 48 inch IIIa-J 2480 A/mm at H gamma IIIa-J + UB
Hazard, Arp and Morton (1979) Arp and Hazard (1980)	UK 48 inch IIIa-J 2480 A/mm at H gamma
Weedman (1983)	Burrel 61/91 cm IIIa-J

As is well know, these surveys have been supplemented with GRISM searches at larger telescopes, so that the technique is often collectively called SLITLESS SPECTROSCOPY (see for instance Osmer, 1980). It is of interest now to ask what the colors are of these objective-prism QSOs: it turns out that most of them are also blue, as expected. It has become good practice now, especially at the UK Schmidt, to obtain both U and B and objective prism plates. Modernization of the Blink machines with TV display following the ideas of J. Bolton (Savage, 1978) is of great help in the examination and intercomparison of the plates. Veron (1983) gives a good discussion of the completeness of the several surveys, variously estimated between 50 and 80%. Here, I add few practical considerations. As already remarked, objective prisms discover QSOs mostly in the range $1.7 < z < 2.4$. The effect of this observational selection is clearly seen in Fig. 5, built using the successive editions of the Asiago Catalogue: in 1966 the Catalog consisted entirely of Quasars, mostly from the 3CR Catalog. One year later, several "interlopers" or Blue Stellar Objects, as they were named at that time, were added and produced a pronounced peak around $z = 2.0$. In 1975 the identification of radiosources from several catalogs (4C, PKS, CT, Ohio etc.) had been very success

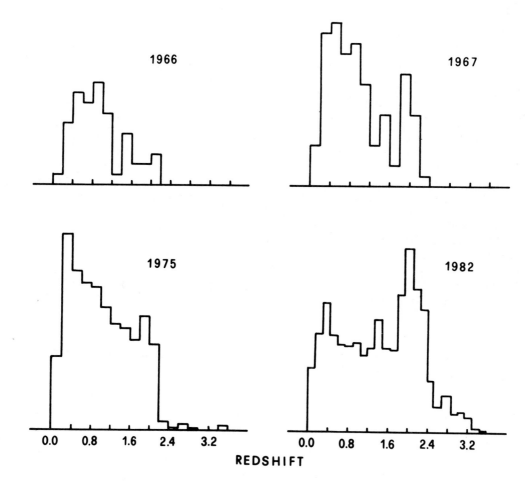

Figure 5 - The distribution of the redshifts in the successive editions of the Asiago Catalog of QSOs.

sfull, and that peak largely reduced. Taday, it dominates entirely the distribution. This observational bias against intermediate redshift could be overcome by a mediumsized Schmidt telescope in Space, capable to investigate in the near UV large fields with an objective prism. Therefore the discovery of intermediate redshift QSOs is at least in principle possible from Space. More difficult is the situation for very high z's, even assuming that those objects do indeed exist. Even the Space Telescope will be of limited help in their discovery, because its field is so limited. Hopefully one will soon discover efficient ways to utilize Schmidts on the ground for this extremely important project. To conclude this Section, it is worth recalling that objective prisms discover QSOs that on average seem one to two magnitudes fainter than quasars. However, great care is needed to discuss this difference; the truth is that the vast majority of QSOs lack good magnitudes. The variability only adds to this unsatisfactory situation.

3. OPTICAL VARIABILITY OF QSOs AND AGNs

In this Section I'll present material mostly obtained with the larger of the two Asiago Schmidts. Indeed, taking advantage of the great collection of plates of the Supernovae Search and of other long-term programs, we have been able to study the variability of QSOs over a period of almost two decades. Occasionaly, films from the smaller Schmidt proved of help, although the short focal length prevents good magnitude estimates. References to the papers published so far can be found in Barbieri et al. (1983). In total we have examined some 2000 plates, investigating both stellar and compact or slightly diffuse objects.

Because most of the material was already available for other programs, the Asiago sample gives a fair idea of the behaviour of the "average" QSO in respect of the optical variability. The largest selection factor is the brightness, due to the limiting magnitude of the plates, around the 19th. Through it however more insidious selection effects can be introduced in the sample, for instance because Quasars tend to be brighter than radio quiet objects, as already remarked in Sect. 2. The magnitudes are usually estimated by eye in respect to surrounding stars. Although this method is perhaps not extremely precise, it has in our opinion a great advantage, namely the possibility to refer the variable at the same time to a fair number of stars. Spurious variations are then easily found, and that is the most critical phase in the entire study. It is difficult in fact to study the variability of QSOs and AGNs, because of their morphology and of their peculiar colors. In the literature there are indeed several cases of spurious variations, especially on photographs taken with short focal length telescopes (e.g. Kinman, 1969). I give two examples to illustrate this point. The first is represented by the anonimous galaxy at 8h+48deg, discovered by G. Romano as a twelwth mag variable "star" (see fig. 6). The galaxy is diffuse, without a bright nucleus, with a complex structure extending several arcsecs. Although we cannot exclude genuine variability, most of the variations, never exceeding 0.6 mag, must be attributed to different conditions of sky transparency, seeing, plate quality etc. On the other hand, we have more than 100 plates of 3C 48, a well known compact galaxy with a much stronger concentration of the light in a stellar nucleus. In this case the variations are contained within 0.4 mag and we would not consider it a variable QSO if not for the photoelectric data by Matthews and Sandage (1963). The Asiago material possesses therefore a good internal consistency and is examined with a cautious eye. Our experience is that a precision of +/-0.15 mag is expected on the average plate, and we disregard variations smaller than 0.3 mag. It is perhaps for this conservative attitude that the Asiago Survey doesn't find such a high number of variable QSOs as other surveys do.

Furthermore, as a consequence of the lack of control on the dates of most of the data points, we content ourselves with a very simple statistical

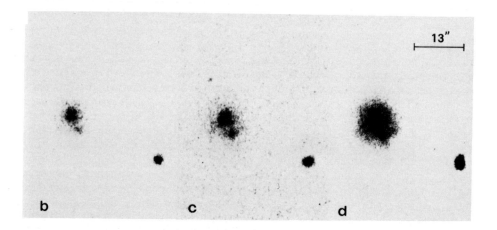

Figure 6 - The anonimous compact galaxy at 8h + 48° discovered as a 12th mag variable "star" on Asiago Schmidt plates. At the bottom, the images of the galaxy obtained with the 182 cm telescope at Cima Ekar (Barbieri et al., 1982)

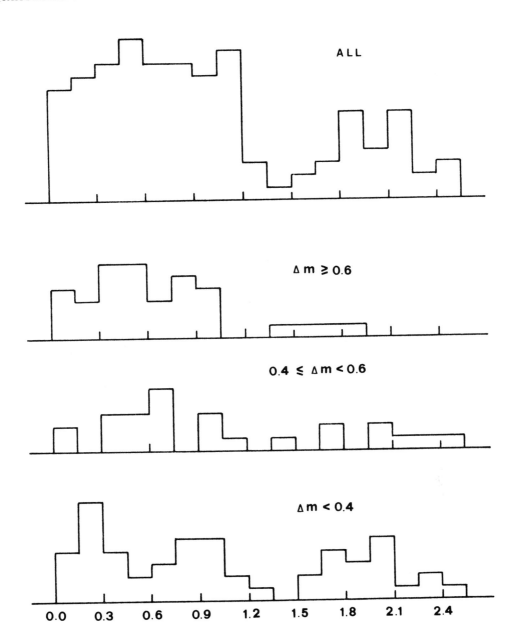

Figure 7 - The Redshift distribution of the Asiago sample. Only one large amplitude variable is found at z>1.8. BL Lac objects are excluded from this figure.

parameter, namely the maximum amplitude DB=Bmin - Bmax observed on the available material.

Let's now examine some results. Firstly, 7 out of 8 BL Lac objects in the sample have been found highly variable. This high percentage clearly separates this sub-class of AGNs from the other QSOs. For those objects with emission-line spectra, Fig. 7 presents the redshift distribution. If we divide the sample in two classes, one with DB <= 0.5 and one with DB >= 0.6 then an intriguing difference is found: there is only one large amplitude variable at z > 1.8, against several non variable objects. This paucity of high redshift variables cannot be due, at least not entirely, to the shortening of the proper time in the rest frame of the QSO: this is shown in Fig. 8 where we have the time coverage for the two classes. Several non variable QSOs have been densely observed in their time frame for almost a decade. Variations like the ones experienced by

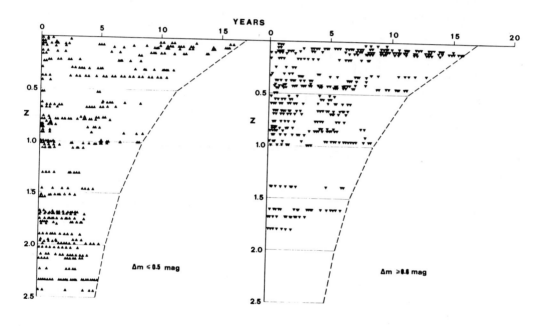

Figure 8 - The coverage in proper-time of highly variable and non variable QSOs in the Asiago sample (BL Lac objects are excluded from this figure).

3C 345 or 3C 446 would not have escaped detection. To a greater extent, the apparent magnitude can be the cause of this effect, because higher redshift objects tend to be fainter (in our sample), with a consequent deterioration of the data. Our conservative attitude raises in this case the detection treshold for genuine variations and we may disregard as plate fluctuations true brightness changes. A case in point is PKS 1116 + 12; it is a 19th mag QSO at z = 2.112 at the plate limit. It has been observed also in the Rosemary Hill Observatory Survey (Pica et al., 1980), and they find an amplitude of more than one mag. But the QSO must be at their very limit too, and we cannot consider it as truly variable. Even taking into account those two effects, a genuine physical property of the high-redshift QSOs cannot be excluded at this moment. Few more years of observations will undoubtely clear this important point. Going back to Fig. 8, a closer inspection would reveal that the high amplitude variables have been on the average more densely observed than the others: taking that into account we conclude that some 15% of QSOs are truly non variable over a period of 5 to 20 years, while at least 50% do change luminosity by 2 to 30 times. With the exception of the already discussed possible correlation with z, the rate of variability does not depend, at least not strongly, on other properties of the objects like radioemission; even radiovariability is not hundred percent correlated with the optical one.

We have in progress a careful analysis of the Asiago data, but this is not the place to present the full results. It is appropriate however in this Colloquium to point out the great amount of work done with our Schmidt and to wish many more years of fruitful researches.

It is my great pleasure to thank Prof. L. Rosino for the constant encouragement and help in this work on QSOs. G. Romano, S. Cristiani and A. Omizzolo have greatly contributed to this presentation.

REFERENCES

Arp, H.C., Hazard, C. 1980, Ap.J. 240, 726
Arp, H.C., Surdej, J. 1982, Astron. & Astrophys. 109, 101
Barbieri, C., Capaccioli, M., Cristiani, S. Nardon, G., Omizzolo, A. 1982, Mem. S.A.It. 53, nr. 3
Barbieri, C., Cristiani, S., Nardon, G., Romano, G. 1983, 24th Liege International Astrophys. Symp.
Barbieri, C., Cristiani, S., Romano, G. 1982, Astron. & Astrophys. 105, 369
Barbieri, C., Erculiani, L., Rosino, L. 1968, Pubbl. Oss. Astr. Padova nr. 143
Braccesi, A. 1967, Nuovo Cimento
Braccesi, A., Formiggini, L., Gandolfi, E. 1970, Astron. & Astrophys. 5, 264
Berger, J., Fringant, A.M. 1977, Astron. & Astrophys. Suppl. 28, 123
Green, R.F. 1976, P.A.S.P. 68, 665
Haro, G., Luyten, W.J. 1962, Boll. Obs. Tonantzintla y Tacubaya nr. 22, 37
Hazard, C., Arp, H.C., Morton, D.C. 1979, Nature 282, 271
Iriarte, B., Chavira, E. 1955, Boll. Obs. Tonantzintla y Tacubaya nr. 12, 33
Kinman, T.D. 1969, Nature vol. 224, 565
Kron, R.G., Chiu, L.T.G. 1981, P.A.S.P. 93, 397
Markarian, B.E. 1967, Astrofizika 3, 55
Matthews, T.A., Sandage, A.R. 1963, Astrophys. J. 138, 30
MacAlpine, G.M., Smith, S.B., Lewis, D.W. 1977, Ap. J. Suppl. 34, 95

Naguchi, T., Maehara, H., Kondo, M. 1980, Ann. Tokio Astr. Obs. vol. 18, pag. 56
Osmer, P.S. 1980, Objects of High Redshift, 78 (Abell and Peebles eds.)
Osmer, P.S., Smith, M.G. 1980, Ap.J. Suppl. 42, 333
Pica, A.J. Pollock, J.T. Smith, A.G. Leacock, R.J. Edwards, P.L. Scott, R.L. 1980, Astron. J. 85, 1442
Richter, N., Sahakian, K. 1965, Mitt. K. Schwarzchild Obs.Tautenburg Nr. 24, Teil I
Rubin, V. et al. 1967, Astrophys. J. 72, 59
Sandage, A.R. 1967, Highlights of Astronomy, 47 (Perek ed.)
Sandage, A.R. Luyten, W.J. 1967, Astrophys. J. 148, 767
Sandage, A.R. Veron, P. 1965, Astrophys. J. 142, 412
Savage, A. 1978, Ph.D. Thesis, University of Sussex
Savage, A., Bolton, J.G. 1979, M.N.R.A.S. 188, 599
Smith, M.G. 1975, Ap.J. 202, 591
Smith, M.G. 1978, Vistas in Astronomy 22, 23
Steppe, H. 1978, Astr. & Astrophys. Suppl. 31, 209
Usher, P.D. 1981, Ap.J. Suppl. 46, 117
Usher, P.D., Warhock, A., Green, R.F. 1983, Astrophys.J. 269, 73
Veron, P. 1983, 24th Liege Internat. Astrophys. Symp.
Weedman, 1983, preprint
Weistrop, D. 1972, A.J. 77, 366

DETECTION OF QSOs NEAR LARGE GALAXIES

A S Pocock[1], J C Blades[2], M V Penston[1], M Pettini[1]
1. Royal Greenwich Observatory,
2. Space Telescope Science Institute

We have presented here some results of our search for quasars near large galaxies. Papers on these results are to be published shortly in Monthly Notices of the Royal Astronomical Society (Mon. Not. R. astr. Soc.)

1. OBJECTIVE

The aim of our programme is to use the absorption spectra of QSOs and related objects (BL Lacs, Seyfert galaxies) as probes of the interstellar media of intervening galaxies. For such a study it is necessary to find new suitably bright QSOs and related objects lying close on the sky to large galaxies because very few are known at present.

Since 1981 we have been carrying out a large scale search programme using UK Schmidt Telescope objective prism plates taken specifically for this purpose. Because we are looking for objects bright enough for high dispersion spectroscopy (> 17.5 mag(B)), short exposure prism plates have been taken for us to a limiting magnitude of ~ 17.5: on sky limited exposures, such bright QSOs are saturated and harder to identify. We have searched several fields and have taken follow-up slit spectra of candidates in the fields of 6 galaxies using the 74 inch reflector at the South African Astronomical Observatory (SAAO). A test of the hypothesis that QSO absorption systems arise in galaxy haloes is best made by looking at the absorption spectrum of background QSOs seen through the outer parts of nearby galaxies.

Recent studies show that nearby spirals can have extents of up to 100-200 kpc in HI.

2. SEARCH PROCEDURE

The spectra of QSOs typically show ultraviolet excess and/or emission lines. The search work is done by inspection of the objective prism plates by eye to look for those objects with ultraviolet excesses and/or emission lines. We have also searched for bright compact blue galaxies selected from the length of the spectra since this gives the required combined measure of blueness and compactness. These bright compact galaxies may also lie behind the foreground galaxies and if they are bright enough they may also be used to probe the interstellar media of foreground galaxies in just the same way as QSOs.

Once an object is selected from the objective prism plate, it is checked against the appropriate direct plate copy from the ESO B/SRC J sky survey to ensure that the candidate is a single object and its long spectrum not attributable to overlapping objects.

The problems encountered during the searching, are worse than in other QSO searches since we are looking for brighter QSOs, and are due to contamination by other types of objects besides QSOs which can also show ultraviolet excess or emission lines.

The types of objects which come into the above catagories are white dwarfs, O, B, A stars, planetary nebulae, cataclysmic variables and M stars (which show apparent emission lines caused by gaps between absorption bands).

3. RESULTS

Follow up low resolution slit spectroscopy of QSO candidates was done on the 74 inch reflector at SAAO using the Reticon Photon Counting System (RPCS) and unit spectrograph with a dispersion of 210 Å/mm to cover the spectral range ~ 3200-7000 Å with a slit of 1.8 x 6 arcseconds orientated E/W. The data has been reduced on the RGO Starlink VAX 11/780 using the SPICA reduction package. Interesting objects identified are:-

11 QSOs magnitudes ~ $16-17_{(B)}$
2 Seyfert 1 galaxies, 1 Seyfert 2 galaxy
Several normal galaxies bright enough to use as probes
M stars (2 of which are thought to be Mira variables)
Planetary nebula
1 cataclysmic variable (shows strong ultraviolet excess on prism plate) see an upcoming paper now in press with Mon. Not. R. astr. Soc. - Echevarria, Pocock, Penston and Blades to be published ~ Nov/Dec 1983.

CaII is the strongest interstellar line in the optical that we expect to see in absorption from an external galaxy accessible from ground telescopes. Other interstellar lines that are strong in ultraviolet and optical will eventually be observed using Space Telescope.

We shall look at the QSOs and Seyfert/bright galaxies at high resolution at the Anglo Australian Telescope (AAT) probing the interstellar medium of the foreground galaxies at different distances from the disk.

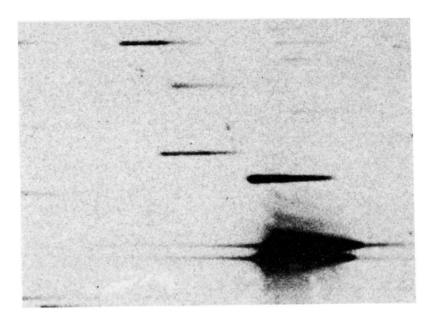

Plate 1

Objective prism spectrum of confirmed QSO 0050-254 (centre)

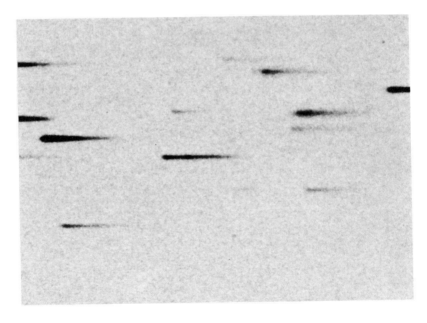

Plate 2

Objective prism spectrum of confirmed QSO 0041-261 (centre)

4. CONCLUSION

The results confirm our ability to select QSO candidates. The point like candidates turned out to be mainly QSOs (~ 25%), white dwarfs, O, B, A stars, M stars, planetary nebula, cataclysmic variable.

It is expected that some of the candidates would be these other objects (ie stars) because of the higher contamination by such objects at the magnitude to which we are working (~ 17.5).

5. FINAL REMARK

Going through Schmidt and Green's Palomar Bright Quasar Survey: we have found that several of these QSOs are close on the sky (with less than 5-10 galaxy diameters) to foreground galaxies, some with known redshifts. For some of these nearby galaxies, we have measured the galaxy redshifts and we shall also use these QSOs to probe the interstellar media of the intervening galaxy in the same way.

The first paper will be submitted to Mon. Not. R. astr. Soc. in the coming months.

DISCUSSION

KUNTH: Since you wish to avoid getting high-z QSOs when aiming to further detect absorptions from intervening galaxies, why do you use objective prism techniques and not other ones such as two colour techniques so as to get lower-z candidates?

POCOCK: We did not think that we would select mainly high-z QSOs by our method. We have, however, found some low-z QSOs and Seyfert galaxies using the objective prism method. It would be ideal to obtain some two colour plates, and combine the two methods.

AN OPTICAL SURVEY FOR VERY FAINT QUASAR CANDIDATES

Donald Hamilton
The Department of Physics
The University of Chicago
Chicago, Illinois, U.S.A.

1. INTRODUCTION

In recent years the development of fast optical telescopes with large corrected fields, of fine-grained, high DQE emulsions such as IIIa-J/F, and the use of rapid scanning microdensitometers and computers, have all made deep photometric studies of large areas of the sky tractable. I am in the process of conducting such a survey using the prime focus cameras of the CTIO and KPNO 4-meter telescopes. Sky-limited plates are taken in four colors defined by filter/emulsion characteristics. Pertinent information about the photometric system is given in Table 1. The large database from this survey of about fifteen areas of the sky is being used to address many problems both extragalactic and galactic in nature. A preliminary report on studies of faint blue stellar objects, quasar candidates, will be presented here for seven of the fifteen fields. Table II lists some information about the fields used for this study.

2. PROCEDURES

The ujfn plates were digitized using the KPNO PDS by raster scanning a 6600 by 6600 pixel area with a 20 micron square aperture in 20 micron steps. At the scale of the 4-meter triplet corrector (18.6 "/mm) our scanned area is no larger than 0.47 sq. degs. Unfortunately, poor placement of the guide probe, nonzero relative displacements between the multicolor plates, and bright stars ultimately reduce the usable area. Because of computer disc space problems, only two-thirds of the images for three out of the seven fields could be reduced.

The digital images are then put through a series of very sophisticated reduction programs collectively known as the Faint Object Classification and Analysis System or FOCAS. Details about FOCAS may be found in Valdes (1982a). A detailed description of the resolution classifier can be found in Valdes (1982b). The advantage of the resolution classifier is that it uses the full image data unlike the r_2 classifier of Kron (1980). Based on my experience, once the proper

Table I: PHOTOMETRIC SYSTEM			
Bandpass	Emulsion	Filter	Typical Exposures
u	IIIa-J*	UG5	160 min
j	IIIa-J*	GG385	50
f	IIIa-F*	GG495	60
n	IV-N†	RG695	50

Notes: * (†) emulsion hypersensitized by baking (soaking) in forming gas ($AgNO_3$ solution)

Table II: FIELD CHARACTERISTICS				
Field	l^{II}	b^{II}	E^*_{B-V}	Useful Area (sq. deg.)
SA 57	66°	86°	0.005	0.46
SA 68:1	111	-46	0.030	0.44
2				0.30
3				0.30
SA 28:2	176	39	0.010	0.41
Hercules 1	90	36	0.015	0.44
2				0.30

Notes: * reddenings derived from Burstein and Heiles (1982)

classification rules have been determined, the resolution classifier is far superior to r_2. The percentage of misclassifications based on r_2 are on the order of 50% at the faint end, but for classifications based on the resolution classifier it is no larger than about 10%. Classification based solely on color also confirms the resolution classifier's efficiency.

3. SELECTION OF QUASAR CANDIDATES

From the large multicolor catalogs I extracted objects which were detected only on the u, j, and f plate material. The additional requirement of an n-detection imposes a bias against blue objects. Although, with the full four-color data one can easily separate nonthermal from stellar continua.

Photometric zero-points were determined generally from photoelectric measurements of faint stars in the fields or by adjusting the zero-point so that the sky surface brightness was some assumed value. I estimate that the zero-points are determined to within approximately 0.5 mag. I am in the process of determining the photometric calibration to much higher precision.

Selection of objects for quasar candidates involves two selection criteria: one of image, and one of color.

3.1 Image Criteria

The determination of whether a particular object is resolved depends on many parameters such as seeing, depth, local graininess, local scattered light, and even wavelength. Also important is how well our noiseless stellar template represents a real stellar image. There are many ways to choose classification criteria. For instance, one could demand that selected objects be unresolved on one, two, or even three of the multicolor plates. Classifications based on two or more colors while improving sample purity also lower the sample from its true size. I feel it is best to decide classification based on only one color, preferably that plate material with the best signal-to-noise ratio, e.g., that of j or f.

Classifications based on the ultraviolet plate material will be less reliable than that based on the redder, higher signal-to-noise bandpasses. The major contamination of the blue star counts will be from faint blue galaxies. Generally these galaxies have compact cores in the near-ultraviolet and blue, but in the red the underlying disc is sufficiently strong to produce a much more extended image. Also, because of atmospheric refraction, ultraviolet images can be elongated, thus reducing resolution sensitivity.

For this quasar survey we have used the image data of the j-plate material and selected only objects with unresolved images. An intercomparison of classifications based on the f-plates with that of the j-plates has not yet been done.

3.2 Color Criteria

Quasar candidates were chosen on the basis of their position in the (u-j) vs (j-f) diagram. This was done for several magnitude intervals. I decided not to count objects brighter than j = 20.5 to avoid any potential saturation problems. A bright-end cut-off this faint ensures that all photometric measurements are on the linear part of the HD curve. The faint-end cutoff was determined by the galaxy completeness limit of about j=24.0. Quasars lie below the subdwarf distribution at about (u-j) = -0.20. Our color criteria are (u-j) < -0.5 and (j-f) < 0.8. Choosing the red (u-j) limit of -0.5 produces a bias against redder quasars. I have corrected for this incompleteness by examining the (U-B) distribution of quasars selected from the radio-selected samples of Schmidt (1968) and of Wills and Lynds (1978). This procedure assumes that there is no color difference between radio- and optically-selected quasars. This correction amounts to about 12%.

3.3 Contamination

Counts of color-selected quasars suffer from three sources of contamination: Galactic stars, narrow emission-line/blue unresolved galaxies, and resolved blue galaxies. Only a spectroscopic survey will eliminate the first two, and good image discrimination will remove most of the third. This resolved galaxy contamination is not the same between fields since the centroid of the galaxy distribution with respect to the stellar locus in two-color diagrams changes from field to field. (These differences are probably indicative of different spatial distributions.) Also, the contamination is a function of the apparent magnitude, since galaxies become progressively bluer at the fainter levels.

The quasar counts of Koo and Kron (1982) suffer from poor image discrimination, as follow-up spectroscopy has confirmed (Koo 1983). Contaminations up to 50% at their faint end are apparently present. In a more recent analysis Koo (1983) presents counts for SA57 corrected for both resolved and unresolved blue galaxies. Corrections for unresolved galaxies are about 10% at j = 22.5 (Koo 1983). The narrow emission-line unresolved galaxies contaminating these faint quasar samples are presumably analogous to B264 (Arp 1970) and B234 (Braccesi et al. 1968).

4. DIFFERENTIAL COUNTS

In Figure 1 I present differential counts (number per square degree per magnitude interval) for this survey averaged over our seven fields. The vertical error bars represent the standard deviation of the seven fields. Also presented are counts from other surveys: Schmidt and Green (1983), Green and Schmidt (1978), Usher (1981), Braccesi et al. (1970) (BFG), Formiggini et al. (1980) (vf), Kron and Chiu (1981), and from Koo and Kron (1982) (KK). I have not corrected our counts for unresolved narrow emission-line objects as these corrections are not known precisely. Also, no corrections for Galactic dust absorption has been made since they are quite small as Table II indicates.

These counts are systematically lower than that of KK. We attribute the predominant difference to a much improved star/ galaxy separation. Also, the data in Figure 1 is an average, whereas the data of KK involves only one field, that of SA68:1. This field appears to yield systematically higher number-counts that the remaining six fields, as indicated in the interfield comparisons of Figure 2. It is also interesting to note that our counts show a brighter turn-over than that of KK.

The differences in the number-counts between various fields might be due to some instrinic property of the field like superclustering or a patchy intercluster medium. Also, such behavior could be due to poorly-determined zero-points, plate characteristics, or nonuniform blue galaxy contamination. The interfield dispersion may be magnitude dependent. If the interfield dispersion is real, it will ultimately limit how well the quasar evolution function will be determined. It might even turn out that the particular form of the evolution function is field-dependent.

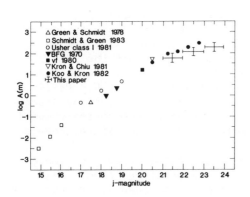 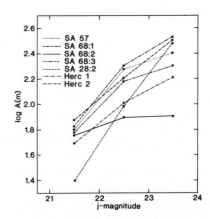

Fig. 1. (left): differential counts; new data represents a mean of the seven fields.
Fig. 2. (right): differential counts for the seven individual fields.

I do not feel that the number-count data, as it currently stands, warrants detailed modelling for evolutionary/luminosity effects. Photometric zero-points will need to be accurately determined. Nevertheless, it is my qualitative impression that pure density evolution can be ruled out because of the steep rise in the number-counts it predicts. It seems entirely possible that pure luminosity evolution might adequately describe the data presented here.

This work was carried our while I was a visitor at Kitt Peak National Observatory. I am grateful to its Director for generous PDS and computer support. I am most appreciative to Mr. Frank Valdes for the invaluable help he has provided me during the last seven months while I was working on the first part of the prime-focus/FOCAS survey. I am also grateful to Mr. Richard Dreiser for his assistance in preparing this manuscript. This work was supported in part by a grant from the National Science Foundation.

REFERENCES

Arp, H.: 1970, Astrophys. J., 162, 811.
Braccesi, A., Lynds, R., and Sandage, A.: 1968, Astrophys. J. Letters, 152, L105.
Braccesi, A., Formiggini, L., and Gandolfi, E.: 1970, Astron. and Astrophys., 5, 264.
Braccesi, A., Zitelli, V., Bonoli, F., and Formiggini, L.: 1980, Astron. and Astrophys., 85, 80.
Burstein, D., and Heiles, C.: 1982, Astron. J., 87, 1165.
Formiggini, L., Zitelli, V., Bonoli, F., and Braccesi, A.: 1980, Astron. and Astrophys. Suppl. Series, 39, 129.
Green, R. F., and Schmidt, M.: 1978, Astrophys. J. Letters, 220, L1.
Koo, D.: 1983, paper presented at the 24th Liege Conference on QSO's and Gravitational Lenses, 21-24 June 1983.
Koo, D., and Kron, R.: 1982, Astron. and Astrophys., 105, 107.
Kron, R., and Chiu, L.-T. G.: 1981, Pub. Astron. Soc. Pacific, 93, 397.
Schmidt, M.: 1968, Astrophys. J., 151, 393.
Schmidt, M., and Green, R. F.: 1983, Astrophys. J., 269, 352.
Usher, P.: 1981, Astrophys. J. Supp. Series, 46, 117.
Valdes, F.: 1982a, "Faint Object Classification and Analysis System", Kitt Peak National Observatory.
Valdes, F.: 1982b, Instrumentation in Astronomy IV, Proc. Soc. of Photo-Opt. Instrum. Eng., 331.
Wills, D., and Lynds, R.: 1978, Astrophys. J. Supp. Series, 36, 317.

ON THE CROSS-CORRELATION OF GALAXIES WITH UVX OBJECTS AND QSOS

B.J. Boyle, T. Shanks and R. Fong
University of Durham, South Road, Durham

ABSTRACT

Recent measurement of UK Schmidt plates have yielded interesting results on the cross-correlation of galaxies with QSOs and ultra-violet excess (UVX) objects. We find that in all Schmidt fields so far analysed there appears to be a significant anti-correlation (at angular scales less than 5') between the positions of galaxies and QSOs and between the positions of galaxies and UVX objects. This anti-correlation appears to be restricted to those galaxies that are found in clusters. These observations can very naturally be explained using a model in which dust within clusters of galaxies obscures the QSOs lying at cosmological distances behind them. This hypothesis may be further corroborated by tentative evidence that the UVX objects and the QSOs appear to be reddened close to clusters of galaxies.

INTRODUCTION

For many years controversy has reigned over the proposed association of low redshift galaxies with high redshift QSOs (Arp 1970, Sulentic 1983). The overwhelming problem with recent statistical tests to assess the significance of these associations has been to find un-biased QSO catalogues. For example, one criticism levied at the recent study of Seldner and Peebles (1979) was that the Burbidge Crowne and Smith (1978) QSO catalogue may well have contained 'ringers'; QSOs that were detected only through their proximity to bright galaxies.

The appearance, in the last five years, of many new unbiased QSO catalogues (at positions of bright galaxies in the field) has now made it possible to test objectively these claims for galaxy/QSO associations, resulting in a far greater degree of confidence in the final results. A great deal of the work done in procuring these new catalogues has focussed on eyeball searches of UK Schmidt objective prism plates for emission-line objects. The area of sky covered by these plates produces catalogues of QSOs whose membership range up to a few hundred, at the limiting magnitude of the search. These large unbiased catalogues are therefore ideally suited for correlation analysis.

We describe here the results of cross correlating these QSO catalogues with the positions of faint galaxies found over the same areas using COSMOS machine measurements of UK Schmidt J plates. A more detailed discussion of these results can be found in Shanks et al. (1983) and Boyle, Fong and Shanks (1984).

RESULTS

Following Seldner and Peebles (1979) we use the 2 point cross-correlation function in analysing the data. The galaxy-QSO cross-correlation function is obtained by centering on each QSO in the catalogue in turn and counting the number of galaxies in an annulus of width, $\Delta\theta$, and angular distance θ from the central QSO. The counts are then compared to those found from a random distribution of test points around the QSO. The 2 point function w_{qg} is then

$$w_{qg} = \frac{\text{average number of galaxies in range } (\theta, \theta + \Delta\theta)}{\text{average number of random points in range } (\theta, \theta + \Delta\theta)} - 1$$

where w_{qg} will be zero from a Poissonian distribution of galaxies around QSOs, positive if galaxies cluster around QSOs and negative if galaxies and QSOs avoid each other.

We have now analysed one field centred at (00h 53, -28°) which contains the South Galactic Pole (SGP), and two other fields at (22h, -18°55') and (1h 12 -35°) for all of which we have both a complete QSO catalogue and a list of galaxy positions from the COSMOS measurement of

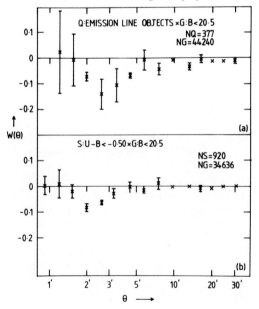

Fig. 1a. Cross correlation of QSOs with galaxies (B<20.5).

Fig. 1b. Cross correlation of UVX objects with galaxies (B<20.5).

FIG 1

UK Schmidt plates. The positions for the QSOs were found from Clowes and Savage (1983), Savage and Bolton (1979), and Savage et al. (1983).

On cross-correlating the samples (Figure 1a) we find a somewhat surprising result. Rather than finding an excess of QSOs around galaxies as Seldner and Peebles had done we see a deficiency of QSOs near the galaxies at the 3σ level. At yet smaller scales <1', there may also be the suspicion of an upturn, but the statistics here are poor and w_{qg} is very noisy. This anti-correlation was originally found on the SGP alone (Shanks et al. 1983). The result was verified on the 22h and 1h 12 fields. The combined result for all 3 fields is shown here, the error bars are derived from the field-to-field variation in the estimator for $w_{qg}(\theta)$ on each field.

On obtaining this result we were immediately concerned that a selection effect was responsible for the observed anti-correlation. One can easily imagine a situation in which emission line objects lying near to galaxies would be rejected from an 'eyeballed' QSO catalogue in case they were merely overlapped galaxy spectra.

To test for this selection effect we procured an automatically detected sample of QSOs. This was achieved by obtaining U and B magnitudes from COSMOS measurements of 50000 stars with $B < 20\overset{m}{.}5$ in three fields; the SGP and 22h fields as previously analysed and one new field centred at (12h 30, 0°).

From Sandage and Luyten (1969) it is well known that, at high galactic latitudes and faint magnitudes, many UVX stars are QSOs. We thus defined a sample of stars with UVX. The criterion for inclusion in the UVX sample was $17.5 < B < 20.0$ and $U-B < -0\overset{m}{.}50$, giving a surface UVX star density of 25 deg^{-2}. It is generally accepted that at $B < 20\overset{m}{.}0$ the density of QSOs is 10-15 deg^{-2} (Veron and Veron 1982). We expect, therefore, that this UVX sample will be contaminated by ordinary galactic stars. This contamination will only serve to decrease any apparent clustering or anti-clustering as galactic stars are Poisson distributed with respect to faint galaxies. From a slit spectrum survey (see Shanks 1983, it appears that the proportion of QSOs in the UVM samples is, indeed, about 40-45%, the remainder being halo stars and white dwarfs.

The cross-correlation of the UVX stars with the galaxies is shown in Fig. 1b. We see a similar result to that found with the QSO sample, although at a lower amplitude (due to the presence of galactic stars). Again this result was first seen on the SGP and was reported in Shanks et al. (1983). It has now been substantiated by the inclusion of two new fields.

Although the same selection effect could not cause both the anti-correlation seen in the QSO/galaxy cross-correlation and in the UVX/galaxy cross-correlation, it is still conceivable that some selection effect in the COSMOS measurement process could be responsible for the observed anti-correlation seen in this case. The cross-correlation of

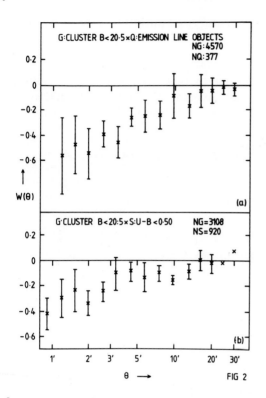

Fig. 2a. Cross correlation of QSOs with cluster members

Fig. 2b. Cross correlation of UVX objects with cluster members.

a large control sample of non-UVX, ordinary galactic stars as detected by the COSMOS machine from their broadband colours did, however, produce a Poissonian result at all scales (Shanks et al. 1983).

To investigate the origin of the anti-clustering in the galaxy sample we divided the galaxies into field and cluster subsamples, using the cluster detection algorithm of Gott and Turner (1977). The clusters were chosen to have a density contrast 8 times that of the background density and to have a membership of at least 6 galaxies. The clusters thus detected ranged in size from small groups to rich Abell-type clusters.

The cross-correlation of these clusters with the UVX stars and the QSOs is shown in Figs. 2a and 2b. The anti-correlation seen at all scales less than 5' in the UVX sample and at scales less than 8' in the QSOs is far more marked than before, (3.5σ with the QSOs and 3σ with the UVX objects) and clearly shows that the clusters are the major cause of the anti-clustering. In contrast the results found with the field galaxies (not shown) are consistent with a Poissonian distribution of QSOs and UVX stars around them.

INTERPRETATION

The simplest explanation of the observed anti-correlation between clusters of galaxies and QSOs is that dust lying in line of sight clusters obscures the QSOs lying at cosmological distances behind them. Evidence for reddening of QSOs in the vicinity of galaxy clusters has recently been claimed by Boyle et al. 1984 who found that QSOs and UVX objects which lay within 2' of a cluster were on average $0\overset{m}{.}1$ redder in U-B than those found elsewhere. If this tentative result is confirmed then it would be a strong indication that the dust absorption model is correct.

On this assumption, we can estimate the amount of absorption needed to produce the observed amplitude of the cluster/QSO cross-correlation w_{qc}. We assume that the dust resides entirely in the galaxy clusters and that, at the faint magnitude limit for the detection of these QSOs the QSOs follow a number magnitude relation of the form

$$n(m_B) \propto 10^{0.6 m_B} \quad \text{(Veron \& Veron 1982)}$$

If we have A_B magnitudes of absorption associated with each cluster then it can easily be shown that

$$w_{qc} + 1 = 10^{-0.6 A_B}$$

giving an A_B of $0\overset{m}{.}35$ for $w_{qc} = -0.40$.

From this value of A_B we derive a reddening of $E(U-B) \simeq 0.1$. This agrees well with the values found from the colours of UVX stars and QSOs close to cluster galaxies, though caution must be urged lest we read too much into the significance of these results. Bogart & Wagoner (1973), again from cross-correlation techniques, proposed an absorption of $A_B = 0\overset{m}{.}5$ associated with Abell clusters, in good quantitative agreement with our results considering our clusters are not as rich as those used by Bogart & Wagoner. The mass of dust associated with this absorption can easily be calculated (Salpeter 1979). Given that our clusters have a typical angular diameter of 6' and are located at redshifts $\simeq 0.15$, we find, on average, $10^{10} M_\odot$ of dust in each cluster.

CONCLUSIONS

It has been shown that, in all fields so far studied QSOs and UVX objects are significantly anti-clustered with respect to galaxies that are members of clusters. Additional evidence from the reddening of QSOs and UVX stars around galaxy clusters, indicates that the most natural and simplest model for explaining these results is one in which QSOs distributed at cosmological distances are obscured by the dust located in intervening, foreground clusters of galaxies.

REFERENCES

Arp, H.C., 1970, Astr. J., 75, 1.
Bogart, R.S., and Wagoner, R.V., 1973, Ap. J., 181, 609.
Boyle, B.J., Fong, R., and Shanks T., 1984, in preparation.
Burbidge, G.R., Crowne, A.M., and Smith, H.E., 1977, Ap. J. Supp. 33, 113.
Clowes, R.G., and Savage, A., 1983, MNRAS, 204, 365.
Gott, J.R., and Turner, E.L., 1977, Ap. J., 216, 357.
Margolis, S.H., and Schramm, D.N., 1977, Ap. J., 214, 339.
Salpeter, E.E., 1977, Ann. Rev. Astron. & Astrophys. 267.
Sandage, A., and Luyten, W.J., 1969, Ap. J., 155, 913.
Savage, A., and Bolton, J.G., 1979, MNRAS, 188, 599.
Seldner, M., and Peebles, P.J.E., 1979, Ap. J., 227, 30.
Shanks, T., 1983, in "Proceedings of the Workshop on Astronomical Measuring Machines", pages 247-252, Royal Observatory, Edinburgh.
Shanks, T., Fong, R., Green, M.R., Clowes, R.G., and Savage, A., 1983, MNRAS, 203, 181.
Sulentic, J.W., 1983, Ap. J. Lett. 265, L49.
Veron, P., and Veron, M.P., 1982, Astron. & Astrophys., 105, 405.

DISCUSSION

J.-L. NIETO: The effect found by Seldner and Peebles (1979) does not hold any longer since Seldner himself and myself have redone the study and showed that the effect was just due to observational biases toward QSOs near galaxies. In our study, however, we found an excess of $m \leq 19$ Lick galaxies at 2° from the QSO, at a quite significant level. Could this be explained by your absorption hypothesis? In addition, what is the meaning of the error bars?

B.J. BOYLE: There is no explanation in the dust model for this excess of galaxies at 2° that you have observed. We have not investigated the properties of the cross correlation function at such large angular scales so I can make no further comment. The error bars used are the r.m.s. field-to-field variation in the values for $\omega(\theta)$ obtained on each field.

THE OPTICAL VARIABILITY OF 3C 446

C. Barbieri, S. Cristiani, G. Romano
Institute of Astronomy, University of Padova

3C 446 is one of the most violently variable quasars.
Its optical variability has been studied for instance by Barbieri et al. (1978), by Pollock et al. (1979) and by Miller (1981). The quasar exibits periods of prolonged activity, with amplitude exceeding 3 mag, and with intraday significant variations. We observed 3C 446 in the last month of Aug. 1983, finding the quasar of unpreceeded brightness.

Three plates were obtained with the 67/92 cm Schmidt telescope on Aug. 4, 9 and 11, and in all three the quasar is seen at B=15.1, the brightest value ever recorded. We have only one previous plate of low quality taken on July 31, and 3C 446 is fainter by approximately 0.7 mag. Finally, on two plates taken the 27th of Aug. with the 182 cm telescope at Cima Ekar we see it again down to B=16.2.

We have drawn in Fig. 1 the light curve from 1967 to 1983. The impression is of an overall increase of brightness. Hopefully other observers will follow 3C 446 in the next months. Spectroscopic coverage of this active phase is also of extreme interest.

REFERENCES

Barbieri, C., Romano, G., Zambon, M.: 1978, Astron. Astrophys. Suppl. 31, 401.
Miller, A.R.: 1981, Astrophys. J. 244, 426.
Pollock, J.T., Pica, A.J., Smith, A.G., Leacock, R.J., Edwards, P.L., Scott, R.L.: 1979, Astron. J. 84, 1658.

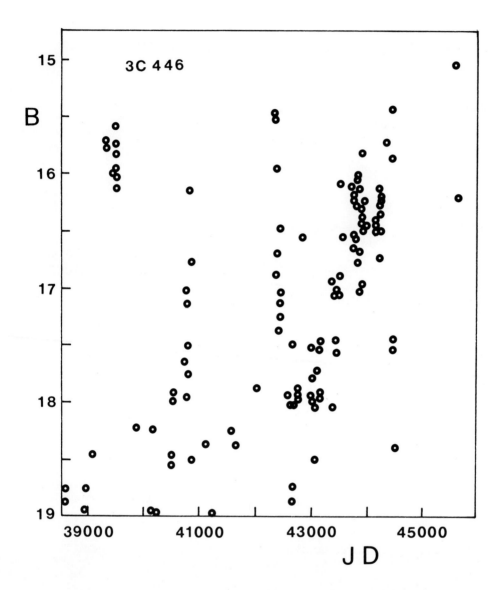

FIGURE 1 - Light curve of 3C 446 from 1967 to 1983.

THE PHYSICAL NATURE OF THE BLUE OBJECTS IN THE FIELD OF BD+15°2469
(VIRGO CLUSTER)

L.Abati Erculiani[1] and H.Lorenz[2]
[1]Istituto di Astronomia, Università di Padova,
 Vicolo dell'Osservatorio, I-35100 Padova, Italy.
[2]Zentralinstitut fur Astrophysik der Akademie der Wissenshaften der DDR, DDR-15 Potsdam, R.-Luxemburg-Str.17a

INTRODUCTION

This is the second Asiago blue objects field studied to determine the physical nature of ultraviolet excess objects (UVX) at high galactic latitudes using low dispersion objective prism plates. As the first of this series (Abati 1982)this work has a double aim :
1. to determine the proportions in the component populations of UVX objects selected by the two-colour method on Schmidt telescope plates;
2. to determine the surface density of quasars down to a limiting magnitude m_{pg} = 18 mag.

The two colour method is an efficient way to find quasars by optical means (Sandage 1965, Braccesi et al.1970).
The majority of quasars with z < 2.2 , as nonthermal continuum sources, shows an ultraviolet excess greater than that of the hot main sequence halo stars, F and G subdwarfs and frequently of white dwarf or blue compact galaxies. Nevertheless, there is a contamination among the UVX quasar candidates selected by colour criteria (Becker 1970, Braccesi 1980, Savage et al.1978). Therefore the separation of the U-B colour index is only partial. Objective prism spectra allow spectral classifications up to a half spectral class (Nandy et al.1977) and additionally show dominant emission or absorption features in the spectra. Thus, without extensive spectral investigations, quasar candidates can be extracted from lists of UVX objects.
Here we investigate the Asiago A_3 UVX objects list (Barbieri and Benvenuti 1974) which cover the magnitude range between 15 and 18 mag in the Virgo cluster region. As the field lies in the Virgo cluster, our nearest neighbour and best studied cluster of galaxies, the surface density of quasars and their location towards the cluster galaxies can be compared with the results obtained in fields outside of clusters of galaxies.
The spectral classification of the UVX objects has been obtained using objective prism plates of the 48" U.K. Schmidt telescope and the 52" Tautenburg Schmidt telescope. The properties of the objective prism spectra of these two telescopes are nearly the same; reciprocal disper-

sion of 2500 A/mm at H_γ and 3500 A/mm at H_β.
The U.K.plates kindly allowed by the Edinburgh Royal Observatory are exposed on Eastman-Kodak IIIa-J hypersensitized emulsion and cover the wavelength range λ 3300-5200 A.
The A_3 objects of the central part of the field (3° x 3°) have been studied also on the Tautenburg objective prism plates. The plates have been exposed both on Kodak 103a-E (wavelength range λ 3300-6700 A) and ORWO ZU 21 (Wavelength range λ 3300-5000 A). The limiting magnitude of the Tautenburg plates is 18.5 mag in the B-band.
All the A_3 UVX objects have been checked on a direct IIIa-J U.K.Schmidt telescope plate, also kindly provided by the Edinburgh Observatory, in order to select the compact galaxies with a starlike appearance on the Asiago Schmidt plates.

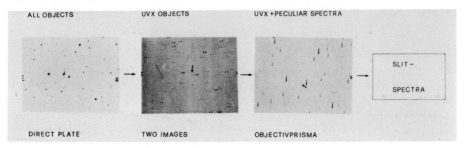

Figure 1 Search line.

RESULTS

277 A_3 UVX objects have been classified as shown in Table 1.

Table 1 - Spectral classification of the A_3 blue objects -

	U-B ≤ -0.2		U-B ≤ -0.4		U-B ≤ -0.7	
	N	%	N	%	N	%
Hot stars	116	43	33	45	6	31
F and G stars	105	39	19	26	0	0
W.D.	7	3	4	5	3	16
Galaxies + G?	20	7	6	8	1	5
Quasars + Q?	22	8	11	15	9	47
Total number	270		73		21	
Double/peculiar	7		5		2	

$\rho_{Q_{max}(Q+Q?)}$ =0.98 Q/deg².; $\rho_{Q_{min}(Q)}$ =0.49 Q/deg².; $\rho_{W.D.}$ =0.31 W.D./deg².

Most of them are stars (43% hot stars, 39% F and G stars, 3% W.D.) 8% are quasars and 7% compact galaxies, unresolved at the scale of the Asiago Schmidt telescope. 7 blue objects have been identified as double stars.
The surface density of W.D. is 0.31 WD/sq.deg.
The surface density of A_3 UVX quasars spreads from 0.49 Q/sq.deg. to 0.98 Q/sq.deg. in good accordance with the one determined from previous works.
Figure 2 shows the distribution of blue objects in the A_3 field as a function of magnitude for different colour index. From this figure (a) one can see that the number of all kinds of objects decreases as the U-B becomes more negative. It is to be pointed out that while the proportion of stars decreases, the proportion of quasars increases (c). Star contamination is lower both at fainter magnitudes and at more negative U-B. In the magnitudes range here considered the number of quasars, at a given U-B, is higher at fainter magnitude. This fact reflects the steep quasar luminosity function. From the same figure (b) it is evident that the proportion of quasars, at a given magnitude, increases at more negative U-B.

BLUE COMPACT GALAXIES

Fifteen A_3 blue objects are identified as galaxies and forty as suspected galaxies in the B.B. list. This means that these objects, which look like stars on the two images plates, have nonstellar appearance on the Asiago direct plates. Two out of the fifteen galaxies lie outside the objective prism field. Eleven out of the thirteen ones inspected show typical galaxies' spectra and two have star like spectra.
All the A_3 blue compact galaxies have been observed recently by Karachentsev and Karachentseva (1982) at the 6m telescope of the URSS during a large investigation of about 100 galaxies in the Virgo cluster. From the radial velocity measured by H and K lines, one can see that 10 out of them are dward emission galaxies, members of the Virgo cluster, and 3 are background galaxies.
The two A_3 blue objects proposed as galaxies in B.B. and here classified as stars do not appear in K.K. galaxies list, and this proves the reliability of the classification obtained from objective prism.
Out of the forty suspected galaxies 8 lie outside the objective prism field, 27 are classified as stars and 5 as galaxies. One of them has been studied by K.K. and estimated as a background galaxies, not connected with the Virgo cluster.

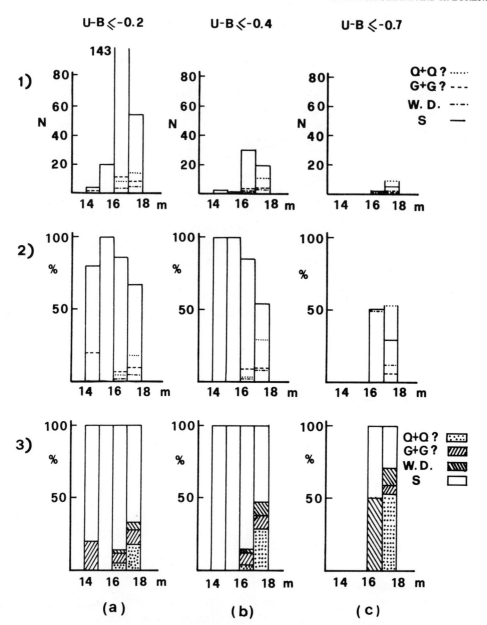

Figure 2.-1) The distribution of A_3 blue objects as a function of magnitude in the subsamples with different U-B.
2) The normalized distribution of A_3 blue objects as a function of magnitude in the subsamples with different U-B.
3) The proportions of A_3 blue objects as a function of magnitude in the subsamples with different U-B.

BLUE OBJECTS AND CLUSTER OF GALAXIES

The location of the A_3 UVX quasar candidates has been checked as regards the bright Virgo cluster galaxies and the more distant cluster of the Zwicky catalogue (Zwicky 1961) in order to inspect the relation quasar-galaxy and quasar-cluster of galaxies.
Nine out of the twenty-two quasar candidates (Q+Q?) are near bright galaxies at distances $r < 1°$ (300 Kpc at the Virgo cluster distance). Four out of the twenty-two quasar candidates are located, in their projection, inside clusters of galaxies classified respectively as Distant (one), Medium Distant (one), and Very Distant (two).
However, since from the objective prism spectra the redshift of quasar candidates cannot be doubtlessly measured, further spectroscopical work at higher dispersion is worthwhile in order to inspect the possibility of physical associations.

PRELIMINARY HIGH DISPERSION SPECTROSCOPICAL RESULTS

Twelve quasar candidates of A_1 and A_3 fields have been investigated at higher dispersion using the UAGS spectrograph attached to the prime focus of the 6m BTA Telescope of the Academy of Sciences of USSR. 10 out of them are quasars with redshift ranging from 0.40 to 2.04. Further spectroscopical work is in progress.

CONCLUSIONS

1) The 276 A_3 blue objects have been separated into their component populations: 85% of them are stars (43% hot stars, 39% F and G stars, 3% white dwarfs) 8% quasars and 7% compact galaxies . 2) The proportions of different kinds of objects change with the magnitude and colour index. In particular the proportion of quasars increases both to fainter magnitudes and more negative U-B. 3) The surface density of UVX quasars to a magnitude limit of B = 18 mag. spreads from 0.49 Q/sq.deg. to 0.98 Q/sq.deg. in good agreement with previous works. 4) Recent investigations on the A_3 blue objects marked as galaxies show that eleven of them are dwarf members of the Virgo cluster. 5) Four out of twenty-two quasar candidates are located in their projection inside distant Zwicky clusters of galaxies. 6) The results here obtained for the A_3 field, which lies in the Virgo cluster, agree, within the statistical uncertainty, with the ones of the A_1 field, outside clusters of galaxies.

ACKNOWLEDGEMENTS

We should like to thank the staff of the U.K.Schmidt Unit and the Tau-

tenburg Observatory of the Academy of Sciences of the G.D.R. for providing to us the objective prism and direct plates.
We are particularly grateful to Dr.M.Hawkins, Dr.I.D.Karachentsev and Dr.G.M.Richter for their helpful comments.

REFERENCES

Abati Erculiani,L.:1982,Astron.Astrophys.Suppl.Ser.48,333.
Barbieri,C.,Abati Erculiani,L. and Rosino,L.:1968,Publ.Oss.PD.N.143.
Barbieri,C. and Benvenuti,P.:1974,Astron.Astrophys.Suppl.Ser.13,269.
Becker,W.:1970,Astron.Astrophys.$\underline{9}$,204.
Braccesi,A.,Formiggini,L. and Gandolfi,E.:1970,Astron.Astrophys.$\underline{5}$,264.
Braccesi,A.,Zitelli,V.,Bonoli,F. and Formiggini,L.:1980,Astron.Astrophys $\underline{85}$,80.
Cheney,J.,E. and Rowan-Robinson,M.:1981,Monthly Notices Roy.Astron.Soc. $\underline{195}$,497.
Green,R. and Schmidt,M.:1982,preprint.
Karachentsev,I. D. and Karachentseva,V. E.:1982 Pism.Astron.Zh.$\underline{8}$,198.
Nandy,K.,Reddish,V.C.,Tritton,K.P.,Cooke,J.A.and Emerson,D.:1977,Monthly Notices Roy.Astron.Soc.$\underline{178}$,63p.
Sandage,A.:1965,Astrophys.J.141,1560.
Savage,A.,Bolton,J.G.,Tritton,K.P.,Peterson,B.A.: 1978,Monthly Notices Roy.Astron.Soc.$\underline{183}$,473.
Zwicky,F.S.:1961,Cat.of galaxies and cluster of galaxies,Caltech.

OPTICAL IDENTIFICATIONS OF RADIO SOURCES WITH ACCURATE POSITIONS USING THE UKST IIIa-J PLATES

Ann Savage
Royal Observatory, Blackford Hill, Edinburgh EH9 3HJ,
Scotland

David L. Jauncey and Michael J. Batty
Division of Radiophysics, CSIRO, Sydney, Australia

S. Gulkis, D.D. Morabito, R.A. Preston
Jet Propulsion Laboratory, California Institute of
Technology, Pasadena, California, U.S.A.

SUMMARY

Three radio identification programmes are described which are drawn from radio samples with accurate radio positions (< 2" arc rms). Optical identifications are being made on the basis of radio-optical positional coincidence alone, without regard to colour or morphology, using the UKST IIIa-J sky survey to a limiting magnitude of 22.5. Some preliminary results are presented.

I. INTRODUCTION

Preliminary optical identification work using the UKST IIIa-J sky survey plate material as it became available and the Parkes 2.7 GHz radio positions (at best accurate to $\sim 10"$) (Savage et al. 1976 and references therein) highlighted the need for accurate radio positions ($\sim 2"$) to minimise the spurious identifications resulting from the high surface density of optical objects on the UKST IIIa-J plates ($\sim 10^{-3}$ per arcsec^2).

This high surface density results from the finer grain of the IIIa-J emulsion and the hypersensitising techniques developed successfully by the UKST team (see Corben et al. 1974 and Sim et al. 1976) which enables these normally slow plates to be useful astronomically, reaching $2^m.5$ fainter than the POSS Whiteoak extension. Additionally the achromat corrector on the UKST means that the image size is well matched over all wavelengths to the emulsion grain size, giving improved resolution and allowing improved morphological classifications (see Savage 1983).

Three major identification programmes are now in progress using the UKST IIIa-J plate material and accurate radio positions. These are described in the following section together with some preliminary results which each programme has produced.

II   THE OPTICAL IDENTIFICATION PROGRAMMES

(a) The Parkes 5.0 GHz deep survey to 30 mJy.

The 5° by 5° region centered 22^h-$18°$ has been surveyed with the Parkes 64 m radiotelescope at 5.0 GHz by Wall et al. (1982). This survey complete to 32 mJy yielded a catalogue of 75 sources, of which 57 had fluxes greater than 30 mJy. These 57 sources have been positioned using the Tidbinbilla Interferometer at a frequency of 2.3 GHz (see Batty et al. 1982 for a description of the performance of the Tidbinbilla Interferometer). A preliminary comparison between the Tidbinbilla and Parkes positions showed a scatter much larger than expected based on the quoted Parkes error. The 6cm position errors appear to be a factor 1.5-2 larger than quoted so that the target diagram appears as in Figure 1.

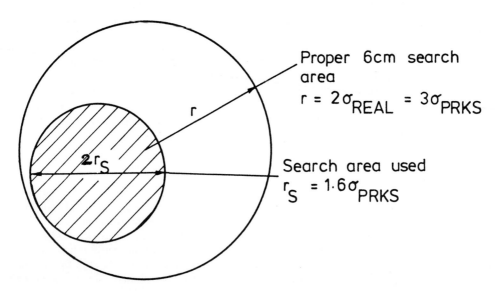

Figure 1. Clear circle, the error circle that should have been searched. Hatched circle, area searched for optical identification using underestimated radio positional errors. This latter area ended up being only one quarter of the area that should have been searched. Thus some 12 misidentifications were made and 24 identifications missed in the work of Savage et al. (1982).

The region searched ends up being only one quarter of the area that should have been searched. Consequently it was expected that few of the published optical identifications would be confirmed and that many of the correct identifications would have been missed.

The identifications using the accurate Tidbinbilla positions have now been completed using UKST IIIa-J, IIIa-F, IV-N and objective prism plates. The original premise is found to be correct; just over 50% (12) of the original identifications claimed by Savage et al. (1982) have been found to be incorrect, and the identification rate has increased from just under 50% (including the spurious identifications) to just over 63%; in all some 24 identifications were wrongly made or excluded by Savage et al.

(b) The Tidbinbilla-UKST Radio Quasar Idenfication Programme

The Tidbinbilla Interferometer is also being used to position all the Parkes 2.7 GHz compact and/or flat spectrum sources south of declination -30°. First results (Jauncey et al. 1982) showed that the Tidbinbilla positions and the UKST survey plates are well matched for the identification of such sources. Anglo-Australian Telescope spectroscopy has confirmed most of the identifications and revealed a redshift of 3.78 for the QSO PKS 2000-330 (Peterson et al. 1982), the highest redshift known so far. For compact sources, the ability to identify such radio sources on the basis of positional coincidence alone, without recourse to colour or morphology, is an important feature of radio identification programmes. Identification criteria such as ultraviolet excess and morphology have in the past provided serious bias in the resulting redshift distributions. Figure 2 demonstrates the bias due to identification procedures based on colour. It shows the error circle due to the Parkes radio position and the original identification with a UVX object, A, although the error circle just touches PKS 2000-330. The small error circle is that from the Tidbinbilla position and it is centered on PKS 2000-330 and excludes object A.

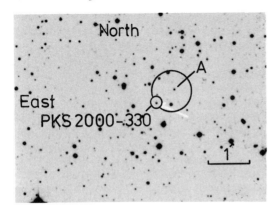

Figure 2. Using UVX criteria as an aid to identification procedures in conjunction with radio source positions with large positional errors, Object A was originally identified as the counterpart of PKS 2000-330. Subsequent acurate radio positioning work uniquely identified the radio source without recourse to colour or morphology criteria.

Additionally, high redshift quasars appear red and have a galaxy type morphology on POSS plates (Savage 1983) because the presence or absence of the strong emission lines, Lyman-α in particular, has a significant effect on the quasar colours and image structure. This effect is compounded by the presence of the Ly-α absorption forest and also of any Lyman limit absorption (see Fig. 3). The density of absorption lines increases with increasing redshift (Peterson 1983) with the result that the integrated continuum magnitudes on either side of the Ly-α emission line differ significantly. For PKS 2000-330 these broad band colours show a $1^m.7$ difference. Thus the most luminous quasar appears as a $17^m.3$ object in the red but drops to $19^m.5$ on the "blue" UKST IIIa-J plates, and we might expect quasars with z > 3.5 to appear at $\sim 20^m.0$ rather than $18^m.0$ on those plates. Interestingly at radio wavelengths the z > 3.0 quasars have been found to have spectra with distinct peaks (see Figure 4).

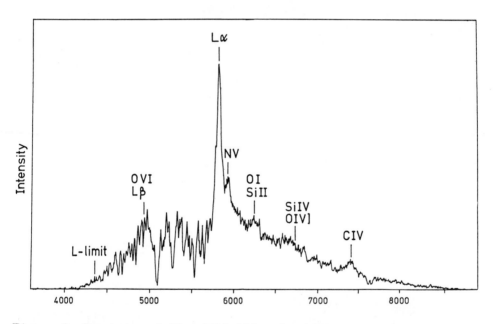

Figure 3. Spectrum of Pks 2000-330, the highest redshift quasars at z = 3.78. The continuum emission drops by about 1.5 magnitudes between the long wavelength side of Ly α and the short wavelength side.

The observed radio flux densities and optical magnitudes of the known z > 3.0 radio quasars show clearly that if similar objects exist at z > 3.5 then they should be well above the existing radio (Parkes 2.7 GHz) and optical (UKST IIIa-J) survey limits. The Tidbinbilla - UKST programme offers a proven method for finding and identifying z > 3.5 quasars with which to test the various cosmological models of differential evolution and galaxy formation.

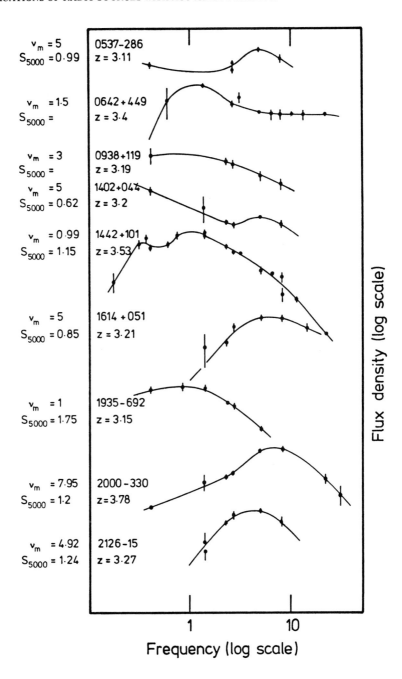

Figure 4. Radio spectra of all the known radio quasars with redshifts greater than 3.0. All show "humps" in their radio spectra, the observed frequency of these humps is given as ν_m on the LHS of the figure.

(c) Optical Identifications of Radio Sources with VLBI Positions.

Very few radio identifications with quasars or bright optically selected quasars are known south of declination -35° and even fewer south of -45°. This VLBI programme of positioning all compact sources with 5.0GHz fluxes > 0.5 Jy (Morabito et al. 1983) has produced some candidates which are brighter than $17^m.0$. These can be used to contribute to the Hipparcos Satellite and Space Telescope complementary astrometry programmes (Duncombe et al. 1982). The NASA/ESA Space Telescope will have the capability of measuring relative positions of 9-17 magnitude stars within its 18 arcmin field of view with an accuracy of 0.002 arcsec rms. The HIPPARCOS instrumental system of bright ($\sim 11^m.0$) stars will produce a solid body reference frame with an overall unknown rotation. Two separate problems with common solutions must be distinguished. One, to determine the rotation of the Hipparcos instrumental system, and two, to tie in the actual coordinates of the Hipparcos system. Using its capability to measure precise angular distances between objects of disparate magnitudes, the ST can be used to tie the Hipparcos system to (in particular) (a) an absolute coordinate system derived from radio interferometric observations using radio sources which have discrete optical counterparts and (b) very distant, and, hence relatively motionless objects such as the QSOs. This VLBI programme will be contributing bright radio QSOs.

The need for such an improved astrometric grid is highlighted by this programme. Residuals from the reference star positions in the solutions for the optical positions were found to increase dramatically with increasing southern declination (i.e. 0.85 at 0° to 2.0 at -80°) and also to increase with epoch: 0.68 (24 fields) for POSS plates epoch 1950.0 to 0.85 (10 fields) for SERC IIIa-J plates epoch 1980.0. Thus in this programme it is now the optical position errors which are dominating and defining the search error circle.

These accurate radio positions have increased the identification rate from 56% to 88%, enabling some identifications to be made in very crowded stellar fields, some subsequently confirmed by AAT spectroscopy. 11% (8 of the 92) of the original identifications have been found to be incorrect.

III CONCLUSIONS

The three accurate radio positioning programmes here and the subsequent optical identification programme undertaken on the UKST IIIa-J plates have demonstrated:

(1) The improvement in the identification rate afforded by the deeper, finer resolution plate material provided by the UKST.

(2) The biasses introduced with identification procedures based on morphology and colour.

(3) The spurious and missed indentifications resulting from identifications using radio positions with large positional errors and high quality deep plate material.

(4) The need for accurate fundamental optical astrometry particularly in the Southern Hemisphere so that identification work based on VLBI positions with an accuracy of $0\rlap{.}''004$ are not limited by the optical position errors of $\sim 2\rlap{.}''0$.

ACKNOWLEDGEMENTS

I would like to thank the staff of the UKST for taking this superb plate material. Thanks also the the Deep Space Network for their assistance with the radio observations. This work is partly supported by NAS7-100.

REFERENCES

Batty, M.J., Jauncey, D.L., Rayner, P.T., and Gulkis, S.: 1982, Astron.J. 87, 938.
Corben, P.M., Reddish, V.C., and Sim, M.E.: 1974, Nature 249, 22.
Duncombe, R.L., Benedict, G.F., Hemenway, P.D., Jefferys, W.H., and Shelus, P.J.: 1982, NASA CP-2244 The Space Telescope Observatory.
Jauncey, D.L., Batty, M.J., Gulkis, S., and Savage, A.: 1982, Astron.J. 87, 763.
Morabito, D.D., Preston, R.A., Slade, M.A., Jauncey, D.L., and Nicolson, G.D.: 1983, Astron.J. 88, 1138.
Peterson, B.A.: 1983, I.A.U. Symp. No. 104, Early Evolution of the Universe and its present Structure, Dordrecht: Reidel (in press).
Peterson, B.A., Savage, A., Jauncey, D.L., and Wright, A.E.: 1982, Astrophys.J.(Lett.) 260, L27.
Savage, A.: 1983, Astron.Astrophys. 123, 353.
Savage, A., Bolton, J.G., and Wright, A.E.: 1976, Mon.Not.R.astr.Soc 175, 517.
Savage, A., Bolton, J.G., and Wall, J.V.: 1982, Mon.Not.R.astr.Soc. 200, 1135.
Sim, M.E., Hawarden, T.G., and Cannon, R.D.: 1976, A.A.S Photo-Bulletin 11, 3.
Wall, J.V., Savage, A., Wright, A.E., and Bolton, J.G.: 1982, Mon.Not.R.astr.Soc. 200, 1123.

GALAXY COUNTS

J. A. Tyson
Bell Laboratories

ABSTRACT

Counts of faint galaxies should reveal any evidence of galaxy luminosity or color evolution, as well as new information on the faint end of the galaxy luminosity function. The FOCAS automated detection and classification software is reviewed, and results of the deep 4m PF photographic survey to 24th magnitude in 23 fields covering 9 sq. degrees are presented. Color-magnitude plots for stars and galaxies are shown, and galaxy color evolution is discussed. Evidence is found for a faint galaxy blue trend at 22-24 J mag. However, the k-correction becomes so severe at redshift ~ 1 that the intrinsically fainter galaxies are emphasized in any magnitude-limited survey. No unambiguous evidence is found for evolution. New 4m limit CCD multi-color data are shown and discussed. The limiting magnitude for detection is 27th J magnitude in 2 hours integration. The data exclude evolution starting at any one epoch for z<10.

1. INTRODUCTION

Hubble recognized the advantage of galaxy counts, as opposed to redshift surveys, as a means of obtaining statistically complete samples of high redshift galaxies, to test for cosmological and evolution effects. In recent years, the limiting magnitude for galaxy count studies has been increased from 19 to 27 B mag. Schmidt telescopes survey large regions but are limited to $B \lesssim 22$ mag for reasonable integration times. Large, low f/number reflectors have pushed beyond this limit into a range of apparent magnitudes dominated by cosmological and evolutionary effects and the faint tail of the luminosity function. Departures of galaxy counts from that expected for a nonexpanding homogeneous euclidean universe ($N \sim dex0.6m$) are dominated by the effect of expansion: the k-correction. Figure 1 shows that the k-correction begins to affect the galaxy counts already at J=17 mag (J≡IIIaJ+GG385), and are dominant by J=21 mag. Plotted are counts from some early surveys, along with early results of our FOCAS (Faint Object Classification and Detection System) 4m PF survey.

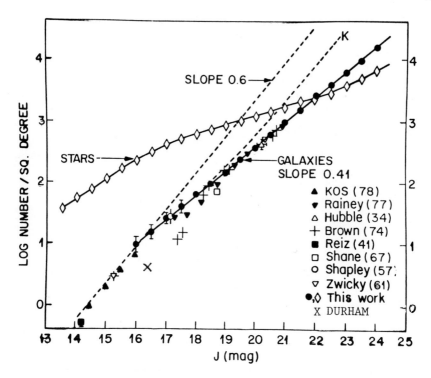

Fig. 1. Galaxy and star counts per square degree to limiting magnitude J, scaled to the north galactic pole. The theoretical curve of slope 0.6 corresponds to the oversimplifications of homogeneity (local density everywhere), flat space-time, and no color-redshift (K) effects. The curve marked K results from making reverse K- and cosmological corrections on the data. Adapted from Tyson and Jarvis (1979).

2. k-CORRECTION DOMINANT

The k-correction dominates all other cosmological or evolutionary effects for all types of galaxies. For example, in figure 2 I show the effects of k-correction on a giant elliptical galaxy. The observed V surface brightness is plotted vs. isophotal radius of the galaxy image, for various redshifts. A photographic surface brightness detection limit of 26 V mag/arcsec^2 implies a nearly stellar image by redshift 1.5, whereas the much lower limit of a CCD detector suggests a 4 arcsec dia resolved image at z=1.5.

I will show some recent CCD data which Pat Seitzer and I have obtained on the CTIO 4m telescope. Although some of the faint galaxies at 26 J mag are probably high redshift, I will argue that it is likely that many are subluminous galaxies at much lower redshift which do not suffer a large k-correction.

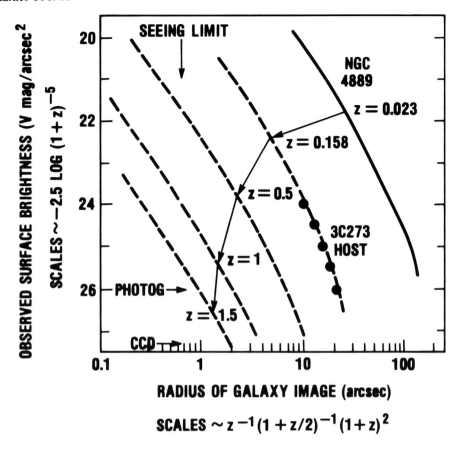

Fig. 2. k-correction dominates the detection limit of even giant elliptical galaxies, suggesting a practical limit of z=1.5 for 1-hour CCD exposure.

Figure 2 is for an unfiltered CCD exposure. If there is a broadband filter or if the detector is a photographic plate then the surface brightness scales like $(1+z)^{-6}$. If one then introduces luminosity evolution $E \sim (1+z)$, then the same scaling as in Fig. 2 is obtained. Thus, considerable evolution cannot prevent a deficit in number counts of these galaxies beyond $z \sim 1$.

Number counts, however, are dominated by spiral galaxies. Assuming Freeman's law we can write the surface luminosity as $B_o \exp(-r/r_o) = (1+z)^4 \alpha \sigma_s/k(z)E(z)$, where σ_s is the night sky surface brightness, α is the fraction of night sky for the outer detection isophote, and k and E are k-correction and luminosity evolution, respectively. This leads to a total "zero-redshift" magnitude correction $K(\alpha,q_o,z)$ given by $K = K_\sigma + K_q$ where $K_\sigma = -2.5 \log \{1-(\alpha \sigma_s/B_o)[1-\ln(\alpha \sigma_s/B_o) -6 \ln(1+z)](1+z)^6\}$ and
$K_q = 5 \log q_o^{-2} \{q_o+(q_o-1)z^{-1}[(1+2q_o z)^{1/2}-1]\}$. Figure 3 shows $K(\alpha,q_o,z)$

plotted vs. redshift z for 2 values of α(1% and 0.4% night sky) and 3 values of q_o(.01,.03,.3). Clearly, aperture or detection isophote effects dominate at the limiting redshifts. However, it is easy to suffer a 3-4 magnitude k-correction for modest redshifts.

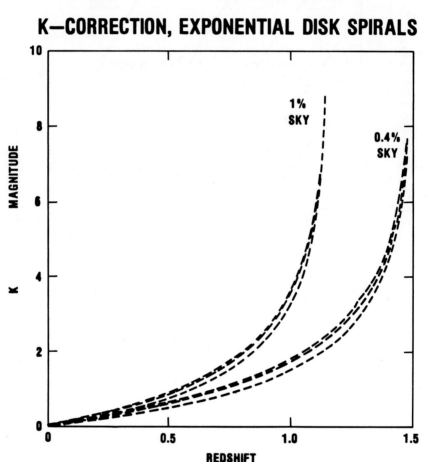

Fig. 3. k-correction to J magnitude for two values of detection limit isophote and three values of q_o (see text), for spiral galaxies.

3. COUNTS OF BRIGHT GALAXIES

Schmidt telescopes play an important role in the detection of evolutionary effects by tying down the galaxy counts at the bright end. For example, the local supercluster may contribute to the counts at $J \simeq 14-16$ mag (see fig. 1), and wide-field surveys going to $J \sim 19$ mag in other directions are needed. Preliminary data from the Durham group (Ellis 1981) from five UKST fields are shown as a cross in figure 1. Clearly, any model of galaxy counts will predict significantly different counts at 24th magnitude depending on which 15th mag count it is normalized to. More complete surveys of 12-19th mag are needed, particularly in the south. Automated reduction classification of the ESO blue and red Schmidt surveys will be very helpful.

4. GALAXY/STAR CLASSIFICATION

Automated classification faint galaxy count surveys are now an international enterprise, although most groups currently report results for only one or two fields. Because of the known clustering of faint galaxies on 1° scales and the open possibility of high-latitude clumped extinction, it is necessary to study tens of fields spaced over the sky. There has also been a trend towards classifiers based only on a few (2 or 3) moments of the intensity. Such classifiers have limited dynamic range, and fail systematically as galaxy brightness approaches sky. Recent faint galaxy count surveys have been reported by Kron (1978,1980), Tyson and Jarvis (1979), Peterson et al. (1979), and Karachentsev (1980). There are several currently in progress, and I will report here on the results of our survey to date. Theoretical predictions incorporating various evolutionary scenarios have been made by Brown and Tinsley (1974), Tinsley (1977,78,80), Bruzual and Kron (1980), Bruzual (1981) and Koo (1981). All the credit for developing luminosity evolution models suitable for direct comparison with observation must go to Beatrice Tinsley. Her schematic models have been an incentive for more extensive galaxy count studies at both bright and fainter limits.

The 7-moment parametric star/galaxy classifier (FOCAS, Jarvis and Tyson 1981) has a range of 17-23 J mag for accurate classification on 1-hour 4m PF J limit plates. An improved version of FOCAS, allowing an additional magnitude of classification at the faint end (Valdes, 1982) has been applied to 23 of our FOCAS photographic fields and two new CCD fields. This new classifier is based on successive convolutions of an ensemble of stellar + various broadened stellar images with every detected object. Goodness of fit is then related to a classification based on the amount of broadening and fraction of broadened image required. A price in computer run-time is paid for this 'infinite-moment' resolution classifier: whereas the 7-moment cluster algorithm classifier takes about 6 hours on a VAX 11/780 for a 4×10^7 pixel 4m PF photographic field to 24th J mag, the new classifier takes 20-30 hours for the same data, and .5-10 hours on a 10^5 pixel CCD image to 26th J mag, depending on field crowding and the number of other users. Figure 4 shows the results for 23 4m limit photographic fields and two recent 4m CCD fields.

5. FAINT GALAXY COUNTS

The best fit to the average of the 23 FOCAS fields for 21 < J < 23 mag is $\log N \sim .432 J$. The dotted lines in figure 4 are the 1σ bounds for field-to-field variations in the photographic data. The two squares are FOCAS galaxy counts based on two 4m PF CCD fields obtained recently by Pat Seitzer at CTIO in a total of 4 hours of integration in J. The limiting magnitude for object detection (50% completeness) is 27 J mag,

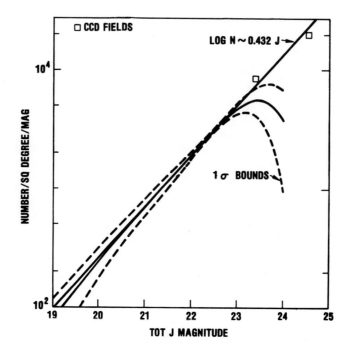

Fig. 4. Differential galaxy counts from 19-25 J mag.

and for photometry and classification 26 J mag. Nearly all objects are found to be blue galaxies. These preliminary CCD data for two fields at 60° galactic latitude are consistent with the continuation of the logN∼.43 J relation to 25 mag and beyond.

6. COMPARISON WITH THEORY

The galaxy count data are compared with theory in figure 5, adapted from Tinsley (1980). Galaxy counts, normalized by logN=0.6 J, are plotted versus total J magnitude. Our original FOCAS results based on 6 high galactic latitude fields are shown as open circles. The results presented in figure 4 are shown as solid circles continuing to J=26 mag. [Our current magnitudes are within 0.2 mag of total due to our low effective surface luminosity threshold of 26.5 mag/sq. arcsec photographic and 28 mag/sq. arcsec CCD]. Tinsley's no-evolution prediction is shown by the solid line intersecting the lower right-hand corner. Other recent no-evolution predictions (based on various mixes of galaxy colors and differing M_J^*'s) by Peterson, et al. (1979), Ellis (1981), and Koo (1981) are shown as dashed lines. Tinsley's three models with evolution are shown as dotted lines (Tinsley 1980). The vertical bar indicates the amount of free adjustment Tinsley estimated between theory and experiment, based on uncertainties at the bright end, problems with magnitudes at the faint end, and allowable range of mixes of galaxy colors input to the theory.

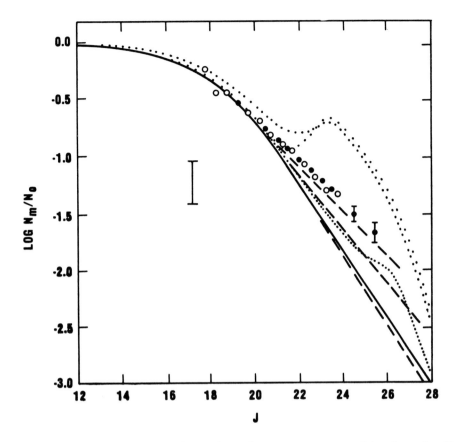

Fig. 5. Observed and predicted number counts normalized to $\log N \sim 0.6\,J$, adapted from Tinsley (1980).

Clearly star-burst evolution models at formation redshifts less than $z \sim 5-10$ are ruled out. Although the data are suggestive of some kind of continuous evolution, the case for evolution is not unambiguous, due to the uncertainties.

7. FAINT GALAXY COLORS

As in faint galaxy counts, the dominant systematic effect in faint galaxy colors is due to the k-correction. M^* galaxies at these faint apparent magnitudes have large k-corrections (see fig. 3) and the sample is systematically biased towards intrinsically fainter 'dwarf' galaxies which are relatively nearby and thus do not suffer as great a k-correction. Since dwarf galaxies are about 0.5 mag bluer (see deVaucouleurs et al. 1981) in J-F than the mean color of a complete sample of nearby galaxies, this means that there will be a blue trend in mean galaxy color with apparent magnitudes for $J \gtrsim 22$ mag. Various burst models of galaxy evolution predict an additional blue trend whose amplitude depends on the assumed formation redshift.

Color-magnitude plots for stars and for galaxies in a typical FOCAS field are shown in figure 6. The well-known sharp edge to the colors of

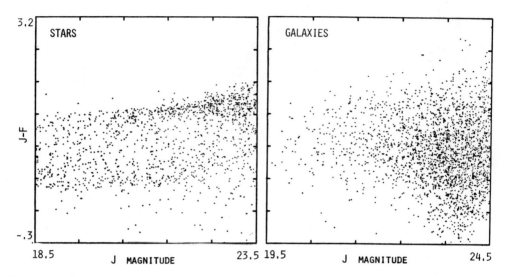

Fig. 6. Color-magnitude diagram for objects classified as stars (left) and galaxies (right) in one FOCAS field.

the m-dwarf stars (upper branch in figure 6, left) can be used as a systematic-error-free calibrator of color: For each half magnitude in each FOCAS field, I measure the difference in J-F color between this red edge to the red dwarf star distribution and the mean of the galaxy colors for the same field. We currently have J,F color data in 12 fields.

The resulting color-magnitude diagram for an average of 12 FOCAS fields and for $20 < J < 24$ mag is shown in figure 7. There is a very significant blue trend in the data, consistent within 2σ with both Tinsley's (error bar) and Bruzual's (dashed curve) mild evolution models. As mentioned above, this is a combination of the k-correction color bias and whatever color evolution may be present. It is easy to account for all of the blue trend with no color evolution, taking the observed luminosity function and a reasonable estimate of the k-correction (1 mag) at J=23.5 mag. Thus, although suggestive, the evidence for evolution in the faint galaxy colors is not unambiguous. What is needed is a redshift for each of these galaxies, an unlikely prospect in the near future.

8. PROBLEMS COMPARING COUNTS WITH THEORY

The problems encountered above in deconvolving any evolutionary effects from galaxy counts and colors divide into three separate areas:

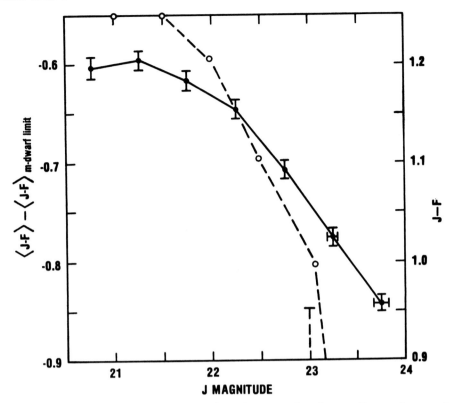

Fig. 7. Color-magnitude diagram for faint galaxies. Some theoretical estimates are also plotted.

8.1. Bright end normalization

Effects of our local superclustering ($J \sim 14$ mag) generate departures from the true cosmic average galaxy counts in this apparent magnitude range.

8.2. Faint end systematics

Aside from occasionally significant differences in definition of magnitude, star/galaxy classification, and multiple object splitting algorithms, there are two known systematics operating in every case which tend to favor counting dwarf galaxies at the faint end: (1) k-correction bias toward the faint end of the luminosity function. (2) High redshift galaxies are very small and are vulnerable to misclassification as stars.

8.3. Theory

In addition to adopting values for the initial mass function, star formation rate, and galaxy formation redshift, the theory must adopt some mix of galaxy types and colors - which affects in turn the k-correction. Also, the absolute magnitude scale M_J^* in the photometric system (J) used for the counts by most observations is not well known, since it has been obtained by conversion from Johnson colors.

Progress in these areas of difficulty will allow difinitive conclusions regarding the type of evolution allowed by the data. We already see that burst models at formation redshifts <5 are ruled out. The bright end normalization problem will go away when automated counts from Schmidt surveys are available over most of the sky. Progress is already being made on the faint end systematics. A dynamic range for the detector/splitter/classifier of at least 6 magnitudes is crucial. It will be helpful to have F-band counts going as faint as the J-band limit, to cover dwarf galaxies in both bands. In summary, galaxy luminosity evolution, if present, is very mild and continuous.

REFERENCES

Brown, G.S. and Tinsley, B.M.: 1974, Ap.J. 194, p.555.
Bruzual, A.G.: 1981, Ph.D. Thesis, University of California, Berkeley.
Bruzual, A.G., and Kron, R.G.: 1980, Ap.J. 241, p.25.
Ellis, R.: 1981, in VIIth Course of the International School of
  Cosmology and Gravitation: Erice, May 11-13.
Jarvis, J.F., and Tyson, J.A.: 1980, A.J. 86, p.476.
Karachentsev, I.D.: 1980, Sov. Astron. Lett. 6, p.1.
Koo, D.C.: 1981, Ph.D. Thesis, University of California, Berkeley.
Kron, R.C.: 1978, Ph.D. Thesis, University of California, Berkeley.
Peterson, B.A., Ellis, R.S., Kibblewhite, E.J., Bridgeland, M.T.,
  Hooley, T., and Horne, D.: 1979, Ap.J. 233, p.L109.
Tinsley, B.M.: 1977, Ap.J. 211, p.621; erratum Ap.J. 216, p.349.
_____ : 1978, Ap.J. 222, p.14.
_____ : 1980, Ap.J. 241, p.41.
Tyson, J.A., and Jarvis, J.F.: 1979, Ap.J. 230, p.L153.
Valdes, F.: 1982, SPIE 331, p.465.
deVaucouleurs, G., deVaucouleurs, A., and Buta, R.: 1981, A.J. 86, p.1429.

DISCUSSION

T. SHANKS: It is well known that magnitude limited samples of galaxies are biased toward bright, high redshift galaxies. Why then do you think your blue galaxy counts are dominated by intrinsically faint galaxies?

A. TYSON: This is not true at these very faint magnitudes, where an M^* galaxy would have a redshift around 0.5, implying a mean k-correction of perhaps 1 magnitude. At that point, for example, the ratio of subluminous galaxies with $M=M^*+2$ to M^* galaxies is near unity in such a faint sample.

FAINT GALAXY NUMBER COUNTS

T. Shanks, P.R.F. Stevenson and R. Fong
Physics Department, University of Durham, England

H.T. Mac Gillivray,
Royal Observatory, Edinburgh.

ABSTRACT

Using photographic data obtained from the U.K. Schmidt and Anglo-Australian telescopes we have determined galaxy number counts in the range $17^m \leq B \leq 23^m$ and $15^m \leq R \leq 22^m$. By comparing our observations with models we find that the blue counts show strong evidence for galaxy luminosity evolution. We find that the red counts are better fitted by non-evolving models and that, here, the effects of the cosmological deceleration parameter, q_o, and luminosity evolution are comparable. We discuss the implications of these results, with particular reference to the prospects of using the red counts to obtain constraints on q_o.

1. INTRODUCTION

The galaxy number-magnitude relation n(m), contains much information on the luminosity evolution of galaxies at high redshift. This information is important because, unlike other probes, n(m) is sensitive to evolution in the absolute luminosity of galaxies as well as to evolution in their colours. Over the years many attempts have been made to determine n(m) but its observed form is still not well established. Here we report on a new attempt to determine n(m) in a blue (b_J) and a red (r_F) passband from photographic data at the South Galactic Pole (SGP). A more detailed account of this work is given by Shanks et al. 1984 (henceforth SSFM).

To interpret galaxy counts it is important to know the form of n(m) at bright as well as faint magnitudes. We therefore began our study by obtaining COSMOS (Stobie et al. 1979) machine measurements of blue and red U.K. Schmidt photographs of a 12 square degree area at the SGP. The measurements defined n(m) in the range $17^m \leq b_J \leq 20^m.5$. We then also obtained COSMOS measurements of AAT photographs of 0.4. square degrees of sky at the centre of the Schmidt photograph. These defined our n(m) relation in the range $20^m.5 \leq b_J \leq 23^m$.

In our photometry we employ isophotal magnitudes with isophotes set at 1-2% of sky for faint galaxies. We have verified our photometry and star-galaxy separation techniques on the UKST plates by comparing with the PDS data of the Oxford group (Fong et al. 1983). MacGillivray and Dodd (1982) have verified our photometry techniques at the faint limits of the AAT plates by comparison with the faint photographic photometry of Carter (1980). A new check on the faint AAT photometry has recently been made possible by the obtaining of CCD photometry on the SGP field by Couch et al. 1984. The comparison of the red photographic and CCD magnitudes is shown in Figure 1 where it can be

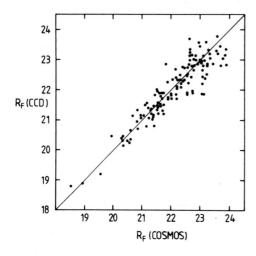

Figure 1.

Comparison between COSMOS photographic galaxy photometry and the CCD photometry of Couch et al., 1984.

seen that there is good agreement in zeropoint, with no evidence of scale errors in our photometry. Similarly good agreement is found for the blue magnitudes, when account is taken of the different photographic and CCD blue passbands. It is interesting to note that the CCD magnitudes were produced using the algorithms of Couch and Newell (1984), which are quite independent of those used to produce the photographic photometry. These algorithms produce pseudo-total magnitudes like those used by Kron (1978) and the absence of any scale error in Figure 1 suggests that our isophotal magnitudes are very close to total magnitudes in the range of interest here.

2. OBSERVED COUNTS

Our observed counts in the blue and red passbands are shown in Figures 2a, b. The circles represent the UKST counts, the triangles the AAT counts. In the fainter magnitude range SSFM made a detailed comparison of these results with other authors. Briefly they found that in the blue there was excellent agreement at faint limits with the counts of Kron (1978), Koo (1981) and Couch and Newell (1984). Less good agreement was found with the counts of Tyson and Jarvis (1979) and Peterson et al. (1979) which seem to be shifted faintwards by $\sim 0.^{m}5$ with respect to our data. It is difficult to believe that

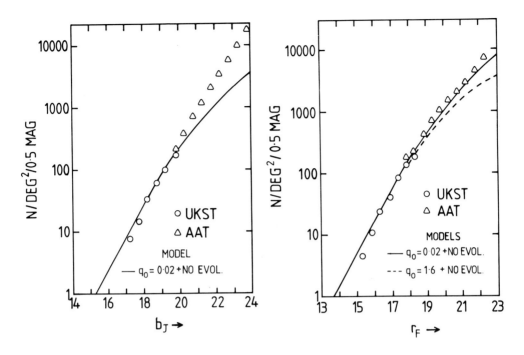

Figure 2. UKST and AAT differential galaxy number counts in the b_J and r_F bands compared to no evolution models.

these are fluctuations due to galaxy clustering because at $b_J \sim 23^m$ galaxies appear relatively unclustered on the sky due to the effect of projection. One explanation for the discrepancy could be the existence of patchy galactic obscuration in some of the fields used by Tyson and Jarvis and Peterson et al. Couch and Newell measured the absorption in each of their widely separated fields from the reddening of field galaxy colours and they found that patchy absorption at lower galactic latitudes accounted for much of the dispersion in their galaxy counts.

In the red band only Kron, Koo and Couch and Newell have produced galaxy counts. When account is taken of the different red passband used by Kron and Koo we find that our faint AAT counts are in good agreement with all of these.

At the brighter magnitudes of the UKST counts, fluctuations caused by galaxy clustering may be a bigger problem. To test how typical the original 12 square degree area of SSFM was, we recently obtained COSMOS measurements on 6 adjoining UKST fields which extends our measured area at the SGP to some 110 square degrees. The resulting b_J galaxy counts in each of these fields is shown together with the original in Figure 3. The excellent agreement among all the fields suggests that the original 12 square degree area was very representative

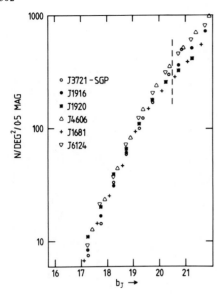

Figure 3

Galaxy counts in 6 separate UKST fields around the SGP. J3721 is the UKST plate number of the original 12 square degree field of SSFM. The dashed line indicates the $b_J = 20^m.5$ limit of reliability for the UKST counts.

in terms of its number count properties.

There are few independent galaxy counts in the range $17^m \leq b_J \leq 20^m.5$ with which our UKST counts can be compared. However SSFM found that our counts seemed to extend naturally from those made at brighter magnitudes by Kirshner et al. (1979) and Zwicky et al. (1961-68).

3. GALAXY COUNT MODELS

Galaxy number count models depend chiefly on the K-corrections, the assumed mix of morphological types and the nearby galaxy luminosity function. Details of the model parameters assumed here are given by SSFM. The red models are better determined than the blue because the blue K-corrections are less reliable at high redshift and because the blue counts contain a much wider spread of morphological types than the red which are dominated by early type E/S0/Sab galaxies. We normalised our models to the UKST counts at bright magnitudes and checked that our assumed morphological mix gave good fits to the observed distribution of galaxy colours.

The solid lines in Figures 2a and 2b are the $q_o = 0.02$, no evolution, standard models of SSFM. It can be seen that in the blue this model underestimates the observed count at $b_J = 23^m$ by more than a factor of 3. The conclusion of SSFM was that the counts therefore implied quite large amounts of galaxy luminosity evolution in the sense that galaxies were brighter in the past.

The no evolution models of Koo (1981) adopted a fainter luminosity function for late type galaxies which decreased the average redshift of galaxies in the model and so produced a steeper number count slope. However this model predicts only a factor of 50 per cent more galaxies than ours at 23rd magnitude in b_J and therefore still requires quite large amounts of evolution to fit the counts. The difference between these two models is a reasonable representation of the uncertainty in the blue model parameters and we conclude that these uncertainties may not be large enough to allow non-evolving models to explain the counts.

In the red band the $q_o = 0.02$ no evolution model of SSFM gives a much better fit to the observations. This difference between the red and blue counts is expected since evolutionary models like those of Bruzual (1981) predict that galaxy luminosity will generally evolve faster at bluer wavelengths. Here the Koo (1981) model is indistinguishable from our own, indicating the smaller uncertainty associated with the red models. The dashed line in Figure 2b is the $q_o = 1.6$ no evolution model which seems to give a worse fit to the observed count. This shows that the effects of q_o on the red counts at these depths can be quite significant. Of course, world models with $q_o = 1.6$ still cannot be rejected on the basis of this result because, with the inclusion of luminosity evolution, such world models can be made to give satisfactory fits to our data.

The q_o dependence of the red counts is interesting because it is different from other cosmological tests which depend on the luminosity redshift relation. In the Hubble diagram, for instance, larger values of q_o mimic evolutionary brightening at high redshift whereas in the number counts it is smaller values of q_o that produce this effect. These almost orthogonal dependencies suggest that it may be possible to use the counts and the Hubble diagram together to obtain quite strong empirical constraints on both q_o and galaxy luminosity evolution. The problem with this approach is that it assumes the evolution of cluster and field early type galaxies is the same and this assumption may be difficult to justify.

An alternative approach to the luminosity redshift relation which overcomes this problem is to determine the average redshift, $\bar{z}$, of galaxies in faint, ($R \sim 21^m$) magnitude limited samples. The advantage of this approach over the Hubble diagram is that it is self-consistent; the same galaxies that define the $n(m)$ relation define $\bar{z}$. The disadvantage with $\bar{z}$ is that much telescope time has to be spent obtaining redshifts for many low redshift, low luminosity galaxies. However the advent of multi-object spectroscopic techniques should make this less of a difficulty and indeed a start has been made on such a project by Koo and Kron at Kitt Peak.

To test the q_o dependence of $\bar{z}$ we have used three models that assume different combinations of q_o and luminosity evolution but all of which produce excellent fits to the red galaxy counts. Table 1 gives the computed $\bar{z}$ for each of these models in a magnitude limited sample with $20\overset{m}{.}5 \leq r_F \leq 21^m$. It can be seen that a relatively small redshift survey should be able to discriminate between the $q_o = 0.02$ and $q_o = 1.6$ models and indeed that there is quite a significant difference between the $q_o = 0.02$ and $q_o = 0.5$ models.

Thus the future prospects for obtaining empirical constraints on both q_o and luminosity evolution using this approach seem encouraging.

Table 1   Model predicted $\bar{z}$ for $20^m.5 \leq r_F \leq 21^m$

Model		$\bar{z}$
$q_o = 0.02$,	$\Delta M = -1z$	0.40
$q_o = 0.5$,	$\Delta M = -1.5z$	0.49
$q_o = 1.6$,	$\Delta M = -2.5z$	0.76

All 3 models give same predicted galaxy count as data.
ΔM is the change in absolute magnitude of galaxies at redshift z.

4. CONCLUSIONS

1. Our blue galaxy counts at the South Galactic Pole show evidence for luminosity evolution in the sense that galaxies seem to have been brighter in the past.
2. Our red galaxy counts show evidence for smaller amounts of evolution than in the blue.
3. The combination of the red counts and faint galaxy redshift surveys may produce empirical constraints on q_o and galaxy luminosity evolution.

REFERENCES

Bruzual, A.G., 1981, Ph.D. Thesis, University of California, Berkeley.
Carter, D., 1980, Mon. Not. R. astr. Soc., 190, 307.
Couch, W.J., and Newell, E.B., 1984 in preparation.
Couch, W.J., Pence, W.D., and Shanks, T., 1984 in preparation.
Fong, R., Godwin, J.G., Green, M.R., and Shanks, T. 1983 "Proc. of the Workshop on Astronomical Measuring Machines" ed. Stobie, R.S., and McInnes, B., Royal Observatory Edinburgh.
Kirshner, N.P., Oemler, A., and Schechter, P.L., 1979, Astr. J. 84, 951.
Koo, D.C., 1981, Ph.D. Thesis, University of California, Berkeley.
Kron, R.G., 1978, Ph.D. Thesis, University of California, Berkeley.
MacGillivray, H.T., and Dodd, R.J., 1982, The Observatory, 102, 141.
Peterson, B.A., Ellis, R.S., Kibblewhite, E.J., Bridgeland, M., Hooley, T., and Horne, D., 1979, Astrophys. J. 233, L109.
Shanks, T., Stevenson, P.R.F., Fong, R., MacGillivray, H.T., 1984, Mon. Not. R. astr. Soc. 206.
Stobie, R.S., Smith, G.M., Lutz, R.K., and Martin, R., 1979, in "Proc. of the International Workshop on Image Processing in Astronomy, Trieste, Italy.
Tyson, J.A., and Jarvis, J.F., 1979, Astrophys. J. 230, L153.
Zwicky, F., Herzog, E., Wild, P., Karpowicz, M., and Kowal C.T., 1961-1968 "Catalogue of galaxies and clusters of galaxies" in 6 vols. Calif. Inst. of Tech., Pasadena.

DISCUSSION

C. CARTER: Have you made any counts on IV-N plates?

T. SHANKS: No, but we do have IVN UKST, AAT photographs and also I CCD frames for calibration and we intend to next make I counts using this material. The advantage of I counts over R counts is that we can probe to deeper redshifts before the quickly evolving ultraviolet part of galaxy spectra is redshifted into the I band.

OBSERVATION OF INTERGALACTIC DUST BY SCHMIDT-TELESCOPES

S. Marx
Central Institute of Astrophysics of the Academy of Science
of the GDR

About 25 years ago Zwicky (1957), Holmberg (1958) and Hoffmeister (1962) found intergalactic dust by observations. Zwicky (1957) explained the deficiency of distant clusters of galaxies by intergalactic dust inside the near clusters. Holmberg (1958) made extensive observations of the Virgo cluster. He found systematic differences of the colour indexes between cluster and field galaxies. Intergalactic dust inside of the Virgo cluster should be the reason of these differences. Hoffmeister (1962) found a clear deficiency of galaxies in the region of Microscopicum. Again the reason should be a large cloud of intergalactic dust.

The possibility of the existence of intergalactic dust has been proved for example by Schmidt (1974, 1975) and Margolis and Schramm (1977). These and other authors, e.g. Karatschensev and Lipovetsky (1968), Crane and Hoffmann (1973), Nandy, Morgan and Reddish (1974), used for the intergalactic dust particles the same parameters which are known for the interstellar dust (density of the particles 1 g cm^{-3}, mean diameter of the particles 10^{-5} cm). The interstellar dust affects the observed colours of the stars, which are behind or inside the dust clouds. Therefore the intergalactic dust inside clusters of galaxies must affect the colours of the galaxies too. Van den Bergh (1975) supposes that the observed colour excess of the cluster galaxies were partly produced by a colour difference between galaxies near the centre and outside of the central region of the cluster. But dust inside the cluster can produce a colour excess too, because the length of the line of sight dependes on the direction of insight in the cluster.

The colour excess was computed by the following assumptions
- the density of the dust is uniform in the volume of the cluster of galaxies
- the mean colour of all galaxies of the cluster are the same for each direction of insight in the cluster
- the reddening of the galaxies depends on the properties of the dust particles in the following way

$$E_{B-V} = R \cdot A_V$$

$$A_V = \frac{1.086 \cdot 3 \cdot D \cdot Q}{4 \quad d \cdot a} \cdot \ell$$

R is taken to be 3, D is the density of the dust inside the cluster, d is the density of the dust particles (1 g cm^{-3}), a is the mean diameter of the dust particles (10^{-5} cm), Q the efficiency factor (2.0) and ℓ the lenght of the line of sight.

Now it is possible to compute the reddening (E_{B-V}) for each galaxy and the mean value for all galaxies of any direction of insight in the cluster, depending on the density of dust.

Table 1 shows the reddening of the galaxies from the centre of to the edge of the cluster for different dust densities. The values of Table 1 are valid for a cluster diameter of 2 Mpc. d is the distance from the centre.

Table 1

reddening in mag

d	10^{-31} g cm^{-3}	10^{-30} g cm^{-3}	10^{-29} g cm^{-3}
0.1 Mpc	0.019	0.186	1.860
0.3	0.019	0.194	1.941
0.5	0.018	0.175	1.750
0.7	0.013	0.132	1.320
0.9	0.008	0.077	0.770

In Figure 1 is illustrated the reddening for clusters of different diameters (1 Mpc $\leq$ D $\leq$ 10 Mpc) and a dust density of 10^{-30} g cm^{-3}.

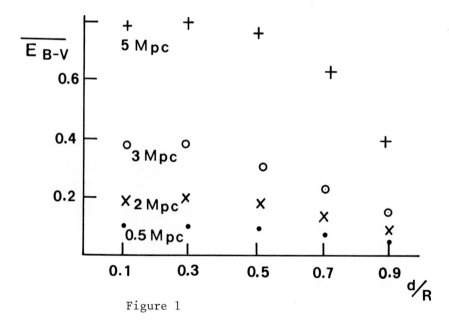

Figure 1

The colour difference of galaxies near the centre and at the edge of the cluster depends on the density of dust inside the cluster. Figure 2 illustrates the logarithm of the colour excess in dependence on the density of dust inside the cluster.

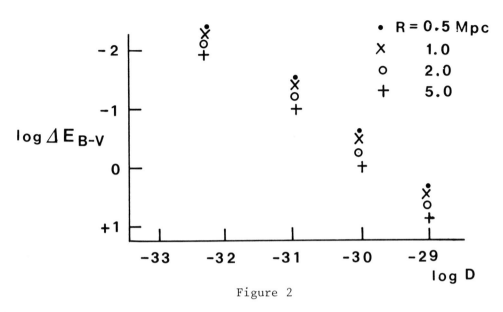

Figure 2

Partly the colour excess of the cluster galaxies can be explained by the effect, supposed by Van den Bergh (1975), partly by dust inside the clusters.

REFERENCES

Crane, P.C., Hoffmann, A. 1973, Astrophys. J. 186, 787
Hoffmeister, C. 1962, Z.f. Astrophys. 55, 62
Holmberg, E. 1958, Meddelande Lund Astr. Obs. Ser. II, No. 136
Karatschensev, I. and Lipovetsky, A. 1968, Astron. Zh. Akad. Nauk. SSR 45, 1148
Margolis, St.H. and Schramm, D.N. 1977, Astrophys. J. 214, 339
Nandy, K., Morgan, D.H. and Reddish, V.C. 1974, Mon. Not. R. Astr. Soc. 169, 19 P
Schmidt, K.H. 1974, Astron. Nachr. 295, 163
Schmidt, K.H. 1975, Astrophys. Space Sci. 34, 23
van der Bergh, S. 1975, Proc. Tercenenary Symp. Roy. Greenwich Obs.
Zwicky, F. 1957, Morphological Astronomy, Springer Verlag

SUPERCLUSTERING AND SUBSTRUCTURES

Guido Chincarini
European Southern Observatory, Garching, West Germany,
and University of Oklahoma

The history of the Universe is infinitely more interesting than the history of the study of the Universe.
                            - Zel'dovich and Novikov (1983) -

PROLOGUE

The understanding we have today of the distribution of galaxies resolves some of the debates which were going on in the seventies. Future data and analysis may allow us to discriminate among cosmological models.

In 1972, Zwicky was convinced, as a result of his work with the Schmidt telescopes (Palomar 18 inch and 48 inch), that the Coma Cluster of galaxies had to be much larger[1] than the 100 minutes of arc estimated by Omer et al. (1965). In his conception of the distribution of galaxies, very large clusters were tenuously connected and superimposed on a uniform background of galaxies.

The problem of the "discrepant" redshift in groups (for instance, the Stephan quintet and Seyfert sextet) was already known to us (see also Zwicky, 1957) and the reality of the Local Supercluster (see de Vaucouleurs, 1983) was not exempt from objections.

Work directed at understanding these, and related cluster problems, (Chincarini and Martins, 1975; Chincarini and Rood, 1975) led to the finding of the segregation of redshifts in cluster and non-cluster regions and to the concept that galaxies, and clusters, were part of very large structures. Regions void of galaxies were detected together with the irregular structures (Chincarini and Rood, 1976; Gregory and Thompson, 1978; Tarenghi et al., 1979; Kirshner et al, 1983). A question asked at a brown-bag lecture at Harvard (1975/1976) and at a meeting in

(1) In <u>Morphological Astronomy</u> Zwicky (1957) estimates a diameter of 320 minutes of arc from the 18-inch Schmidt observations and of at least 12° from the 48-inch Schmidt data.

Rome by Rees (1976) on the statistical significance of the distribution in the redshift space, was better answered only later (Chincarini, 1978).

Fundamental surveys and studies existed which had been undertaken with the aim of understanding the distribution of galaxies. I refer in particular to the survey of clusters by Abell (1958), to the catalogue by Zwicky and collaborators (1961-1968) and, above all, to the careful counts of galaxies by Shane and Wirtanen (1967), and the related studies by Neyman, Scott and Shane (1953) and Scott, Shane and Swanson (1954). On this material, Peebles (1980 and references therein) developed the powerful "machinery" of the autocorrelation functions aiming 1) to understand the distribution of matter in the Universe and its implications and 2) to discriminate among cosmological models.

The IAU Tallin Symposium (1978) marks the acceptance of the new findings and stimulates further work. The Crete IAU symposium (1982) reflects the gain in new knowledge at all wavelengths allowing a broad discussion and intercomparison between theoretical and observational findings. By the time of the Crete meeting, among others, the extensive CfA survey by Davis et al. (1982) had also been completed.

The fast enrichment of knowledge we achieve through the work of capable scientists makes it exciting to be even a small part of all this. The flourishing of excellent work, the involvement of highly capable astronomers, and the recent publication of good reviews (see for instance Oort, 1983) make it difficult, and perhaps unnecessary, to make a new review of the field at this time.

In this lecture, therefore, I prefer to discuss only a few topics and show some preliminary results of the work in progress. The analysis I present is preliminary in the sense that the sample and statistics must be refined, and in a few cases we see improvements which must be made and are in progress. More important, the comparison with models is only in its infancy. It is appropriate, nevertheless, to discuss the work in progress at this time and location since new lines of developments may be suggested.

INTRODUCTION

What is meant by supercluster depends on the time and on the author. In all cases, however, it identifies a large agglomerate of something, clusters of galaxies, and/or galaxies, and in this sense it is synonymous with large scale structure. Its meaning is generally defined by the context in which it is used and it may be appropriate to avoid, at this time, a sharp definition and rather to leave it floating. It is part of the research which is in fact in progress to determine the statistical properties of these structures (nomen est numen). As an operational definition the reader may identify it with the concept of "clouds" as empirically defined by Shane and Wirtanen (1967), extended, however, to three dimensions.

Third order clustering is not yet observed and II order clustering is not meant to be characterized by clusters of clusters of galaxies, in the sense, for instance, that the Coma cluster is a cluster of galaxies. It is more accurate, perhaps, to talk of groups, up to 10-20 members, of clusters. Clusters are clumped as shown by Abell (1958) and, to mention only the latest work, by Bahcall and Soneira (1983a). The latter authors in fact show that clusters of galaxies are correlated over separations of about 100 h^{-1} Mpc (h = H_o/100) and define structures comparable in size to the one defined by the galaxies.

The observed distribution of galaxies is very clumpy and the agglomerations define large structures (no boundary has yet been found) of very irregular forms with density peaks which coincide with the clusters of galaxies. Such structures seem to be connected to each other and never seem to be isolated. This picture was proposed, based on some early observational evidence, by Chincarini and Rood (1980), suggested by the theory of Zel'dovich and colleagues (1978) and in a somewhat different form proposed by Einasto et al. (1978). In such a picture it would be strange to detect an isolated cluster of galaxies.

If positive density fluctuations form from a homogeneous medium we expect (as the a posteriori logic suggests) negative density fluctuations, that is, regions where the density of galaxies is very low. Such regions have been observed (see, for instance, Chincarini and Rood, 1976; Tifft and Gregory, 1976; Chincarini, 1978; Kirshner et al., 1982). The perhaps unexpected result was that, so far, no galaxy has been observed in such voids. In this way, an upper limit can also be put on the density of an eventual "uniform background" of galaxies. The voids themselves assume a high cosmological significance because they characterize the high order correlation functions and allow tests between numerical models and observations.

Statistics on the voids may soon be available. The topology is stable and favoured by the passage of time because the positive and negative density fluctuations act in the same direction and tend to group the matter (for numerical and analytical discussion see Peebles, 1982, and Salpeter, 1983). One of the problems to be solved, observationally, is the determination of the density enhancement above which we have the formation of bound groups and clusters and below it unbound density enhancements. It is as yet somewhat unclear whether unbound groups, or clusters, have been observed (see, however, Gott et al., 1973).

The study of a fair sample of the Universe allows a determination of the density parameter Ω which is unaffected by non-observed mass in galaxies (flat rotation curves detected by Bosma, 1978, and Rubin et al., 1982) and by stability problems in clusters. The most recent determinations give Ω = 0.1-0.3 (Davis and Peebles, 1982; Bean et al., 1983).

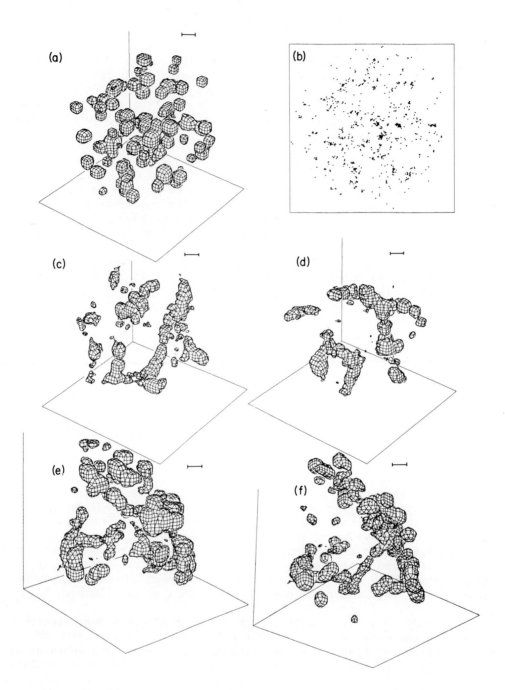

Figure 1. Comparison between the space distribution of numerical models and observations. (a) and (b) Poisson model, (c) and (d) pancake models (adiabatic) and (e) and (f) Center for Astrophysics Northern Survey. (From Frenk, White and Davis, Ap.J. 271, 417).

Is the above picture reflecting a hierarchical distribution? Do we have evidence of third order clustering? In fact, is the large correlation length estimated for clusters of galaxies an indication of III order clustering?

Shandarin (1983) and Frenk et al. (1983) show that in the adiabatic models, the correlation function $\xi(r)$ and the visual appearance of the structures match better than in other models (Poisson-isothermal) the observed distribution of galaxies (Figure 1). A large coherence length is theoretically demanded not only by models developing from primordial adiabatic fluctuations, but also by models in which the mass distribution in the Universe is dominated by neutrinos with non-zero rest mass (Bond, Efstathiou and Silk, 1980). In the very low density regions, voids, generated by adiabatic numerical models, galaxies may be unable to form. While the statistical reality of voids is not in doubt, we must be cautious (Peebles, 1983) in seeking their physical interpretation, since these can also be produced in a hierarchical process. Always following Peebles, the existence of voids does not mean that there is a reason to think that $\xi(r)$ has been underestimated on large scales. If the voids were produced by a physical process operating in a coherent way over scales ~ 100 h^{-1} Mpc, the process would have had to have operated in a peculiar way, leaving $|\xi| \ll 1$ on this scale. For arguments in this direction see also Soneira and Peebles (1978) and Bean et al. (1983). On the other hand, as we have seen, clusters give a larger correlation length.

In summary, we would like to evolve models to match the observed distribution, that is to go from a more homogeneous and isotropic past to the observed clumpy distribution (an expanding Universe is unstable against growth of departures from homogeneity and isotropy). To do this, we need to understand what we observe. Such an understanding will finally shed light also on the problem of the formation and evolution of galaxies as well.

The visual appearance may not be enough to discriminate between models. The autocorrelation functions are statistical descriptors able to average over the complexities of the structure to evidence the basic properties; they are, however, not sensitive to topological details which may be important. It seems worthwhile, therefore, to look into some properties of the structures (substructures) and possibly define some parameters and/or characteristics which allow a close comparison with models.

A WAY TO SELECT STRUCTURES: THE PERCOLATION (OR DENDROGRAM) ALGORITHM

The percolation algorithm in its most sophisticated developments is used in various branches of science and especially in solid state physics. It was imported into astronomy by Materne (1978) and used in selecting groups and applied to superclusters to define membership in the embedded clusters by Materne and others (see Appendix of Tarenghi et al.,

1980). Gerola and Seiden (1978) used a similar technique in the stochastic study of the formation of spiral arms. Zel'dovich et al. (1982) and Shandarin (1983) demonstrate its usefulness in discriminating among numerical models and observations of the large scale structures.

In brief: The objects are considered "connected" if their separation is smaller than a preselected parameter R_i. A structure, let us say, at a level i, is the ensemble of all the connected points of the sample (Figure 2). The definition is independent of symmetry or smoothing and the computer follows the structures as we would site the details of a crack in a wall or water suddenly spreading in a dry creek. In a cristal an electric current would follow the path defined by the impurities. The method is equivalent to selecting a structure with density (number of galaxies per square degree or per Mpc^3) above a certain level.

While the matter is almost straightforward in two dimensions, complications arise when a magnitude limited sample is used. A minor and easily treatable inconvenience is that the presence of cluster virial velocities will cause spurious separations. A more difficult problem to treat is the distortion introduced into the geometry by a sample limited by apparent magnitude. This is especially true for large values of the parameter R_i which allow to probe larger regions of space. Due to the fact that we are probing at different absolute magnitudes and different distances, we ficticiously change the density of galaxies.

To take this into account we may use a value of R_i which is a function of the distance x. Since $N(x) \propto D(x) \Gamma (\alpha+1, L/L^*)$ where $N(x)$ is the density $D(x)$ corrected by the effect of the magnitude limited sample and

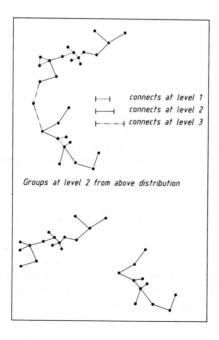

Figure 2. Connected ensembles at different levels for various levels of the parameter R. On the bottom of the figure the two structures isolated at the 2nd level.

$L/L^* = \text{dex}\left[0.4\left(M^1 - m_1 + 5\log x + 25\right)\right]$ we can write:

$$R(x) \simeq 2\left(1/N(x)\right)^{1/3}$$

or

$$R(x_1)/R(x_2) \simeq \left(\frac{D(x_1)}{D(x_2)}\right)^{1/3} \left(\frac{\Gamma(\alpha+1, L_1/L^*)}{\Gamma(\alpha+1, L_2/L^*)}\right)^{1/3}$$

There is naturally a simpler way and this has been used also by Einasto, Klypin, Saar and Shandarin (1983). That is, after an estimate of the cluster volume one can remove the virial velocities from the sample and assign distance velocities according to a "reasonable" model. Furthermore, instead of using an apparent magnitude limited sample, an absolute magnitude limited sample can be used. In this case, however, only part of the data are used. Here I prefer to use the whole sample.

STRUCTURES AND SUBSTRUCTURES

An analysis of the ESO/Uppsala catalogue showed the Hydra-Centaurus supercluster as the most prominent structure of the southern hemisphere. Such a structure may be connected to a filament evidenced by Moody et al. (1982) in their analysis of the Shane and Wirtanen catalogue, their filament N.13. The growth of the 3-dimensional clustering, as a function of the parameter R is reproduced in Figures 3 and 4 for the main structures in the region of Coma/A1367 and Perseus/Pisces.

The structures are rather extended in one dimension (note that the boundaries are defined by the region of the sky selected) and of the order ⩾ 100 Mpc while, especially in the Perseus-Pisces supercluster, the width and depth are of the order of 15-20 Mpc. Galaxies are, however, still "connected" in a redshift range of 4000-5000 km/sec. In addition to the main structure (or main structures when detected from a larger sample), substructures are also evidenced. Some of these have been reproduced in Figure 5.

For the main structure and a set of substructures with at least 5 members we have measured the length and width. In this case we called length the sum of the separations connecting the "first" point of the structure to the "last" connected point (see Figure 2), and width either the mean separation or the r.m.s. of the separation of the rest of the points from the segmented line defining the length. Following an analysis similar to the one by Zel'dovich et al. (1983), we also constructed the multiplicity function for the substructures isolated for various values of the parameter R in order to follow the growth of the structures as a function of R.

Comparison with a random distribution of points is needed to see a) whether the detected substructures are statistical fluctuations and b) whether or not some of the statistical properties of the observed filaments differ from those generated as fluctuations of a random ensemble. The same analysis, therefore, which was used for the observed redshift sample was applied to a sample of 2000 random points simulating a magnitude limited sample of objects in a volume of space similar to the observed volume. That is, we have the additional constraint

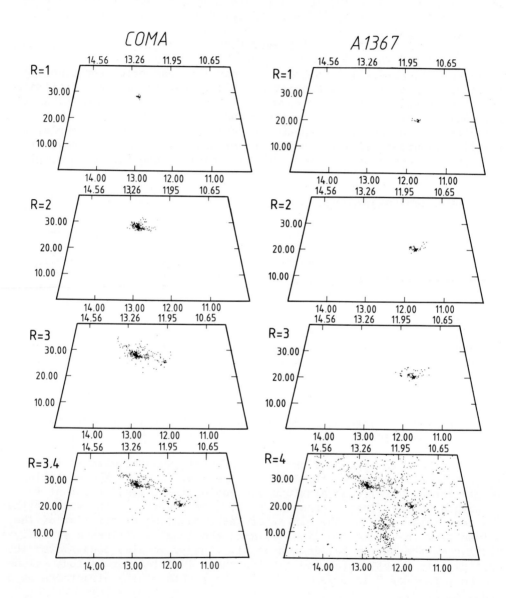

Figure 3. The growth (percolation in 3 dimensions) of the Coma/A1367 supercluster as a function of the parameter R. The fast connection to Virgo is partly spurious due to the use of an apparent magnitude limited sample. (R.A. in hours, declination in degrees.)

$$N(v_i,v_j) = \int_{x_i}^{x_j} x^2 \, D(x) \, \Gamma \, (\alpha+1, \, L/L^*) \, / \int_0^\infty x^2 \, D(x) \, \Gamma \, (\alpha+1, \, L/L^*) \, .$$

The substructures detected in the random sample do not differ noticeably from the real structures. A sample of the former is reproduced in Figure 6. In Figure 7 we have the distribution thickness/length (thickness and length as defined above) at some value of the parameter R, both for the observations and for the random sample. The distributions are very similar, a matter which may be only partly due to our definition of length and width. A better discriminator seems to be the maximum length of the connected region as a function of the parameter R (Zel'dovich et al., 1982).

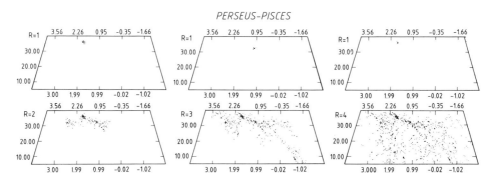

Figure 4. Same as in figure 3 for the Perseus-Pisces sample. Note the filament extending toward Pegasus.

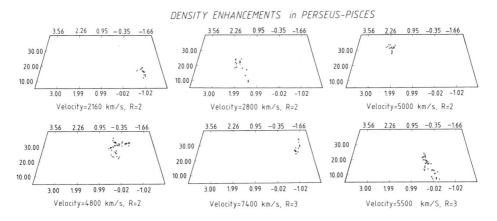

Figure 5. Substructures detected by the percolation algorithm in the Perseus-Pisces sample. The approximate mean velocity of the group is also given.

What characterizes a real structure, however, is its stability to variations of the parameter R. A set of structures detected in the random sample at some value of R disappears in the background for a different value of R, at which value a new set of substructures will appear. Structures which are "stable" in the observed sample will preserve their identity as a function of R. To some extent it is similar to the "stability" of a density perturbation. Differences between the random and observed samples are also detected in the variations of the multiplicity function as a function of values of the parameter R (Figure 8). In Figure 8 such variations have been reproduced for the Coma extended region, the random sample and the Perseus/Pisces sample. At large values of R in all cases we form a main "agglomerate" and some distribution of smaller

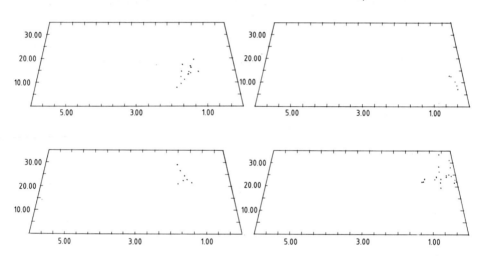

Figure 6. Substructures detected by the percolation algorithm in a random (apparent magnitude limited) sample of 2000 points.

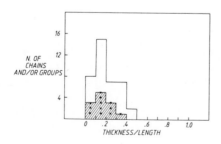

Figure 7. Distribution of the ratio thickness/length in the observed and random sample obtained for substructures (with more than 5 members) at some value of the parameter R.

SUPERCLUSTERING AND SUBSTRUCTURES

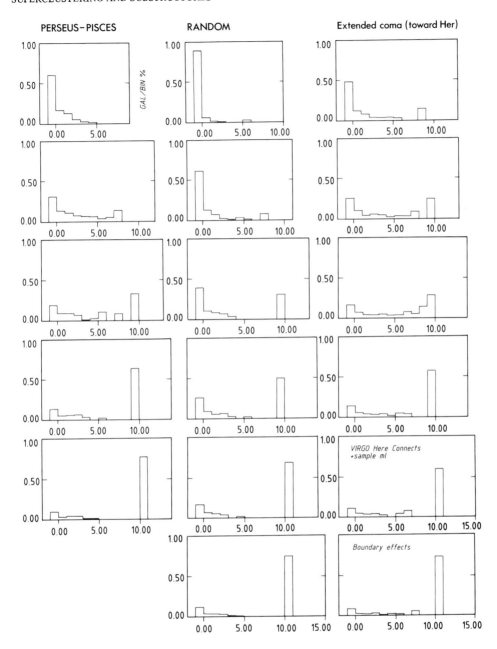

Figure 8. Variation of the multiplicity function, % of galaxies per ensemble of various richness (richness in bin of $\log_2 N$), as a function of the parameter R. The random sample does not form intermediate clustering and the substructures are unstable (lose their identity) to variations of the parameter R. All the substructures have been detected, and the number of members counted, using the percolation algorithm. At large R, with the samples used, we have boundary effects and the connection (in part spurious) to Virgo.

groups. The main difference consists in the early development, where the random sample never develops a sizable number of intermediate richness clusters.

The conclusion is that the majority of the observed substructures of the real world are "stable" and are not statistical fluctuations. They are not due to the unknown mechanism by which the eye picks out textures and patterns. Such structures can be isolated and their topology, and, perhaps, kinematics, measured. In a simple way, it is a matter of contrast as expected.

An extreme example given in Figure 9 is the filament selected in the Pegasus region, extending between 5° and 25° declination. The tip of this

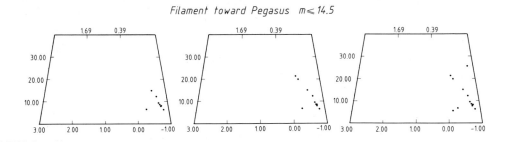

Figure 9. The filament in Pegasus, m ≤ 14.5.

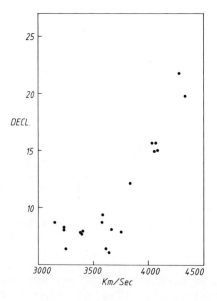

Figure 10. The filament in Pegasus in the declination-redshift plane.

very narrow filament points away from us as can be seen from Figure 10 where the objects are plotted in a declination redshift diagram.

Such substructures may be difficult to generate in isothermal (Poisson) models and may finally argue against a hierarchical Universe.

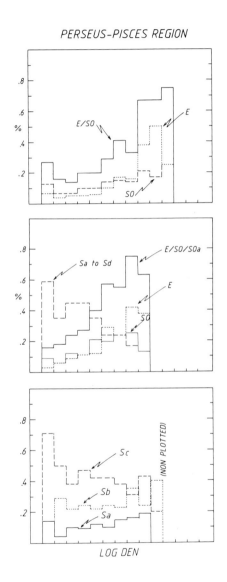

Figure 11. Distribution of types in the Perseus-Pisces sample as a function of surface density of galaxies.

MORPHOLOGICAL SEGREGATION

Giovanelli, Haynes and Chincarini (1983) noticed that in the region of the Perseus-Pisces galaxies are somewhat segregated according to their morphological type (see fig. 16 in Oort, 1983).[2] The distribution of galaxies is indicative of a correlation between morphological type and density. Such correlation is illustrated in Figure 11.

The result, which is being tested in other regions of space, stresses the fact that type segregation is a phenomenon which extends to very low density regions of space. In the sample considered here, the density ranges, for galaxies with m ≤ 14.5, from a maximum of about 3.7 galaxies per square degree (note that the Perseus cluster, A426, is outside the sample area) to a density of about $1.7 \, 10^{-3}$ galaxies per square degree. Such segregation is visible in various low density filaments (substructures) as well, where a preponderance of spiral is observed, a fact evidenced also by Focardi et al. (1983) in other samples.

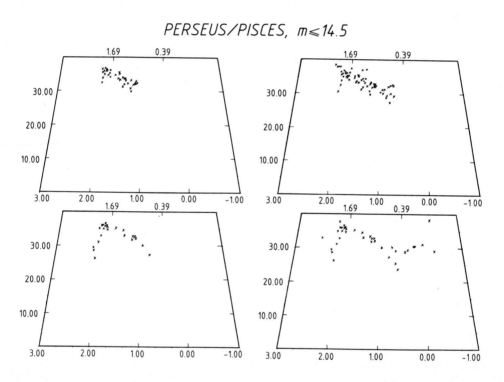

Figure 12. The main structure detected, using the percolation algorithm, in a magnitude limited subsample, m ≤ 14.5, of elliptical and spiral galaxies. The pattern of the elliptical galaxies (similar to the pattern of spirals) cannot be generated by galaxies which evaporate from clusters. If a merging mechanism of formation is at work it should act locally and during the collapse phase (if any) of the filaments.

(2) A similar effect was evidenced by Tarenghi et al. (1980) in an early version of the Hercules paper. Because of the low statistical significance of the effect the statement was modified following the referee's comments. Compare, however, their figure 11, top and bottom.

As has been mentioned already this is suggestive of a formation mechanism in which the morphological type is somewhat conditioned by its environment with little, if any, evolution along the Hubble sequence. The merger mechanism (see, for instance, Silk and Norman, 1981) is very attractive. However, it is doubtful that ellipticals in the low density regions may be the result of cluster evaporation, as is demonstrated by their distribution in Figure 12. Are such patterns understood in the framework of a hierarchical Universe? We are eager to proceed with our work and further simulations, confident to gain further understanding on this matter.

ACKNOWLEDGEMENTS

I am grateful to S.F. Shandarin for discussions we had in Crete on the percolation algorithm. My appreciation goes to L. Woltjer and G. Setti for the comfortable environment at ESO and to J. Manousoyannaki who helped during the first steps of the programming. I am indebted to P. Bristow and C. Stoffer for their skillful and patient typing of the camera ready manuscript.

Part of this work is being supported by the Research Council of the University of Oklahoma and by NSF Grant AST 82-00727.

REFERENCES

Abell, G.O.: 1958, Ap.J. Suppl. 3, p. 211.
Bahcall, N.A., and Soneira, R.M.: 1982, Ap.J. 262, p. 419.
Bean, J., Efstathiou, G., Ellis, R.S., Peterson, B.A., Shanks, T., and Zou, Z.L.: 1983, in Early evolution of the Universe and its present structure, eds. G.O. Abell and G. Chincarini, D. Reidel Publishing Company.
Bond, J.R., Efstathiou, G., and Silk, J.: 1980, Phys. Rev. Letters 45, p. 1980.
Bosma, A.: 1978, dissertation, Rijksuniversiteit te Groningen, Holland.
Chincarini, G.: 1978, Nature 272, p. 515.
Chincarini, G., and Martins, D.: 1975, Ap.J. 196, p. 335.
Chincarini, G., and Rood, H.J.: 1975, Nature 257, p. 294.
Chincarini, G., and Rood, H.J.: 1976, Ap.J. 206, p. 30.
Chincarini, G., and Rood, H.J.: 1980, Sky and Telescope 59, p. 364.
Davis, M., and Peebles, P.J.E.: 1983, in press, Ap.J. (Center for Astrophysics, preprint series).
Davis, M., Huchra, J.P., Latham, D.W., and Tonry, J.: 1982, Ap.J. 253, p. 423.
de Vaucouleurs, G.: 1983, Proceedings, Colloquium on Groups and Clusters of Galaxies, held in Trieste.
Einasto, J., Klypin, A., Saar, E., and Shandarin, S.F.: 1983, Tallin Preprint A-4.
Focardi, P., Marano, B., and Vettolani, P.: 1983, COSPAR/IAU Symposium, Rojen, Bulgaria.

Frenk, C.S., White, S.D.M., and Davis, M.: 1983, Ap.J. 271, p. 417.
Gerola, H., and Seiden, P.E.: 1978, Ap.J. 223, p. 129.
Giovanelli, R., Haynes, M.P., and Chincarini, G.: 1983, in preparation.
Gott, J.R., Wrixon, G.T., and Wannier, P.: 1973, Ap.J. 186, p. 777.
Gregory, S.A., and Thompson, L.A.: 1978, Ap.J. 222, p. 784.
Joeveer, M., and Einasto, J.: 1978, in The Large Scale Structure of the Universe, eds. M.S. Longair and J. Einasto, D. Reidel Publishing Company.
Kirshner, R.P., Oemler, A., Schechter, P.L., and Schectman, S.A.: 1983, in Early evolution of the Universe and its present structure, eds. G.O. Abell and G. Chincarini, D. Reidel Publishing Company.
Materne, J., 1978, Astr. Ap., 63, 401.
Moody, E.A., Turner, E.L., and Gott, J.R.: 1983, Princeton Observatory Preprint No. 37.
Neyman, J., Scott, E.L., and Shane, C.D., 1953, Ap.J. 117, p. 92.
Omer, C.G., Page, T.L., and Wilson, A.G.: 1965, A.J. 70, p. 440.
Oort, J.H.: 1983, in press, Annual Review of Astronomy and Astrophysics.
Peebles, P.J.E.: 1980, The Large-Scale Structure of the Universe, Princeton University Press, New Jersey.
Peebles, P.J.E.: 1983, in The Origin and Evolution of Galaxies, eds. B.J.T. Jones and J.E. Jones, D. Reidel Publishing Company.
Peebles, P.J.E.: 1982, Ap.J. 257, 438.
Rubin, V.C., Ford, W.K., and Thonnard, N.: 1980, Ap.J. 238, p. 471.
Salpeter, E.E.: 1983, in Early evolution of the Universe and its present structure, eds. G.O. Abell and G. Chincarini, D. Reidel Publishing Company.
Scott, E.L., Shane, C.D., and Swanson, M.D.: 1954, Ap.J. 119, p. 91.
Shandarin, S.F.: 1983, in The Origin and Evolution of Galaxies, eds. B.J.T. Jones and J.E. Jones, D. Reidel Publishing Company.
Shane, C.D., and Wirtanen, C.A.: 1967, Publ. Lick Obs. 22, part 1.
Silk, J., and Norman, C.: 1981, Ap.J. 247, p. 59.
Soneira, R.M., and Peebles, P.J.E.: 1978, Ap.J. 211, p. 1.
Tarenghi, M., Tifft, W.G., Chincarini, G., Rood, H.J., and Thompson, L.A.: 1979, Ap.J. 234, p. 793.
Tifft, W.G., and Gregory, S.A.: 1976, Ap.J. 205, p. 696.
Zel'dovich, Ya.B.: 1978, in The Large Scale Structure of the Universe, eds. M.S. Longair and J. Einasto, D. Reidel Publishing Company (and references therein).
Zel'dovich, Ya.B., Einasto, J., and Shandarin, S.F.: 1982, Nature 300, p. 407.
Zel'dovich, Ya.B., and Novikov, I.D.: 1983, in The Structure and Evolution of the Universe, ed. G. Steigman, The University of Chicago Press.
Zwicky, F.: 1957, Morphological Astronomy. Springer-Verlag Publishing Company.
Zwicky, F.: 1972, private communication.
Zwicky, F., Herzog, E., Wild, P., Karpowicz, M., and Kowal, C.T.: 1961-1968, Catalogue of Galaxies and Clusters of Galaxies, 6 Vol., Pasadena, Calif.: California Institute of Technology.

COSMOLOGY WITH THE SPACE SCHMIDT TELESCOPE- GALAXY COLORS AND COLOR DISTRIBUTIONS

GUSTAVO BRUZUAL A.
Centro de Investigaciones de Astronomía (CIDA),
A.P. 264, Mérida 5101-A, Venezuela.

ABSTRACT

Galaxy spectral evolutionary models are used to compute the following quantities in optical and UV bandpasses: (a) galaxy color, luminosity, two-color diagrams, and surface brightness profiles as functions of redshift; and (b) galaxy counts, and color and redshift distributions as functions of apparent magnitude. These predictions may be used as a guide to prepare and interpret observing runs with the Space Telescope, the Space Schmidt Telescope, and the Starlab Observatory.

INTRODUCTION

In this paper I report some results from a series of investigations carried out by the author (Bruzual 1981, 1983 a, 1983 b, 1983 c). In these papers parametric models for the spectral evolution of galaxies are developed using the evolutionary synthesis technique. The direct result from the synthesis program is the prediction of the evolution in time of a galaxy spectral energy distribution (s.e.d.). The model s.e.d.'s make it possible to predict observational quantities of cosmological interest that will serve as a guide in the preparation for and the interpretation of observing runs with telescopes from space (Space Telescope, Space Schmidt Telescope, Starlab Observatory, etc.). These predictions are subject to the limitations imposed upon the models by the simplifying assumptions underlying these models. Some of these assumptions are the following: (1) Galaxies can be treated as closed systems. (2) Chemical evolution is not important (for our purposes) after the stars in galaxies are formed. The models assume solar composition throughout. (3) The star formation rate (SFR) is a smooth function of time (independent of stellar mass) which determines the spectral and luminosity evolution of a galaxy. (4) The initial mass function (IMF) is a simple function of the stellar mass (independent of galaxy age) of the same general form as the IMF observed in the solar vicinity. (5) The effects of gas and dust on galaxy spectra can be neglected on a first approximation. The validity of these assumptions lies in the ability of the models to reproduce the observations.

Under these assumptions at a given galaxy age a model is specified by the SFR and the IMF. To relate time and redshift a cosmological model must be used. For reasons of space no more details about the spectral evolutionary models will be given here. The reader is referred to the papers cited above for further details.

RESULTS

Model predictions have been computed in the UBV photoelectric system, the U^+J^+FN photographic system, and in four gaussian shaped UV bands centered at wavelengths 1400, 1700, 2200 and 2700 A. The magnitudes corresponding to the UV bands will be denoted 14, 17, 22, and 27, respectively. A complete definition of the photometric systems is given in Bruzual (1983 c). A summary of the most relevant predictions follows.

(a) Color Evolution.

Color versus redshift lines have been computed for most of the color combinations in the systems mentioned above. As an example Fig. 1a shows the behaviour of the color 22-27 with redshift. The quantities next to each line refer to the values of the parameters used in the SFR and the IMF. $H_o=50$, $q_o=0$ in this figure.

(b) Two-Color Diagrams as a Function of Redshift.

The loci in the 17-22 versus 22-27 two-color diagram expected to be occupied by galaxies at z=0, 0.2, 0.5, 1.0, 1.5, and 2 are shown in Fig. 1b, for the $H_o=50$, $q_o=0$ cosmology. The hatched area indicate the region of the diagram where stars are expected. This kind of diagram should be helpful in identifying high redshift objects which subsequently can be studied spectroscopically.

(c) Luminosity Evolution.

The dependence on Z of the 27 magnitude is shown in Fig. 1c for the same models and cosmology as in Fig. 1a. The reader is referred to Bruzual (1983 b) for details about

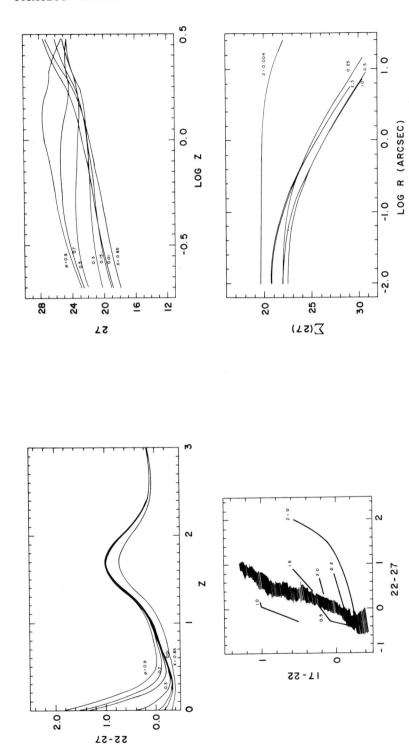

Figure 1.- (a) Behaviour of the color 22-27 versus redshift for several representative models described in Bruzual (1983 c). (b) Loci in the 17-22 versus 22-27 two-color diagram expected to be occupied by galaxies at Z=0, 0.2, 0.5, 1.0, 1.5 and 2. The hatched area indicate the region of the diagram where stars are expected. (c) Dependence on Z of the 27 magnitude. (d) Surface brightness profiles in the 27 magnitude for a $\mu = 0.7$ model at the redshifts indicated next to the curves. For the four figures it was assumed that $H_o=50$, $q_o=0$, and galaxy age = 16 Gyr.

the absolute magnitude assigned to each galaxy s.e.d.

(d) Surface Brightness Profiles.

Figure 1d shows the surface brightness profile in the 27 magnitude for a mildly evolving s.e.d. ($\mu = 0.7$). The shape of the profile is preserved at any Z. The vertical displacement for a given Z is determined by the luminosity evolution of the given model in the specific emitted wavelength. In the case shown the galaxy is brighter for Z=2 and 3, than at z=1. For bluer galaxies ($\mu = 0.01$) the same effect takes place irrespective of evolution (just due to the K correction term).

(e) Galaxy Counts and Galaxy Colors and Redshift Distributions.

With the information presented so far it is possible to predict galaxy number counts as a function of apparent magnitude in any desired bandpass (Fig. 2a). The derived color and redshift distributions can also be computed as a function of apparent magnitude. Figure 2b shows the 22-27 color distribution as a function of apparent magnitude. Figure 2b shows the 22-27 color distribution as a function of apparent B magnitude. Figure 2c shows the distribution of log Z for the same models. The reader is referred to Bruzual (1983b) for further detail.

CONCLUSIONS

Under the assumptions named in the introduction, it would seem safe to conclude that in a first approximation no unexpected results are anticipated. Faint galaxy counts are predicted to increase as expected from an extrapolation of ground-based observations. Color distributions as a function of apparent magnitude would seem to be have as hinted from data already available (Bruzual and Kron 1980, Koo 1981). Stars and galaxies can be differentiated, in principle, by their position in the UV two-color diagrams. As expected, galaxies with widely different UV luminosities at the present epoch may come from systems that looked equally bright in the past. The surface brightness profiles will, in general, behave as expected with galaxy redshift, even though for distant enough galaxies the spectral evolution (or just the K correction) in some bands may dominate the redshift term.

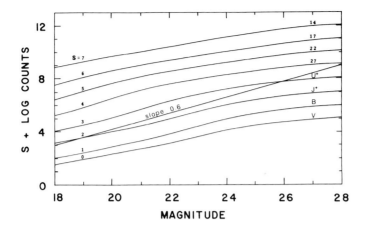

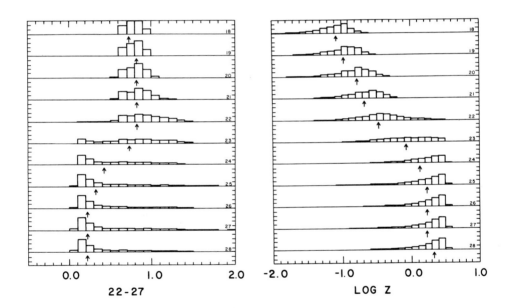

Figure 2.- (a) Galaxy number counts as a function of apparent magnitude. (b) 22-27 color distribution as a function of apparent B magnitude for B in the range from 18 to 28 magnitude. The arrows point to the median color of the distribution. (c) Distribution of log Z as a function of apparent magnitude. The arrows point to the median value of log Z. In all cases $H_o = 50$, $q_o = 0$, and galaxy age = 16 Gyr.

REFERENCES

Bruzual A., G. 1981, Ph.D. thesis, University of California,
    Berkeley.
Bruzual A., G. 1983 a, Ap. J., 273 (in press).
Bruzual A., G. 1983 b, Ap. J. Suppl., 53 (in press).
Bruzual A., G. 1983 c, Rev. Mexicana Astr. Ap., 8. pp.63-81.
Bruzual A., G., and Kron R.G. 1980, Ap. J., 241, p. 25.
Koo, D.C. 1981, Ph.D. thesis, University of California,
    Berkeley.

OPTICAL DESIGN WITH THE SCHMIDT CONCEPT

1. GROUND-BASED DEVELOPMENT
2. THE SPACE SCHMIDT PROJECT FOR THE 1990'S ?

G. Lemaître

Observatoire de Marseille
Place Le Verrier, 13248 Marseille, France

The basic principle of the camera described by Bernard Schmidt in 1932 is that a single concave mirror with a stop at its center of curvature has no unique axis and therefore yields equally good images at all points of its yield. The field is curved, and to correct the spherical aberration produced by the mirror, Schmidt introduced, in the stop at the center of curvature of the mirror, a thin non spherical corrector plate of glass. Around 1930 at the Hamburg Observatory and in spite of many difficulties, it was Schmidt's genius as an optician that succeeded - after several judiciously interpreted trials - in figuring a corrector plate by elastic relaxation, and thus demonstrated the optical performance of this new generation of instruments.

For such critical photographic work as sky surveys by the Palomar, ESO and UK Schmidt telescopes, an attempt was made to match the size of the residual aberrations to that of the resolving power of the emulsion used. A third order theory has been treated by Strömgren (1935) for minimizing the chromatism, by Carathéodory (1940) for the determination of the on-axis stigmatism and by Linfoot (1949) for the optimization and balancing of the aberrations in the field. The theory of elasticity applied to the original Schmidt method of figuring aspherical plates has been treated by Couder (1940).

The length of a Schmidt telescope is twice that of the focal length. At low F-ratio the concave mirror would be much larger than the corrector plate and the length of the telescope prohibitive. Ratios between F 2.7 and F 3.5 were mainly adopted for large classic (refractive) singlet corrector Schmidt, but it has been shown that the chromatism of even a fused silica plate hardly limits the optical performance of the system. Solutions for reducing this effect will be presented below. New developments, such as all-reflective design and spectrographic cameras using aspherical reflective gratings, need the treatment of a more elaborate theory of aberrations than the third order one. New methods of figuring refractive or reflective corrector, and aspherical gratings by elastic relaxation are very efficient if used with plane polishing.

1. GENERAL DESIGN

A large variety of systems can be designed according to the principle outlined by Schmidt of locating both the stop and the corrector at the center of curvature C of the mirror. This is the case with refractive or, more recently reflective cameras working with object at infinity or at a finite distance and also new spectrographic cameras using aspherical reflective gratings.

1.1 WAVEFRONT FIGURES AT THE CENTER OF CURVATURE OF A SPHERICAL MIRROR

For determining the correction to be done in C with corrective element, let us consider, as in Figure 1, a point source I that will become the stigmatic image point of the system with the corrector. The shape of the corrective element can be readily defined from that of the wavefront $Z_w(r)$ passing point C after reflection at P on the mirror whose radius of curvature is R. The Gaussian focus G of the mirror alone for an object at infinity is located midway between segment CS. Let $GI = M \times R/2$ be the distance from the point source I to the point G. The adimensional parameter M is not necessarily small and takes any negative values between 0 and -1 for Schmidt cameras working at finite distance, as is shown in the case of Figure 1. With the object at infinity, M is positive and relatively small, and point I is close to point G - between G and S.

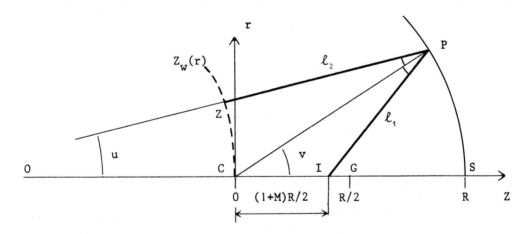

Figure 1. Reflection at a spherical mirror.

The shape of the wavefront $Z_w(r)$ can be obtained from the condition of a constant light path $2\ IS + IC$ on the axis, so that for a point P of the mirror and for positive lengths $IP = \ell_1$ and $PZ = \ell_2$

$$\ell_1 + \ell_2 = (3 - M)R/2 \tag{1}$$

In a cylindrical coordinate system the wavefront $Z_w(r)$ can be expressed by the even polynomial series

$$Z_w = \sum_{n=1} A_{2n} \, r^{2n}/R^{2n-1} \qquad (2)$$

For large spectrographic cameras having an F-ratio as fast as F1, high order terms of eq. (2) will be necessary to obtain an accurate correction and also as starting parameters for solving the elasticity problem, mainly in the case of aspherical gratings - working at infinity - where deformable masters, used for replication, are free of the thickness distribution.

A set of relations can be deduced from the geometrical properties of Figure 1, where u and v are respectively the angles of segments ZP and CP with the Z axis. These are

$$\begin{aligned} Z_w &= R \cos v - \ell_2 \cos u \\ r &= R \sin v - \ell_2 \sin u \\ \ell_1 \cos u &= R \cos v - (1+M)\,R/2 \\ \ell_1^2/R^2 &= (5 + 2M + M^2)/4 - (1+M)\cos v \end{aligned} \qquad (3)$$

An extremely laborious work using numerous series expansions and many operation on these has been carried out from relations (1) and (3) for the determination by identification with the coefficients A_{2n} of eq. (2). This analysis would be too long to be presented here but could be published later. For the five first coefficients, the result is

$$A_2 = \frac{M}{1+M}, \qquad A_4 = -\frac{1}{4}\frac{1 - M - M^2 - 3M^3}{(1+M)^3} \qquad (4)$$

$$A_6 = -\frac{3}{8} + 2M - 4M^2 + \ldots, \quad A_8 = -\frac{45}{64} + 4M - \ldots, \quad A_{10} = -\frac{193}{128} + \ldots,$$

In the case of a camera working with an object at finite distance, the adimensional parameter M is also the magnification ratio, i.e. $M = -CI/CO = -SI/SO$. As a verification, the first two coefficients of relations (4) give a spherical wavefront for the particular cases where $M = +1$ and $M = -1$, corresponding respectively to point I in S and point I in C.

1.2 WAVEFRONT FIGURE FOR AN OBJECT AT INFINITY

In this section and the following, only the case of object at infinity is to be considered so that the point I is close to G. For I at G, parameter $M = 0$ and coefficient $A_2 = 0$; the aspherical wavefront does not presents curvature in C. When optimizing the corrector for reducing off-axis aberrations and possibly on-axis sphero-chromatism for refractive elements, one will need a coefficient A_2 of sign opposite to that of the higher coefficients. These wavefronts shows an inflexion point, and for a radial distance r_o the propagation is parallel to the Z axis. If r_m is the radius of the clear aperture beam, the zone of parallel propagation can be characterized by a a ratio as

$$a = r_o^2/r_m^2 \tag{5}$$

The ratio is defined by $F = R/2r_m$, since parameter M is necessarily small. From the derivation of the two first terms in eq. (2), one obtains as a first approximation

$$M \simeq a/2^5 F^2 \tag{6}$$

For $a = 1$, the wavefront is parallel to the Z axis for $r_o = r_m$. As will see, optical optimizations lead us mainly to study cases where $a < 1$ and $a > 1$.

1.3 DESIGN OF VARIOUS CORRECTIVE ELEMENTS

Refractive or reflective correctors and self-corrective gratings can be used in Schmidt's design. To avoid full obstruction of the camera, a tilt of the corrector is necessary if there is a reflection. Let us consider a cylindrical coordinate system Z, r, θ, linked to the corrector, with the r, θ plane being tangent to it at its vertex C, the Z axis being positive toward the spherical mirror and $\theta = 0$ in the symmetry plane of the camera. The general figure $Z(r,\theta)$ of the corrective element is of the form

$$Z = \Sigma\, B_{m,2n}\, r^{2n} \cos m\theta\, /\, R^{2n-1}, \tag{7}$$

and will be defined from that of the wavefront Z_w as defined in eq. (2) and (4) by the following relation

$$B_{m,2n} = s\, \mu\, T_{m,2n} \cdot A_{2n}, \tag{8}$$

where s is a field optimization factor very close to unity (see below), μ is a constant depending upon the mounting and the nature of the corrector and $T_{m,2n}$ are coefficient appearing for tilted correctors. Table 1 lists the values of μ and ratios $B_{m,2n}/A_{2n}$ for different types of corrector. For a refractive plate the rays are assumed to emerge from the aspherical surface that faces the concave mirror. The case where the reflective corrector is not tilted leads to full obstruction on-axis but can be usable off-axis. This family belongs to the class of centred systems. The case where the reflective corrector is tilted by an angle i has been listed with $t = \frac{1}{2} \sin^2 i$. This is also valid for Littrow-mounted reflective gratings. These families belong to the class of non-centred systems : the figure of the corrector is of bi-axial symmetry. For spectrographs, self-corrected reflective gratings have a rotational symmetry if mounted in normal diffraction ($\beta = 0$).

At the edge of the field, 1) the corrective element gives a stronger correction than when working on-axis. This being due to the inclination φ of the incident beams, and 2) the stop is circular but the cross-section of the off-axis beams is elliptical. Thus the corrective element increases the correction in the sagital section of these beams. This two off-axis effects produce degradation of the astigmatism type. For field optimization it is necessary to choose, if possible, an s value slightly under unity to under-correct the aspherical element.

TABLE 1. Value of μ and $B_{m,2n}/A_{2n}$ for different types of corrector

CORRECTOR TYPE	REFRACTIVE PLATES Correcting for index n	REFLECTIVE MIRRORS without tilt, 100% obstruction on-axis	REFLECTIVE MIRRORS OR GRATINGS IN LITTROW tilt angle = i	REFLECTIVE GRATINGS incident angle = α diffraction angle β = 0
μ	$-1/(n-1)$	$1/2$	$1/2 \cos i$	$1/(1 + \cos \alpha)$
$B_{0,2}/A_2$	$-s/(n-1)$	$s/2$	$s(1-t)/2 \cos i$	$s/(1 + \cos \alpha)$
$B_{2,2}/A_2$	0	0	$-st/2 \cos i$	0
$B_{0,4}/A_4$	$-s/(n-1)$	$s/2$	$s(1-2t+\frac{3}{2}t^2)/2 \cos i$	$s/(1 + \cos \alpha)$
$B_{2,4}/A_4$	0	0	$-2st(1-t)/2 \cos i$	0
$B_{0,6}/A_6$	$-s/(n-1)$	$s/2$	$s(1-3t)/2 \cos i$	$s/(1 + \cos \alpha)$

2. REFRACTIVE CAMERAS

2.1 OFF-AXIS ABERRATIONS AND CHROMATISM OF A SINGLET PLATE

To determine the size of off-axis images at an angle φ from the axis, spot diagrams have been obtained with 73 rays in the stop. There are 1, 16, 24 and 32 rays for semi-apertures of 0, $r_m/2$, $3 r_m/4$ and r_m, respectively, (see Figure 2 at the top of the curves). First, different values of the parameter M, and then a from eq. (6), are considered with s = 1. Off-axis images are plotted on the sphere of center C and radius $(1+M)R/2$ that gives the on-axis stigmatism. The two-dimensional size of these images is denoted by ℓ_r and ℓ_t, in the radial and tangential directions respectively. These spot-diagrams are shown at the top line of Figure 2, the center of the field being in the vertical direction of the Figure. Except for a = 4/3, these images can be improved by focusing by the quantity Δf. The sign of Δf is positive for a focusing toward the mirror. One then obtains images inscribed in a circle of minimum diameter d. These spot diagrams are shown on the second line of Figure 2. Variations of the quantities ℓ_r, ℓ_t, d and Δf with the parameter a are given as a function of the quantities F, φ

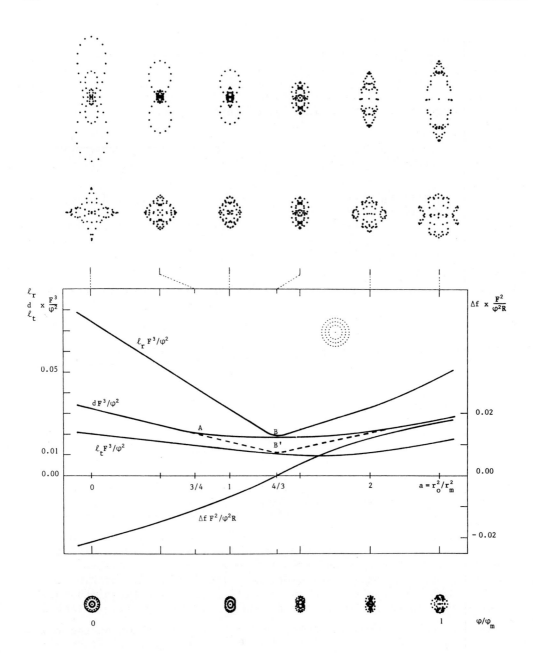

Figure 2. Off-axis residual aberrations of refractive Schmidts

and R. As results, for a = 3/4, i.e. a null power zone at $r_o = 0.866\, r_m$, the diameter of the largest image is

$$d = 0.020\, \varphi^2/F^3 \qquad (9)$$
$$s = 1, \quad a = 3/4 \quad \text{i.e.} \quad M = 3/128\, F^2$$

(cf. point A, Fig. 2). For a = 4/3 the null power zone is at $r_o = 1.155\, r_m$, out of the clear aperture and $d = 0.0185\, \varphi^2/F^3$ (cf. point B, Fig. 2) which is the minimum for s = 1.

In the range 1 < a < 5/2 it is possible to improve the resolution by undercorrecting the plate. If φ_m is the semi-field angle to optimize, the best parameter of under-correction is $s = \cos^2 \varphi_m$. After focusing, the size of the off-axis blur at $\varphi = \varphi_m$ is shown by the dotted-line of Figure 2. The best resolution is again for a = 4/3 (cf. point B', Fig. 2)

$$d = 0.011\, \varphi_m^2/F^3 \qquad (10)$$
$$s = \cos^2 \varphi_m, \quad a = 4/3 \quad \text{i.e.} \quad M = 1/24\, F^2$$

Spot-diagrams of this optimization are shown on a spherical surface for different field values φ/φ_m at the bottom line of Figure 2. For instance, at F3 and for $2\varphi_m = 5°$ the diameter of the images according to eq. (9) and (10) do not exceed 0.29 and 0.16 arcsec respectively.

Unfortunately, for singlet refractive plates, the chromatic variation of spherical aberration is much larger than off-axis aberrations. To minimize the effect of this variation, the slope of the correcting plate must be chosen as weak as possible in absolute value. This criterium, which corresponds to the old rule stated by Kerber in 1886 to optimize blue and red wavelengths of composite lenses, is translated here by a = 3/4, giving a correcting plate having opposite slopes at $r_o = r_m/2$ and $r_o = r_m$. If the plate is designed for correcting the wavelength λ_o with refractive index n_o (on-axis stigmatism), at the wavelength λ with n as the corresponding refractive index, the angular diameter of the on-axis image (c.f. Bowen, 1960) is

$$d_\lambda = 1/128\, F^3 \nu \qquad (11)$$

where $\nu = (n_o - 1)/(n_o - n)$. For an F3 telescope and a plate in fixed silica correcting for $\lambda_o = 405$ nm, at $\lambda = 320$ nm or at $\lambda = 656$ nm the absolute value of ν is $\nu = 36$ and eq. (11) gives $d_\lambda = 1.6$ arcsec.

Finally if d_φ is the diameter on off-axis image given by eq. (9) and d_T that of the seeing, the image size in the field of a singlet plate Schmidt cannot be better than

$$D < d_\varphi + d_\lambda + d_T = 20\cdot 10^{-3}\, \varphi^2/F^3 + 7.8\cdot 10^{-3}/F^3\nu + d_T \qquad (12)$$

If d_T is 1 arcsec, singlet plate Schmidts designed at F3 with $2\varphi = 5°$ in the wavelength range 320 – 656 nm have residual aberrations of at least 3 arcsec. Improving this resolution by factor 2 would lead in the same conditions to an aperture of F 4.75 and then a very long tube.

2.2 ADDITION OF A MENISCUS FOR CORRECTING THE RED REGION

The predominant effect of the sphero-chromatic aberration of a singlet corrector plate can be divided by a factor two by adding a meniscus just before the focus only when working in the red region of the spectrum. The full spectral range of the telescope is then shared by blue and red, each range corresponding to an equal variation Δn of the refractive index of the plate. The plate is designed alone with $a = 3/4$ for correcting the blue; this gives an under-correction in the red which is compensated by adding a monocentric meniscus. Both mirror and meniscus have a common center of curvature so that no off-axis aberrations are added by the meniscus. If N is its mean index in the red, its thickness is roughly equal to

$$e = \frac{1}{8} \frac{\Delta n}{n-1} \frac{N^3}{N^2-1} R \qquad (13)$$

The image size in the field for this design is given in eq. (12) where now the number ν is twice as great as for the singlet plate of the previous section if the full spectral ranges (red + blue) are the same. For instance, an F3 telescope of 1 meter clear aperture having a plate and a meniscus in fused silica and extended spectral range such as 320 to 440 nm in the blue and 440 to 1000 nm in the red leads to a meniscus of 65 mm thickness.

An in-convenience of this design is to use two plate-holders of slightly different curvature and collimation. The red plate scale is smaller than that of the blue plate (0.8% smaller for the preceding example). Advantages of this solution (compared to that described in following section) are the high transparency of a singlet plate in the UV and also of having only one aspherical surface to figure. Also, filters can be inserted into the meniscus for cutting off the shorter wavelengths without additional glass-air surfaces.

2.3 DOUBLET PLATE CORRECTOR

Another possibility for correcting the sphero-chromatism aberration of a single plate is to design a doublet plate corrector. If λ and λ' are the two wavelengths for which the instrument is corrected and n and n' the refractive indices, one can define the number $\nu = (n_o - 1)/(n - n')$ for each glass as ν_1 and ν_2. Let us suppose that ν_2 corresponds to the most dispersive glass. If ψ_1 and ψ_2 are the powers of the plates for a given value of the a ratio, and ψ that of a singlet plate correcting for index n_o, the condition of a chromatism is

$$\psi_1/\nu_1 + \psi_2/\nu_2 = 0 \quad \text{and} \quad \psi_1 + \psi_2 = \psi \qquad (14)$$

giving the power of the components as

$$\psi_1 = \frac{\nu_1}{\nu_1 - \nu_2} \psi \quad \text{and} \quad \psi_2 = \frac{-\nu_2}{\nu_1 - \nu_2} \psi, \qquad (15)$$

the correcting plate of most dispersive glass being divergent. Assuming that the secondary chromatism is negligeable, one can choose the ratio a = 4/3 for a better off-axis correction of the two plates (cf. resolution given by eq. (10)), and then the null power zones are out of the clear aperture. The figure of the two plates can be written by using the parameter s of eq. (8) as

$$s_1 = \frac{\nu_1}{\nu_1 - \nu_2} \cos^2 \varphi_m \quad \text{and} \quad s_2 = -\frac{\nu_2}{\nu_1 - \nu_2} \cos^2 \varphi_m . \quad (16)$$

For instance a Schmidt at F3 achromatized at λ = 365 nm and λ' = 588 nm by using Schott glass like crown UBK7 ($\nu_1 \simeq 27$) and light flint LLF1 ($\nu_2 \simeq 18$) have plates of equivalent asphericity of F2.06 for the crown and of F2.35 for the flint which is divergent at its vertex. Such correcting plates were built for the 1.2 m, F3 UK Schmidt in Siding Spring (Australia) while another doublet is under completion for the 1 m, F3 ESO Schmidt in La Silla (Chile).

3. REFLECTIVE TELESCOPES

All-reflecting Schmidts are particularly well adapted for carrying out sky surveys from space (Fig. 3). Below, optical performance will be compared between centred systems working off-axis and non centred systems. The best result is in favour of the latter.

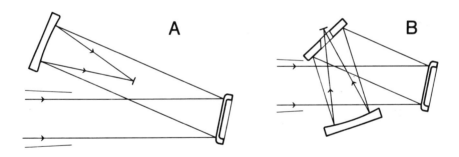

Figure 3. Reflective Schmidts

3.1 CENTRED SYSTEMS OPTIMIZED OFF-AXIS

Let us consider, first, a full obstructed centred system. As was done for refractive telescopes, and for a parameter s = 1, off-axis images have been plotted, as a function of the a ratio, on the sphere of center C and radius $(1+M)R/2$ that gives the on-axis stigmatism. The size of these images is denoted by ℓ_r out ℓ_t in the radial and tangential directions respectively (see spot-diagrams at the top line of Fig. 3). With the same conventions as for § 2.1 the diameter of the best images d and defocussing Δf are also given for the spot-diagrams on the second line of Figure 4. The result is that for a = 3/2, i.e. a null power zone at $r_o = 1.22\ r_m$, the diameter of the off-axis blur is close to the minimum. A careful interpolation between the considered values of a seems to give the smallest blur for $r_o = 1.10\ r_m$, which corresponds to $a = \sqrt{2}$. If φ_m is the semi-field angle to be optimized, the best parameter of under-correction is $s = \cos \varphi_m$. After focussing, the size of the off-axis blur at $\varphi = \varphi_m$ is shown by the dotted line of Figure 4. For $a = 3/2$ the best resolution is $d = 0.011\ \varphi_m^2/F^3$. Spot-diagrams of this optimization are shown on a spherical surface for different field values φ/φ_m on the bottom line of Figure 4.

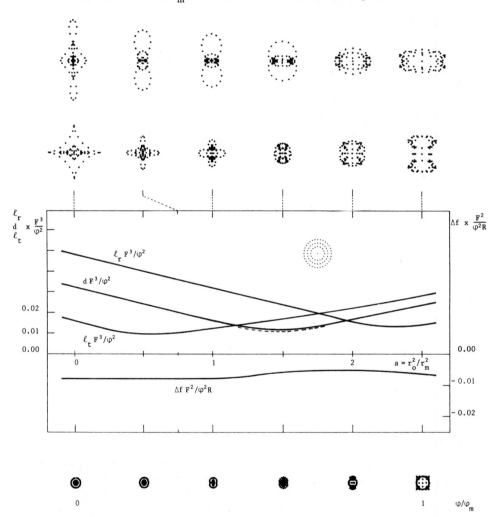

Figure 4. Off-axis residual aberrations of centred reflective Schmidts

To make such a two-reflection telescope practicable and then avoid the obstruction, the angle of incidence of the principal beam must be at least

$$i = \varphi_m + 1/4\, F \tag{17-1}$$

When adding a holed folding mirror (three-reflection telescope) for a better detector access to the focus, this angle can be kept as

$$i = \varphi_m + 7/16\, F \tag{17-2}$$

Relations (17-1) and (17-2) hold for beams having circular cross-section. For both of the preceding mountings, the resolution for the most off-axis beam of inclination $i + \varphi_m$ is

$$d_c = 0.011\, (i + \varphi_m)^2/F^3 \tag{18}$$
$$s = \cos(i + \varphi_m)\, , \quad a = 3/2 \quad \text{i.e.} \quad M = 3/64\, F^2$$

This resolution is the same as that found for a monochromatic refractive camera (cf. eq. (10)).

3.2 NON CENTRED SYSTEMS

For a non-centred system the mounting is the same as for the preceding section. Relations (17) hold, but the shape of the correcting mirror is not a surface of revolution. The level lines of this surface are homothetical ellipses as defined by eq. (7) and (8) and coefficients given in the forth column of Table 1, so that stigmatism is achieved at the center of the field.

The best resolution is again determined by ray tracing with respect to the a ratio. The result is that the null power zone is slightly under $a = 3/2$ as for the preceding section. An under-correction of the corrector does not improve the performance. For the two points at the edge of the field and in the symmetry plane of the telescope, the largest blur is that which corresponds to the largest distance from the Z axis of the corrector. The angular diameter of that image is found as

$$d_{NC} = 0.012\, \varphi_m\, (\tfrac{3}{2} i + \varphi_m)/F^3 \tag{19}$$
$$s = 1\, , \quad a = 3/2$$

Figure 5 gives spot diagrams in the field of a design with $a = 1$. For this ratio the image blurs are 1.44 times larger than for eq. (19) but the corrector can be readily figured by the elastic relaxation technique using a built-in edge at the null power zone.

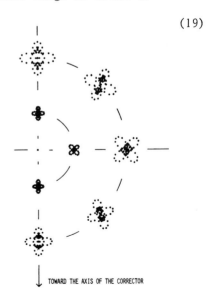

Figure 5. Off-axis residual aberrations in the field of non-centred reflective Schmidts for $a = 1$

3.3 GAIN OF NON-CENTRED SCHMIDTS

For an image-blur having larger size than that of the detector pixel the gain in magnitude of the non-centred design over the centred system is readily determine from eq. (18) and (19). Considering with eq. (17-2) the folded three-reflection design, one obtains

$$\Delta m = 2.5 \log (d_C^2/d_{NC}^2) = 5 \log \frac{11}{18} \frac{(2\varphi_m + 7/16\,F)^2}{\varphi_m(\varphi_m + 7/16\,F)} \qquad (20)$$

As instance, at F3 with a semi-field $\varphi_m = 2.5°$, the gain in magnitude is $\Delta m = 2.7$, and the non-centred design gives, at the edge of the field, an image blur equal or smaller to 1.2 arcsec (G. Lemaître, 1976, 1979, 1980). This performance is very promising for carrying-out the construction of a 2 meter focal length space Schmidt telescope for an UV Survey. Proposals submitted to NASA (J. Wray et al., 1982) could be retained in the future.

4. SCHMIDT SPECTROGRAPHS WITH REFLECTIVE ASPHERICAL GRATINGS.

In a slit spectrograph using reflective gratings, one of the difficulties encountered is the obstruction of the incident beam by the camera optics. To avoid this, the camera optics are usually designed to be placed at a distance of several times the collimator diameter from the gratings. It is then necessary to provide camera optics having an aperture substantially larger than that of the collimator beam if severe losses due to vignetting are to be avoided at the edges of spectra. These optics, much larger than the grating size, have increased asphericities, and for cameras of fast F ratios the figuring of such optics becomes crucial.

Aspherical gratings have been produced by the elastic relaxation method These gratings lead to more nearly ideal mountings from the standpoints of a small number of surfaces, of a fast F ratio and of a wide field. An example of the procedure is the following. 1) A plane grating is duplicated on a flexible optical blank when in a state of zero stress. 2) The flexible blank is re-duplicated while submitted to a state of stress giving the required asphericity by flexure.

Four arrangements having the same focal length, F ratio and field of view are shown in figure 6. The first mounting is obtained with a classical refractive schmidt camera (a). In (b) a doublet corrector camera is used with all surfaces spherical and one type of glass. The doublet corrector can be replaced by a Maksutov or a Bouwers meniscus, but the optical positioning is kept approximately identical. One method of avoiding the difficulty of these large correctors is to place the corrector parallel to the grating and practically in contact

with it (Bowen, 1952). The light
then passes the corrector plate
twice, once before and once after
diffraction at the grating (c).
It can be shown that in this case,
the figure of the twice-trough
plate is reduced by the factor
$(1+\cos \alpha)/\cos \alpha$ from that of the
regular once-through plate (a),
where α is the angle of incidence
on the plate and the grating. The
chromatic aberrations are essentially the same as for the conventional corrector plate.
With aspherical gratings
(Lemaître, 1977) these chromatic
aberrations are entirely suppres-
sed and the problem of four refrac
tions in (c) vanishes. Table 1
shows that for mountings (d) the
figure of the corrective grating
is reduced by the factor $(1+\cos\alpha)/(n-1)$ from that of the regular once
through plate (a). For a grating
having symmetry of revolution the
axis of the camera is taken
nearly normal to the grating ($\beta=0$).
The coefficients defining the
figure of the grating according
to eqs (7) and (8) are given in
the fifth column of Table 1. The
residual off-axis aberrations are
smaller than those given by eq.
(18) for centred reflective
Schmidts in which the angle i is
taken equal to zero.

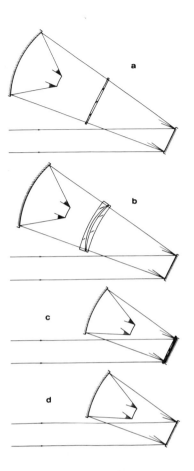

Fig.6 - Comparison of four spectrographic mountings using the same
reflective grating.

The reflective normal diffraction mounting is surely the highest
performance design of all those based on the Schmidt concept. For
cases where the angle of diffraction β is non-zero , such
grating have only a bi-axial symmetry, as for the Littrow mount, the
aspherical coefficients of which, are given in the fourth column of
Table 1. In order to obtain a more accessible focus, folding mirrors
have been placed between the gratings and the camera mirror. These
holed mirrors can fold either both incident and diffracted beams, or
only the latter.

5. TRENDS FOR SCHMIDTS IN ASTRONOMY

Because of their wide field of view, Schmidt cameras have found many uses in astronomy (detection, identification, classification). The main optical designs are the followings :

- photography for sky surveys and selected areas.

- field spectrography with single prisms.

- field spectrography with a two position bi-prism for measuring the radial velocities of stars and nebulas. An accuracy of 4 km/s is obtained (Fehrenbach, 1981).

- Schmidt cameras and their modifications especially for slit spectrographs on large telescopes.

Ground-based Schmidt telescopes are susceptible to differential refraction effects in the field. For this reason, larger field Schmidts could be built only if the size of the corrector were increased and the focal length kept not very different than usual in order to decrease the exposure time. Critical definition of such refractive Schmidts, as given in section 2 by eq. 10, is valid only for a doublet lens corrector having a null secondary chromatism. Under this assumption a Schmidt at F2.2 will have a fifth order astigmatism blur of 1 arcsec at the edge of a field of 7.5° in diameter.

A space Schmidt telescope could surely bring more spectacular discoveries at a price of less optical research than those needed for a very large ground based Schmidt. In 1963, Henize proposed to NASA the use of a Schmidt telescope in space for an UV sky survey.

A two-mirror reflective Schmidt was presented by Epstein (1967) and a folded version was developped by Wray (1969). These were designed with the centred system concept used off-axis.

Fifth order astigmatism can be reduced by using non-centred systems e.g. tilted correctors that have bi-axial symmetry. Korsch (1974) and Lemaître (1976) have given the figure of such correctors for a zonal-ratio $a = 0$. Schroeder (1977) has improved the performance by taking $a = 3/4$. Optimization of the corrector zonal-ratio led finally to choose $\sqrt{2} < a < 3/2$ (Lemaître, 1979) for obtaining the best optical performance. For these values the null power zone is out of the clear aperture. The corresponding size of residual aberration is given in section 3.2 by eq. 19 and the gain over centred systems in section 3.3.

The folding mirror is holed, as for modern spectrographic cameras, in order to adapt a large curved cathode detector. Cathodes of 10 cm in diameter are now in operation in different observatories and have a resolution better than 10 μm with a permanent magnet focussing. A 20 cm detector as needed for the space Schmidt can readily be made (Carruthers, 1979 ; Griboval, 1983 ; Servan, 1983) but the cost is not negligeable for working with an automatic film transport.

At F 3.2 with a 5° field, a resolution of 1 arcsec is achieved. The feasibility of such a complete instrument has been demonstrated and has led, for its construction, to the proposed joint American-Italian-French Schmidt Telescope Project (Wray, 1982).

      the S S T   in the 90's
          will be surely   for the S T
                 a good fellow

REFERENCES

Bowen, I.S.: 1952, *Ap. J.* 116, 1.
          1960, *Telescopes, Stars and Stellar Systems*, Vol. I, ed. Kuiper, p. 58.
Caratheodory, C.: 1940, *Hamburg. Math. Einzelschr.*, 28.
Carruthers, G.R.: 1979, *Advances in Electronics and Electron Physics*, 52, 283.
Couder, A.: 1940, *Comptes-Rendus*, Paris, 210, 327.
Epstein, L.C.: 1967, *Publ. Astron. Soc. Pacific*, 79, 132.
Fehrenbach, C., Burnage, R.: 1981, *Astron. Astrophys. Suppl. Series*, 43, 297.
Griboval, P.: see this Colloquium.
Henize, K.G.: 1963, *NASA Proposal*, Northwestern Univ.
Kerber, see in Chretien, H.: 1958, *Le Calcul des Combinaisons Optiques*, Ed. Sennas, Paris, 346.
Korsch, D.: 1974, *Appl. Optics*, 13, 2005.
Lemaître, G.: 1976, *J. Opt. Soc. Am.* 66, 12, 1334.
              1977, *Astron. Astrophys.* 59, 249.
              1979, *Comptes Rendus*, 288 B, 297.
              1980, *Comptes Rendus*, 290 B, 171.
              1981, *Current Trends in Optics*, ICO 12 Conference, Taylor et Francis, London 131.
Linfoot, E.H.: 1945, *Proc. Phys. Soc.*, London, 57, 209.
           1949, *M.N.R.A.S.* 109, 279.
Linfoot, E.H., Wolf, E.: 1949, *J. Opt. Soc. Am.* 39, 752.

Schmidt, B.: 1932, Mitt. Hamburger Sternw., Bergedorf, 7, 36, 15.
Schroeder, D.J.: 1977, *Appl. Optics*, 17, 141.
Servan, B.: 1976, *IAU Colloquium n° 40*, Paris, 1.
Stromgren, B.: 1935, Vierteljahrsschr. Astr. Gessellsch., 70, 65.
Wray, J.D., O'Callaghan, F.G.: 1969, SPIE, Santa Barbara.
Wray, J.D., Smith, H.J., Henize, K.G., Carruthers, G.R.: 1982, SPIE, 332, 141.

IMPLEMENTATION AND USE OF WIDE FIELDS IN FUTURE VERY LARGE TELESCOPES

J. R. P. Angel
Steward Observatory, University of Arizona

ABSTRACT

The full potential of the next generation of larger telescopes will be realized only if they have well instrumented large fields of view. Scientific problems for which very large ground-based optical telescopes will be of most value often will need surveys to very deep limits with imaging and slitless spectroscopy, followed by spectroscopy of faint objects taken many at once over the field. Improved instruments and detectors for this purpose are being developed. Remotely positioned fibers allow the coupling of light from many objects in the field to the spectrograph slit. CCD arrays, operated in the TDI or drift scan mode, will make large area detectors of high efficiency that may supercede photographic plates. An ideal telescope optical design should be based on a fast parabolic primary, have a field of at least 1° with achromatic images < 0.25 arcseconds and have provision for dispersive elements to be used for slitless spectroscopy and compensation of atmospheric dispersion over the full field. A good solution for a general purpose telescope that can satisfy these needs is given by a three element refractive corrector at a fast Cassegrain focus. A specialized telescope dedicated to sky surveys, with better image quality and higher throughput than presently available, might be built as a scaled up Schmidt with very large photographic plates. Better performance in most areas should be obtained with a large CCD mosaic detector operated in the drift scan mode at a telescope with a 2-mirror reflecting corrector.

1. INTRODUCTION

Telescopes with apertures considerably larger than 4m diameter are being planned by a number of groups in the U.S.A. and elsewhere. A new generation of 8m class instruments, some making use of single glass honeycomb mirrors and alt-azimuth mounts, can be expected. One or two instruments with multiple primary mirror elements to achieve even larger area are planned, such as the US NNTT and ESO's VLT. The new telescopes should be built to realize the best quality imaging possible through the atmosphere and to match the most efficient detectors and instruments. In this paper we will consider the wide field capability that is needed and can be realized in these telescopes, and also future types of specialized survey instruments that will best support them.

Spectroscopy of faint objects will be one of the major tasks for the largest telescopes, and it is for this work that a wide field is most valuable. Even with a very large telescope, a long time is required to obtain spectra of objects at or below the sky limit, but recently techniques have been developed to obtain aperture or slit spectra of dozens of objects at once, taken from all over the field of view. This goes to the heart of a science in which no experiments are possible, but understanding has to be built up from observations of many objects.

In addition to slit spectroscopy of multiple objects, a wide field is valuable for slitless spectroscopy and for direct imaging. Most of the present generation of 4m telescopes take advantage of wide field prime focus correctors with photographic plates to make low resolution grism surveys. An interesting aspect of slitless spectroscopy is that the large telescope acts efficiently as its own survey instrument. The limiting magnitude for low resolution spectra projected against the sky background is similar to that of higher resolution spectra obtained through multiple apertures in the focal plane.

Direct imaging remains an area where very large telescopes will have an important role and should exploit the best seeing. With an 8m telescope the sky signal in photometric bands is about 1000 photon/sec from one square arcsecond. When the seeing is excellent, sampling with pixels of 0.1 - 0.15 arcseconds is desirable. Mapping to 1% of the night sky in these pixels will thus still require long exposures with the best detectors.

What field of view should we strive for in a telescope to be used in these ways? No hard and fast answer is possible, because of the great diversity of programs undertaken with optical telescopes. However, once it is accepted that equipment for identifying and simultaneous spectroscopy of many objects, of order 100 at a time, will be used, then there is considerable advantage in having the widest possible field. Consider the study of gravitationally bound systems, such as open and globular clusters, dwarf galaxies, galaxies and clusters of galaxies. The nearest examples of all these types, those that can be studied in the most detail, have a diameter of a degree or considerably more. It is clearly advantageous if entire systems can be studied in one or only a few fields of multiobject spectroscopy.

A second general class of observation is of randomly distributed objects, such as the local disc stellar population, halo stars, distant galaxies and quasars. There is no natural angular scale for studying these objects, but there is a relationship between number in a given field and apparent magnitude. Suppose one wishes to study many isotropically distributed objects of certain luminosity and the telescope is to be equipped to take a certain number of objects at once from a field of angular diameter θ. The apparent magnitude that must be reached then varies as $10/3 \log\theta$, thus a doubling of the field diameter allows one to work almost exactly 1 magnitude brighter. At cosmological distances this simple relationship breaks down due to the effects of redshift, evolution and geometry, and each type of object must be treated explicitly. Thus, for bright quasars the number density increases much more strongly with magnitude than would be expected for an isotropic distribution. But by 21st magnitude, where the density is ~ 100/square degree, the rate of increase is not very different from the isotropic case above.

Our conclusion from these considerations is that the highest scientific productivity of very large telescopes operating in the optical spectrum will be achieved by giving them the widest possible fields of view. The actual size to be built turns on the practicality of making instruments and detectors for large fields, optical design solutions with the required image quality and scale, and compatibility of wide field use with other applications, particularly in the thermal infrared.

2. INSTRUMENTS AND DETECTORS

Aperture or slit spectra of multiple objects in the field are currently obtained either by making a drilled aperture plate at the spectrograph entrance, or relaying light from the focal plane to the spectrograph entrance with fibers. Aperture plates are the easier to implement and have been used at the Mayall and Hale telescopes (Dressler and Gunn 1983). However, the method cannot be adapted to large fields and is restricted to fairly low dispersion. Fiber coupling, in use at the Steward Observatory and AAT telescopes, overcomes these limitations and is the method of choice for very large telescopes.

A fiber is located at the position of each object in the focal plane, and all the fibers are brought into a line at the entrance slit of the spectrograph. In this way a detector with 1000 x 1000 large pixels can record linear spectra of moderate dispersion of around 100 objects at once. Higher dispersion spectra require large numbers of pixels in the form of mosaic detectors in a single large spectrograph, or multiple smaller spectrographs each handling a subset of the total number of spectra. Existing multiple fiber instruments use plates with drilled holes to locate fibers correctly in the focal plane (Hill et al. 1980, Gray 1983). The next generation will incoporate some type of mechanism so fibers can be remotely positioned under computer control (Hill, Angel and Scott 1983). This is not only a convenience; fibers in a permanent set can be prepared with more care, and equipped with microlenses to allow the most efficient coupling.

Two telescope requirements for getting the most out of multi-fiber spectroscopy are as follows. First, the images should be achromatic. Most current wide field correctors were designed for imaging in photometric bands, when refocusing from band-to-band is not a serious defect. However, if silicon detectors covering 0.3 - 1µ are used in the spectrograph, this range of wavelengths should all come to a sharp focus in the same plane. Second, assuming the seeing is good and the corrector is achromatic, the spread of the image from dispersion in the atmosphere will be quite pronounced, even for objects well above the horizon. A system that incorporates prism elements to balance out atmospheric dispersion is thus very desirable, so that small round apertures with the maximum sky rejection can be used.

One aspect that is not critical for fiber spectroscopy is scale or focal ratio. With the aid of microlenses, which can be contacted to a fiber end with no loss, efficient coupling can be made to fast or slow foci (Hill, Angel and Richardson 1983). Another point is that there is no special difficulty in instrumenting very large fields with fibers, particularly if the fibers are remotely postioned. For Schmidt size fields with fixed fibers, differential refraction during the exposure is significant (Watson 1983), but remote positioning allows correction to be made during an exposure. A practical consideration is the need for access around the field for remotely positioned fibers. Optical designs like the Schmidt or Paul have 'trapped' foci, with little room for actuators unless folding mirrors are used.

Turning our attention to slitless spectroscopy and imaging, the big question is the detector to be used for large telescopes. In the best seeing, when images of 1/3 arcsecond can be recorded, pixel sizes of 0.1 - 0.15 arcseconds should be used, i.e., $\sim 10^9$ per square degree. The ideal detector would combine the large area and convenience of photographic plates with the linearity, dynamic range and high sensitivity of CCDs. Many CCDs used together are needed, or more sensitive emulsions. In fact, we can anticipate that both may be available within the next decade, the time scale for development of the new big telescopes.

We have seen at this meeting the continuing strength of photographic methods, particularly when the most sensitive plates are analyzed with the new generation of high speed microdensitometers. The primary deficiency of photography is its low detective quantum efficiency (DQE), currently $\sim 2\%$ for the best hypersensitized IIIaJ plates. However, advances in speed continue to be made, with Kodak's new T grains holding the promise of a factor two increase in DQE for astronomical plates (Millikan 1984). The large plate scale of very big telescopes may allow still further advances, since larger grains can be tolerated. Presently a focal ratio of f/3 is considered ideal for photography, yielding optimum exposure for sky limited images in about an hour. Future increases in emulsion speed should allow for somewhat slower foci or shorter exposures.

Charge coupled devices (CCDs) offer higher efficiency, linearity and dynamic range than photographic plates, with near unity DQE over a wide spectral range obtained in the best devices. Single devices are now widely used for imaging relatively small fields. Two techniques are now being developed to realize wide field coverage with CCDs: the deployment of many devices in a mosaic and the drift scan method. The wide field camera on space telescope uses four devices to sample a 2.6 arcminute square field with 0.1 arcsecond pixels. A similar configuration has just been developed at Palomar with 0.3" sampling over a 8' square field (Gunn, Westphal and Danielson 1983). In the drift scan or TDI method, the CCD is clocked slowly and continuously as the star field moves in synchronization with the charge image across the face of the chip. The deepest optical images yet made, reaching 26th magnitude in R, have been made this way (McKay 1983). Wide field surveys are also now being undertaken by this method. McGraw et al. (1982) describe a 1.8m transit instrument which will use two CCDs to scan a strip 8 arcminutes wide, of area 20 square degrees per night, to about the same limiting magnitude as a IIIaJ Schmidt plate.

The advantages of the TDI method are that seamless large area images can be realized from a mosaic of separated small devices, and that

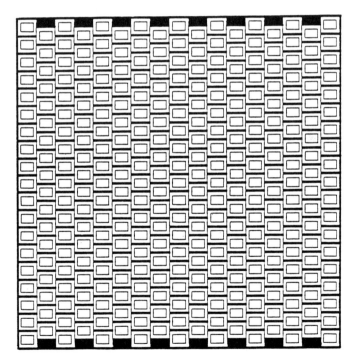

Figure 1. A possible configuration of CCDs of aspect ratio 2:3, suitable for drift scan imaging of a large field. The vertical repeat spacing is just less than twice the height of the active area, so that a horizontal scan leaves no gaps.

flat fielding becomes a fairly tractable one dimensional problem. A mosaic configuration suitable for TDI imaging of large fields is shown in Figure 1. The devices here are packed with a filling factor of 1/3. The DQE of such an array would thus be about 25%, if the individual device quantum efficiency is 75%. If devices are used in this way, there may be no compelling advantage in using very large individual chips. In some ways it is better for them not to be too large. If the telescope is drifted across the star field, image distortion will result in blurring if the individual devices are too large.

A critical design parameter in using CCD arrays for imaging is to match the pixel size and image scale. Present devices used for astronomy have pixels in the range 15 - 30μ. Taking 22μ as representative, we find for sampling at 0.1 arcseconds a focal length of 45.4m is needed, for 0.15 arc seconds, 30.3m. Thus a 7.5m telesope should operate in the range f/4 - f/6. Another critical parameter is device cost. The present cost for premium quality devices in unit quantity is approximately 1¢/pixel. At this price enough devices for a 10^9 pixel array with 1/3 filling factor would cost \$3M, and a detector system with resolution comparable to that of a Schmidt plate would then be comparable in cost to the large primary mirror that feeds it. We can anticipate improved yield and technology over the next few years may result in an order of magnitude reduction in cost, making the cost of large arrays not disproportionate, even for smaller telescopes.

A concern commonly expressed at the prospect of large CCD arrays is difficulty of data handling and reduction. The spectacular results reported at this meeting by the Edinburgh and Cambridge groups should help allay some of these fears. Digital processing of full Schmidt plates, $\sim 10^9$ pixels, is already now being successfully undertaken with modest sized computers and tape data storage. When the new large telecopes are operating we can expect more powerful computers will be readily available, and also, as Grosbol (1984) has pointed out here, low cost laser disc storage of data.

If, as we can hope, the exposure and recording of large electronic images becomes manageable and routine, then the possibility arises for serendipitous imaging. When a single object is scheduled for extensive spectroscopic or polarimetric study, the surrounding field can be recorded and used for other projects. As an example, for many cosmological studies the exact area covered is not as important as the quality and depth of images. In conditions of good seeing, the serendipity mode will be of special value.

One final aspect to be considered in optical design is provision for the dispersing elements. Correction for atmospheric dispersion requires a way to disperse each object in the field by an amount equal to that produced by the atmosphere, but in the opposite direction. Dependence on an wavelength must be the same as the atmosphere's and with magnitude proportional to the tangent of zenith distance. Systems to make such corrections have been built for speckle interferometry, and make use of a pair of counter-rotating prisms. Slitless spectra will in general require higher dispersion, either by prism or grism elements. Both functions should be accomplished with elements that are a small fraction of the primary diameter, if they are to be practical for very large telescopes.

3. OPTICAL DESIGNS FOR LARGE TELESCOPES

From the discussion above, it would appear that electronic or photographic detectors can be realized for fields of at least 1°, and that multiobject fiber systems could be made for even larger fields. Ideally, the wide field should be achieved with a primary mirror that is rather fast, f/2 or faster, to keep down telescope length and enclosure size. Also, if the telescope is to be reconfigured for operation in the infrared, it is valuable to have a parabolic or nearly parabolic primary figure, so that small Cassegrain secondaries will give an acceptable field of view at long focal ratio and without refractive correctors. A summary of the reqirements to be met by an ideal optical system is given in Table 1:

Table 1.

field of view	1° or more
aberration	$\lesssim 0.15$ arcseconds
achromatism	over range 0.3 - 1μ
correction for atmospheric dispersion	
dispersion for slitless spectroscopy (prism and/or grism)	
scale	4.5 - 7 arcseconds/mm
primary figure	$\lesssim$ f/2, parabolic

These are tough specifications, not met by any existing telescope. The two designs that come closest, and that we can take as points of departure, are the Dupont 2.5m telescope, a Ritchey Chretien that achieves a 2° field at f/7.5 with a single element Gascoigne corrector (Bowen and Vaughan 1973a) and the CFHT prime focus corrector, 1° at f/4.2 (Fouéré et al. 1982).

The general idea of Ritchey Chretien optics is to get a wide field of good resolution by using two reflecting surfaces to maximum advantage, requiring the primary to be hyperbolic instead of parabolic. A refractive element can then be used to fine tune the image, to flatten the field or reduce astigmatism to get more field. This approach has several deficiencies. Firstly, if a reasonably fast focus is required,

F $\lesssim$ 6, then the primary figure departs quite strongly from a parabola. Secondly, the insertion of a single refractive corrector element compromises achromaticity. In the Dupont telescope, for example, the corrector has to be repositioned to correct different wavelength bands. Thirdly, the introduction of glass prism or grism elements for either dispersion correction or slitless spectroscopy will cause some aberration, especially at the faster focal ratios (Bowen and Vaughan 1973b).

If a design is to take advantage of the inherent achromaticity and good correction of reflecting optics, it would seem that a three-mirror system should be used. Wide field corection can be achieved without the need for any refractive element. If the primary and secondary together are made afocal, then dispersing elements can be introduced between the secondary and tertiary without compromising aberrations or achromaticity. The third element and focal surface must be tilted if there is deviation.

An early analytic solution for a two-mirror corrector with a parabolic primary yielding extremely low aberrations was given by Paul (1935), an example of which is shown in Figure 3a below. Even with a primary of focal ratio f/1, an optimized design described by Angel, Woolf and Epps (1982) gave 0.2 arcsecond images over a 1° field. The Paul configuration being built for McGraw's transit telescope has a 1.8m f/2.2 primary, and a 1° field of view. It will incorporate for slitless spectroscopy a zero deviation prism whose diameter is 1/3 that of the primary.

A disadvantage for general purpose telescopes is the inconvenience of the focal position, half way up the tube with very little room for instruments. A scheme was devised by Woolf et al. (1982), to access the focus with multifiber probes but is not simple and the field is limited to 40' by obscuration. A modification of the three-mirror design to produce an accessible focus just above the secondary has been developed by Epps and Takeda (1983). In this case though, the primary must be hyperbolic if high quality imaging is to be realized, and again the field is limited by obscuration. The most practical design for a versatile 3 element system, given by Epps (1983a), places the tertiary mirror behind the primary, and the focus a little above the primary vertex. A steering mirror gives several accessible focal stations below the primary.

Central obscuration is the dominant problem in three-reflector systems. Aberrations at wide field angles can be very well controlled, but a solution for a big telescope that gives a full 1° field and acceptable obscuration requires large auxiliary optics and will be somewhat inconvenient for access and reconfiguring with low IR background.

Let us return then to the prime focus type of solution, in which wide field achromatic imaging relies heavily on the use of refractive elements, and the particular primary figure is not very important. The idea that three refractive elements of the same material could be used to obtain achromatic correction, originally recognized by Sampson (1913), was realized in the wide field correctors designed for both hyperbolic and parabolic primaries (Wynne 1968, Faulde and Wilson 1973). The recent CFHT corrector of Wynne and Richardson (Fouéré et al. 1982) gives high quality images over a wide field. Operating at the prime focus of an f/3.78 parabola, it yields an unvignetted field of 46' at focal ratio f/4.20, with images better than 0.7" over a total field of 55'. It incorporates a grens as part of the interchangeable third corrector element, allowing slitless spectroscopy over the full field with high image quality. The CFHT system comes close to achieving many of the features of Table 1, including a parabolic figure on the primary, but fails in that the primary is rather slow (f/3.8). What happens if we use a three-element corrector with an f/2 or faster primary? This question is now being studied by Epps for the California group. It appears at the present time that, even with the field restricted to 30 arcminutes, good images over the full optical range need two alternative corrector configurations, with refocusing of different broad wavelength bands (Epps 1983a). The question of dispersion over the field is unresolved, but aberrations produced by prisms and grisms in the fast beam may be a problem, especially if good images are to be preserved.

If correction at a fast prime focus cannot meet our needs, a more favorable approach is to use multi-element refractive correctors at a Cassegrain focus derived from a fast primary. Wynne (1973, 1983) has shown that even two element correctors can give remarkably good performance over a 1° field. Three element correctors are potentially even more powerful, and indeed a recent solution by Epps, Angel and Anderson (1983) described below satisfies all the requirements of Table 1. Starting with an f/2 primary, a relatively fast classical Cassegrain focus, f/5.3, is formed by a secondary with about 1/3 the primary diameter. The focus could be used as is for small field work, but aberrations outside a 5 arcminute field become larger than an arcsecond. Coma is the same as for an f/5.3 parabola, astigmatism is greater. Correction of this focus is made with three fused silica elements, optimized to give minimum aberration over the wavelength range 0.33 to 1μ and over a 1° field. The design, which has a flat focal plane at f/6, was made with the inclusion of counter rotating zero deviation prisms placed after the second element, where the beam is narrowest and has least divergence. These prism elements are of FK5 and LLF2 glass, a pair that matches atmospheric dispersion quite well, and allows transmission down to 3300Å.

Ray tracing of the design at different field angles and wavelengths, from 0.33μ to 1μ <u>with no refocusing</u>, shows spot diagrams mostly having 100% of the rays within 0.25 arcseconds (Figure 2). Only the spots of extreme wavelengths at the edge of the field fall outside this limit. The prisms introduce no significant additional aberration, and compensate up to 60° zenith distance, reducing atmospheric dispersion over the full wavelength range by an order of magnitude, more for

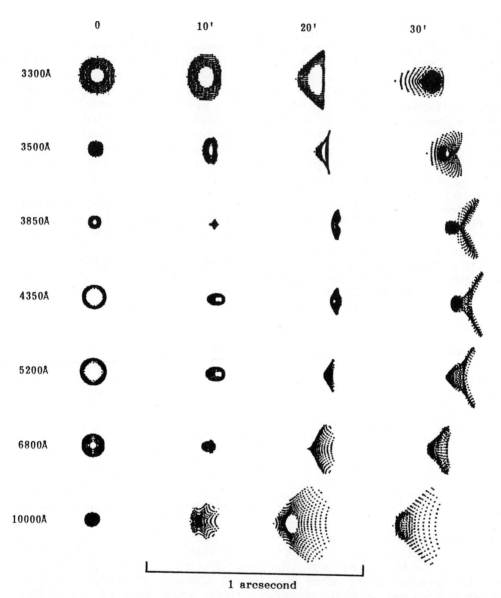

Figure 2. Spot diagrams for the corrected Cassegrain focus of Epps, Angel and Anderson (1983), all computed for the same flat focal plane. The horizontal registration is accurate, showing the small residual chromatic difference of magnification. Summed over all wavelengths, 0.33μ–1μ, 100% of the energy lies in a 0.17 arcsecond circle on axis, within 0.29 arcsecond at 20' radius and 0.46 arcseconds at the edge of the 1° diameter flat field. Atmospheric dispersion over this wavelength range is 4 arcseconds at 60° zenith distance, and can be reduced to 0.4 arcseconds by the prisms built into the corrector. On the scale of this drawing, the field diameter is 200m.

reduced bandwidth. Scaled for a 7.5m primary, the completely unvignetted field of 1° requires a secondary 2.48m of diameter and corrector elements of 1.19m, 0.86m and 0.97m. The field diameter is 0.8m and the scale is 222µ/arcsecond.

Epps' optimization of this f/6 system has given a solution that combines extraordinary achromaticity, image quality, spectral range and field size. The design is evolving. We plan next to try for a slightly faster final focus, explore the performance to 2µ and look at dispersive elements for slitless spectroscopy.

4. SPECIALIZED SURVEY INSTRUMENT

Our discussion so far has been restriced to the implementation of wide fields in very large general purpose telescopes. Now let us consider briefly two directions that might be taken by specialized survey instruments with larger aperture and better image quality than present day Schmidts. For survey instruments having the same spatial resolution, the time taken to survey a large area of sky to a given limiting magnitude inversely is proportional to $(A\Omega)_{eff}$, the product of telescope area, the field of view and the DQE of the detector. This figure of merit can be reexpressed as

$$(A\Omega)_{eff} = \pi a/4F^2 \times DQE$$

where a is the detector area and F is the telescope focal ratio. Its value for an f/2.5 Schmidt telescope with 35 cm square photographic plates is 3 cm^2, assuming a DQE of 2%.

In order to achieve higher spatial resolution and a higher figure of merit we consider the performance of a larger Schmidt with better emulsions, say a DQE of 4%. With a clear aperture of 2.5m, and operating at f/2.5 one could achieve spatial resolution of ~ 0.5 arcseconds, at least in moderately narrow wavelength bands and with moderately short exposures, so as to minimize variable distortion from differential refraction. The figure of merit would be 24 cm^2, assuming 70 cm square plates. The telescope tube would be 12.5m in length and a primary of ~ 4m diameter would be needed (Figure 3b). If the images are to be treated digitally and not to be degraded, the telescope would have to be operated in conjunction with a large plate scanner, having 0.25 arcsecond pixels (6µ).

As an alternative, we can consider a shorter, fatter telescope, a 6m f/2 primary in a three-mirror sytem, such as that shown in Figure 3a, operated with CCDs in the TDI mode. In order to be specific let us take hypothetical CCDs not unlike those widely used today, with a DQE of 75% and 400x600 pixels each 25µ square, subtending 1/4 arcsecond on the sky. It follows then that F at the corrected focus must be 3.44, and $(A\Omega)_{eff}$ = n x .075 cm^2, where n is the number of CCDs operated in the focal plane. A figure of merit of 25 cm^2 requires the use of 330 chips. Packed with a filling factor of 1/3, the required field of view in the focal plane would be 1.1 square degrees. The array shown in Figure 1 represents this concept. It has 323 devices, and would measure 37 cm square.

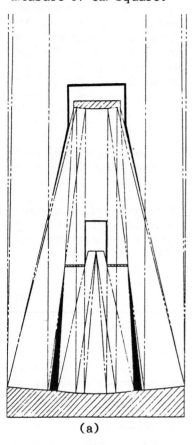
(a)

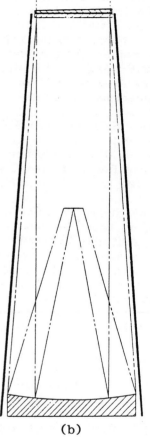
(b)

Figure 3. Two large survey telescopes drawn to the same scale, (a) shows a Paul 3 mirror telescope in which the f/2 primary and tertiary surfaces are figured on the same 6m blank (Arganbright 1983). The field is 1.45° diameter at f/3.44 and the image size ~ 0.1 arcseconds (Epps 1983b), (b) shows a classical Schmidt of 2.5m clear aperture with 8.5° diameter field at f/2.5.

The construction and operation of such a large detector array may seem prohibitively difficult but it should be contrasted with the construction and operation of appropriate dark room facilities and a plate scanner than can accurately scan 10^{10} 6µ pixels over a 70 cm plate in a reasonable period. The CCD approach offers linear data with high dynamic range and an immediate digital record. The Paul telescope has considerably better optics and wider spectral range (0.3 - 2.2µ). Its aperture is larger than the Schmidt's and it would be much more powerful for intensive study of restricted fields much less than 36 square degrees.

5. SOME CONCLUSIONS

Perhaps the most important conclusion is that the need for wide fields of excellent imaging can be realized in optical designs based on fast primaries. Predictably though, the good optical solutions require relatively large auxiliary mirrors and lenses. This need should be of concern when telescopes with very large primary mirrors are contemplated, particularly if rapid reconfiguration of the telescope for low background IR operation is envisaged. A specific solution to this problem is offered by the MMT design reported by Lynds et al. (1983). Four 2.5m Cassegrain secondaries, which work with four 7.5m primaries as individual wide field telescopes, are mounted in a single large structure across the top of the telescope. When this structure is rotated by 45° the secondaries and their supports move out of the light path, to be replaced by small chopping secondaries and beam combining mirrors.

Another point we have considered is application of the above designs to telescopes in space. Paul optics are an obvious choice for wide field imaging, given their achromaticity and excellent images. However, the three element Cassegrain corrector has high enough performance to merit consideration for a larger ST successor that can do multiobject spectroscopy as well as imaging. Large and small field correctors could be used interchangeably at the same basic Cassegrain focus, to give either the wide field at 0.2 arcsecond resolution, or a smaller field at the diffraction limit.

Whether in space or on the ground, the problems of transmitting, recording, accessing and calibrating images each of 10^9 pixels is formidable. Still more of a challenge is to use computers efficiently to help us digest all this information. As we plan and develop new very large telescopes with wide fields and the highest resolution, the experience gained from Schmidt telescopes will be invaluable.

I am indebted to Harland Epps for sharing his insights into much of this material with me. This work is supported by NASA under grant NAGW-121.

REFERENCES

Angel, J. R. P., Woolf, N. J., and Epps, H. W.: 1982, SPIE, Proc. 332, 42.
Arganbright, D.: 1983, Steward Observatory Report.
Bowen, I. S. and Vaughan, A. H.: 1973a, Applied Optics, 12, 1430.
Bowen, I. S. and Vaughan, A. H.: 1973b, Pub. Ast. Soc. Pac., 85, 174.
Dressler, A. and Gunn, J.: 1983, Ap. J., 270. 7.
Epps, H. W. and Takeda, M.: 1983, Annals of the Tokyo Observatory, in press.
Epps, H. W.: 1983a, Report on Optical Design prepared for CTIO, October 1983.
Epps, H. W.: 1983b, private communication.

Epps, H. W., Angel, J. R. P. and Anderson, E.: 1983, in preparation.
Faulde, M. and Wilson, R. N.: 1973, Astron. and Astrophys., 26, 11.
Fouére, J. C., Lelievre, G., Lemonier, J. P., Odgers, G. J., Richardson, E. H. and Salmon, D. S.: 1982, in "Instrumentation for Astronomy with Large Optical Telescopes", C. H. Humphries ed., Reidel.
Gray, P. M.: 1983, Proc. SPIE 374 and 444, in press.
Grosbol, P.: 1984, Proc. IAU Colloquium #78, Astronomy with Schmidt-type Telescopes, M. Capaccioli, ed., this volume.
Gunn, J., Westphal, J. and Danielson, G.: 1983, in preparation.
Hill, J. M, Angel, J. R. P., Scott, J. S., Lindley, D. and Hintzen, P.: 1980, Ap. J. (Letters), 242, L69.
Hill, J. M., Angel, J. R. P. and Scott, J. S.: 1983, Proc. SPIE, 380, in press.
Hill, J. M., Angel, J. R. P. and Richardson, E. H.: 1983, Proc. SPIE 445, in press.
McGraw, J. T., Stockman, H. S., Angel, J. R. P., Epps, H. and Williams, J. T.: 1982, Proc. SPIE, 331, 137.
McKay, C.: 1983, Proc. SPIE 445, in press.
Millikan, A. G.: 1984, Proc. IAU Colloquium #78, Astronomy with Schmidt-type Telescopes, M. Capaccioli, ed., this volume.
Paul, M.: 1935, Rev. D'Opt. 14, 169.
Sampson, R. A.: 1913, Obs., 36, 248.
Watson, F. G.: 1983, Edingburgh Astronomy preprint.
Woolf, N. J., Angel, J. R. P., Antebi, J., Carleton, N., Barr, L. D.: 1982, Proc. SPIE 332, 79.
Wynne, C. G.: 1983, private communication.
Wynne, C. G.: 1968, Ap. J., 152, 675.
Wynne, C. G.: 1973, Mon. Not. R. Astr. Soc., 163, 357.

SCHMIDT ASTRONOMY AND THE SPACE TELESCOPE

Barry M. Lasker
Space Telescope Science Institute
Homewood Campus
Baltimore, Maryland

ABSTRACT: The Space Telescope is described from the viewpoint of its interaction with Schmidt telescopes, specifically with regard to the planning and interpretation of observations and to the construction of the Guide Star Selection System.

I. INTRODUCTION

It is entirely appropriate that this gathering of astronomers who are in one way or another concerned with the applications of Schmidt telescopes should devote significant attention to the Space Telescope (ST). Indeed, I can think of no better example of the complementarity of greatly differing techniques in optical astronomy than one finds in the interplay between Schmidt telescopes and the ST. Obviously, Schmidts have played and will continue to play enormous roles in the identification of the problems (as well as the specific objects) that we expect to study with the ST. Additionally, examination of the surprises that we must reasonably (even conservatively) expect to encounter with the ST will require the use of Schmidts both for immediate follow-up and for the planning of subsequent observations. Finally, the large number of guide stars necessary to point and stabilize the ST can best be located with astrometric techniques based on the use of all-sky Schmidt surveys.

II. ST INSTRUMENTATION AND CAPABILITIES

The concepts, goals, and capabilities of the ST have been described in detail in a number of generally available collections (e.g. Hall 1982; Longair and Warner 1979; Macchetto, Pacini, and Tarenghi 1979). The advances that may be expected to come from the ST originate in its wide spectral coverage (i.e. 10000 A to 1150 A), from its outstanding resolution (better than 0.06 arc-sec [FWHM] at 6330 A, correspondingly higher in the ultraviolet), from its operation as a long-lived (15 year) observatory that is accessible (by open solicitation and peer review) to the entire astronomical community, and from its versatility (which arises from the availability of six observing instruments and from the possibility of space-shuttle revisits for instrument repair or replacement or even for ground refurbishment of the entire telescope).

The telescope itself, which is a 2.4-meter Ritchey-Chrétien instrument with a focal-plane scale of 3.58 arc-sec/mm, has pointing stability commensurate with its optical quality and is capable of integrating sufficiently to reach magnitudes as faint as 28. The initial complement of science instruments for ST, which are described in considerable detail in the references cited above, consists of the Wide Field/Planetary Camera, the development effort for which is led by J. A. Westphal at the California Institute of Technology; the Faint Object Camera, by F. Macchetto at the European Space Agency; the High Resolution Spectrograph, by J. C. Brandt at the Goddard Space Flight Center (NASA); the Faint Object Spectrograph, by R. J. Harms at the University of California at San Diego; and the High Speed Photometer, by R. C. Bless at the University of Wisconsin. Additionally, a sixth configuration, which is achieved by using one of the three Fine Guidance Sensors of the ST as an observing instrument, is under development by the Astrometry Team, led by W. H. Jefferys at the University of Texas.

In preparation for the launch of the ST in 1986, the instrumentation described above (as well as the supporting ground systems) is presently in advanced stages of preparation. The institutional basis for the operation of the ST will be provided by the Space Telescope Science Institute (ST ScI), which is responsible for defining the science program of the ST, for supporting all scientific (and some operational) aspects of ST observing and data reduction, and for maintaining a scientifically useful archive of ST data (see Hall 1982 for details). It is expected that the ST ScI will call for ST observing proposals early in 1985, and so now is an excellent time for each of us to be planning such proposals.

III. ST COMPLEMENTARITY WITH SCHMIDT TELESCOPES

When Schmidts, large reflectors (e.g. the 4-meters), and the ST are considered together (i.e. with respect to resolution, to scale, or to field size), the evident hierarchy, namely Schmidt: large reflector: ST, suggests that a primary mode of doing science is to use the Schmidt as a discovery instrument for the large reflector, which then in turn serves as a discovery instrument for the ST. This surely will be a useful research mode in certain cases, for example, those involving crowded fields (like globular clusters) or compact diffuse objects (e.g. specific small structures in nebulae). However, to base research procedures on this three layer hierarchy above is overly simplistic, both because the ST is so dramatically different from either kind of ground-based telescope and because time on large reflectors is so scarce.

Specifically, if we look at the ratios of resolution, scale, or field size for the case Schimdt-to-large-reflector and for the case large-reflector-to-ST, we see that the first set of numbers is much smaller than the second, which is a quantitative way of showing that from the vantage point of ST performance, Schmidts and large reflectors are somewhat similar. Therefore, we may realistically expect to find classes of problems for which the Schmidt alone is an entirely adequate discovery instrument for ST observing, for example the ST observations of a QSO discovered with a Schmidt, possibly one equipped with an objective prism. These lines of reasoning suggest that ST programs will frequently rely directly on Schmidts much as other modes of research do, namely as a discovery and planning instrument (cf. the early paper in this volume), and that this usage mode is likely to be quite heavy.

The availability of time on large reflectors is already insufficient to meet the needs of the astronomical community, and it may be expected that pressures generated by the ST will make this situation worse. Furthermore, even for such telescopes equipped with large scale cameras (scales of the order of 6 to 12 arc-sec/mm and fields of the order of one degree), the speed of the photographic process and the scheduling constraints are such that the amount of direct photographic work being done with such configurations is decreasing, while an approximately compensating amount of direct imaging is being done with small field electronic cameras (CCDs and the like). This practice is yet another factor that makes the Schmidt telescopes a major wide field resource for supporting ST observing.

One may also observe that while the competition for Schmidt observing time is heavy, the oversubscription is still considerably less than for the large reflectors.

While my next remark is clearly unnecessary for the Schmidt specialists here, for the more general audience who may tend to think of the Schmidts in terms of the published standard surveys, I wish to point out yet again the versatility of the Schmidt telescope. It can be configured for direct photography or for (slitless) spectrographic work at a variety of resolutions (ranging from that just sufficient to identify emission line objects to that adequate to do MK spectral classification). A large variety of filters are generally available, and an important development of the last decade is the existence of narrow-band interference filters covering fields several degrees in size. Finally, image growth can be controlled by exposure time so as to obtain very nicely exposed images of specific phenomena; for example, a short H-alpha exposure will produce an excellent image of elephant trunks (in H II regions) that are totally blackened on the standard surveys.

IV. THE GUIDE STAR SELECTION SYSTEM

One of the most important interactions between the ST and Schmidt astronomy lies in the efforts associated with the Guide Star Selection System (GSSS). The ST contains an inertial system (based on gyroscopes and reaction wheels) sufficient to point to within about one arc-min of a desired target and to hold there with a stability of the order of 0.1 arc-sec per second (of time). Obviously, such performance is sufficient only for rough pointing; and the more precise pointing required by the excellent imaging properties of the ST (see above) will be obtained by reference to off-axis guide stars.

The required measurements on the guide stars are made with the Fine Guidance Sensors (FGS), which are optical interferometers that furnish error signals to the pointing and control system of the ST. The area accessible to each FGS (69 square arc-min) is sufficiently small that relatively faint stars ($9 < V < 14.5$) will routinely need to be used as guide stars. Furthermore, the possibility that the pointing uncertainty cited above (about 1 arc-min) may lead to the acquisition of an incorrect guide star necessitates the use of photometric information from the FGSs to confirm the correctness of the acquisitions; to support this checking, the GSSS must predict the magnitudes of guide stars to a precision of 0.4 magnitudes. Finally the requirement that GSs be astrometrically sufficient for reliable (3 sigma success probability) positioning of targets in the smallest acquisition aperture (2 arc-sec on the High Resolution Spectrograph) leads to the requirement that the GS-target separations be known to within 0.33 arc-sec. The performance of the GSSS at the galactic poles, where there are relatively few appropriate guide stars, has been addressed by Soneira and Bahcall (1981), who show that the ST system performance goal of 85 per-cent successful acquisitions is generally achievable.

The approach adopted to meeting the above requirements is based on photographic photometry and astrometry using Schmidt plates. The properties of the plate material is given in Table 1. The measuring devices, which are Perkin-Elmer 2020G PDS scanning microdensitometers, accomodate plate sizes up to 50 cm (square) and have scanning speeds up to 200 mm/sec. The nominal precision of these machines as manufactured is 5 microns, which is insufficient to achieve the required astrometric precision for plate scales of 67 arc-sec/mm; and so we are increasing the precision to 1 micron by installing a new optical encoding system based on Hewlett-Packard (5501A) laser interferometers in place of the original encoders. This and other needed PDS modifications are described in Kinsey (1983).

The astrometric error budget adopted allocates 0.25 arc-sec of the 0.33 arc-sec total to plate measuring, centroiding, and reduction to an astrometric reference catalog; the remainder is allocated to proper motion. The most relevant experiments available to date are those conducted on Palomar and SRC plates of the Praesepe astrometric standard region (Russell 1978; also private communication) by Jane Russell at the

ST ScI. The results of these experiments, which were analyzed with the "Yale" centroider (e.g. Chiu and van Altena 1979; also van Altena private communication) and a standard astrometric reduction program, support pair separations within the 0.25 arc-sec allocation; and additional experiments, currently in progress, are dedicated to verifying that this performance is achievable over the entire sky.

Table 1

Schmidt Survey Material Used in the GSSS

SRC-J Survey
- GG 395 + IIIaJ on SRC Schmidt;
- 606 plates covering area south of -20° on 5° centers.
- mean epoch 1976
- copies of SRC A grade exposures

SRC-J Survey Equatorial Extension
- GG 395 + IIIaJ on SRC Schmidt
- 288 plates covering equator through -15° on 5° centers.
- mean epoch 1981
- copies of SRC A or B grade exposures[*]

Palomar Quick V
- Wratten #12 + IIaD
- 583 plates covering +6° through +90° on same 6° centers as original Palomar Survey
- mean epoch 1982
- original plates; 20 min (nominal) exposure on "best effort" basis; limiting magnitude about V=19.

[*]Mixture of A and B grade plates being adjusted to meet GSSS development schedule.

The performance of the GSSS near the edges of Schmidt plates is of particular concern, as important objects surely will lie there; and we must even expect cases in which ST targets and their guide stars are on adjacent plates. The generous overlap that exists in the southern surveys (e.g. 6.6 degree plates on 5 degree centers for the SRC J surveys) makes this a minimal problem there; however, for the Palomar material (6.6 degree plates on 6 degree centers), astrometric overlap solutions will routinely be required.

As I indicated above, proper motions are a particular concern for the GSSS. Table 2, which is based on proper motions of 0.019 arc-sec per year at low latitudes and 0.036 arc-sec per year at high latitudes (cf. adopted from the Radcliffe catalog of proper motions; also Harrington, private communication), on Gaussian combination with an 0.25 arc-sec measuring error (including reduction to the sky), and the mean plate epochs given in Table 1, shows that GSSS performance in significantly large areas of the southern sky will be out of specification (0.33 arcsec) by 1990, i.e. about five years after ST launch. Note that the similar problem for the northern sky does not become serious until significantly later (circa 1995) because we shall be able to incorporate the new Palomar sky survey, scheduled to begin in 1984, into the GSSS.

Table 2

EFFECTS OF CUMULATIVE PROPER MOTION
Combined with Measuring and Transofrmation Error of 0."25

Survey	Mean Epoch	Assumed Proper Motions*	GSSS Performance 1985	1990	1995
SRC	1976	0."019/yr	0."303	0."365	0."439
		.036	.409	.563	.728
SRC-Ext	1981	.019	.261	.303	.365
		.036	.289	.409	.563
Palomar	1983	.019	.253	.283	.338
		.036	.260	.355	.499

*Upper and lower entries correspond to low and high galactic latitudes, respectively.

The long operational life of the ST requires that the GSSS take steps to assure performance within the ST specifications well into the twenty-first century. While enough exciting new things are happening in astrometry that creating a detailed 20 year plan at this time would be foolish, it is clear that, at least in the short run, that matter of maintaining GSSS performance is best pursued by obtaining recent epoch Schmidt plates (i.e. ones for which proper motions may be neglected).

One photoelectrically measured photometry sequence (generally six stars covering the required range, measured in B and V, and having a precision of at least 0.05 mag) per Schmidt survey plate center will be used to calibrate the photographic photometry that the GSSS must do to support the correct identification and acquisition of guide stars. These data, which are a combination of sequences form the literature and new observations, are already about 85% complete; and we do intend to present the complete set of sequences as a published paper.

V. THE GSSS STAR CATALOG

The logical processes of planning and scheduling ST observing sequences requires the use of large quantities (say, months) of approximate guide star information early in each ST planning cycle, i.e. before detailed and precise GSSS results could possibly be made available. Therefore, the GSSS Star Catalog is being constructed to support this ST planning requirement. The catalog will be nearly complete over the set of possible guide stars, i.e. complete to 15th mag at high and intermediate galactic latitudes and at lower latitudes magnitude-limited at about 13.5 or 14th mag, corresponding to a surface density of about 1000 stars per square degree. About 20,000,000 entries are expected in the catalog, each with a precision of about 0.5 to 1.0 arc-sec in relative position (about 3 arc-sec absolute) and 0.6 mag in brightness.

The catalog will be constructed from so called "coarse scans" (entire plates scanned with the PDSs at low resolution, generally 50 microns, corresponding to 3 arc-sec pixels) of the plate material described in Table 1. This work, which has already begun using prototype software developed at the ST ScI as well as that provided by our colleagues (particularly the COSMOS group at Edinburgh), is to be completed before ST launch, and we expect to publish the catalog in a scientifically usable form as soon thereafter as possible.

VI. RECOMMENDATIONS

Our gathering here is a unique forum for generating ideas related to the use of Schmidt telescopes, and especially to making new Schmidt surveys. Therefore, I wish to take this opportunity to indicate three directions that, if adopted, would benefit all of us by best supporting the ST, in the first two cases by improving the precision and throughput of the GSSS, in the third case by the general support of ST research:

1. Make a new southern survey with a mean epoch of about 1990 so as to provide a set of southern plates with small accumulated proper motions, for soon after that time the proper motions accumulated in the original southern survey will cause GSSS errors that present an operational problem for the ST.

2. Make new surveys on 5 degree centers so as to minimize the necessity of performing astrometric overlaps to plan ST observations, for such overlaps reduce GSSS throughput (and possibly precision).

3. Adopt usage and scheduling policies for Schmidt telescopes such that significant and convenient amounts of observing time are available for individual observing programs dedicated to planning ST programs and to interpreting ST results.

It is, of course, painfully clear that item 3, astronomical research, is in priority contention with items 1 and 2, new surveys. This is presumably a theme that will arise many times at this conference, and I hope that the special needs of ST will be considered as it is resolved.

VII. FUNCTIONAL RELATIONS TO OTHER PROGRAMS

The interactions between the Schmidt-ST activities that we have been considering and other programs of astronomical instrumentation will generally be driven by specific research projects; nevertheless, an overview is evident. In addition to its obvious uses for stellar statistics, the GSSS Star Catalog may be useful in supporting various all-sky astrometric programs, of which the Hipparcos/Tycho mission, discussed elsewhere in this volume, is one possibility. Additionally, the engineering and astronomical constraints that have led to the requirements for a GSSS for the ST will be pertinent to other planned or proposed astronomical satellites, and so not only the GSSS Star Catalog but also some functional details of the GSSS design may be useful in these newer telescopes.

REFERENCES

Chiu, L.T., and van Altena, W.F. 1979, in 'Image Processing in Astronomy, G.Sedmak, M.Capaccioli, R.J.Allen eds., Trieste Observatory.

Hall, D.N.B., editor, 'The Space Telescope Observatory', Special Session of Commission 44 at I.A.U. 18th General Assembly, Patras, Greece, 1982, NASA CP-2244.

Kinsey, J.H. 1983, in 'Astronomical Microdensitometry Conference', D.Klinglesmith ed., Goddard Space Flight Center, in press.

Longair, M.S., and Warner, J.W., editors, 'Scientific Research with the Space Telescope', I.A.U. Colloquium No. 54, Princeton, 1979, NASA CP-2111.

Macchetto, F., Pacini, F., Tarenghi, M., editors, 1979, ESA/ESO Workshop on Astronomical Uses of the Space Telescope, Geneva, 1979.

Russell, J.L. 1978, in 'Modern Astrometry', I.A.U. Colloquium No. 48, F.V.Prochazka and R.H.Tucker eds., University Observatory, Vienna.

Soneira, R.M., and Bahcall, J.N. 1981, 'Guide Star Probabilities', NASA Contractor Report No. 3374.

DISCUSSION

R.D. CANNON: Dr.Lasker has referred to the desiderability of the UK Schmidt Telescope being used to re-survey the southern sky in the 1990's. While many people feel that this would be a very valuable project for many reasons, I should make it clear that the scientific programme of UKST is determined by the British telescope time assignment committees, and that it will be necessary to make the case that re-surveying the southern sky represents the best scientific use of the telescope at that time.

GROUND BASED DEVELOPMENTS - REDUCTION SYSTEMS

P. J. Grosbøl
European Southern Observatory
Karl-Schwarzschildstr. 2, D-8046 Garching, FRG.

ABSTRACT

The reduction of Schmidt plates is divided into scanning the plates, processing the data, and analyzing the results. At least two new measuring machines for analyzing Schmidt plates are being designed which will increase the total throughput to over 10 plates per day. The developments of microprocessors and mass storage open new possibilities for faster and cheaper processing of these large amount of data. Finally, the introduction of DataBase Management and Table Data systems will make it easier for astronomers to analyze the extracted informations. With these new developments the aim to make a detailed analysis the plates at the same rate at which they are obtained can be achieved within this decade.

1. INTRODUCTION

The reduction of Schmidt plates can be divided into the three subsequent steps, namely : digitizing the plates, locating and classifying objects on them, and finally analyzing these data in order to obtain astronomical results. The main difference in analyzing Schmidt plates compared with data from other area detectors is the amount of data. A 30 cm * 30 cm photographic plate contains approximately 1 Gpixel being equivalent to 5000 standard CCD frames or 400 images from the Space Telescope Wide Field Camera. Since several such plates are taken per night the total amount of data coming from Schmidt telescopes is orders of magnitudes larger than that obtained from other astronomical image detectors. As a result only a small fraction of the Schmidt plates available have been studied fully by means of digital technichs. Thus, the main aim for the future development of reduction systems for this type of data is simply to analyze this material properly at the same rate at which they are obtained. Several recent advances in hardware and software technics have made it possible to achieve this goal within the present decade. These new technics and their impact on the three reduction steps are discussed below.

2. MEASURING MACHINES

The first step in the reduction of a Schmidt plate is to digitize its photographic information. Important properties of this process are the range of densities which can be measured correctly, and the rate with which it can be done. Estimates of these two quantities assuming a 10 micron squared aperture are shown in Figure 1 for a number of measuring machines. Due the large differences between the machines the data given should only be regarded as an indication of their performances within an order of magnitude.

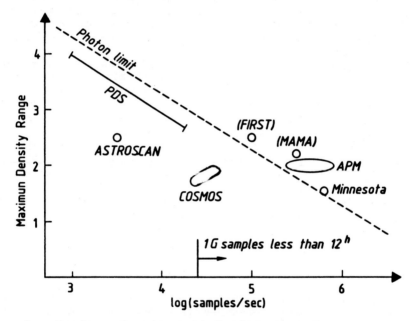

Figure 1 : Maximum density range as a function of sample rate for measuring machines (see text for explanation).

The two machines written in parentheses (i.e. FIRST in ESO-Garching and MAMA in Paris) are being designed and constructed at present. In the diagram the line 'Photon Limit' gives the possible location of a single aperture microdensitometer with a numeric aperture of 0.167 and a standard halogen lamp as light source assuming a sampling time of 0.1 times the sample interval and a maximum density deviation of 0.01. The position of the line will be shifted if a different set of parameters is chosen. Two of the four machines placed over the line are using lasers for illumination (i.e. APM and Minnesota) while the other two have array detectors (i.e. FIRST and MAMA). The maximum density range for these machines is limited rather by stray light than by photon statistics.

Two scanning modes are normally used for digitizing information from Schmidt plates, namely : search mode where objects over a large area of the plates are found, and raster scan mode where objects with known positions are measured. The sample rate is of prime importance for search mode whereas a lower maximum density can be accepted because the majority of the objects will be faint. A scanning time of 12 hours for one Schmidt plate (i.e. approx. 25 K samples/sec) is a practical limit at which it becomes feasible to analyze many plates. The fast array detector or laser scanner machines are built for this type of work but they are only marginally suited for detailed measurements of brigther objects. That can be done in raster scan mode where the limited number of object makes it more important to achieve high densities than to scan fast. This is of special interest for the newer emulsions (e.g. IIIa-J) with their very high saturation densities. Due to stray light problems it can only be done on machines having a single matched set of preslit and aperture (e.g. PDS).

Including the two machines being designed the total throughput of Schmidt plates analyzed in search mode will soon become larger than 10 per day. The production of deep exposures from Schmidt telescopes is of the same order. The measurement of plates will therefore not be the limiting factor in the reduction. It should, however, be noted that the geographic distribution of these machines is far from uniform. Also concerning the study of single bright objects an increasing amount of single aperture microdensitometers (e.g. PDS) is becoming available.

3. PROCESSING OF DATA

Only the processing of the large amount of data produced in search mode pose a problem whereas data from raster scan mode normally are so few that they easily can be handled by standard image processing systems (e.g. IHAP or MIDAS from ESO, GIPSY from Groningen, IPPS or FOCAS from KPNO, AIPS from NRAO etc.). The algorithms for detection and classification of objects on Schmidt plates are well known (see e.g. Bijaoui,1979 and Stobie,1980). The main problem is to process the data at the same rate at which they are digitized (i.e. larger than 100 KHz) in order to reduce them before they are further analyzed. The present high speed measuring machines are using special hardware to performe these computations. However, the recent development of fast 16-bit microprocessors (e.g. MC68000 from Motorola and Z8002 from Zilog) makes in possible to used such devices. Most of the operations needed for object detection and classification can be done using 16-bit integers. A comparison of three standard processors is shown in Table 1 for this type of operations.

Table 1: Relative performance of different processors for 16-bit integer operations.

Processor	Compiler	Optimized	Rel. performance
MC 68000	C	No	1.0
MC 68000	Assembler	Yes	2.1
HP 1000F	Fortran	No	1.0
VAX 11/780	Fortran	Yes	1.8

It can be seen that microprocessors such as MC68000 have a performance which is equivalent to that of standard mini-computers. The large difference between assembler and C code is due to an optimized use of the many internal registers in assembler. Further, the microprocessor systems are at least an order of magnitude cheaper than the mini-computers mentioned in Table 1. Since both multi-task operating systems (see Tinnon,1982) and cross-compilers for high level languages are available the use of microprocessors for detection and classification algorithms is more cost efficient also when development costs are included.

The increased capacity of mass storage will soon make it feasible to store the original digitized data directly. The storage density on optical devices has exceeded that of Schmidt plates and will in the coming years reach approx. 100 times higher value (Ohr,1983). When such densities are obtained a distribution of optical digital disks containing measured data from plates would be possible instead of reducing the data before making them available. This would have two advantages, namely : that the reductions could be done at the astronomers home institute, and that different routines optimized for the individual application could be used. By spreading the processing task to more institutes the load on the individual installation would further be made smaller.

4. ANALYSIS OF EXTRACTED DATA

The result after processing the plates is a set of data describing each of the objects found. In order to extract physical informations from these data their internal statistical properties have to analyzed, and they must be compared with other available data. Most present image processing systems do not provide such facilities to the user who therefore must do this time consuming task for his individual data. It is, however, possible to make a system which operates on table with heterogeneous data instead of images (see Ponz and Grosbøl, 1982). By defining commands which performe statistical operations and tests on the tables the user can analyze the internal distributions of his data in an efficient way. The second problem is to find other available data related to the material and to compare it with the data obtained. The bibliographic search for references and data is very normally very slow. This type of information is now becoming available in data bases at different centers. Access to these data through DataBase Management Systems will also greatly improve the efficiency of final analysis. Especially, when the data found can be directly transfered to the table processing system.

5. CONCLUSION

The measurement and processing of the data from Schmidt plates will soon be able to keep up with the incoming material. The main challenge for the reduction systems is to provide flexible tools for the users to analyze their final data. First then it will be possible for the individual astronomer to obtain result from this vast amount of information in a limited time. The availability of such reduction facilities would also interest more people in doing research on Schmidt plates.

ACKNOWLEDGEMENTS

I would like to thank Dr. M. Cullum for a number of interesting discussions on the optical performance of measuring machines.

REFERENCES

Bijaoui,A.: 1979, Image Processing in Astronomy, Eds. G.Sedmak, M.Capaccioli, R.J.Allen, Trieste, pp. 173-185.
Ohr,S.: 1983, Electronic Design Vol.31 No.17 , pp. 137-146.
Ponz,J.D.,Grosbøl,P.: 1982, Tables in MIDAS, ESO-Garching.
Stobie,R.S.: 1980, SPIE Vol.264, pp. 208-212.
Tinnon,J.: 1982, Electronics Vol.55 No.8 ,pp. 137-140.

SUMMARY

L. Woltjer
European Southern Observatory
Karl-Schwarzschild-Strasse 2
D-8046 Garching b. München

In summarizing this meeting, let me first turn to questions of instrumentation and reduction under the following headings:

Current Schmidt Telescopes. Cannon, West, Birkle and others have stressed the great difficulties in getting a Schmidt telescope to function at the necessary performance level - that is to the limits imposed by differential refraction and similar effects. With many Schmidt telescopes, much engineering work remains to be done.

Future Large Telescopes. Angel stressed the need for a relatively large field (~1 square degree) in the future very large telescopes (NNTT, VLT, etc.). It is, however, not evident that such telescopes should be based on the Schmidt concept.

Space Schmidt. There is wide agreement on its usefulness; its feasibility was discussed by Lemaître, but funding remains a problem.

Detectors. CCD detectors may be superior in efficiency but are still very small, and electronography encounters many difficulties. It is clear that in the coming ten years photographic plates will continue to be the principal detectors in Schmidt telescopes, and the potential improvements in photographic materials outlined by Millikan are therefore most welcome.

Auxiliary Instrumentation. Fibers may perhaps have a role, if it is desired to do slit spectroscopy with a Schmidt. But the competition with large telescopes will remain severe. The Fehrenbach double prism seems to be an interesting accessory for a Schmidt telescope. The reported 4 km/sec accuracy for radial velocity measurements would be more than adequate in many galactic structure applications. One may wonder if similar methods could be used at much fainter magnitudes for quasars and galaxies. The 2000 km/sec accuracy for galaxies with a simple prism reported by Clowes seems to be just a factor of four or so lower than what one would like.

Measuring and Processing. This seems to be the real bottleneck for many Schmidt users. The availability of fast measuring machines in the U.K. has led to a prodigious output of data. It was reported that such machines will also become available in France and at ESO. Image processing and the automatic analysis of images and spectra also are of much importance but one will have to keep track with great care of what it really is what one measures "automatically".

Archiving. During this decade, the sky surveys will certainly continue to be distributed on film and glass. From Grosbøl's projections, however, it is clear that the next decade may be different.

Intermediate Steps. Problems of calibration have been stressed by several authors. Gilmore noted the non-Gaussian distribution of errors which may lead to the erroneous discovery of "interesting" objects.

The photographically enhanced pictures made by Malin are most impressive. Most of us would believe that making use of a measuring machine + image processing or using a CCD detector, the same enhancement ought to be possible. However, the CCD pictures shown here still look very fade compared to Malin's. De Vaucouleurs discussed the limits of photographic photometry: accuracies of $0.^m02$ for stellar magnitudes and surface brightness measurements at the level of 28^m/sq arc second seem to be possible, but are achieved by few.

The separation of the images of stars and galaxies is essential in much of the Schmidt work at faint magnitudes, and Tyson and others described how this may be done. Much checking will be needed to ascertain the full reliability of these methods.

Incompleteness. This has been discussed long ago by van Gent and others for variable star searches. Quasars show surprisingly large incompleteness on spectral plates, and from Kinman's discussion of emission line galaxies, the situation there seems to be still worse. A full understanding of the causes would be of much importance.

Turning now to the scientific results reported at the meeting, let me summarize these - without aiming at anything like completeness - under the following headings:

Stellar Spectra. The Michigan surveys produce much valuable material. After decades of dissatisfaction with the precision of the HD classification, it is good to see a reclassification on the MK system taking place. Also the searches for peculiar stars by Bidelman will generate finding lists for much work at large telescopes.

Variables. Much work was reported to be in progress at Asiago on supernovae by Rosino and on quasars by Barbieri. The latter work needs a long time base, and its continuation would seem to be of much importance. Of particular interest are the quasars with a relatively bright level most of the time, interrupted by occasional sharp declines. The variable quasars studied by Hawkins suggest that most faint quasars are intrinsically also rather faint. The RR Lyrae found in the same programme will be important for galactic structure, especially so if more radial velocities become available.

Galactic Structure. The monumental work reported by Becker should give much information on the distribution of stars at intermediate latitudes. Ishida's results on the distribution of giants also contribute to outlining the distribution of visible matter. Bahcall, however, concluded that near the sun about half of the matter is "invisible". Kron outlined some of the important programmes for the future: searches for very low metal stars, the luminosity function of the galactic halo, and the finding of more low luminosity stars and degenerates.

Magellanic Clouds. While many surveys have been made, Westerlund indicated that much is still to be done in searching for variables, Hα objects and dust clouds. E. and M. Kontizas reported initial studies on the shape and content of clusters; much follow up will be needed to lower magnitudes with CCDs at large telescopes.

Quasars. Numerous searches are under way with a variety of techniques.

Other Active Galaxies. Fairall reported on searches for Seyferts on the basis of the appearance of galaxies, while Kinman reported spectroscopic searches for emission line galaxies. Takase and his associates studied uv excess galaxies. The follow up with large telescopes is very necessary, and Kachikian showed examples of several Markarian objects being just parts of larger galaxies, H II regions being much in evidence.

Astrometry. Murray's report dealt with the determination of proper motions and average parallaxes on Schmidt plates. Rather good positions in the south seem to be coming along, while in the north the situation is less good owing to obsolete catalogues. Obviously, the astrometric situation is of much concern, in connection with the Space Telescope as discussed by Lasker, and also in connection with the Hipparcos astrometry satellite.

It seems, in fact, rather clear that at the moment in several ways the southern hemisphere is favoured: some Schmidt telescopes are located there under dark skies; the sky surveys are more recent and deeper. The result is evident at the present meeting where probably more than half of the newer work reported pertained to the southern hemisphere.

INDEX OF AUTHORS

Abati Erculiani L.	475	Freeman K.C.	261
Adorf H.-M.	423	Gilmore G.	77
Andrews A.D.	279	Griboval P.J.	173
Angel J.R.P.	549	Grosbol P.J.	571
Azzopardi M.	351	Guibert J.	165
Bahcall J.N.	241	Gulkis S.	481
Balazs L.G.	237, 269	Gutierrez L.	177
Barbieri C.	443, 473	Hadley B.W.	73
Batty M.J.	481	Hamilton D.	461
Beard S.M.	401, 405	Hasegawa T.	283
Becker W.	247	Hassan S.M.	295
Bidelman W.P.	273	Hawkins M.R.S.	121
Birkle K.	203	Hazard C.	395
Bisiacchi G.F.	177	Hewett P.C.	137
Blades J.C.	457	Irwin M.J.	89, 137, 147, 395
Boyle B.J.	467	Ishida K.	257
Brandt J.C.	233	Iye M.	215
Bridgeland M.T.	89, 137	Jauncey D.L.	481
Brown D.S.	185	Jia X.Z.	173
Bruzual A.G.	527	Jockers K.	237
Bunclark P.S.	89, 137, 147	Kelly B.D.	401, 405
Burnage R.	291	Khachikian E.Y.	427
Campusano L.E.	433	Kibblewhite E.J.	89, 137
Cannon R.D.	25	Kinman T.D.	409
Capaccioli M.	379, 393	Klinglesmith D.A.	155, 233
Carignan C.	385	Kodaira K.	383
Carter D.	389	Kohoutek L.	311
Cawson M.G.M.	89	Kontizas E.	347, 359, 363
Charvin P.	165	Kontizas M.	347, 355, 359, 363
Chincarini G.	511	Kowal C.T.	229
Clowes R.G.	107	Kron R.G.	315
Cooke J.A.	401, 405	Lasker B.M.	563
Crane P.C.	99	Le Gall J.-Y.	197
Creze M.	325	Lelievre G.	379, 393
Cristiani S.	473	Lemaitre G.	533
Davoust E.	379, 393	Lorenz H.	475
Dawe J.A.	181, 193	MacGillivray H.T.	125, 405, 499
de Vaucouleurs G.	367	Maehara H.	169, 439
Dialetis D.	363	Major J.V.	185
Dodd R.J.	125, 405	Malagnini M.L.	133
Dunlop C.N.	185	Malin D.F.	57
Emerson D.T.	401, 405	Malyuto V.	287
Fairall A.P.	397	Marx S.	507
Fehrenbach C.H.	291	Maury A.	141
Firmani C.	177	McCarthy M.F.	37
Fong R.	467, 499	McGee B.	279

McMahon R.	137, 395
Melnick J.	396
Morabito D.D.	481
Murray C.A.	217
Niedner M.B.	233
Nieto J.-L.	379, 393
Noguchi T.	439
Ogura K.	283
Ohtani H.	331
Okamura S.	383
Paresce F.	177
Parker Q.A.	405
Penston M.V.	457
Pesch P.	53
Pettini M.	457
Pocock A.S.	457
Preston R.A.	481
Prokakis T.	363
Pucillo M.	133
Rahe J.	233
Ratnatunga K.	261
Robin A.	325
Romano G.	473
Roser H.-J.	423
Rosino L.	301
Ruiz E.	177
Rupp S.W.	155
Saisse M.	197
Salas L.	177
Sanduleak N.	53
Santin P.	133
Savage A.	481
Sedmak G.	133
Seitter W.C.	159
Shanks T.	467, 499
Sicuranza G.L.	133
Sim M.E.	143
Stevenson P.R.F.	499
Stoclet P.	165
Takase B.	439
Terlevich R.	395
Torres C.	433
Tsvetkov M.K.	207
Turon-Lacarrieu C.	225
Tyson J.A.	489
Watanabe M.	383
Watson F.G.	181
West R.M.	13
Westerlund B.E.	333
Westpfahl D.J.	265
Wirth A.	129
Woltjer L.	3, 577
Woszczyk A	211
Yamagata T.	169

INDEX OF NAMES

Aaronson M.	35, 356-357, 383-384
Abati Erculiani L.	49-50, 448, 455, 475, 480
Abell G.O.	7, 9, 11, 13, 20, 23, 29, 126, 128, 456, 512-513, 525-526
Acker A.	344
Adam G.R.	26, 34
Adorf H.-M.	206, 423-424, 426
Agnelli G.	155-158
Aksnes K.	232
Alcaino G.	48, 50
Alksne Z.	50
Alksnis A.	46, 50
Allen C.W.	220-221, 223
Allen D.A.	35, 70, 314, 390, 392
Allen R.J.	106, 140, 334, 377, 570, 575
Amieux G.	43, 50
Anderson E.	557-558, 562
Anderson J.H.	219, 223, 318, 323
Ando H.	216
Andrews A.D.	279-280, 282, 341, 343
Angel J.R.P.	184, 549, 551-552, 556-558, 561-562, 577
Antebi J.	562
Aoyagi Y.	216
Apt J.	35
Arakelian M.A.	411, 421
Arganbright D.	560-561
Argue A.N.	218, 223
Arhipova V.P.	7, 11
Aritome H.	216
Arp H.C.	7, 29, 34, 375, 377, 389, 392, 396, 427, 430, 447, 449
	455, 463, 465, 467, 472
Arpigny C.	237, 239
Askins B.S.	58, 70
Aspin C.	184
Athanassoula E.	71
Azzopardi M.	337, 340, 343, 351-353
Baade W.	47-48
Babcock H.W.	69-70
Bacchus P.	201
Bahcall J.N.	241, 246, 315, 320, 322, 325, 327, 368, 377, 566, 570
	579
Bahcall N.A.	513, 525
Baker J.G.	8, 11
Balazs L.G.	237, 269-271
Balzano V.A.	413, 417, 419, 421
Banse K.	106
Bappu M.K.V.	39, 50

Barbieri C.	217-218, 223, 317, 322, 412, 421, 443, 447, 451-452 455, 473, 475, 480, 578
Barbon R.	62, 70, 367, 377, 381-382
Barkhatova K.A.	295, 299
Barnard E.E.	233, 235
Barnes J.V.	241, 246
Barr L.D.	562
Barrow J.	34
Bartaya R.	43, 49, 51
Bartolini C.	50
Bateson F.M.	344
Batty M.J.	481-482, 487
Baum W.A.	57-58, 66, 70, 209
Bautz L.P.	406, 408
Bean J.	513, 515, 525
Beard S.M.	113, 119-120, 128, 140, 401-403, 405, 408
Beck H.	41, 50
Becker W.	3, 247, 297, 299, 318, 322, 475, 480, 579
Beer A.	38, 51, 214, 345
Benacchio L.	223, 377, 381-382
Benedict G.F.	487
Benvenuti P.	442, 475, 480
Berger J.	317, 322, 447, 455
Bernacca P.L.	377
Bertiau F.C.	44, 50, 317, 323
Bianchini A.	308
Bidelman W.P.	38, 42, 45, 50, 52, 273-274, 276-277, 392, 578
Bijaoui A.	353, 573, 575
Birkle K.	203, 577
Bisiacchi G.F.	177, 180
Blaauw A.	8, 11, 38, 51, 246, 271, 323-324
Black D.L.	145
Blades J.C.	457-458
Blair M.	85, 87
Blanco B.M.	47, 50, 353
Blanco V.M.	24, 38-40, 46-47, 50-51, 83, 87, 341, 343, 345, 351 353, 361, 417, 421
Bless R.C.	564
Bogart R.S.	471-472
Bohannan B.	341, 343
Bohme D.	222-223
Bohuski T.J.	398
Bok B.J.	49, 283, 285
Boksenberg A.	119, 426
Bolton J.G.	109, 120, 413, 415, 421, 433, 437-438, 449, 456, 469 472, 480, 487
Bond H.E.	48, 50, 276
Bond J.R.	515, 525
Bonoli F.	465, 480
Bosma A.	513, 525

INDEX OF NAMES

Bowen I.S.	539, 545, 547, 555-556, 561
Boyle B.J.	467-468, 471-472
Bozyan E.P.	175, 370, 377
Braccesi A.	433, 437-438, 446-447, 455, 463-465, 475, 480
Brand P.W.J.L.	28, 35, 60, 71
Brandt J.C.	155, 158, 233-235, 564
Breysacher J.	343, 352-353
Bridgeland M.T.	89, 97, 137, 140, 498, 504
Brooks J.W.	34
Brosche P.	384
Brown D.S.	185
Brown G.S.	374, 377, 490, 493, 498
Brshaft M.	158
Bru P.	219, 223
Bruch A.	163
Bruck M.T.	337-338, 343, 355-357, 360
Brunet J.P.	345
Bruzual A.G.	493, 496, 498, 503-504, 527-530, 532
Budell R.	163
Bunclark P.S.	52, 89-90, 95, 97, 137, 140, 147, 154
Buonanno R.	46, 51, 331, 334
Burbidge G.R.	467, 472
Burenkov A.N.	430
Burnage R.	43, 50, 227, 291-294, 547
Burstein D.	87, 462, 465
Burton W.	421
Bus S.J.	33-34
Bushouse H.	412, 421
Buta R.	375, 377, 498
Butcher H.R.	410, 421
Butler C.J.	43, 50, 339, 344
Campbell A.W.	68, 70, 73, 75, 193, 196
Campbell B.	46, 287, 290
Campusano L.E.	433, 436, 438
Cannon A.J.	42
Cannon R.D.	13, 25-26, 28, 31, 34-35, 71, 73, 75, 108, 119, 142
	375, 377, 426, 487, 570, 577
Capaccioli M.	60-62, 70, 106, 128, 140, 223, 334, 368, 371-374, 376-3
	379, 381-382, 393-394, 408, 455, 562, 570
	575
Caratheodolis C.	533, 547
Cardon P.	46
Carignan C.	385, 387
Carleton N.	562
Carlson C.	180
Carrasco L.	180
Carruthers G.R.	333, 345, 547-548
Carter D.	28, 35, 60, 62-63, 70-71, 176, 376-377, 389-390, 392
	406, 408, 500, 504-505
Casini C.	421, 427, 430, 442

Caulet A. 387
Cawson M.G.M. 33, 89, 91, 96-97
Cayrel R. 287
Charvin P. 165
Chavira E. 50, 305, 308, 317, 323, 447, 455
Chen J. 155-158
Cheney J.E. 480
Chin G. 282
Chincarini G. 175, 511-513, 523, 525-526
Chiu L.T.G. 446, 455, 464-465, 567, 570
Chretien H. 547
Christensen C. 287, 290
Christian C.A. 83, 87, 266-267, 424, 426
Chromey F.R. 266-267
Chun M.S. 347, 349
Ciatti F. 39, 50
Clark D.H. 70
Clements E.D. 223
Clowes R.G. 31-32, 107-108, 112, 115-116, 119, 128, 406, 408
424, 426, 435, 438, 472, 577
Colgate S.A. 155, 158
Collinder P. 295, 299
Conti P.S. 314
Contopoulos G. 246, 271
Cooke J.A. 32, 35, 110-113, 115, 119-120, 128, 139-140, 401
403, 405-406, 408, 480
Corbally C. 50
Corben P.M. 220, 222, 224, 481, 487
Cordwell C.S. 283, 285
Cormack W.A. 184
Corsi C.E. 331, 334
Corwin H.G. 29, 34-35, 111-113, 119, 392
Cosmovici C. 308
Couch W.J. 500-501, 504
Couder A. 533, 547
Coupinot G. 427, 430
Cowley A.P. 39, 42, 51
Coyne G.V. 46, 50
Coyte E. 68, 196
Crabtree D.R. 345
Craine E.R. 331, 334
Crane P.C. 87, 99, 106, 135, 384, 507, 509
Creze M. 325, 327
Cristiani S. 455, 473
Crowne A.M. 467, 472
Cruikshank D.P. 260
Cullum M. 575
Curtis H.D. 273
D'Odorico S. 376-377
Da Costa G.S. 363-364

INDEX OF NAMES

Dahn C.C.	323
Dainty J.C.	192
Danezis E.	347, 349, 357
Danguy T.	167
Danielson G.	553, 562
Danziger I.J.	34
Davenhall C.	87
Davidson K.	421
Davies R.D.	33-34, 335, 342, 344
Davis M.	369, 377, 512-514, 525-526
Davis Philip A.G.	227, 260, 354
Davis R.J.	327
Davoust E.	379, 381-382, 393-394
Dawe J.A.	24, 27, 31, 33-34, 61, 68-70, 73, 75, 109, 119-120 145, 181-184, 193, 196, 369, 371, 377
De Biase G.	382
De Graeve E.	46-47
de Jager C.	97
de Jong T.	9, 11
de Loore C.W.H.	343
de Vaucouleurs A.	29, 34, 376-377, 381-382, 392, 498
de Vaucouleurs G.	29, 34, 60-61, 70, 155, 157-158, 176, 334, 336, 338 344, 367-377, 379, 381-384, 387, 389, 392, 495, 498 511, 525, 578
de Vegt C.	218, 223
Dermendjian D.	373, 377
Dettmar R.	161
Deutschman W.A.	327
Di Martino D.	308
Di Serego S.	421
Di Tullio G.	308
Dialetis D.	363
Dibai E.A.	411, 421
Dickman R.L.	285
Dieckvoss W.	218, 224
Dimitroff G.Z.	3, 11
Dixon M.E.	338, 344
Dodd R.J.	25, 29, 34-35, 113, 119-120, 125-126, 128, 405-406 408, 500, 504
Dopita M.A.	71
Doroshenko V.T.	411, 421
Downes A.J.B.	50
Doyle J.C.	43, 50
Dressler A.	551, 561
Drilling J.S.	43, 50
Duemmler R.	161
Duerr R.	331, 334
Duflot M.	294, 340, 344
Dufour R.J.	335, 344, 387
Dumoulin B.	19-20, 23

Dunbar R.S.	34
Duncombe R.L.	486-487
Dunlop C.N.	185
Eberhard G.	368
Ebner H.	218, 223
Echevarria J.	458
Edwards P.L.	456, 473
Efstathiou G.	515, 525
Eggen O.J.	243, 246, 315, 322
Egret D.	158
Einasto J.	438, 513, 517, 525-526
Ekers R.D.	28, 35
Elliott D.A.	171
Elliott K.H.	30, 33-34, 60, 70, 344
Ellis R.S.	34-35, 492, 494, 498, 504, 525
Elson R.A.W.	35
Emerson D.T.	35, 50, 113, 119-120, 128, 140, 401, 403, 405, 408 426, 480
Epps H.W.	341, 343, 556-562
Epstein L.C.	546-547
Ershkovich A.I.	234-235
Esipov V.F.	411, 421
Evans D.S.	377, 387
Evans L.	356-357
Faber S.M.	48, 85, 87
Fabian A.C.	390, 392
Fairall A.P.	29, 397-399, 411, 421
Fall M.S.	364
Faulde M.	557, 562
Feast M.W.	356-357
Fehrenbach C.H.	40, 43, 50, 227, 290-294, 339-340, 343-344, 352-354 546-547, 577
Feige J.	317, 322
Feigelson E.	377
Feldman F.R.	55-56
Feldman L.H.	145
Fellgett P.B.	57, 70
Fenkart R.P.	299, 318, 320, 322
Festou M.	239
Firmani C.	177, 180
FitzGerald M.P.	266-268
Florsch A.	340, 344
Flower D.R.	344
Focardi P.	524-525
Fong R.	34-35, 320, 324, 438, 467-468, 472, 499-500, 504
Ford V.L.	338, 344
Ford W.K.	526
Forman W.	381-382
Formiggini L.	433, 437-438, 447, 455, 464-465, 480
Fort B.P.	391

INDEX OF NAMES

Forti J.	49-50
Fouere J.C.	555, 557, 562
Freedman W.	95
Freeman K.C.	261, 338, 344, 347, 349, 385-387, 491
Frenk C.S.	363-364, 514-515, 526
Fresneau A.	220, 223
Fried D.L.	192
Fringant A.M.	317, 322, 447, 455
Frogel J.A.	35, 270-271, 361
Frye R.L.	274, 276
Fuenmayor F.J.	46, 50
Fujita Y.	46, 50
Furenlid I.	191-192
Gallagher J.	412, 421
Gallet R.M.	377
Gandolfi E.	433, 437-438, 447, 455, 465, 480
Ganz R.	223
Gaposchkin S.	339, 344-345
Garrison R.	51
Gascoigne S.C.B.	335, 344, 356-357
Geffert M.	159
Georgelin Y.M.	410, 421
Georgelin Y.P.	410, 421
Gerola H.	516, 526
Geyer E.H.	239, 349, 364
Giguere P.T.	237, 239
Gilmore G.	77, 79-87, 108, 320-321, 323, 325-327, 403, 578
Gilmozzi R.	60, 70
Giovanelli R.	523, 526
Gisler G.R.	410, 419, 421
Glaspey J.	337-338, 345
Gliese W.	223-224
Godwin J.G.	320, 323, 504
Goerigk W.	161
Golev V.K.	208-209
Gonzales G.	50, 305, 308
Gonzalez L.E.	436-437
Good A.	145
Gooding R.A.	335, 344
Gordon M.A.	285
Goss W.M.	28, 34-35
Gott J.R.	470, 472, 513, 526
Goudis C.	336, 344
Graham J.A.	339, 344
Graham J.R.	71
Gray P.M.	181-182, 184, 551, 562
Green M.R.	438, 472, 480, 504
Green R.F.	8, 11, 317, 323, 418, 421, 447, 455-456, 460, 464-465
Greenberg J.M.	334
Gregory S.A.	511, 513, 526

Grenon M.	227
Gresham M.	323
Grewing M.	226
Griboval D.	175
Griboval P.J.	173, 175-176, 372, 377, 547
Grosbol P.J.	106, 554, 562, 571, 574-575, 578
Grossenbacher R.	274, 276
Guarnieri A.	50
Guibert J.	165
Gulkis S.	481, 487
Gunn J.	551, 553, 561-562
Gutierrez L.	177, 180
Guyenne T.D.	227
Guzzi L.	308
Hack M.	345
Hadley B.W.	27, 29, 73
Haist G.M.	58, 71
Hall D.N.B.	563-564, 570
Hamajima K.	45, 50, 260, 442
Hamilton D.	420-422, 461
Hanel A.	239
Harding P.	411, 422
Harms R.J.	564
Haro G.	9, 45, 50, 265-267, 305, 308, 316-317, 323, 412, 418-41 422, 438, 447, 455
Harrington J.P.	567
Hartkopf W.I.	243, 245-246
Hartley M.	26, 30-31, 34
Harvey G.M.	223
Hasegawa T.	283, 285
Hassan S.M.	295, 299
Haug U.	39, 50-51, 223, 422, 424-426
Hawarden T.G.	28, 34-35, 60, 71, 487
Hawkins M.R.S.	25, 29, 35, 82, 121-123, 355, 357, 480, 579
Hayakawa S.	258, 260
Hayes D.S.	260
Hayford P.	295, 299
Haynes M.P.	523, 526
Haynes R.F.	28, 35
Hazard C.	395, 449, 455
Heckman T.M.	419, 422
Heckmann O.	203, 206, 218-219, 224
Hecquet J.	427, 430
Hedin B.	346
Heidmann J.	421, 427, 430, 442
Heiles C.	462, 465
Helin E.F.	34, 230, 232
Helon G.	97
Hemenway P.D.	487
Henize K.G.	34, 44, 51, 335, 341-342, 344, 351, 353-354, 546-548

INDEX OF NAMES

Herbig G.H.	285
Herbst W.	285, 316, 323
Herd J.T.	184
Herman J.	45
Herzog A.D.	331, 334
Herzog E.	11, 399, 504, 526
Hesser J.E.	285, 357
Heudier J.-L.	23, 34, 76, 120, 145, 167, 192, 377, 408
Hewett P.C.	80, 83, 87, 89-90, 137, 396
Hewitt A.	171
Hidayat B.	45, 50-51, 260
Hilditch R.W.	241, 246
Hill G.	241, 243-244, 246
Hill J.M.	181, 184, 551-552, 562
Hill R.	274
Hiltner W.A.	58, 70, 209
Hintzen P.	184, 412, 422, 562
Hippelein H.	420, 422
Hoag A.A.	426
Hobbs R.W.	155, 158
Hodge P.W.	335-339, 344-345, 360
Hoffmann A.	507, 509
Hoffmeister C.	507, 509
Hog E.	226
Holmberg E.	507, 509
Hooley T.A.	97, 498, 504
Hoover P.S.	345
Hopp U.	349
Horne D.	97, 498, 504
Houk N.	39, 42, 51, 274-276
Hubble E.P.	320, 323, 489-490
Huchra J.P.	411-412, 417, 419, 422, 525
Huebner W.F.	237, 239
Humason M.L.	317, 323, 409-410, 422
Hume W.	155, 158
Humphreys R.M.	340, 344
Humphries C.H.	183-184, 562
Hunstead R.W.	217, 224
Hunter D.	421-422
Ibanez M.	345
Ichikawa T.	50, 260
Iijima T.	257, 260
Illingworth G.	331, 334
Inoue M.	442
Ionson J.A.	234-235
Iriarte B.	317, 323, 447, 455
Irvine N.J.	275-276
Irwin M.J.	89, 97, 137, 140, 147, 395
Ishida K.	50, 257, 260, 442, 579
Isserstedt J.	295, 299

Iwanowska W.	212-214
Iye M.	215
Jacoby G.H.	342, 344
Jaidee S.	317, 323
Jarvis J.F.	133, 135, 490, 493, 498, 500-501, 504
Jauncey D.L.	481, 483, 487
Jefferys W.H.	487, 564
Jeneralczuk X.	213-214
Jensen E.B.	386-387
Jia X.Z.	173, 176
Jimsheleishvili G.	288
Jockers K.	237, 239
Joeveer M.	526
Johanssen K.	43, 51
Johnson H.L.	266-267, 297, 299
Johnson H.M.	335-336, 338, 344
Johnson P.G.	336, 344
Jones B.F.	323
Jones B.J.T.	315, 318, 323, 526
Jones C.	381-382
Jones J.E.	526
Jones R.C.	57, 71
Jones W.B.	367, 369, 377
Jugaku J.	260
Kahn F.	422
Kalnajs A.	387
Karachentsev I.D.	477, 480, 493, 498, 507, 509
Karachentseva V.E.	477, 480
Karpowicz M.	11, 399, 504, 526
Kawakami H.	216
Kawara K.	258, 260
Kayser S.	411, 422
Kazarian M.A.	412, 422, 427, 430, 442
Keel W.C.	410, 420-422
Keenan P.C.	37-38, 42, 46-47, 293
Kellman E.	37
Kelly B.D.	113, 119-120, 128, 140, 401, 403, 405, 408
Kennicutt R.C.	420, 422
Kent S.M.	420, 422
Khachikian E.Y.	427, 430-431, 579
Kharadze E.	43, 48-49, 51
Kibblewhite E.J.	89, 97, 137, 140, 498, 504
King D.F.	375, 377
King D.J.	28, 35
King I.R.	48, 87, 347-349, 374, 377
Kinman T.D.	39, 50-51, 318, 323, 409, 412, 417, 422, 451, 455, 578-5
Kinsey J.H.	566, 570
Kirshner N.P.	502, 504, 511, 513, 526
Kjar K.	87, 106, 135, 239, 384
Kleinman D.E.	35

INDEX OF NAMES

Klemola A.R.	318, 323
Klinglesmith D.A.	155, 163, 233, 235, 570
Klypin A.	517, 525
Kobayashi Y.	260
Kodaira K.	171, 215-216, 383-384
Kogure T.	171
Kohoutek L.	9, 11, 45, 51-52, 206, 311, 314
Kondo M.	317, 323, 442, 456
Kontizas E.	347, 349, 356-357, 359-360, 363, 365, 579
Kontizas M.	347, 349, 355-357, 359-361, 363-364, 579
Koo D.C.	463-465, 493-494, 498, 500-504, 530, 532
Kormendy J.	368, 373, 375-377, 379, 382, 384
Korovyakovskiy Y.P.	427, 430
Korsch D.	546-547
Koutchmy S.	239
Kovalevsky J.	201, 226
Kowal C.T.	11, 229-232, 399, 417, 422, 504, 526
Kox H.	224
Kozasa T.	260
Kreidl T.J.	426
Kron G.E.	337, 344
Kron R.G.	81, 87, 315, 446, 455, 461, 463-465, 493, 498, 500-501 503-504, 530, 532, 579
Krug P.A.	43, 51
Kruszewski A.	106, 134-135
Kuiper G.P.	38, 51, 547
Kukarkin B.V.	308
Kun M.	46, 51
Kunkel W.E.	345
Kunth D.	409, 417-418, 422, 460
Kurtanidze O.M.	46, 51
Kutner M.L.	279-280, 282, 285
Lacroute P.	201, 219, 223
Lampton M.	180
Lamy P.L.	239
Landolt A.U.	43, 50, 266-267
Larson R.B.	384
Lasker B.M.	335, 344, 563, 570, 579
Latham D.W.	377, 525
Lauberts A.	7, 11, 22-23, 106
Laustsen S.	5, 11
Le Gall J.-Y.	197, 201, 225
Leacock R.J.	456, 473
Lee S.G.	46
Leighton R.B.	257, 260
Lelievre G.	379, 381-382, 393-394, 562
Lemaitre G.	201, 533, 544-547, 577
Lemonier J.P.	562
Lentes F.T.	163, 239
Lequeux J.	353

Levialdi S.	135
Lewis D.W.	417, 419, 421-422, 455
Liebert J.	317, 323-324
Liller W.	232
Lindblad B.	38
Lindegren L.	201
Lindley D.	184, 562
Lindsay E.M.	337, 339, 341-345, 351, 353-354
Linfoot E.H.	533, 547
Lipovetsky V.A.	412, 422, 507, 509
Loden L.O.	277
Longair M.S.	438, 526, 563, 570
Longmore A.J.	28, 35
Longo G.	381-382
Lorenz H.	49-50, 475
Lorre J.	375, 377
Lucke P.B.	338, 344-345
Lundstrom I.	314
Lutz R.K.	140, 504
Luyten W.J.	9, 11, 221-222, 224, 265, 267, 316-318, 320-321, 323-3 438, 446-447, 455-456, 469, 472
Lynden-Bell D.	315, 322
Lynds R.	463, 465, 561
Lynga G.	317, 323, 337-338, 345
MacAlpine G.M.	55-56, 417, 421-422, 449, 455
MacConnell D.J.	45-46, 48, 51, 273-274, 276, 345, 354
MacGillivray H.T.	29, 34-35, 87, 109, 113, 119-120, 125-126, 128, 336 345, 365, 405-406, 408, 499-500, 504
Macchetto F.	563-564, 570
Machnik D.E.	285
Madore B.F.	7, 29, 34, 95
Maehara H.	45, 47, 51, 169-171, 317, 320, 323, 439, 442, 456
Maffei P.	307-308
Mahaffey C.T.	421
Major J.V.	185
Malagnini M.L.	133-135
Malin D.F.	27-29, 35, 57-58, 60, 62-63, 65-68, 70-74, 76, 91 141, 376-377, 379, 381, 389-390, 392, 578
Malkan M.	31, 35
Malyuto V.	287-290
Mammano A.	39, 50
Manchester R.N.	28, 34
Manefield G.A.	34
Manousoyannaki J.	525
Maran S.P.	155, 158
Marano B.	525
Marchal J.	353
Marchant J.C.	57, 71
Marei M.	299
Margolis S.H.	472, 507, 509

INDEX OF NAMES

Marin M.	175
Markarian B.E.	7, 11, 396, 412, 418, 420, 422, 427, 448-449, 455
Marsden B.G.	232
Marsoglu A.	338, 343, 357, 360
Martin N.	345, 352, 354
Martin R.	140, 504
Martinez J.	175
Martinez R.E.	46, 51
Martins D.	511, 525
Marx S.	507
Materne J.	515, 526
Mathewson D.S.	336, 345
Matsumoto T.	260
Matthews T.A.	451, 455
Maurice C.	345
Maury A.	141
Maury A.C.	42
Mavridis L.N.	340, 345
Mayall N.U.	42, 409-410, 422
Maza J.	304
McCarthy M.F.	24, 37, 39-40, 44, 47, 50-52, 286, 341, 343, 353, 417, 422
McCuskey S.W.	8, 38, 43, 49, 51, 248, 270-271, 316, 318, 323
McGee B.	279
McGraw J.T.	553, 556, 562
McHardy I.M.	398-399
McInnes B.	87, 119-120, 128, 408, 504
McKay C.	553, 562
McMahon R.	137, 395
McNeil R.	48, 51
Mclean I.S.	182, 184
Meaburn J.	30, 33-34, 336, 344-345
Mebold U.	35
Meier D.L.	47
Meinel A.B.	335
Melnick J.	106, 395-396
Merrill J.E.	46
Metcalfe N.	27, 34, 68, 70, 109, 120, 196, 369, 371, 377
Mianes P.	345
Middelburg F.	106
Middlehurst B.	38, 51
Mikami T.	257, 259-260
Mikolajewski M.	214
Miller A.R.	473
Miller R.H.	376-377
Miller W.C.	57, 71
Miller W.J.	145
Millikan A.G.	18, 23, 57, 71, 141, 145, 552, 562, 577
Minkowski R.L.	13, 20, 23, 40, 51, 415, 422
Mizuno S.	171

Moffat A.F.J.	83, 156, 158, 266-267, 295, 297, 299
Moody E.A.	517, 526
Moore E.P.	155, 158
Moore S.	317, 323
Morabito D.D.	481, 486-487
Morales-Duran C.	318, 323
Moreno A.	52
Moreno H.	52, 436
Morgan D.H.	32, 34-35, 119, 507, 509
Morgan W.W.	37-38, 42, 46-47, 49, 266-267, 293, 322, 406, 408 411, 422
Morton D.C.	43, 51, 87, 222, 319-320, 322, 449, 455
Moss C.	418, 420-422
Mould J.R.	32, 35, 356-357, 383-384
Mulholland D.	176
Mullan D.J.	342, 345
Muller A.B.	15, 18, 23
Munch G.	420, 422
Murakami H.	260
Murdin P.G.	70
Murray C.A.	163, 217, 220, 222-224, 579
Murray K.M.	58, 71
Muzzio J.C.	46, 49, 51
Nail V.McK.	335, 345
Namba S.	216
Nandy K.	31, 34-35, 46, 48, 51-52, 110, 119-120, 475, 480, 507 509
Nanni G.	155, 158
Nardon G.	455
Nassau J.J.	38, 42, 46-47, 51, 248, 340
Natriashvili R.S.	51
Natriashvili V.V.	51
Neckel T.	295, 299
Nelles B.	239
Neugebauer G.	257, 260
Newell E.B.	81, 87, 374, 377, 500-501, 504
Neyman J.	512, 526
Nicholson W.	163
Nicolson G.D.	487
Niedner M.B.	155, 158, 233-235
Nieto J.-L.	368, 377, 379, 381-382, 392-394, 472
Nilson P.	7, 11
Nishi K.	216
Nissen P.	287, 290
Noguchi T.	317, 320, 323, 439, 441-442, 447, 456
Nordstrom B.	267
Norman C.	525-526
Norris M.V.	339, 344
Novikov I.D.	511, 526
Nulsen P.E.J.	390, 392

INDEX OF NAMES

O'Callaghan F.G.	548
Obitts D.L.	377
Ochsenbein F.	158
Odgers G.J.	562
Oemler A.	504, 526
Ogura K.	46, 51, 283, 285-286
Ohashi M.	442
Ohman Y.	38
Ohr S.	574-575
Ohtani H.	331, 334
Okamura S.	78, 87, 171, 367, 377, 383-384
Oke J.B.	401, 403
Okuda H.	258, 260
Olander N.	345-346
Omer C.G.	511, 526
Omizzolo A.	455
Onaka T.	216
Oort J.H.	93, 241, 243-246, 512, 524, 526
Opal C.B.	175, 370, 377
Orsatti A.	49-51
Osborn W.	345
Osman A.M.I.	344
Osmer P.S.	8, 11, 396, 449, 456
Ostriker J.P.	315, 323
Otten L.B.	50
Ounnas C.	106, 353
Pacini F.	563, 570
Page T.L.	237, 239, 333, 345, 526
Paresce F.	177, 180
Parker Q.A.	32, 113, 120, 128, 403, 405-406, 408
Parthasarathy M.	39, 50
Pasian E.	135
Paul M.	556, 560, 562
Payne-Gaposchkin C.	339, 345
Peach J.V.	320, 323
Pease F.G.	412, 422
Peebles P.J.E.	456, 467-469, 472, 512-513, 515, 525-526
Pegrin Y.	345
Pelt J.	287-290
Pence W.D.	504
Penhallow W.S.	274, 276
Penston M.V.	457-458
Perek L.	45, 52, 456
Perryman M.A.C.	225, 227
Pesch P.	45, 48-49, 52-53, 56, 317, 324, 351, 354, 417, 422
Peterson B.A.	35, 155-158, 411, 422, 480, 483-484, 487, 493-494 498, 500-501, 504, 525
Petrossian A.R.	427, 429-431, 442
Pettini M.	457
Philip A.G.D.	39, 49, 52, 340-342, 345

Phillipps S.	34, 320, 324
Pica A.J.	455-456, 473
Piccioni A.	50
Piccirillo J.	374, 377
Pickering E.C.	38, 222
Pickering J.	83, 87
Pinto G.	223
Pittella G.	155, 158
Platais I.	47, 52
Plaut L.	418, 422
Pocock A.S.	457-458, 460
Pollock J.T.	456, 473
Ponz J.D.	106, 574-575
Poulakos C.	317, 323-324
Preston G.W.	9, 40
Preston R.A.	481, 487
Prevot L.	340-341, 345
Prochazka F.V.	223, 570
Prokakis T.	363
Pucillo M.	133-135
Pye J.P.	398-399
Quinn P.J.	63, 71, 391-392
Racine E.	83, 87
Racine R.	157-158
Radford G.A.	243, 245-246
Raharto M.	260
Rahe J.	155, 158, 233, 235
Ratnatunga K.	261
Rayner P.T.	487
Rebeirot E.	340-341, 345, 352, 354
Recillas Cruz E.	43, 52
Reddish V.C.	25, 34-35, 119-120, 480, 487, 507, 509
Rees M.	379, 382, 512
Reid I.N.	81, 83-84, 86-87, 320-321, 323, 325-327, 403
Reif K.	35
Reipurth B.	60, 71, 283, 285
Requieme Y.	227
Richardson E.H.	214, 552, 557, 562
Richer H.B.	340, 345
Richter G.M.	480
Richter L.	317, 323
Richter N.	317, 323, 447, 456
Richter W.	9, 11
Richtler T.	161, 364
Robin A.	325, 327, 345
Rodgers A.W.	334-335, 342-343, 345-346, 411, 417-418, 422
Roemer E.	232
Roharto R.	50
Roman N.	422
Romanishin W.	385-387

INDEX OF NAMES

Romano G.	451, 455, 473
Rood H.J.	408, 511, 513, 525-526
Roosen R.G.	155, 158
Rose A.	57, 71
Rosenbush D.	275-276
Roser H.-J.	206, 423
Rosino L.	301, 303, 307-309, 317, 322, 455, 480, 578
Rousseau J.	340, 345
Rowan-Robinson M.	480
Rubin V.C.	317, 323, 382, 447, 456, 513, 526
Ruffini R.	377
Ruiz E.	177, 180
Rupp S.W.	155
Ruprecht J.	295, 299
Russell J.L.	219, 224, 566, 570
Russell K.S.	34
Saar E.	517, 525
Sahakian K.A.	427, 430-431, 447, 456
Saio H.	315, 324
Saisse M.	197, 202, 225
Saito T.	331, 334
Sakka K.	171
Salas L.	177, 180
Salmon D.S.	562
Salpeter E.E.	97, 471-472, 513, 526
Salter C.J.	50
Sampson R.A.	557, 562
Sandage A.R.	7-8, 11, 57, 71, 315, 318, 322, 324, 375, 377, 401, 403
	409-410, 422, 444, 446-447, 451, 455-456, 465
	469, 472, 475, 480
Sandig H.U.	222-223
Sanduleak N.	39, 45, 48-49, 52-54, 56, 317-318, 324, 340-342, 345
	351, 353-354, 417, 422
Sanford P.W.	46
Santangelo M.	268, 334, 365
Santin P.	133-135, 382
Sarazin C.L.	406, 408
Sargent W.L.W.	409, 411-412, 417, 422, 427, 431
Sasaki T.	171
Sato S.	260
Savage A.	29, 32, 34, 52, 109, 116, 119-120, 412, 422, 426, 433
	435, 437-438, 449, 456, 469, 472, 475, 480-484, 487
Sawyer D.L.	316, 323
Scaddan R.J.	192
Schalen C.	49
Schechter P.L.	504, 526
Schectman S.A.	9, 40, 526
Scheuer H.-G.	162-163
Schilbach E.	320, 324
Schild R.E.	327

Schiller S.	48, 52
Schmidt B.	3, 37, 333, 345, 533-534, 548
Schmidt K.H.	239, 507, 509
Schmidt M.	8, 11, 38, 51, 246, 271, 323-324, 460, 463-465, 480
Schmidt-Kaler T.	295, 297, 299, 339, 345
Schneider S.E.	91, 97
Schnell A.	317, 323
Schramm D.N.	472, 507, 509
Schroeder D.J.	546, 548
Schuster H.-E.	13, 18, 23
Schwarzschild M.	394
Schweizer F.	63-64, 71, 381-382, 386-387, 389-390, 392
Scott J.S.	184, 551, 562
Scott R.L.	456, 473, 512, 526
Searle L.	361, 427, 431
Secchi A.	46
Seddon H.	34
Sedmak G.	106, 133-135, 140, 334, 377, 382, 570, 575
Seiden P.E.	516, 526
Seitter W.C.	46, 159
Seitzer P.	490, 493
Seldner M.	467-469, 472
Serrano A.	180
Servan B.	547-548
Setti G.	8, 11, 525
Sexton J.	337, 344
Shandarin S.F.	515-517, 525-526
Shane C.D.	46, 490, 512, 517, 526
Shanks M.G.	29
Shanks T.	35, 184, 320, 324, 392, 435, 438, 467-470, 472, 498-499 504-505, 525
Shapley E.H.	337, 345, 490
Shaw R.	192
Shelus P.J.	487
Shepherd W.M.	219, 224
Shimizu M.	442
Shiukashvili M.	288
Shoemaker E.M.	10-11, 34, 230, 232
Sicuranza G.L.	133-135
Silk J.	515, 525-526
Sim M.E.	18, 23, 28, 34, 120, 143, 145, 167, 408, 481, 487
Simien F.	293-294
Sinclair M.W.	34
Sion E.M.	317, 324
Sivan J.-P.	410, 421
Skedd D.	34
Slade M.A.	487
Slettback A.	45
Smak J.	344
Smiriglio F.	46, 48, 51-52

INDEX OF NAMES

Smith A.G.	456, 473
Smith H.A.	381-382
Smith H.E.	467, 472
Smith H.J.	175, 548
Smith L.F.	342-343, 346
Smith M.G.	32, 34-35, 39, 52, 119, 140, 409, 415, 422, 426, 443, 448-449, 456, 504
Smith S.B.	455
Smolinski J.	212, 214
Sneath P.H.A.	130-131
Snyder L.E.	285
Sokal R.R.	130-131
Soneira R.M.	241, 246, 315, 320, 322, 325, 327, 513, 515, 525-526, 566, 570
Staller R.F.A.	9, 11, 318, 324
Standen P.R.	27, 35
Steigman G.	526
Stenholm B.	314
Stenquist E.	38
Stepanian D.A.	412, 422
Stephenson C.B.	38-39, 45, 47-48, 52, 54, 56
Steppe H.	317, 324, 447, 456
Stevenson P.R.F.	499, 504
Stewart G.C.	390, 392
Stewart J.M.	270-271
Stobie R.S.	87, 109, 119-120, 128, 137, 140, 170-171, 408, 499, 504, 573, 575
Stock J.	43, 52, 299, 340, 345
Stockman H.S.	562
Stoclet P.	165
Stothers R.	270-271
Straizys V.	221, 224
Strand K.A.	23
Strewinsky W.	15
Strittmatter P.A.	323
Strobel A.	212-214
Strom S.E.	385-387
Stromgren B.	315, 324, 533, 548
Struble M.F.	408
Sulentic J.W.	380-381, 467, 472
Surdey J.	447, 455
Sviderskiene Z.	221, 224
Swanson M.D.	512, 526
Swierkowska S.	214
Swings P.	237, 239
Takase B.	171, 412, 419, 421-422, 439-440, 442, 579
Takeda M.	556, 561
Talbot R.J.	386-387
Tamazian V.S.	442
Tanaka W.	216

Tapia S.	333-334
Tarenghi M.	62, 70, 427, 430, 511, 515, 524, 526, 563, 570
Taylor K.N.R.	29, 35, 375, 377
Terebizh V.Y.	411, 421
Terlevich R.	395-396
Terzian Y.	97
Teuber D.	159, 163
Thaddeus P.	282
The P.S.	48, 50, 318, 324
Thomas N.G.	426
Thompson L.A.	511, 526
Thonnard N.	526
Thuan T.X.	315, 323, 376-377
Tifft W.G.	338, 345, 356-357, 513, 526
Tinnon J.	574-575
Tinsley B.M.	87, 384, 493-496, 498
Tomita Y.	334
Tonry J.	525
Toomre A.	390
Torres C.	433, 436, 438
Toth I.	271
Treanor P.J.	417, 422
Trevese D.	155, 158
Tritton K.P.	26-27, 29-30, 34-35, 43, 51, 120, 222, 319-320, 322 375, 377, 480
Tritton S.B.	28, 34-35, 59
Trumpler R.J.	295, 299
Tsvetkov M.K.	207-209
Tucholke H.-J.	159, 161, 163
Tucker K.D.	282, 285
Tucker R.H.	223-224, 570
Tully R.B.	7
Turlo Z.	214
Turner E.L.	470, 472, 526
Turon-Lacarrieu C.	225-227
Twarog B.A.	327
Typek J.	214
Tyson J.A.	133, 135, 489-490, 493, 498, 500-501, 504, 578
Upgren A.R.	274, 276, 318, 324
Urch I.H.	344
Usher P.D.	438, 446-447, 456, 464-465
Uyama K.	260
Vaghi S.	202
Valdes F.	461, 465, 493, 498
van Altena W.F.	567, 570
van Leeuwen F.	163, 219
van Rhijn P.J.	316, 324
van Woerden H.	7, 11, 35
van de Hulst H.C.	334
van den Bergh S.	336, 345, 507, 509

INDEX OF NAMES

van der Hucht K.A.	314
van der Lans J.	11
Vasilevskis S.	161, 163
Vaucher B.	424, 426
Vaugham A.H.	555-556, 561
Veron M.P.	469, 471-472
Veron P.	239, 443-444, 447, 449, 456, 469, 471-472
Vettolani P.	525
Vidal J.-L.	379, 382
Vignato A.	155, 158
Vigneau J.	337, 340, 343, 352-353
Vitrichenko V.A.	209
Vogt N.	295, 297, 299
Voigt H.H.	299
Volkmer C.C.	163
Vorontsov-Velyaminov B.A.	7, 11, 412, 422
Wackerling L.	44, 52
Wade C.M.	345
Wagoner R.V.	471-472
Waldhausen S.	46, 51
Walker M.F.	335, 345, 355, 357
Wall J.V.	413, 415, 421, 482, 487
Wallace P.T.	26, 35, 119, 426
Wannier P.	526
Warhock A.	456
Warner J.W.	563, 570
Wasilewski A.J.	417, 419, 421-422
Watanabe M.	169, 171, 216, 383-384
Watson F.G.	181-184, 552, 562
Watson W.	33
Weedman D.W.	398, 449, 456
Wehinger P.A.	237, 239
Weiberger R.	317, 324
Weistrop D.	320, 324, 447, 456
Welch G.A.	344
Welin G.	46, 52
Wesselink A.J.	336
West R.M.	11, 13, 19-20, 23-24, 34, 46, 51, 76, 134-135, 192 207, 209, 287-288, 290, 577
Westerlund B.E.	290, 333, 335-338, 340-346, 579
Westpfahl D.J.	265-268
Westphal J.A.	553, 562, 564
White R.A.	411, 422
White S.D.M.	514, 526
Whitmore B.C.	130-131, 382
Whitney C.A.	345
Whittle M.	418, 421-422
Wielen R.	241, 244, 246, 270-271
Wikierski B.	214
Wild P.	11, 309, 399, 504, 526

Willerding E.	162
Williams G.A.	417, 422
Williams J.T.	562
Williams T.B.	394
Wills A.J.	343
Wills D.	463, 465
Wills R.	344
Wilson A.G.	526
Wilson O.C.	288, 290
Wilson R.N.	11, 557, 562
Wirtanen C.A.	512, 517, 526
Wirth A.	129
Wischnjewsky M.	437
Wisniewsky W.	46, 50
Wolf E.	547
Wolstencroft R.D.	25, 35
Woltjer L.	3, 8, 11, 13, 525, 577
Wood R.	339, 346
Woolf N.J.	556, 561-562
Woolley R.	270-271
Woszczyk A	211, 214
Wray J.D.	544, 546-548
Wright A.E.	487
Wright F.W.	335, 337, 339, 344
Wrixon G.T.	526
Wyckhoff S.	237, 239
Wynn-Williams C.G.	260
Wynne C.G.	26, 35, 557, 562
Wyse R.F.G.	315, 322-323
Yamagata T.	169, 171
Yoshii Y.	315-316, 324
Yoss K.M.	243, 245-246
Yutani M.	320, 323
Zaleski L.	212
Zambon M.	421, 473
Zealey W.J.	25, 35, 60, 71
Zel'dovich Y.B.	511, 513, 516-517, 519, 526
Zitelli V.	465, 480
Zou Z.L.	155-158, 525
Zug R.	295, 297, 299
Zwicky F.	7, 11, 304, 317, 322-324, 396-397, 399, 411, 479-480 490, 502, 504, 507, 509, 511-512, 526

INDEX OF ASTRONOMICAL OBJECTS

A0136-0801	381
Adonis	231
Alpha Persei	219-220
Amor	10, 229
Anon 8h+48	452
Apollo	10, 229-231
Arp 299	414
Aten	229
B 234	463
B 264	463
BD +44 493	276
Barnard 5	284
Barnard 34	284
Barnard 145	284
Barnard 157	284
Barnard 161	284
Barnard 163	284
Barnard 164	284
Barnard 227	284-285
Barnard 335	284
Barnard 343	284
Barnard 361	284
Barnard 362	284
Barnard 367	284
Bok 17B	262
Bok 44B	262
Bok 45B	262
Bok 98F	262
Bok 215B	262
Bok 267B	262
Bok 268B	262
Bok 307F	262
Bok 310F	262
CG 22	60
CI Cyg	214
Carina dwarf spheroidal galaxy	28
Centaurus A	63
Chiron	230-231
Clusters: 0003.5-35	407
Clusters: Abell 118	126
Clusters: Abell 140	112, 126
Clusters: Abell 141	126
Clusters: Abell 155	126
Clusters: Abell 426	524
Clusters: Abell 1367	517-518

Clusters: Abell 2197	418
Clusters: Abell 2198	418
Clusters: Abell 2199	418
Clusters: Abell 2670	112
Clusters: Centaurus	66
Clusters: Coma	511, 513, 517-518, 521
Clusters: Fornax	91, 126
Clusters: Indus 2133-5732	111-113
Clusters: Indus 2151-5805	113
Clusters: Microscopium	507
Clusters: Pegasus	522
Clusters: Virgo	65, 91, 475, 479, 507, 518, 521
Coalsack	46, 330-331
Comet 1967n	214
Comet 1968c	214
Comet Austin	237-238
Comet Halley	155, 231, 233-235
Comet Kohoutek	234
Comet Taylor	231
ESO 113-IG45	398
ESO 263-G13	398
Epsilon Cha	275
Fields: Ara	46
Fields: BD +15 2469	475
Fields: Cassiopeia	47, 307
Fields: Cepheus	46, 48
Fields: Coma	303
Fields: Crux	46
Fields: Cygnus	208
Fields: Herc 1	464
Fields: Herc 2	464
Fields: Monoceros	46
Fields: Orion	208
Fields: SA 28.2	464
Fields: SA 57	463-464
Fields: SA 68.1	464
Fields: SA 68.2	464
Fields: SA 68.3	464
Fields: Sagitta	307
Fields: Ursa Major	383
Fields: Virgo	303, 383
Fields: X UMa	444
Fornax A	63
Galactic Poles: North	318, 320, 417, 433, 490
Galactic Poles: South	48, 59-60, 81-82, 86, 116, 125-127, 222, 320 325-326, 397, 405, 417, 433-437, 468-469, 504
Galaxy: Anon 8h+48	451
Gould belt	270
Gum Nebula	10, 60, 283, 336
HB 457	312

INDEX OF ASTRONOMICAL OBJECTS

HD 8783	275
HD 33599	275
HD 81410	275
HD 104237	275
HD 168785	276
HD 184738	46
Haro 0049.5+01	415-416, 419
He 2-442	313-314
Hidalgo	230
Horsehead	306
IC 1613	385
IC 3370	63-64
IC 4329	390
IZw 18	396, 427
Iades	305
Il Hya	275
J XIII (Leda)	231
Jupiter	231
K 3-50	311
K 3-62	313-314
K 3-67	313-314
KUG 0025-103	440
KUG 0239+345	440
KUG 0935+407	440
KUG 1047+332	440
KUG 1626+413	440
KUG 2257+157	440
KUG 2259+157	440
Kazarian 5	427
Leo Group	60-61
Local Supercluster	511
Lynds 1225	284
Lynds 1622	284-285
M 11	337
M 15	412
M 31	127, 155, 157, 375
M 33	94-95, 127, 385-387
M 41	337
M 81	95
M 83	386
M 101	95, 414
M 2-50	313-314
Magellanic Clouds	31, 33, 47, 49, 93, 336, 375
Magellanic Clouds: Large	333, 347, 355, 364, 386
Magellanic Clouds: Small	333, 347, 349, 351, 359, 363, 386
Magellanic Clouds: Small/Bar	338
Magellanic Clouds: Small/Wing	338
Magellanic Stream	43
Markarian 5	427
Markarian 7	427-429

Markarian 8	427
Markarian 12	427
Markarian 35	178-179
Markarian 38	427
Markarian 59	427
Markarian 71	427
Markarian 94	427
Markarian 104	427
Markarian 111	427
Markarian 116	427
Markarian 256	427
Markarian 306	427
Markarian 307	427, 431
Markarian 319	431
Markarian 325	427
Markarian 557	413, 419
Markarian 665	431
Markarian 691	431
Markarian 710	427
Markarian 739	427
Markarian 804	431
Markarian 848	427
Markarian 984	427
Markarian 1118	431
Mich III-283	415-416, 419
Mich III-286	415-416
Mich III-295	415-416, 419
Mich III-296	415-416, 419
Milky Way	46, 212, 221, 248, 252, 295, 306, 311-312, 329
Monoceros OB1	283-285
Monoceros R1	283-285
Monoceros region	305
N Cyg 75	214
N Del 67	214
N Her 63	214
NGC 103	296
NGC 121	356
NGC 152	359-360
NGC 247	385-387
NGC 300	95, 385-387
NGC 456	338, 352
NGC 460	352
NGC 465	338, 352
NGC 584	394
NGC 602	337
NGC 654	296
NGC 663	296
NGC 669	296
NGC 1023	420
NGC 1097	25, 28

INDEX OF ASTRONOMICAL OBJECTS

NGC 1316	63, 389-390
NGC 1344	62-63, 390-391
NGC 1510	28
NGC 1512	28
NGC 1999	306
NGC 2264	283, 285, 305
NGC 2362	266
NGC 2367	296, 299
NGC 2383	296, 299
NGC 2384	295-296, 298-299
NGC 2403	95, 126
NGC 2483	266
NGC 2685	381-382
NGC 2865	390
NGC 3115	43
NGC 3370	63
NGC 3379	60-62, 91, 368, 371-372, 374, 376, 379, 381
NGC 3384	61-62, 381
NGC 3389	62
NGC 3923	62-63, 390-391
NGC 4036	381
NGC 4374	380-382
NGC 4406	380-381
NGC 4486	368
NGC 4643	64-65
NGC 4672	65-66
NGC 5018	390-391
NGC 5128	28, 390
NGC 5291	28
NGC 6522	47
NGC 6702	393-394
NGC 7531	28
NGC 7793	385-387
Orion Trapezium	279, 306
Orion complex	305
Orion nebula	67, 279-280, 306
PHL 938	424, 426
PHL 6780	419
PKS 1116+12	454
PKS 2000-330	483-484
Pleiades	208, 219, 305
Pluto	217
Praesepe	219, 305, 566
Puppis window	265-266, 268
QSO 0041-261	459
QSO 0050-254	459
R 4419	105
Ross 368	180
SAO 166722,3	59
SMC-SNR N. 49	336

SMC-SNR N. 55A	336
SMC-SNR N. 63A	336
SMC-SNR N. 132D	336
SS 433	39, 45
Saturn	231
Sculptor dwarf irregular galaxy	5
Sculptor group	385-386
Scutum region	257, 259
Seyfert sextet	511
Stephan quintet	511
Superclusters: Hydra-Centaurus	517
Superclusters: Indus	113, 401
Superclusters: Local	511
Superclusters: Perseus-Pisces	517, 519-521, 523-524
The Galaxy	92, 241, 247, 257, 266, 269, 317
The Galaxy: anticenter	318
The Galaxy: center	31, 46, 318
Tololo 3	409
Trojans	230
Uranus	231
V 1016 Cyg	312
V 1057 Cyg	312
V 1329 Cyg	311
V801 Cen	275
Vela SNR	10
WR 132	313-314

INDEX OF SUBJECTS

APM	89, 137, 147
Absorption	8, 126, 297, 318, 386, 471
Algorithm: detection	470
Algorithm: search	420
Associations	46, 305
Associations: galaxy/QSO	467
Asteroids	10, 229
Astrometry	173, 217, 225, 566
Atlasses	19, 26
Autocorrelation function	512
Automated detection	107, 121, 133
Automated measuring machines	77-78, 171, 185, 571
Automated procedures	263, 406
BL Lac objects	8, 454
Blue objects	475
Blue stragglers	356
Bok globules	45, 60, 283
C-C diagram	297
C-M diagram	92, 279, 297, 355
CCD	10, 22, 33, 40, 169, 215, 297, 391, 489, 553
CO clouds	285
COSMOS	79, 109, 121, 125, 137, 401, 405, 468, 499
Catalogues	7, 197, 226
Chromaticity	200
Classification: automated	129, 287, 493
Classification: galaxies	383
Classification: images	81, 129, 133, 461
Classification: parameters	130
Classification: spectral	41, 273, 476
Classification: stars	287
Clusters of galaxies	7, 29, 107, 113, 126, 406, 507, 513
Clusters of galaxies: morphology	523
Clusters of galaxies: numerical models	514
Clusters of galaxies: stability	522
Co-occurrence matrix	134
Cometary globules	29, 60, 283
Comets	155, 214, 230, 233, 237
Comets: ion column density	237
Comets: plasma tails	233
Cosmology	511, 527
Counts: galaxies	464, 489, 499, 527
Counts: galaxies/models	502
Counts: stars	79, 315, 329, 347
Counts: stars/in MC	335
Dark clouds	46, 316, 329

Dark clouds: extinction	329
Data analysis	102, 165
Data compression	167
Deceleration parameter	499, 528
Density enhancements	513
Density fluctuations	513
Density moment sum method	156
Density parameter	513
Density wave theory	391
Detectors	57, 173, 177, 551
Detectors: linear	181
Detectors: resolution	177
Discoveries	3
Dust clouds	28, 59
Dust: intergalactic	507
Dust: interstellar	467
ESO facilities	99
Electronographic camera	173
Emission line objects	311, 467
Emulsions: T-grain	141, 552
Emulsions: contrast	141
Emulsions: gold spot disease	28, 143
Emulsions: granularity	141, 185
Emulsions: reciprocity failure	68
Emulsions: sensitivity	196
Emulsions: speed	142
FAST	197
FOCAS	461, 489
Fabry-Perot interferometry	177
Fourier transform	131, 189, 200
GIOTTO	233
Galactic astronomy	247
Galactic structure	8, 257, 315, 325
Galactic winds	390
Galaxies	7, 28, 53, 60, 110, 129, 133, 171, 333, 367
	379, 383, 385, 389, 401, 405, 409, 439, 457
	489, 507
Galaxies: HII regions	395, 441
Galaxies: Haro	413
Galaxies: M/L	386
Galaxies: Markarian	411, 427
Galaxies: N-body simulations	390
Galaxies: Seyfert	397, 410, 427
Galaxies: UV excess	427, 439, 467
Galaxies: Zwicky	420
Galaxies: active nuclei	427, 443
Galaxies: blue	54, 463
Galaxies: blue compact	396, 476
Galaxies: bright nucleus	397
Galaxies: bulges	385

INDEX OF SUBJECTS

Galaxies: classification	383
Galaxies: colour excess	507
Galaxies: colour maps	385
Galaxies: colours	411, 495, 527, 530
Galaxies: compact	411, 458
Galaxies: contamination	463
Galaxies: counts	489, 499, 527
Galaxies: differential counts	464
Galaxies: discrepant redshifts	511
Galaxies: disks	65, 385, 390
Galaxies: distribution	125, 418
Galaxies: early type	379, 389, 393
Galaxies: emission line	54, 409
Galaxies: evolution	489, 527
Galaxies: extension	61
Galaxies: interactions	379, 441
Galaxies: internal absorption	386
Galaxies: isophotes	126, 161
Galaxies: luminosity evolution	499
Galaxies: luminosity function	489, 530
Galaxies: mass distribution	385
Galaxies: mergers	63
Galaxies: models of counts	502
Galaxies: morphology	440, 502
Galaxies: nearby	95
Galaxies: optical variability	398
Galaxies: photometry	126, 383, 428
Galaxies: polar rings	381
Galaxies: light profiles	62, 383
Galaxies: redshifts	32, 109, 401, 405
Galaxies: rotation	428
Galaxies: shell	62-63, 389
Galaxies: standard	368
Galaxies: starburst nuclei	413
Galaxies: substructures	381
Galaxies: surveys	397, 440
Galaxies: young	395
Grant machine	102
Grid photography	368
Grism	40, 216, 341, 352, 434, 449
Guide Star Selection System	28, 563
HII regions	54, 312
HII regions: compact	342
HIPPARCOS	197, 218, 225, 288, 294, 486
HIPPARCOS: consortia	226
Herbig-Haro objects	306
Holographic grating	216
Hubble flow	396
Hydrogen peroxide	143
IHAP	102

IIIa-F	5, 44, 73, 143
IIIa-J	5, 29, 44, 57, 66, 68, 71, 73, 108, 143, 191 481, 552
IRAS	31, 48
IUE	49
Image processing	573
Images	185
Images: analysis	89
Images: automated analysis	137
Images: automated detection	169
Images: blending	330
Images: classification	81, 129, 133, 461
Images: detection of faint	57, 91
Images: multiple	135
Images: profiles	148
Images: sizes	79
Images: stellar	147
Images: unresolved	81
Interloopers	449
International Halley Watch	155, 231, 233
Interstellar matter	244, 409, 457
Isophotes: ellipticity	363
Isophotes: twisting	63, 365
Kelvin-Helmholtz instability	234
Kodak IV-N	30
LAST	183
Large Scale Phenomena Network	233
Line ratios	213, 287, 417
Low luminosity objects	122
Luminosity function: MC	359
Luminosity function: disk	246
Luminosity function: galaxies	489, 530
Luminosity function: halo	325
Luminosity function: stars	9, 86, 259, 315
M.A.M.A. project	165
Magellanic Clouds	33, 333
MC: charts	335
MC: clusters	337, 359, 363
MC: luminosity function	359
MC: objects	333
MC: star counts	335
MC: surveys	333
MEPSICRON	177
MIDAS	103
Magnitude scale	82
Mass storage	571
Matter: hierarchical distribution	513
Metal abundances	261
Microdensitometers	78, 89, 159, 165, 370
Microprocessors	571

INDEX OF SUBJECTS

Microspots	143
Missing mass	9, 241
Missing mass: candidates	244
Modulation transfer function	189
Multeplicity function	517
Neutrinos	515
Novae	46
Objective-prism	26, 41, 107, 137, 205, 227, 237, 258, 273, 279, 283, 287, 291
Objective-prism: targets	39
Oort's constants	93
Optical fibers	33, 115, 181, 552
Optronics S3000 (FIRST)	101
PDS	101, 159, 261, 566
PLANET-A	233
Parallaxes: secular	93
Parallaxes: trigonometric	222
Peculiar objects	56
Photography: adjacency effect	368
Photography: calibration	18, 26, 74, 79, 147, 155, 367
Photography: colour	67
Photography: contrast	19, 69
Photography: faint images	73
Photography: future	69
Photography: high contrast	74
Photography: masking	58, 66, 73
Photography: sensitization	18, 26, 40, 58, 371
Photography: signal enhancement	73
Photography: superposition	66
Photography: uniformity	68
Photometry	173, 332
Photometry: K-correction	421, 490, 502
Photometry: automated	77
Photometry: calibration	82, 147, 150
Photometry: colour equation	85
Photometry: errors	84
Photometry: photographic	355
Photometry: standards	82
Photometry: stars	161, 295
Photometry: three colour	247
Photometry: threshold	83
Photometry: two colour	475
Planetary nebulae	311
Planetary satellites	229
Point Spread Function	189, 373
Poisson-Boltzman equation	243
Population synthesis	325
QSOs	8, 31, 39, 107, 121, 396, 409, 427, 433, 443, 457, 461, 467, 473, 475
QSOs: automated detection	115

QSOs: discoveries	443
QSOs: evolution	464
QSOs: multicolour photometry	443
QSOs: observational selection	449
QSOs: optical variability	443, 473
QSOs: radio emission	443
Quantum sensitivity	190
Radio sources	8, 390, 481
Radio sources: accurate positions	481
Radio sources: morphology	481
Radio sources: optical identifications	481
Rapid Selenium toner	145
Ray tracing	193
Rayleigh scattering	373
Reddening	122, 318
Reduction systems	571
Reduction techniques	99
Reflection nebulosities	374
Ringers	467
SMC: clusters	347, 355
SMC: dinamical parameters	347
SMC: emission line objects	353
SMC: members	351
SMC: nebulosities	351
SMC: stars	352
SMC: structural parameters	348
SMC: tidal radius	347
ST/Schmidt complemetarity	564
STARLAB	527
STARLINK	89, 110
Satellites	231
Second Cape Photographic Catalogue	218
Seeing	15, 185, 220, 372, 393
Selected Areas	43, 223
Sky background	57, 78, 112, 420
Slitless spectroscopy	237, 423, 449, 552
Solar System	229
Space Telescope	527, 563
Spectrographs	544
Spectrophotometry	423
Star clusters: open	295
Star/galaxy separation	133, 320, 464
Stars	37, 44, 77, 133, 212, 241, 247, 257, 265, 273, 279, 283, 287, 291, 301, 315, 325, 329, 333, 351, 356
Stars counts	161
Stars: A type	269
Stars: Mira	307

INDEX OF SUBJECTS

Stars: Sr-Eu-Cr	275
Stars: age distribution	271, 315
Stars: blue	55
Stars: calibration	267
Stars: carbon	46
Stars: classification	259, 287
Stars: colours	315
Stars: contamination	359, 469
Stars: counts	79, 161, 315, 329, 347
Stars: counts/MC	335
Stars: counts/authomated methods	329
Stars: degenerate black dwarfs	9
Stars: density distribution	247, 269, 280, 325-326
Stars: early type	49
Stars: emission line	55, 312
Stars: flare	208, 279, 305
Stars: formation	269, 283, 329, 386, 395, 442, 498
Stars: horizontal branch	360
Stars: kinematics	269, 315
Stars: late type	47, 241, 279
Stars: luminosity function	9, 86, 259, 315
Stars: metal abundance	9, 251, 315
Stars: number counts	258
Stars: peculiar	273
Stars: photometry	77
Stars: profiles	155
Stars: proper motions	9, 92, 222, 320, 568
Stars: radial velocities	40, 43, 161, 291
Stars: red giants	257
Stars: reddening line	249
Stars: sequences	297
Stars: standard	287
Stars: transverse velocities	321
Stars: types of	262
Stars: variable	306
Stars: velocity dispersion	270
Statistical astronomy	77, 92
Stellar populations	315
Stellar reference frame	218
Stellar statistics	315
Strasbourg Stellar Data Center	218, 226, 287
Summary of the meeting	577
Superassociations	427
Superclusters	7, 511
Superclusters: substructures	511
Supernovae	301
Supernovae: search	93, 302
Supernovae: search/Asiago	451
Surface photometry	367, 383, 385
Surface photometry: errors	367

Surface photometry: sky level	369
Surface photometry: zero point	371
Surveys	3-4, 13, 25, 389
Surveys: 2-micron	258
Surveys: Case-Hamburg	45
Surveys: ESO/SRC	27
Surveys: ESO/Uppsala	22
Surveys: Michigan	42
Surveys: O-B stars	265
Surveys: QSO/Asiago	446
Surveys: QSOs	425
Surveys: Schmidt	225, 301, 337, 340, 567
Surveys: UKST equatorial	29
Surveys: completeness	277, 437, 489
Surveys: elements	14
Surveys: emission objects	44, 424
Surveys: faint galaxies	125
Surveys: feasibility	6
Surveys: intruments	559
Surveys: methods	409
Surveys: multicolour	447
Surveys: near IR	30
Surveys: objective-prism	9, 31, 38, 42, 53, 212, 261, 265, 291, 311, 339 352, 395, 401, 405, 412, 449, 457, 467, 476
Surveys: optical	461
Surveys: photometric	279, 461
Surveys: quantitative	423
TD1	553
TYCHO	218
Table data	571
Telescopes: AAT	115
Telescopes: Asiago Schmidt	301
Telescopes: Bulgarian Schmidt	207
Telescopes: Burrell Schmidt	54
Telescopes: CFH	382, 391, 393, 557
Telescopes: CNRS-Liege Schmidt	292
Telescopes: Calar Alto Observatory	203, 423
Telescopes: ESO 1-m Schmidt	14
Telescopes: ESO 3.6-m	5
Telescopes: Schmidt	3, 41
Telescopes: Southern Schmidt	334
Telescopes: Space Schmidt	10, 34, 451, 527, 544
Telescopes: Space Telescope	486, 527, 563
Telescopes: Torun 60/90 Schmidt	211
Telescopes: UK 1.2-m Schmidt	26, 108, 195
Telescopes: aberrations	537
Telescopes: all reflective	199, 225, 541
Telescopes: astrographs	247
Telescopes: corrector	18, 26, 109, 185, 536

Telescopes: field	6
Telescopes: flat field	193
Telescopes: focal surface	217
Telescopes: new generation	549
Telescopes: optical design	533, 555
Telescopes: radial distortion	219
Telescopes: resolution	379, 393-393
Telescopes: vignetting	193, 369
The Galaxy: disk	242, 259
The Galaxy: halo	123, 261
The Galaxy: halo stars	320
The Galaxy: kinamatics	261
The Galaxy: massive halo	241
The Galaxy: model	242
The Galaxy: rotation	92
The Galaxy: spheroid	84, 242
Trans-Plutonian planets	230
UKIRT	31
UKSTU	181
UVSAT	215
UVX objects	8, 49, 396, 412, 433, 457, 475, 483
Unseen material	244
VEGA	233
VLBI	486
Variable objects	82, 121, 212

ASTROPHYSICS AND SPACE SCIENCE LIBRARY

Edited by

J. E. Blamont, R. L. F. Boyd, L. Goldberg, C. de Jager, Z. Kopal, G. H. Ludwig, R. Lüst,
B. M. McCormac, H. E. Newell, L. I. Sedov, Z. Švestka, and W. de Graaff

1. C. de Jager (ed.), *The Solar Spectrum, Proceedings of the Symposium held at the University of Utrecht, 26–31 August, 1963*. 1965, XIV + 417 pp.
2. J. Orthner and H. Maseland (eds.), *Introduction to Solar Terrestrial Relations, Proceedings of the Summer School in Space Physics held in Alpbach, Austria, July 15–August 10, 1963 and Organized by the European Preparatory Commission for Space Research*. 1965, IX + 506 pp.
3. C. C. Chang and S. S. Huang (eds.), *Proceedings of the Plasma Space Science Symposium, held at the Catholic University of America, Washington, D.C., June 11–14, 1963*. 1965, IX + 377 pp.
4. Zdeněk Kopal, *An Introduction to the Study of the Moon*. 1966, XII + 464 pp.
5. B. M. McCormac (ed.), *Radiation Trapped in the Earth's Magnetic Field. Proceedings of the Advanced Study Institute, held at the Chr. Michelsen Institute, Bergen, Norway, August 16–September 3, 1965*. 1966, XII + 901 pp.
6. A. B. Underhill, *The Early Type Stars*. 1966, XII + 282 pp.
7. Jean Kovalevsky, *Introduction to Celestial Mechanics*. 1967, VIII + 427 pp.
8. Zdeněk Kopal and Constantine L. Goudas (eds.), *Measure of the Moon. Proceedings of the 2nd International Conference on Selenodesy and Lunar Topography, held in the University of Manchester, England, May 30–June 4, 1966*. 1967, XVIII + 479 pp.
9. J. G. Emming (ed.), *Electromagnetic Radiation in Space. Proceedings of the 3rd ESRO Summer School in Space Physics, held in Alpbach, Austria, from 19 July to 13 August, 1965*. 1968, VIII + 307 pp.
10. R. L. Carovillano, John F. McClay, and Henry R. Radoski (eds.), *Physics of the Magnetosphere, Based upon the Proceedings of the Conference held at Boston College, June 19–28, 1967*. 1968, X + 686 pp.
11. Syun-Ichi Akasofu, *Polar and Magnetospheric Substorms*. 1968, XVIII + 280 pp.
12. Peter M. Millman (ed.), *Meteorite Research. Proceedings of a Symposium on Meteorite Research, held in Vienna, Austria, 7–13 August, 1968*. 1969, XV + 941 pp.
13. Margherita Hack (ed.), *Mass Loss from Stars. Proceedings of the 2nd Trieste Colloquium on Astrophysics, 12–17 September, 1968*. 1969, XII + 345 pp.
14. N. D'Angelo (ed.), *Low-Frequency Waves and Irregularities in the Ionosphere. Proceedings of the 2nd ESRIN-ESLAB Symposium, held in Frascati, Italy, 23–27 September, 1968*. 1969, VII + 218 pp.
15. G. A. Partel (ed.), *Space Engineering. Proceedings of the 2nd International Conference on Space Engineering, held at the Fondazione Giorgio Cini, Isola di San Giorgio, Venice, Italy, May 7–10, 1969*. 1970, XI + 728 pp.
16. S. Fred Singer (ed.), *Manned Laboratories in Space. Second International Orbital Laboratory Symposium*. 1969, XIII + 133 pp.
17. B. M. McCormac (ed.), *Particles and Fields in the Magnetosphere. Symposium Organized by the Summer Advanced Study Institute, held at the University of California, Santa Barbara, Calif., August 4–15, 1969*. 1970, XI + 450 pp.
18. Jean-Claude Pecker, *Experimental Astronomy*. 1970, X + 105 pp.
19. V. Manno and D. E. Page (eds.), *Intercorrelated Satellite Observations related to Solar Events. Proceedings of the 3rd ESLAB/ESRIN Symposium held in Noordwijk, The Netherlands, September 16–19, 1969*. 1970, XVI + 627 pp.
20. L. Mansinha, D. E. Smylie, and A. E. Beck, *Earthquake Displacement Fields and the Rotation of the Earth, A NATO Advanced Study Institute Conference Organized by the Department of Geophysics, University of Western Ontario, London, Canada, June 22–28, 1969*. 1970, XI + 308 pp.
21. Jean-Claude Pecker, *Space Observatories*. 1970, XI + 120 pp.
22. L. N. Mavridis (ed.), *Structure and Evolution of the Galaxy. Proceedings of the NATO Advanced Study Institute, held in Athens, September 8–19, 1969*. 1971, VII + 312 pp.

23. A. Muller (ed.), *The Magellanic Clouds. A European Southern Observatory Presentation: Principal Prospects, Current Observational and Theoretical Approaches, and Prospects for Future Research. Based on the Symposium on the Magellanic Clouds, held in Santiago de Chile, March 1969, on the Occasion of the Dedication of the European Southern Observatory.* 1971, XII + 189 pp.
24. B. M. McCormac (ed.), *The Radiating Atmosphere. Proceedings of a Symposium Organized by the Summer Advanced Study Institute, held at Queen's University, Kingston, Ontario, August 3–14, 1970.* 1971, XI + 455 pp.
25. G. Fiocco (ed.), *Mesospheric Models and Related Experiments. Proceedings of the 4th ESRIN-ESLAB Symposium, held at Frascati, Italy, July 6–10, 1970.* 1971, VIII + 298 pp.
26. I. Atanasijević, *Selected Exercises in Galactic Astronomy.* 1971, XII + 144 pp.
27. C. J. Macris (ed.), *Physics of the Solar Corona. Proceedings of the NATO Advanced Study Institute on Physics of the Solar Corona, held at Cavouri-Vouliagmeni, Athens, Greece, 6–17 September 1970.* 1971, XII + 345 pp.
28. F. Delobeau, *The Environment of the Earth.* 1971, IX + 113 pp.
29. E. R. Dyer (general ed.), *Solar-Terrestrial Physics/1970. Proceedings of the International Symposium on Solar-Terrestrial Physics, held in Leningrad, U.S.S.R., 12–19 May 1970.* 1972, VIII + 938 pp.
30. V. Manno and J. Ring (eds.), *Infrared Detection Techniques for Space Research. Proceedings of the 5th ESLAB-ESRIN Symposium, held in Noordwijk, The Netherlands, June 8–11, 1971.* 1972, XII + 344 pp.
31. M. Lecar (ed.), *Gravitational N-Body Problem. Proceedings of IAU Colloquium No. 10, held in Cambridge, England, August 12–15, 1970.* 1972, XI + 441 pp.
32. B. M. McCormac (ed.), *Earth's Magnetospheric Processes. Proceedings of a Symposium Organized by the Summer Advanced Study Institute and Ninth ESRO Summer School, held in Cortina, Italy, August 30–September 10, 1971.* 1972, VIII + 417 pp.
33. Antonin Rükl, *Maps of Lunar Hemispheres.* 1972, V + 24 pp.
34. V. Kourganoff, *Introduction to the Physics of Stellar Interiors.* 1973, XI + 115 pp.
35. B. M. McCormac (ed.), *Physics and Chemistry of Upper Atmospheres. Proceedings of a Symposium Organized by the Summer Advanced Study Institute, held at the University of Orléans, France, July 31–August 11, 1972.* 1973, VIII + 389 pp.
36. J. D. Fernie (ed.), *Variable Stars in Globular Clusters and in Related Systems. Proceedings of the IAU Colloquium No. 21, held at the University of Toronto, Toronto, Canada, August 29–31, 1972.* 1973, IX + 234 pp.
37. R. J. L. Grard (ed.), *Photon and Particle Interaction with Surfaces in Space. Proceedings of the 6th ESLAB Symposium, held at Noordwijk, The Netherlands, 26–29 September, 1972.* 1973, XV + 577 pp.
38. Werner Israel (ed.), *Relativity, Astrophysics and Cosmology. Proceedings of the Summer School, held 14–26 August, 1972, at the BANFF Centre, BANFF, Alberta, Canada.* 1973, IX + 323 pp.
39. B. D. Tapley and V. Szebehely (eds.), *Recent Advances in Dynamical Astronomy. Proceedings of the NATO Advanced Study Institute in Dynamical Astronomy, held in Cortina d'Ampezzo, Italy, August 9–12, 1972.* 1973, XIII + 468 pp.
40. A. G. W. Cameron (ed.), *Cosmochemistry. Proceedings of the Symposium on Cosmochemistry, held at the Smithsonian Astrophysical Observatory, Cambridge, Mass., August 14–16, 1972.* 1973, X + 173 pp.
41. M. Golay, *Introduction to Astronomical Photometry.* 1974, IX + 364 pp.
42. D. E. Page (ed.), *Correlated Interplanetary and Magnetospheric Observations. Proceedings of the 7th ESLAB Symposium, held at Saulgau, W. Germany, 22–25 May, 1973.* 1974, XIV + 662 pp.
43. Riccardo Giacconi and Herbert Gursky (eds.), *X-Ray Astronomy.* 1974, X + 450 pp.
44. B. M. McCormac (ed.), *Magnetospheric Physics. Proceedings of the Advanced Summer Institute, held in Sheffield, U.K., August 1973.* 1974, VII + 399 pp.
45. C. B. Cosmovici (ed.), *Supernovae and Supernova Remnants. Proceedings of the International Conference on Supernovae, held in Lecce, Italy, May 7–11, 1973.* 1974, XVII + 387 pp.
46. A. P. Mitra, *Ionospheric Effects of Solar Flares.* 1974, XI + 294 pp.
47. S.-I. Akasofu, *Physics of Magnetospheric Substorms.* 1977, XVIII + 599 pp.

48. H. Gursky and R. Ruffini (eds.), *Neutron Stars, Black Holes and Binary X-Ray Sources*. 1975, XII + 441 pp.
49. Z. Švestka and P. Simon (eds.), *Catalog of Solar Particle Events 1955–1969. Prepared under the Auspices of Working Group 2 of the Inter-Union Commission on Solar-Terrestrial Physics*. 1975, IX + 428 pp.
50. Zdeněk Kopal and Robert W. Carder, *Mapping of the Moon*. 1974, VIII + 237 pp.
51. B. M. McCormac (ed.), *Atmospheres of Earth and the Planets. Proceedings of the Summer Advanced Study Institute, held at the University of Liège, Belgium, July 29–August 8, 1974*. 1975, VII + 454 pp.
52. V. Formisano (ed.), *The Magnetospheres of the Earth and Jupiter. Proceedings of the Neil Brice Memorial Symposium, held in Frascati, May 28–June 1, 1974*. 1975, XI + 485 pp.
53. R. Grant Athay, *The Solar Chromosphere and Corona: Quiet Sun*. 1976, XI + 504 pp.
54. C. de Jager and H. Nieuwenhuijzen (eds.), *Image Processing Techniques in Astronomy. Proceedings of a Conference, held in Utrecht on March 25–27, 1975*. XI + 418 pp.
55. N. C. Wickramasinghe and D. J. Morgan (eds.), *Solid State Astrophysics. Proceedings of a Symposium, held at the University College, Cardiff, Wales, 9–12 July 1974*. 1976, XII + 314 pp.
56. John Meaburn, *Detection and Spectrometry of Faint Light*. 1976, IX + 270 pp.
57. K. Knott and B. Battrick (eds.), *The Scientific Satellite Programme during the International Magnetospheric Study. Proceedings of the 10th ESLAB Symposium, held at Vienna, Austria, 10–13 June 1975*. 1976, XV + 464 pp.
58. B. M. McCormac (ed.), *Magnetospheric Particles and Fields. Proceedings of the Summer Advanced Study School, held in Graz, Austria, August 4–15, 1975*. 1976, VII + 331 pp.
59. B. S. P. Shen and M. Merker (eds.), *Spallation Nuclear Reactions and Their Applications*. 1976, VIII + 235 pp.
60. Walter S. Fitch (ed.), *Multiple Periodic Variable Stars. Proceedings of the International Astronomical Union Colloquium No. 29, held at Budapest, Hungary, 1–5 September 1976*. 1976, XIV + 348 pp.
61. J. J. Burger, A. Pedersen, and B. Battrick (eds.), *Atmospheric Physics from Spacelab. Proceedings of the 11th ESLAB Symposium, Organized by the Space Science Department of the European Space Agency, held at Frascati, Italy, 11–14 May 1976*. 1976, XX + 409 pp.
62. J. Derral Mulholland (ed.), *Scientific Applications of Lunar Laser Ranging. Proceedings of a Symposium held in Austin, Tex., U.S.A., 8–10 June, 1976*. 1977, XVII + 302 pp.
63. Giovanni G. Fazio (ed.), *Infrared and Submillimeter Astronomy. Proceedings of a Symposium held in Philadelphia, Penn., U.S.A., 8–10 June, 1976*. 1977, X + 226 pp.
64. C. Jaschek and G. A. Wilkins (eds.), *Compilation, Critical Evaluation and Distribution of Stellar Data. Proceedings of the International Astronomical Union Colloquium No. 35, held at Strasbourg, France, 19–21 August, 1976*. 1977, XIV + 316 pp.
65. M. Friedjung (ed.), *Novae and Related Stars. Proceedings of an International Conference held by the Institut d'Astrophysique, Paris, France, 7–9 September, 1976*. 1977, XIV + 228 pp.
66. David N. Schramm (ed.), *Supernovae. Proceedings of a Special IAU-Session on Supernovae held in Grenoble, France, 1 September, 1976*. 1977, X + 192 pp.
67. Jean Audouze (ed.), *CNO Isotopes in Astrophysics. Proceedings of a Special IAU Session held in Grenoble, France, 30 August, 1976*. 1977, XIII + 195 pp.
68. Z. Kopal, *Dynamics of Close Binary Systems*, XIII + 510 pp.
69. A. Bruzek and C. J. Durrant (eds.), *Illustrated Glossary for Solar and Solar-Terrestrial Physics*. 1977, XVIII + 204 pp.
70. H. van Woerden (ed.), *Topics in Interstellar Matter*. 1977, VIII + 295 pp.
71. M. A. Shea, D. F. Smart, and T. S. Wu (eds.), *Study of Travelling Interplanetary Phenomena*. 1977, XII + 439 pp.
72. V. Szebehely (ed.), *Dynamics of Planets and Satellites and Theories of Their Motion. Proceedings of IAU Colloquium No. 41, held in Cambridge, England, 17–19 August 1976*. 1978, XII + 375 pp.
73. James R. Wertz (ed.), *Spacecraft Attitude Determination and Control*. 1978, XVI + 858 pp.

74. Peter J. Palmadesso and K. Papadopoulos (eds.), *Wave Instabilities in Space Plasmas. Proceedings of a Symposium Organized Within the XIX URSI General Assembly held in Helsinki, Finland, July 31–August 8, 1978.* 1979, VII + 309 pp.
75. Bengt E. Westerlund (ed.), *Stars and Star Systems. Proceedings of the Fourth European Regional Meeting in Astronomy held in Uppsala, Sweden, 7–12 August, 1978.* 1979, XVIII + 264 pp.
76. Cornelis van Schooneveld (ed.), *Image Formation from Coherence Functions in Astronomy. Proceedings of IAU Colloquium No. 49 on the Formation of Images from Spatial Coherence Functions in Astronomy, held at Groningen, The Netherlands, 10–12 August 1978.* 1979, XII + 338 pp.
77. Zdeněk Kopal, *Language of the Stars. A Discourse on the Theory of the Light Changes of Eclipsing Variables.* 1979, VIII + 280 pp.
78. S.-I. Akasofu (ed.), *Dynamics of the Magnetosphere. Proceedings of the A.G.U. Chapman Conference 'Magnetospheric Substorms and Related Plasma Processes' held at Los Alamos Scientific Laboratory, N.M., U.S.A., October 9-13, 1978.* 1980, XII + 658 pp.
79. Paul S. Wesson, *Gravity, Particles, and Astrophysics. A Review of Modern Theories of Gravity and G-variability, and their Relation to Elementary Particle Physics and Astrophysics.* 1980, VIII + 188 pp.
80. Peter A. Shaver (ed.), *Radio Recombination Lines. Proceedings of a Workshop held in Ottawa, Ontario, Canada, August 24-25, 1979.* 1980, X + 284 pp.
81. Pier Luigi Bernacca and Remo Ruffini (eds.), *Astrophysics from Spacelab*, 1980, XI + 664 pp.
82. Hannes Alfvén, *Cosmic Plasma*, 1981, X + 160 pp.
83. Michael D. Papagiannis (ed.), *Strategies for the Search for Life in the Universe*, 1980, XVI + 254 pp.
84. H. Kikuchi (ed.), *Relation between Laboratory and Space Plasmas*, 1981, XII + 386 pp.
85. Peter van der Kamp, *Stellar Paths*, 1981, xxii + 155 pp.
86. E. M. Gaposchkin and B. Kołaczek (eds.), *Reference Coordinate Systems for Earth Dynamics*, 1981, XIV + 396 pp.
87. R. Giacconi (ed.), *X-Ray Astronomy with the Einstein Satellite. Proceedings of the High Energy Astrophysics Division of the American Astronomical Society Meeting on X-Ray Astronomy held at the Harvard-Smithsonian Center for Astrophysics, Cambridge, Mass., U.S.A., January 28–30, 1980.* 1981, VII + 330 pp.
88. Icko Iben Jr. and Alvio Renzini (eds.), *Physical Processes in Red Giants. Proceedings of the Second Workshop, held at the Ettore Majorana Centre for Scientific Culture, Advanced School of Agronomy, in Erice, Sicily, Italy, September 3–13, 1980.* 1981, XV + 488 pp.
89. C. Chiosi and R. Stalio (eds.), *Effect of Mass Loss on Stellar Evolution. IAU Colloquium No. 59 held in Miramare, Trieste, Italy, September 15–19, 1980.* 1981, XXII + 532 pp.
90. C. Goudis, *The Orion Complex: A Case Study of Interstellar Matter*, 1982 (forthcoming).
91. F. D. Kahn (ed.), *Investigating the Universe. Papers Presented to Zdeněk Kopal on the Occasion of his retirement, September 1981.* 1981, X + 458 pp.
92. C. M. Humphries (ed.), *Instrumentation for Astronomy with Large Optical Telescopes, Proceedings of IAU Colloquium No. 67.* 1982 (forthcoming).
93. R. S. Roger and P. E. Dewdney (eds.), *Regions of Recent Star Formation, Proceedings of the Symposium on "Neutral Clouds Near HII Regions - Dynamics and Photochemistry", held in Penticton, B.C., June 24–26, 1981.* 1982, XVI + 496 pp.
94. O. Calame (ed.), *High-Precision Earth Rotation and Earth-Moon Dynamics. Lunar Distances and Related Observations*, 1982, xx + 354 pp.
95. M. Friedjung and R. Viotti (eds.), *The Nature of Symbiotic Stars*, xx + 310 pp.
96. W. Fricke and G. Teleki (eds.), *Sun and Planetary System*, xiv + 538 pp.
97. C. Jaschek and W. Heintz (eds.), *Automated Data Retrieval in Astronomy*, xx + 324 pp.
98. Z. Kopal and J. Rahe (eds.), *Binary and Multiple Stars as Tracers of Stellar Evolution*, 1982, XXX + 504 pp.
99. A. W. Wolfendale (ed.), *Progress in Cosmology*, 1982, VI + 360 pp.
100. W. L. H. Shuter (ed.), *Kinematics, Dynamics and Structure of the Milky Way*, 1983, XII + 392 pp.
101. M. Livio and G. Shaviv (eds.), *Cataclysmic Variables and Related Objects*, 1983, XII + 352 pp.

102. P. B. Byrne and M. Rodonò (eds.), *Activity in Red-Dwarf Stars*, 1983, XXVI + 670 pp.
103. A. Ferrari and A. G. Pacholczyk (eds.), *Astrophysical Jets*, 1983, XVI + 328 pp.
104. R. L. Carovillano and J. M. Forbes (eds.), *Solar-Terrestrial Physics*, 1983, XVIII + 860 pp.
105. W. B. Butron and F. P. Israel (eds.), *Surveys of the Southern Galaxy*, 1983, XIV + 310 pp.
106. V. V. Markellos and Y. Kozai (eds.), *Dynamical Trapping and Evolution on the Solar System*, 1983, XVI + 424 pp.
107. S. R. Pottasch, *Planetary Nebulae*, 1984, X + 322 pp.
108. M. F. Kessler and J. P. Phillips (eds.), *Galactic and Extragalactic Infrared Spectroscopy*, 1984, XII + 472 pp.
109. C. Chiosi and A. Renzini (eds.), *Stellar Nucleosynthesis*, 1984, XIV + 398 pp.